Pseudomonas

Model Organism, Pathogen, Cell Factory

Edited by
Bernd H. A. Rehm

WILEY-VCH Verlag GmbH & Co. KGaA

The Editor

Prof. Dr. Bernd H. A. Rehm
Institute of Molecular BioSciences
Massey University
Private Bag 11222
Palmerston North 4442
New Zealand

Library of Congress Card No.: applied for

British Library Cataloguing-in-Publication Data
A catalogue record for this book is available from the British Library.

Bibliographic information published by the Deutsche Nationalbibliothek
Die Deutsche Nationalbibliothek lists this publication in the Deutsche Nationalbibliografie; detailed bibliographic data are available on the Internet at <http://dnb.d-nb.de>.

Composition Thomson Digital, Noida, India
Printing Betz-Druck GmbH, Darmstadt
Bookbinding Litges & Dopf GmbH, Heppenheim
Cover design Adam Design, Weinheim

Printed in the Federal Republic of Germany
Printed on acid-free paper

ISBN: 978-3-527-31914-5

Contents

Pseudomonas. Model Organism, Pathogen, Cell Factory. Edited by Bernd H.A. Rehm

ISBN: 978-3-527-31914-5

Preface

The genus *Pseudomonas* represents a diverse group of medically, environmentally and biotechnologically important bacteria. Pseudomonads are characterized by an enormous metabolic capacity, which is reflected by their capability to (i) adapt to diverse and challenging environments, (ii) degrade recalcitrant compounds, and (iii) synthesize a variety of low-molecular-weight compounds as well as biopolymers.

In this book a selection of scholarly *Pseudomonas* review papers is presented by internationally recognized researchers who have been active in the field over many years.

The aim of this book is to provide a concise and comprehensive survey of all relevant aspects related to the genus *Pseudomonas*. The first chapter introduces us to *Pseudomonas* genomics by comparing about 20 genomes from various pseudomonads, emphasizing the vast genomic diversity in this group of bacteria. The species *P. aeruginosa*, an opportunistic human pathogen causing chronic infections in the lungs of cystic fibrosis (CF) patients, is of particular clinical relevance. Molecular mechanisms of adaptation to the unique CF lung environment as well as pathogenicity mechanisms of this species are discussed in two chapters. *P. aeruginosa* has emerged as a model organism for biofilm studies. One chapter has been dedicated to describing the recent advances in understanding the complex processes of biofilm formation. The ability to thrive in diverse environments requires complex regulatory networks implementing synthesis of a number of signal molecules, motility and chemotaxis, which have been thoroughly outlined in chapters depicting the respective individual aspects. Since various pseudomonads show particular interactions with plants not only as pathogens, but also causing promotion of plant growth, one chapter is dedicated to comprehensively describing the adaptation of pseudomonads to the particular plant environment. Plasmids and bacteriophages are important contributors to the genetic diversity found in pseudomonads, and recent advances in these fields are concisely summarized in two separate chapters.

The book closes with a chapter comprehensively outlining the use and potential applications of pseudomonads in biotechnology. Applications in bioremediation,

Pseudomonas. Model Organism, Pathogen, Cell Factory. Edited by Bernd H.A. Rehm

ISBN: 978-3-527-31914-5

bioconversion, and as a cell factory for the production of low-molecular-weight compounds, biopolymers and recombinant proteins are discussed.

Bernd H.A. Rehm
Palmerston North January 2008

List of Contributors

Tim T. Binnewies
Technical University of Denmark
Center for Biological Sequence Analysis
BioCentrum-DTU
2800 Lyngby
Denmark

Wulf Blankenfeldt
Max-Planck Institute of Molecular Physiology
Otto-Hahn-Str. 11
44227 Dortmund
Germany

Alexander Boronin
Russian Academy of Sciences
Institute of Biochemistry and Physiology of Microorganisms
142290 Pushchino
Russia

M. Cámara
University of Nottingham
Institute of Infection, Immunity and Inflammation
University Park
Nottingham NG7 2RD
UK

Oana Ciofu
University of Copenhagen
Institute of International Health, Department of Bacteriology
Blegdamsvej 3B
2200 Copenhagen
Denmark

Tim Conibear
University of Southampton
School of Biological Sciences
Southampton SO17 1BJ
UK

S. A. Crusz
University of Nottingham
Institute of Infection, Immunity and Inflammation
University Park
Nottingham NG7 2RD
UK

S. P. Diggle
University of Nottingham
Institute of Infection, Immunity and Inflammation
University Park
Nottingham NG7 2RD
UK

Pseudomonas. Model Organism, Pathogen, Cell Factory. Edited by Bernd H.A. Rehm

ISBN: 978-3-527-31914-5

J. F. Dubern
University of Nottingham
Institute of Infection,
Immunity and Inflammation
University Park
Nottingham NG7 2RD
UK

Richard C. Draper
University of Otago
Department of Biochemistry
PO Box 56
Dunedin 9054
New Zealand

A. Franks
National University of Ireland (UCC)
Microbiology Department
Cork
Ireland

M. P. Fletcher
University of Nottingham
Institute of Infection,
Immunity and Inflammation
University Park
Nottingham NG7 2RD
UK

Carsten Friis
Technical University of Denmark
Center for Biological Sequence Analysis
BioCentrum-DTU
2800 Lyngby
Denmark

Anthony S. Haines
University of Birmingham
School of Biosciences
Birmingham B15 2TT
UK

Susse Kirkelund Hansen
Technical University of Denmark
Infection Microbiology Group
BioCentrum-DTU
2800 Lyngby
Denmark

Susanne Häussler
Helmholtz Zentrum für
Infektionsforschung GmbH
Department of Cell Biology
Inhoffenstrasse 7
38124 Braunschweig
Germany

S. Heeb
University of Nottingham
Institute of Infection,
Immunity and Inflammation
University Park
Nottingham NG7 2RD
UK

Kirsten Hertveldt
Katholieke Universiteit Leuven
Department of Biosystems
Kasteelpark Arenberg 21
3001 Leuven
Belgium

Niels Høiby
Rigshospitalet
Department of Clinical Microbiology
Juliane Maries Vej 22
2100 Copenhagen
Denmark

Francesco Imperi
University "Roma Tre"
Department of Biology
Viale G. Marconi 446
00146 Rome
Italy

Randall T. Irvin
University of Alberta
Department of Medical Microbiology
and Immunology
Edmonton
Alberta T6G 2H7
Canada

Lars Jelsbak
Technical University of Denmark
Infection Microbiology Group
BioCentrum-DTU
2800 Lyngby
Denmark

Helle Krogh Johansen
Rigshospitalet
Department of Clinical Microbiology
Juliane Maries Vej 22
2100 Copenhagen
Denmark

Jeevan Jyot
University of Florida
Department of Medicine Infectious
Diseases
1600 SW Archer Road
Gainesville, FL 32610
USA

Junichi Kato
Hiroshima University
Department of Molecular Biotechnology
Hiroshima 739-8530
Japan

Kristoffer Kiil
Technical University of Denmark
Center for Biological Sequence Analysis
BioCentrum-DTU
2800 Lyngby
Denmark

Sylvia M. Kirov
University of Tasmania
School of Medicine (Pathology)
Hobart, Tasmania 7001
Australia

Staffan Kjelleberg
University of New South Wales
School of Biotechnology and
Biomolecular Science
Sydney, NSW 2052
Australia

Janosch Klebensberger
University of New South Wales
School of Biotechnology and
Biomolecular Science and Center for
Marine Biofouling and Bio-Innovation
Sydney, NSW 2052
Australia

Irina A. Kosheleva
Russian Academy of Sciences
Institute of Biochemistry and
Physiology of Microorganisms
142290 Pushchino
Russia

Iain L. Lamont
University of Otago
Department of Biochemistry
PO Box 56
Dunedin 9054
New Zealand

Rob Lavigne
Katholieke Universiteit Leuven
Department of Biosystems
Kasteelpark Arenberg 21
3001 Leuven
Belgium

G. L. Mark
National University of Ireland (UCC)
Microbiology Department
Cork
Ireland

Carsten Matz
Helmhotz Center for Infection Research
Division of Microbiology
38124 Braunschweig
Germany

Dmitri V. Mavrodi
Washington State University
Department of Plant Pathology
Pullman, WA 99164-6430
USA

Diane McDougald
University of New South Wales
School of Biotechnology and
Biomolecular Science and Center for
Marine Biofouling and Bio-Innovation
Sydney, NSW 2052
Australia

Karla A. Mettrick
University of Otago
Department of Biochemistry
PO Box 56
Dunedin 9054
New Zealand

S. H. Miller
National University of Ireland (UCC)
Microbiology Department
Cork
Ireland

Claus Moser
Rigshospitalet
Department of Clinical Microbiology
Juliane Maries Vej 22
2100 Copenhagen
Denmark

F. O'Gara
National University of Ireland (UCC)
Microbiology Department
Cork
Ireland

Reuben Ramphal
University of Florida
Department of Medicine Infectious
Diseases
1600 SW Archer Road
Gainesville, FL 32610
USA

Bernd H. A. Rehm
Massey University
Institute of Molecular BioSciences
Private Bag 11222
Palmerston North 4442
New Zealand

Scott A. Rice
University of New South Wales
School of Biotechnology and
Biomolecular Science and Center for
Marine Biofouling and Bio-Innovation
Sydney, NSW 2052
Australia

Ute Römling
Karolinska Institutet
Department of Microbiology, Tumor
and Cell Biology,
Box 280
17 177 Stockholm
Sweden

Matt Shirley
University of Otago
Department of Biochemistry
PO Box 56
Dunedin 9054
New Zealand

Federica Tiburzi
University "Roma Tre"
Department of Biology
Viale G. Marconi 446
00146 Rome
Italy

Christopher M. Thomas
University of Birmingham
School of Biosciences
Birmingham B15 2TT
UK

Linda S. Thomashow
Washington State University
USDA-ARS Root Disease and Biological
Control Research Unit
Pullman, WA 99164-6430
USA

Tim Tolker-Nielsen
Technical University of Denmark
Center for Biomedical Microbiology
2800 Lyngby
Denmark

David W. Ussery
Technical University of Denmark
Center for Biological Sequence Analysis
BioCentrum-DTU
2800 Lyngby
Denmark

Paolo Visca
University "Roma Tre"
Department of Biology
Viale G. Marconi 446
00146 Rome
Italy

Jeremy S. Webb
University of New South Wales
School of Biotechnology and
Biomolecular Sciences
Sydney NSW 2052
Australia

Hanni Willenbrock
Technical University of Denmark
Center for Biological Sequence Analysis
BioCentrum-DTU
2800 Lyngby
Denmark

P. Williams
University of Nottingham
Institute of Infection,
Immunity and Inflammation
Center for Biomolecular Sciences
Nottingham NG7 2RD
UK

Lei Yang
Technical University of Denmark
Infection Microbiology Group
BioCentrum-DTU
2800 Lyngby
Denmark

1
Comparative Genomics of *Pseudomonas*

Kristoffer Kiil, Tim T. Binnewies, Hanni Willenbrock, Susse Kirkelund Hansen, Lei Yang, Lars Jelsbak, David W. Ussery, and Carsten Friis

1.1
Introduction

The genus *Pseudomonas* covers one of the most diverse and ecologically significant groups of bacteria. Members of the genus are found in large numbers in a wide range of environmental niches, such as terrestrial and marine environments, as well as in association with plants and animals. This almost universal distribution of *Pseudomonas* suggests a remarkable degree of genomic diversity and genetic adaptability.

Pseudomonas aeruginosa is a motile Gram-negative rod-shaped bacterium. It differs from other members of the *Pseudomonas* genus because of its potential pathogenicity for human beings and other mammals. *P. aeruginosa* participates in infections in immunocompromised individuals, such as patients suffering from AIDS, cancer, burn wounds and cystic fibrosis (CF) [1–3]. *P. aeruginosa* infections are normally difficult to eradicate.

P. aeruginosa is a generalist with a diverse metabolic competence. It can utilize simple small molecules as well as complex organic compounds as carbon sources and is capable of growing aerobically as well as anaerobically using NO_3 or arginine as respiratory electron acceptors. *P. aeruginosa* grows over a wide range of temperatures, from common environmental temperatures up to 42°C, with an optimum temperature at 37°C.

The first completed genome of *P. aeruginosa* is the genome of PAO1, published in 2000 [4]. It is 6.3 Mbp and contains 5570 open reading frames. A significant number (8.4%) of *P. aeruginosa* genes are predicted to be involved in regulation, which at the time of publication of the genome was the largest fraction of regulators among sequenced bacterial genomes. The large genome size and remarkable gene complexity of *P. aeruginosa* enable it to adapt and thrive in different environmental conditions and hosts. In addition to PAO1, PA14 as well as nine other clinical *P. aeruginosa* strains are currently available either as completed genomes or as unfinished sequences. Genome comparison analysis of PAO1 and five different clinical *P. aeruginosa* isolates (four of them are CF isolates) shows that about 80% of

Pseudomonas. Model Organism, Pathogen, Cell Factory. Edited by Bernd H.A. Rehm

ISBN: 978-3-527-31914-5

the PAO1 genome is conserved among other strains [5]. More than 30 relatively large regions of the PAO1 genome are found to be nonconserved in the five clinical strains. Those regions include phages, the pyoverdine biosynthesis locus, genes encoding a putative type I secretion system and a putative restriction modification system.

P. aeruginosa is an opportunistic pathogen capable of producing a wide variety of virulence factors, including lipopolysaccharides, flagellum, type IV pili, proteases, exotoxins, pyocyanin, exopolysaccharides, type III secretion, etc. Many of the extracellular virulence factors have been shown to be regulated by quorum-sensing signals [6,7]. Biofilm formation can also be regarded as a virulence factor [8]. Acute infections with *P. aeruginosa* can be life threatening, resulting in severe tissue damage and septicemia (i.e. bloodstream invasion). Although the pathogenicity of *P. aeruginosa* is typically characterized by a high level of toxin production [9], several important exceptions to this are known, particularly in the cases of certain chronic infections of *P. aeruginosa*. These include infections of CF patients, where *P. aeruginosa* develops genetic adaptations during long-term persistence, in which virulence factors are normally selected against [10–12]. This shows *P. aeruginosa* is capable of choosing distinct strategies for different types of infections.

Another remarkable and unfortunate character of *P. aeruginosa* is its tolerance to many antimicrobial drugs. It has a number of inherent antibiotic-resistance mechanisms that include an AmpC β-lactamase that can be induced by β-lactams, which makes it inherently resistant to cephalothin and ampicillin [13]. *P. aeruginosa* also has efflux pumps such as MexAB–OprM, making it impermeable to many antibiotics. MexAB–OprM removes β-lactams, chloramphenicol, fluoroquinolones, novobiocin, as well as various dyes and detergents [14]. Finally, it has been shown that through mutation, *P. aeruginosa* is capable of developing resistance to antibiotics that the strain is not inherently resistant to, such as aminoglycosides and colistin [15].

1.1.1 Other Species of *Pseudomonas*

In addition to *P. aeruginosa* PAO1, several other *Pseudomonas* genomes have been sequenced (Table 1.1). *P. putida*, for instance, is a ubiquitous bacterium frequently isolated from soil (particularly polluted soil), the rhizosphere and water. *P. putida* is a paradigm of a metabolic versatile saprophytic soil bacterium and the best characterized strain, KT2440 [16,17], has become a model bacterium worldwide – both in laboratory studies and for the development of biotechnological applications. Some of these include bioremediation of contaminated sites [18,19], biocatalysis for the production of useful chemicals [20], and the potential development of new biopesticides and plant growth promoters as a plant rhizosphere protective agent. Strain KT2440 is a plasmid-free derivative of the original isolate, designated *P. arvilla* strain mt-2 [21], the natural host of the archetype TOL plasmid pWW0 [16], and subsequently reclassified *P. putida* mt-2 [22]. It has maintained its ability to survive in its natural environment and is the first Gram-negative soil bacterium that has been certified as a biosafety host for expression of foreign genes [23]. The genome of strain KT2440 was sequenced in 2004 [24]; the sequence information is continuously

Table 1.1 Feature overview of currently sequenced *Pseudomonas* genomes.

Organism	Genome size (bp)	No. contigs[a]	G + C content (%)	Percent coding (%)	No. genes[b]	Accession [reference]
P. aeruginosa 2192	6 826 253	82	66.2	85	5546	NZ_AAKW00000000 [32]
P. aeruginosa C3719	6 146 998	124	66.5	86	5065	NZ_AAKV00000000 [32]
P. aeruginosa LES	6 601 757	2	66.3	80	5326	[33]
P. aeruginosa PA7	6 663 529	147	66.4	84	5309	NZ_AAQE00000000 [32]
P. aeruginosa PACS2	6 492 423	1	66.3	86	5317	NZ_AAQW00000000 [32]
P. aeruginosa PAO1	6 264 404	–	66.6	89	5568	AE004091 [32]
P. aeruginosa UCBPP-PA14	6 537 648	–	66.3	89	5892	CP000438 [32]
P. entomophila L48	5 888 780	–	64.2	88	5134	CT573326 [32]
P. fluorescens Pf-5	7 074 893	–	63.3	88	6137	CP000076 [32]
P. fluorescens PfO-1	6 438 405	–	60.5	89	5736	CP000094 [32]
P. mendocina ymp	5 072 807	–	64.7	89	4594	CP000680 [32]
P. putida F1	5 925 059	120	61.9	88	4735	NZ_AALM00000000 [32]
P. putida GB-1	6 053 195	30	62.0	90	4890	NZ_AAXR00000000 [32]
P. putida KT2440	6 181 863	–	61.5	86	5350	AE015451 [32]
P. putida W619	5 748 550	62	61.5	89	4624	NZ_AAVY00000000 [32]
P. stutzeri A1501	4 567 418	–	63.9	89	4128	CP000304 [32]
P. syringae pv. *phaseolicola* 1448A	5 928 787	–	57.9	84	4984	CP000058 [32]
P. syringae pv. *syringae* B728a	6 093 698	–	59.2	87	5089	CP000075 [32]
P. syringae pv. *tomato* str. DC3000	6 397 126	–	58.3	85	5470	AE016853 [32]

[a] The numbers correspond to the number of contigs for unassembled genomes, while a dash signifies a completely assembled genome.

[b] For assembled genomes this is the number of proteins annotated in the official NCBI release; for the unassembled genomes this is the number of predicted genes from the EasyGene runs.

providing new insight into the biology of an adaptable and metabolic versatile group of soil bacteria, and it is facilitating the increased use of this organism for biotechnological purposes.

P. syringae is a plant pathogen that can infect a wide range of plant species [25]. *P. syringae* strains are assigned to different pathovars on the basis of their distinctive pathogenicity and ability to infect different plant species. More than 50 different pathovars exists and completed genome sequences of *P. syringae* strains from three different pathovars (*tomato, syringae* and *phaseolicola*) are available for analysis [26–28] (Table 1.1). *P. syringae* pv. *tomato* is the causal agent of bacterial speck disease on tomato and *Arabidopsis*, while *P. syringae* pv. *phaseolicola* and *P. syringae* pv. *syringae* cause halo blight and brown spot disease on bean plants, respectively.

Other *Pseudomonas* species are nonpathogenic plant-associated bacteria that exhibit plant growth-promoting properties [29]. For example, *P. fluorescens* Pf-5 and *P. fluorescens* PfO-1 are commensal bacteria that colonize plant surface environments and produce various secondary metabolites that suppress the growth of soil-borne plant pathogens. These bacteria have potential roles as biological disease control agents in agricultural settings. *P. stutzeri* strain A1501 is able to colonize and infect rice roots, and is widely used as a rice inoculant in China [30]. It can fix nitrogen and may provide rice plants with fixed nitrogen and hence promote plant growth.

P. entomophila is an entomopathogenic bacterium that is highly pathogenic for a variety of insects from different orders [31].

1.1.2
Obtaining Sequence Data on *Pseudomonas*

In total, 19 *Pseudomonas* genomes were downloaded from online resources: 18 from The National Center for Biotechnology Information (NCBI) [32] and one was produced by the Pathogen Sequencing Group at the Sanger Institute [33]. An overview of the genomes is given in Table 1.1. Of the 19 genomes, 11 were completely sequenced and assembled at the time of download, and were obtained with traditional NCBI annotation describing the location of genes within the genomes. The remaining eight genomes were, however, under assembly at the time and thus had little in the way of any official annotation. To identify the position of genes in these genomes we applied the EasyGene [34] method for gene finding in bacterial genomes.

1.2
Pan/Core Genome of *Pseudomonas*

With the availability of an increasing amount of fully sequenced *Pseudomonas* genomes it becomes possible to conduct an investigation of genetic characteristics defining the *Pseudomonas* genus. For instance, what fraction of the genome is conserved throughout all sequenced *Pseudomonas*?

The "core genome" of *Pseudomonas* is defined as containing those genes that are present in all strains of *Pseudomonas*, i.e. the minimum amount of genes required for a bacterium to be considered part of the *Pseudomonas* genus. In contrast the "pan-genome" is defined containing any gene present in any strain of *Pseudomonas*. A histogram of the sizes of the core and pan-genomes of *Pseudomonas* similar to that previously done for *Escherichia coli* [35] is shown in Figure 1.1. Figure 1.1 was constructed by continuously adding to a list of genes, beginning with the whole genome of *P. aeruginosa* 2192 and then examining other *Pseudomonas* genomes,

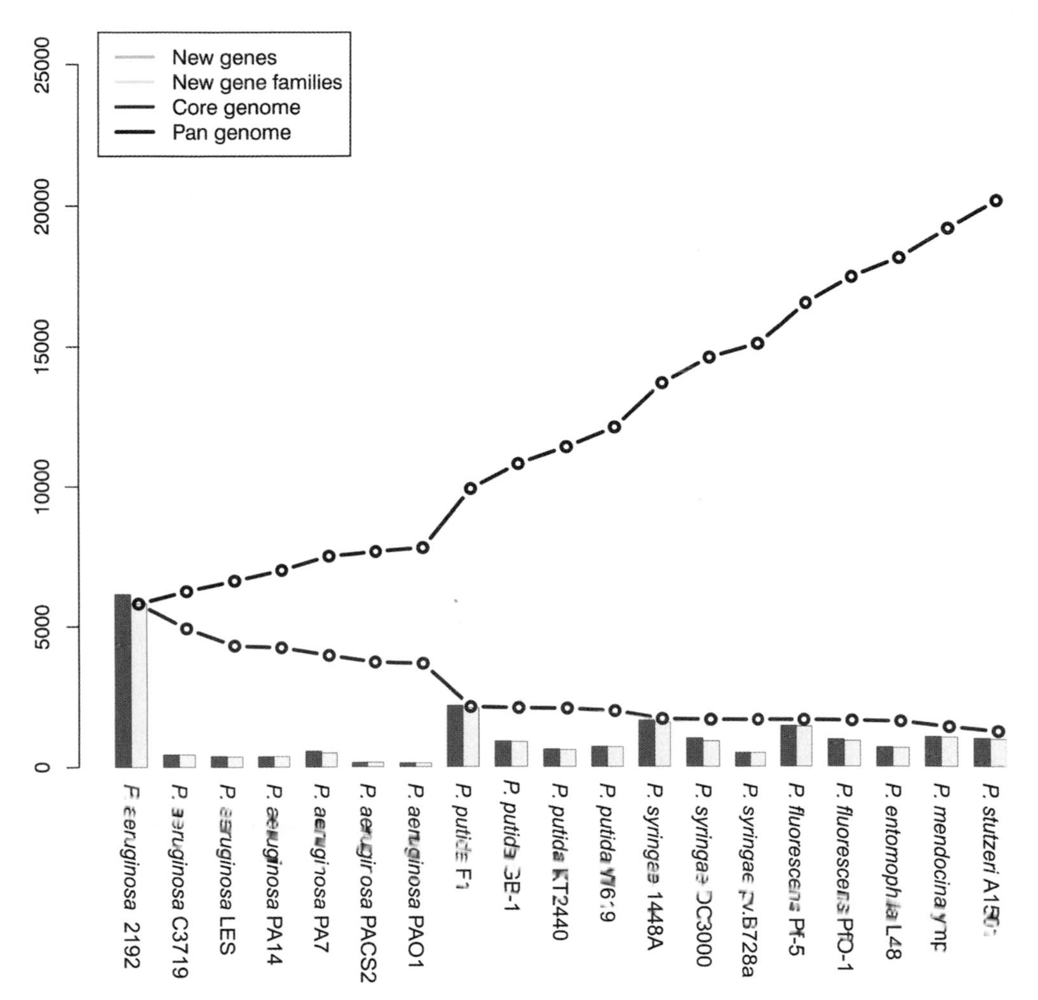

Figure 1.1 *Pseudomonas* pan/core genome. The core genome is defined as comprising all genes present in all sequenced genomes of *Pseudomonas*, while the pan-genome encompasses all genes present in any strain. The figure shows how the numbers of genes decline and increase in the two with each consecutive release of a new sequenced *Pseudomonas* genome. Each genome is represented by two bars: one solid giving the number of new proteins not found in any of the previous strains (i.e. any of the strains to the left) and one light grey representing the number of unique protein families. The genomes are presented in an order roughly equal to evolutionary distance from *P. aeruginosa* 2192.

adding any genes not already present in the list, in an order based on evolutionary distance to *P. aeruginosa* 2192. To determine whether a given gene is present in an organism or not, the "50/50" rule was used on a BLAST alignment [36] at the protein level. The "50/50" rule states that to be considered conserved across two organisms at least 50% of the length of the protein should show a minimum of 50% sequence identity. The number of genes not present in any of the previous strains is shown for each strain with a black column. Another column shows the number of new gene families, illustrating that the new genes are not just functional orthologs, but indeed do add new functionality.

While the core genome is obviously restrained in size, the pan-genome is theoretically infinite or at least can be considered to be very large. As more and more strains within the *P. aeruginosa* species are sequenced and added to Figure 1.1, the number of genes in the core genome slowly drops until it reaches what appears to be a stable level after four or five genomes have been sequenced describing the *P. aeruginosa* core genome. When one begins to add non-*P. aeruginosa* genomes to the plot the core genome starts to drop again, now stabilizing at a significantly lower level, which describes the core genome for the whole *Pseudomonas* genus. Of course, the exact values for the sizes of both the *P. aeruginosa* and the *Pseudomonas* genus core genomes are subject to change as even more fully sequenced *Pseudomonas* genomes become available. In particular, the genus core genome is almost certain to drop further as new strains become sequenced, not hitherto represented in the core genome.

Analysis of the pan/core genomics of an organism is of great importance towards describing at the genetic level what constitutes a taxonomic group, e.g. a "*Pseudomonas* bacteria". There is a considerable difference (about 10-fold) in the size between the core and pan-genomes, and even just within *P. aeruginosa* the pan-genome is twice the size of the core genome, with many of the genes unique to one particular strain not being merely functional orthologs. While perhaps a trivial observation, it nevertheless has a great impact towards determining whether a given gene exists in *Pseudomonas* or not. One should be careful before making statements to that effect or at the very least take the trouble to specify precisely in which particular isolate one believes the gene to be present or absent. The genetic diversity of bacteria is quite vast even at the species level and much more so at the genus level.

1.3 Phylogeny of *Pseudomonas*

In comparative genomics phylogeny is almost always a principal component of the analysis – either as the actual subject of study or as the underlying structure, explaining most of the data. In the study of pan-genomics phylogeny can be used to explain a lot of the data. There are several ways to assess the phylogeny of a given group of strains or species. One of the oldest and most common ways is by comparing the sequence of the 16S ribosomal subunits. Figure 1.2 is an example of such a tree.

The tree in Figure 1.2 is made by first extracting the 16S rRNA genes, using the hidden Markov model-based RNAmmer program [37], then aligning them using

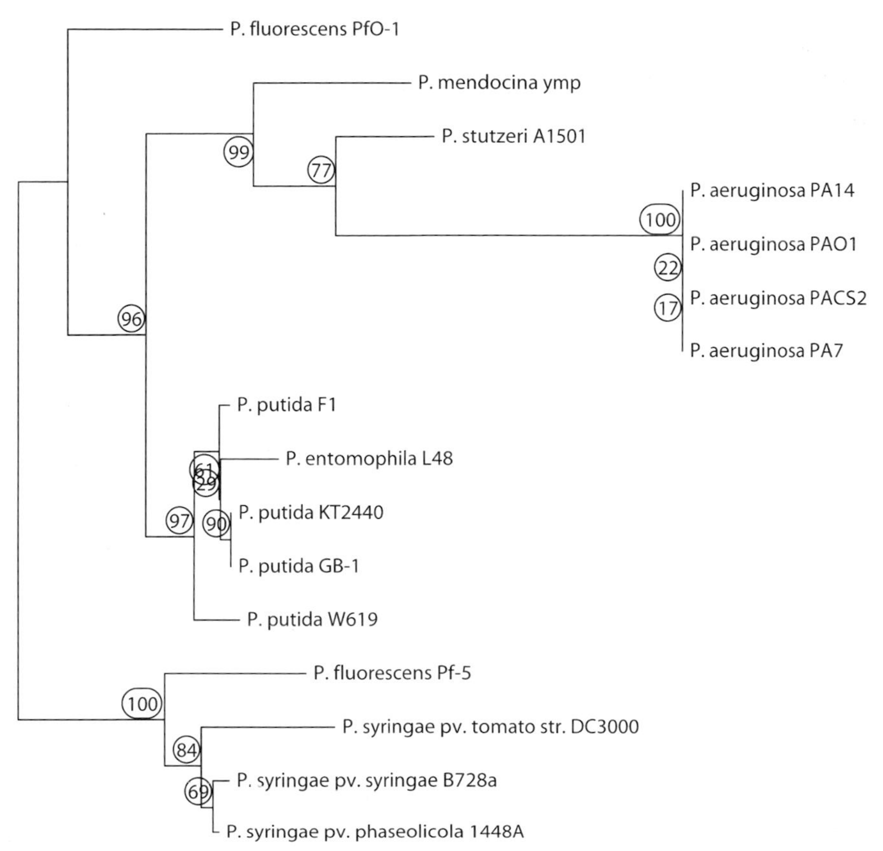

Figure 1.2 16S phylogenetic tree of the *Pseudomonas* genus. The tree is based on similarities in the sequence of 16S ribosomal subunits and shows the evolutionary distance between these. Study of the 16S ribosomal subunit is a common way of establishing evolutionary distance, but it is not the only way and nor is it guaranteed to give the best results.

ClustalW [38]. Using the BIONJ algorithm the phylogeny is reconstructed from the multiple alignment [39]. This last step was performed using the phylowin program [40]. To get a sense of the robustness of the tree it is also bootstrapped. From the tree we can derive some simple conclusions. The four *P. aeruginosa* are very closely related and it seems that *P. aeruginosa* is a more "well-defined" species than many of the other *Pseudomonas*, at least from a 16S perspective. Also, *P. stutzeri* and *P. mendocina* seems to be quite distinct from the other species, while *P. entomophila*

L48 is quite closely related to the *P. putida* strains. In fact the 16S rRNA of *P. putida* W619 is less similar to that of the other *P. putida* strains compared to that of *P. entomophila* L48. Even more interesting, it is remarkable that while *P. fluorescens* PfO-1 has a distance to the other species that is comparable to that seen among the other *Pseudomonas* species, *P. fluorescens* Pf-5 is so similar to the *P. syringae* strains that it might be mistaken for one if the taxonomists were not there to say otherwise.

When confronted with this kind of data, the proper question to ask is whether the 16S rRNA comparison actually tells us what we want to know? To address this question we will turn to a comparison of the σ factors of our *Pseudomonas* strains. The σ factors, the essential ones at least, happen to have some of the same traits that make the 16S rRNA a good phylogenetic marker: it interacts with numerous genes, making changes relatively infrequent and uptake by horizontal gene transfer improbable; also the household σ factor σ^{70} is ubiquitous.

Studying the RpoD subtree on Figure 1.3, we see again that the two *P. fluorescens* strains cluster with the *P. syringae* strains and that *P. entomophila* L48 clusters with the *P. putida* strains. We also see, however, that the resolution is very poor on the RpoD subtree. If we instead look at the FliA and RpoH clades, which have higher resolutions, we see that the two *P. fluorescens* strains and *P. entomophila* L48 split out from the *P. syringae* and *P. putida* clades, although they still seem to be closely related.

1.4
Blast Atlas of *Pseudomonas* Genomes

The "Blast atlas" is a visualization method to show a vast amount of data in one plot by taking a reference genome and mapping the conservation of each protein-encoding gene along the chromosome. An example of a BLASTatlas is given in Figure 1.4, using *P. aeruginosa* UCBPP-PA14 as the reference genome, compared to 18 other *Pseudomonas* genomes, as well as the UniProt database (outer circle in black). The atlas is constructed largely as described previously [41], with the sole refinement being the introduction of a continuous color scale to more accurately describe the BLAST hits.

As for Figure 1.1, we used the "50/50" rule (see above) as a requirement for what constitutes a conserved protein. However, in this case, since we are visualizing BLAST results along a sequence, it is possible to display more information from the BLAST report. In this context, any protein which is well conserved across the *Pseudomonas* genus will appear as a strongly colored band on the atlas, while proteins that are weakly conserved or present only in the reference strain and absent in the

→

Figure 1.3 Phylogenetic tree of the *Pseudomonas* genus based on σ factor similarities. As σ factors generally show high degrees of conservation and σ^{70} in particular, is also ubiquitous, it – like the 16S ribosomal subunit – is useful as a measure of evolutionary distance. While it can be said to be generally neither more nor less accurate than the 16S tree of Figure 1.2, it provides an alternate view.

72 species , 168 sites (global gap removal)

BIONJ Method

Observed divergence

0.081

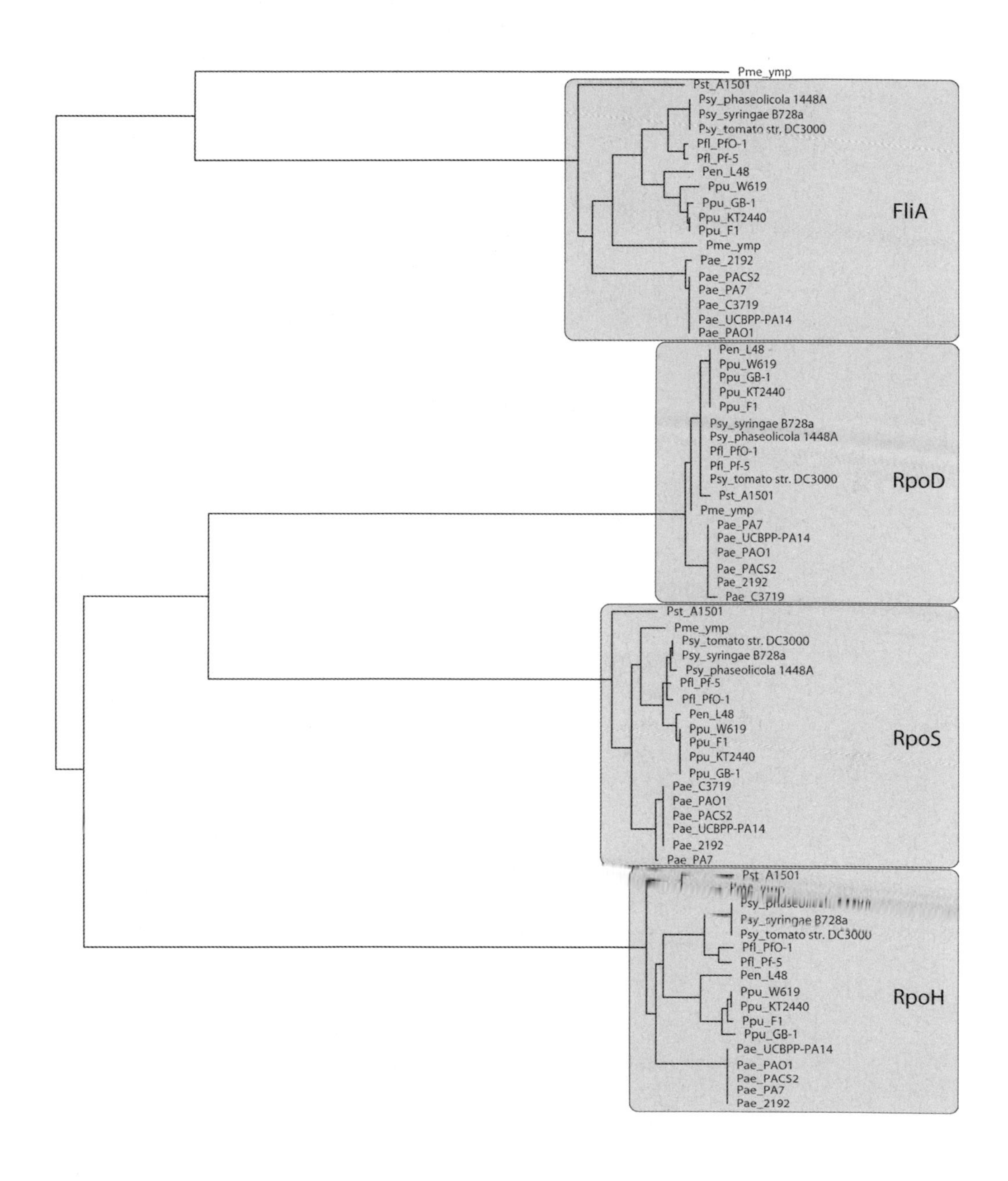

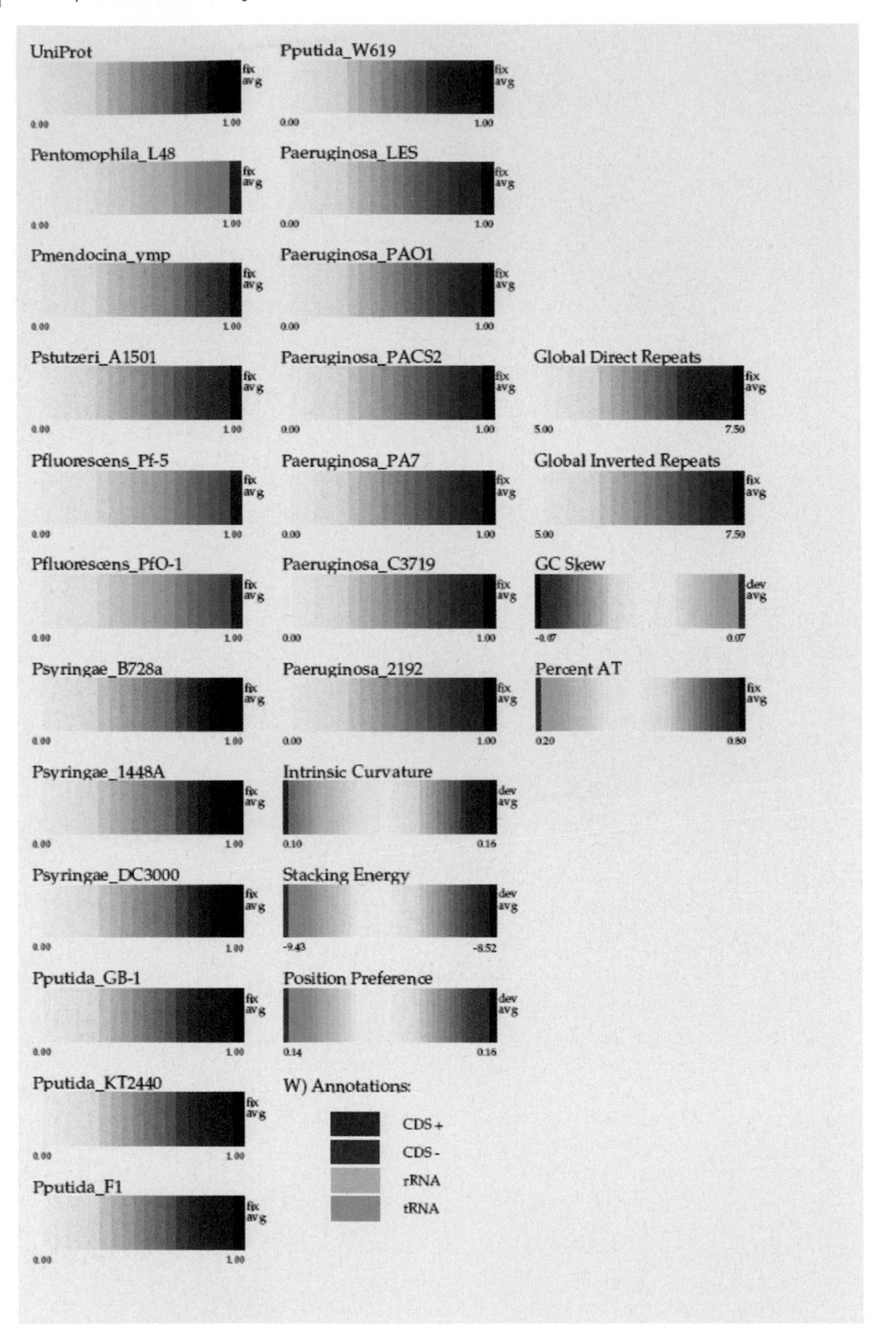
UniProt
fix
avg
0.00
1.00
Pputida_W619
Pentomophila_L48
Paeruginosa_LES
Pmendocina_ymp
Paeruginosa_PAO1
Pstutzeri_A1501
Paeruginosa_PACS2
Global Direct Repeats
5.00
7.50
Pfluorescens_Pf-5
Paeruginosa_PA7
Global Inverted Repeats
Pfluorescens_PfO-1
Paeruginosa_C3719
GC Skew
dev
-0.07
0.07
Psyringae_B728a
Paeruginosa_2192
Percent AT
0.20
0.80
Psyringae_1448A
Intrinsic Curvature
0.10
0.16
Psyringae_DC3000
Stacking Energy
-9.43
-8.52
Pputida_GB-1
Position Preference
0.14
0.16
Pputida_KT2440
W) Annotations:
CDS +
CDS -
rRNA
tRNA
Pputida_F1
Center for Biological Sequence Analysis
http://www.cbs.dtu.dk/
GENOME ATLAS

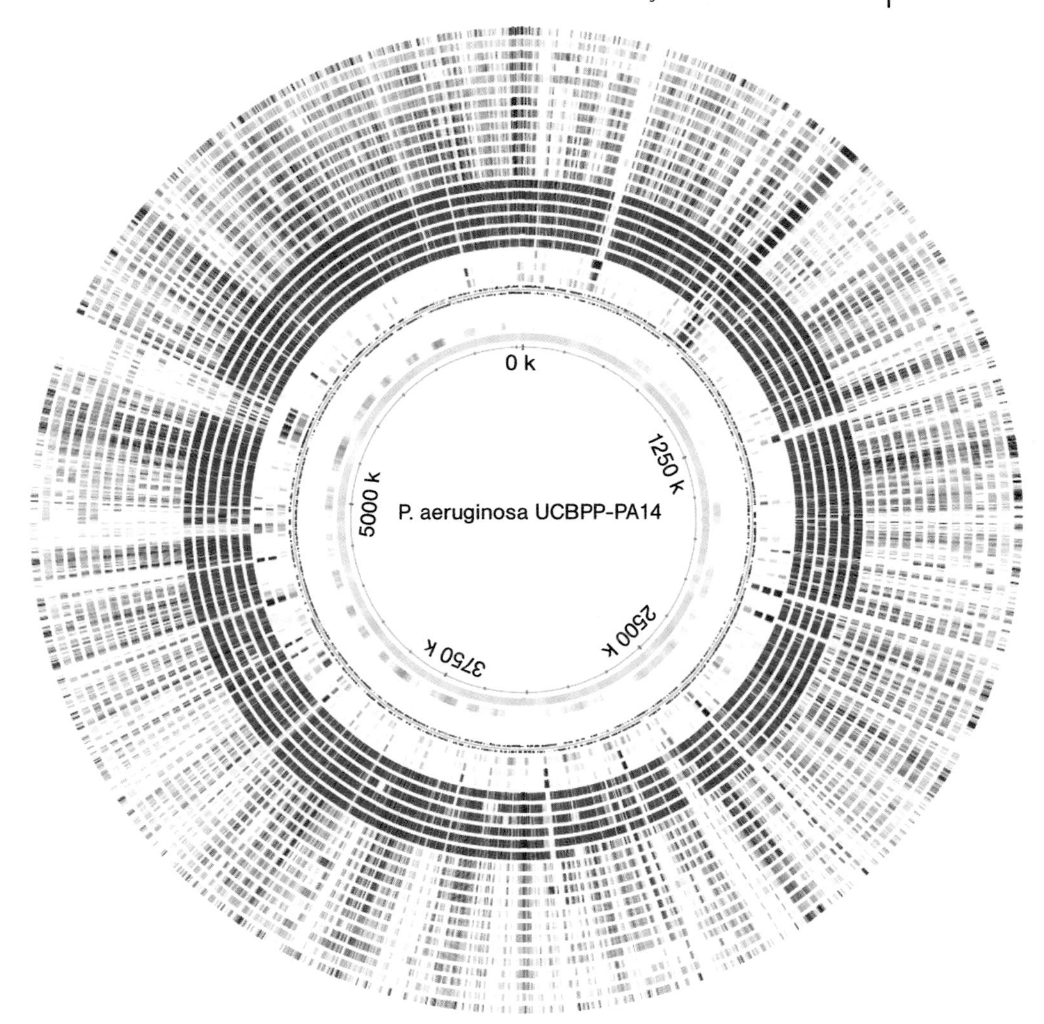

Figure 1.4 Blast atlas of *P. aeruginosa* UCBPP-PA14 versus 18 other *Pseudomonas* strains. The innermost circles show properties related to the base composition of the UCBPP-PA14 genome as well as the locations of the genes. Starting from the red colored rings and outwards, the colors represent BLAST identities to other *Pseudomonas* genomes. In general the evolutionary distance from UCBPP-PA14 increases as one moves from the innermost circles to the outermost, e.g. the bright red circles all represent various *P. aeruginosa* species, while the blue show alignments to *P. putida* strains, etc. The strength of the color bands in each circle indicates the degree of conservation in the other strains of that area of the UCBPP-PA14 genome. BLAST alignments are performed at the amino acid level and only for proteins. Thus, anything not being translated (e.g. rRNA) has not been aligned and will thus be colored grey. A larger and zoomable version of the atlas is available online: http://www.cbs.dtu.dk/services/GenomeAtlas/suppl/zoomatlas/.

other proteomes will result in weak bands or outright gaps in the alignments circles. However, because the BLAST alignments are done at the protein level, gaps will also appear for genomic regions not being translated, e.g. rRNA islands. Thus, while weak and strong bands can be interpreted directly, it is prudent to cross-reference any complete gap with the genomic annotations which are also given on Figure 1.4 before drawing any conclusions either way.

The innermost circle is the AT content, with the turquoise color indicating that this genome is GC rich. The second innermost circle is deep red representing regions that are more than three standard deviations AT-rich than the chromosomal average and dark turquoise regions similarly represent GC-rich regions in the chromosome. The second innermost circle is the GC skew (the bias of Gs towards one strand or the other, averaged over a 10 000-bp window). The third circle shows the Global repeats as they represent the best match of a 100-bp piece of DNA, centered at a given position in the chromosome, when searched against the entire chromosome. The fourth circle represents the genes, with blue for protein-encoding genes oriented clockwise and red coding for genes on the other strand (counterclockwise). The three lanes before the individual proteomes start are DNA structural features, based on the physico-chemical properties of the DNA helix. From innermost to outermost they are the "Position Preference" (a measure related to the rigidity of the DNA [42]), the "Stacking Energy" (which states the energy needed to de-stack, i.e. melt, the DNA [43]) and, finally, the "Intrinsic curvature" (which describes the DNA helix preference towards bending) [44]. It is also possible to zoom into specific regions of interest and a web-based "zoomable atlas" for *P. aeruginosa* UCBPP-PA14 can be found online, as well as for several other bacterial genomes (http://www.cbs.dtu.dk/services/GenomeAtlas/suppl/zoomatlas/).

One thing which is characteristic in distinguishing pathogens from nonpathogens is the presence of pathogenic gene islands. Since the Blast atlas is adept at identifying genetic regions in the reference genome that show either an abnormally high or low degree of conservation, it should be able to pick up on the presence of pathogenic islands. For the same reason, the *P. aeruginosa* UCBPP-PA14 strain was chosen to serve as the reference genome as this strain is famous for its high degree of virulence and is known to be much more virulent than, for example, *P. aeruginosa* PAO1 [45].

1.4.1 Region 5 243 000–5 361 000

The atlas of UCBPP-PA14 reveals many regions of potential interest. For example, a large region around 5250 kbp shows only partial similarity to other *P. aeruginosa* genomes and practically no similarity to more distantly related *Pseudomonas* (see the zoom in Figure 1.5). The region also shows some interesting structural properties and an examination of the annotations reveals that the majority of the genes present are located on the direct strand. The region is the previously described pathogenicity island PAPI-1 which is a cluster of more than 100 genes [46]. Some of the genes within PAPI-1 are homologous to known genes with virulence functions in other human and plant bacterial pathogens, and mutations in a number of these PAPI-1

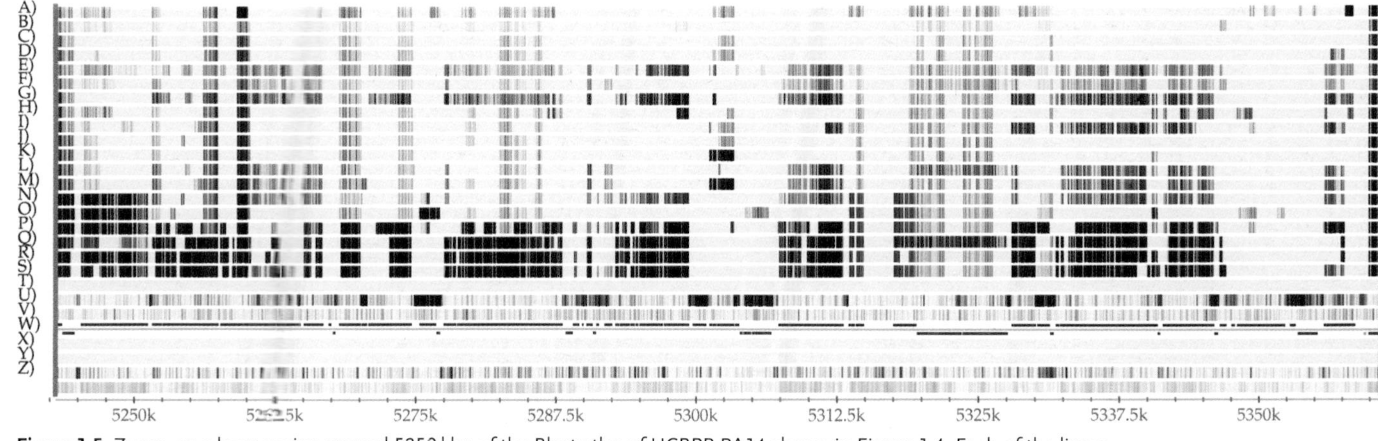

Figure 1.5 Zoom on a large region around 5250 kbp of the Blast atlas of UCBPP-PA14 shown in Figure 1.4. Each of the linear colored bars in this plot reflects one of the circles in the circular atlas; for further explanation of the colors, refer to the description in Figure 1.4 and the description in the main text. This region shows little similarity to other *P. aeruginosa* genomes and less similarity to more distantly related *Pseudomonas*. The region is the previously described pathogenicity island PAPI-1, which is a cluster of more than 100 genes, some of which are homologous to known genes with virulence functions in other human and plant bacterial pathogens.

genes results in the attenuation of PA14 virulence in both plant and animal infection models [46]. PAPI-1 is located next to a $tRNA^{Lys}$ gene, which presumably provides an attachment site for integration of the island after its acquisition.

It was recently experimentally demonstrated that PAPI-1 is a mobile genetic element [47]. The island can be excised from the PA14 genome, transferred and integrated into the genomes of *P. aeruginosa* strains such as PAO1 and other strains that do not harbor the island naturally [47]. Although the molecular mechanism of PAPI-1 transfer remains incompletely understood, PAPI-1 carries several genes predicted to encode proteins that could function in a conjugative transfer mechanism [47]. As the PAPI-1 island carries several virulence determinants and is capable of spreading among *P. aeruginosa* strains it could potentially contribute to the evolution of variants with enhanced pathogenicity due to an increased virulence gene repertoire.

Hybridization studies and microarray genomotyping of different *P. aeruginosa* strains of diverse environmental origin have showed the presence of PAPI-1 sequences in a significant fraction of the strains, although most these isolates appear to carry only a portion of the island [45,46]. This pattern of PAPI-1 sequence distribution, which is also evident in Figure 1.5, is probably due to progressive accumulation of mutations or deletions of genes that are either unnecessary, deleterious or provide no selective advantages to the recipient after acquisition of the island.

1.4.2 Region 713 000–785 000

Another region revealed by the Blast atlas, this time for its high degree of conservation, is found around 713 kbp (see the zoom in Figure 1.6). This region contains 36 genes involved in "Translation, ribosomal structure and biogenesis". These are mainly ribosomal proteins. The region also contains a ribosomal RNA operon and several tRNAs. The region obviously plays an important role in transcription and translation in the organism, and the high degree of conservation is understandable.

Another observation in line with the biological significance of this island is found if one looks to the bars displaying the structural parameters: "Intrinsic curvature", "Stacking Energy" and "Position Preference". These not only reveal the unusual base composition of the rRNA genes (which must in turn give rise to unusual DNA structures), but also show characteristic structures in the intergenic regions between the genes. These are related to promoter activity since the DNA must destack and unwind for the RNA polymerase to bind, thus initiating transcription. The characteristic structures facilitate this, and are indicators of strong promoters and high levels of expression for the associated genes.

1.5 Functional Categories

Are there differences, on a broad level, between the distributions of functional genes in the various different species of *Pseudomonas*? This question can now be evaluated

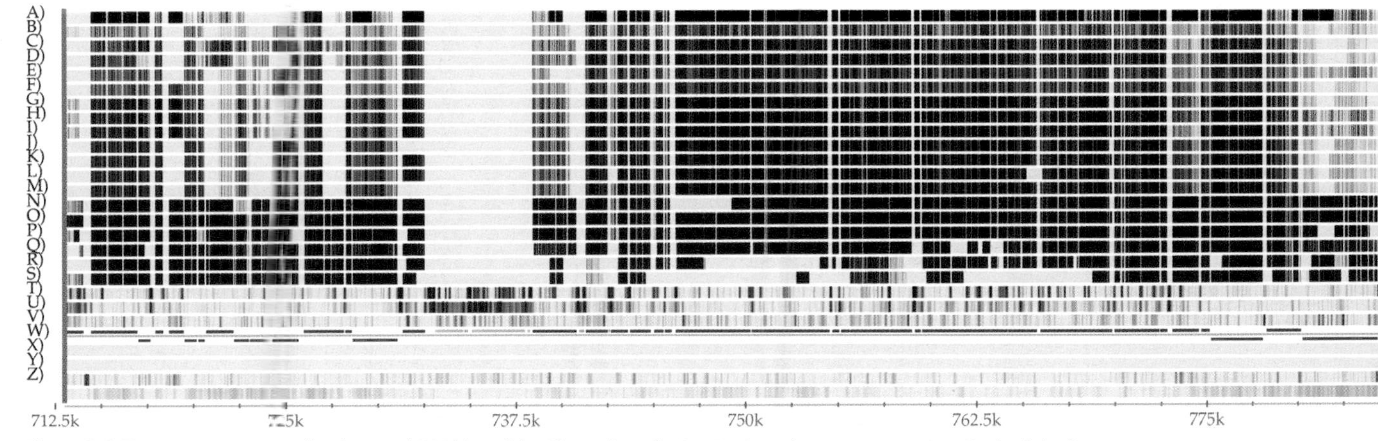

Figure 1.6 Zoom on a genetic island around 750 kbp of the Blast atlas of UCBPP-PA14 shown in Figure 1.4. Each of the linear colored bars in this plot reflects one of the circles in the circular atlas; for further explanation of the colors, refer to the description in Figure 1.4 and the description in the main text. The island is identified by its high degree of conservation across the *Pseudomonas* genus, and contains 36 genes related to translation, ribosomal structure and biogenesis. The region also contains a rRNA operon (visible by the light blue bars in the annotation line) as well as several tRNAs. Notice that the apparent lack of conservation for the rRNA operon is an artifact of the BLAST alignment (which is only performed for translated sequences) and not an accurate representation of the biology.

based on the set of genomes sequenced to date. For each genome, the fraction of genes in each of several different functional categories was determined. A bar diagram of the distribution of Clusters of Orthologous Gene (COGS) functional category assignments [48] obtained from the Integrated Microbial Genomes (IMG) system [49] for *Pseudomonas* is shown in Figure 1.7. At the time of writing the IMG contained assignment information on 16 of the 19 genomes given in Table 1.1. The number of genes for each strain which could be found varied from the 3700 (the lowest) identified for *P. mendocina* YMP to 4811 (the highest) for *P. aeruginosa* 2192. In general, the COG assignments cover roughly 75% of the proteins for each organism in Table 1.1.

As can be seen from Figure 1.7, the great majority of genes (around 90%) have some form of assignment, while approximately 10% have been labeled *function*

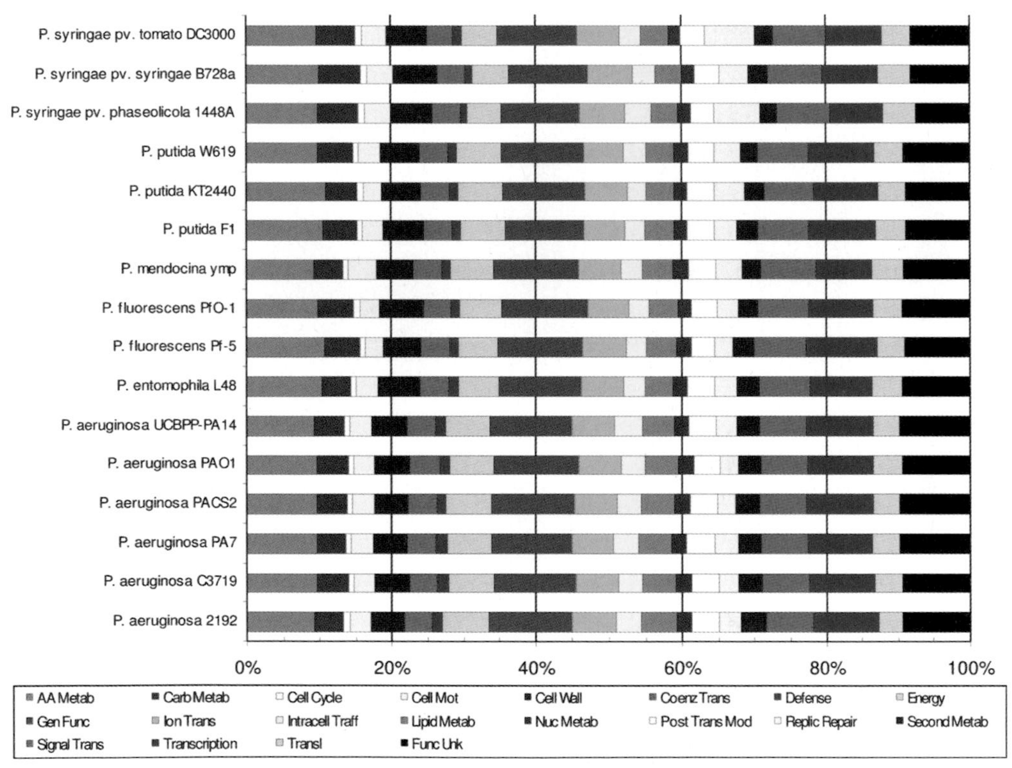

Figure 1.7 The percentage of proteins assigned to COG categories by organism. The figure shows data for the 16 species of *Pseudomonas* found in the IMG database (see text). Although on the whole the functional breakdown of the various strains is very similar, the similarity appears slightly higher within each species than observed when comparing across species.

unknown. (An assignment of *function unknown* is different from a protein not having a COG assignment. A protein assigned *function unknown* is known to exist, although its function is unknown. A protein having no assignment might not be a protein at all!) On the whole the organisms show a high degree of functional similarity at this broad level, although a slightly higher difference is observed when comparing across species relative to within species. It appears, based on these results that there are no major differences on a broad level of distribution of functional categories across all the *Pseudomonas* genomes.

1.6
Codon Usage and Expression

Elucidation of the physical and biological properties of highly expressed genes as well as the examination of codon usage preferences have been addressed by a number of studies [50–52]. Figure 1.8 demonstrates how the average codon usage in

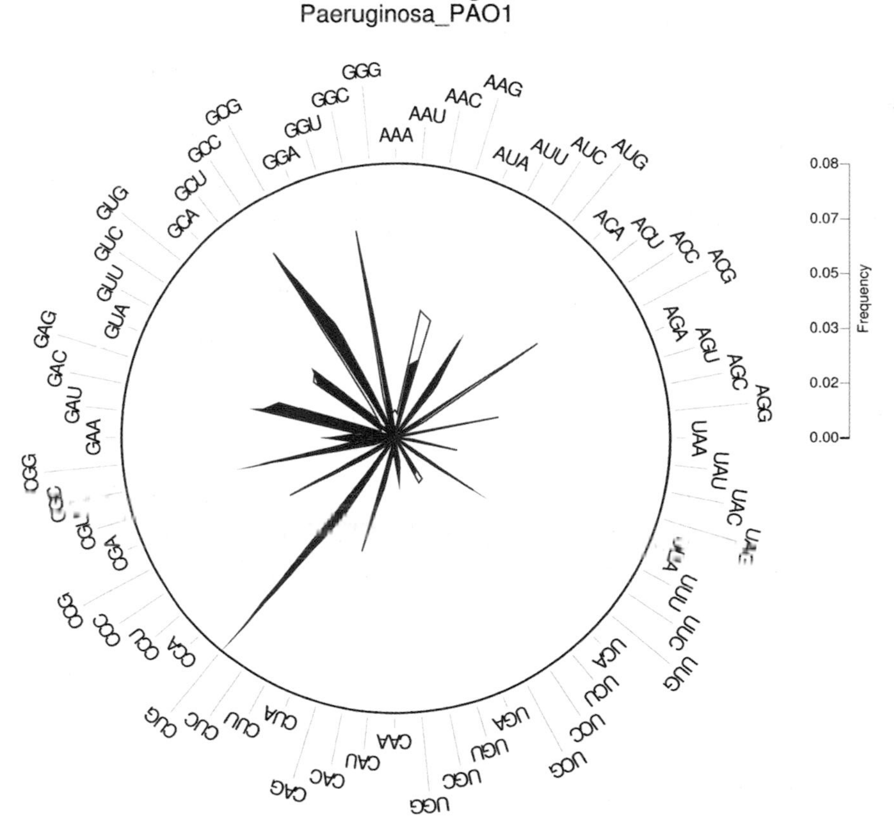

Figure 1.8 Codon usage in *P. aeruginosa* PAO1. Genomic codon usage (blue) versus highly expressed genes (red).

P. aeruginosa PAO1 differs from the codon usage preference of highly expressed genes. This so-called codon usage bias represents an evolutionary strategy to modulate gene expression and hence mathematical formulations of the codon usage bias have been widely used to predict gene expression on a genomic scale [1].

While the first genomes examined in this way – *E. coli* and *Saccharomyces cerevisiae* – provided strong evidence of high translational codon usage bias [52], recent studies report bacterial genomes with little codon usage bias [53,54], *often* these species have genomes with extreme AT or GC content. Thus, the dominating codon bias in *P. aeruginosa* PAO1 is not translational. Here, the codon usage bias is much more correlated with GC content than with translational bias [54]. However, a translational codon adaptation index (tCAI) may still be useful for estimation of gene expression levels and prediction of highly expressed genes in *P. aeruginosa* PAO1 (Figure 1.9). The codon bias in other *Pseudomonas* species may be more or less translational than in PAO1 (Table 1.2).

Obviously, the tCAI measure is only able to predict highly expressed proteins (translated genes) since this measure is based on codon usage bias. Therefore, this method does not consider tRNAs, ribosomal RNAs and other noncoding RNAs. However, transcription of DNA is also highly influenced by DNA stacking and flexibility. Consequently, on a more global scale, gene expression may be regulated by specific promoters that are sensitive to DNA structural properties [2,55]. For example, the "Position Preference" measure (Figure 1.10) is a DNA structural measure that was originally derived for eukaryotes using chicken DNA and is a trinucleotide model of nucleosome-positioning patterns. It reflects the preference of a given trinucleotide for being found in a region where the DNA minor groove faces

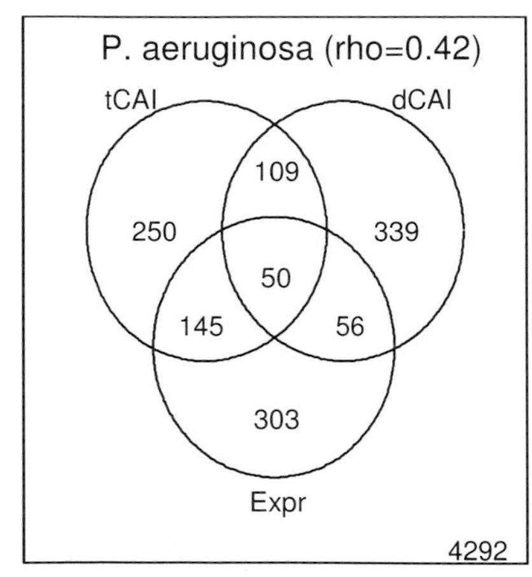

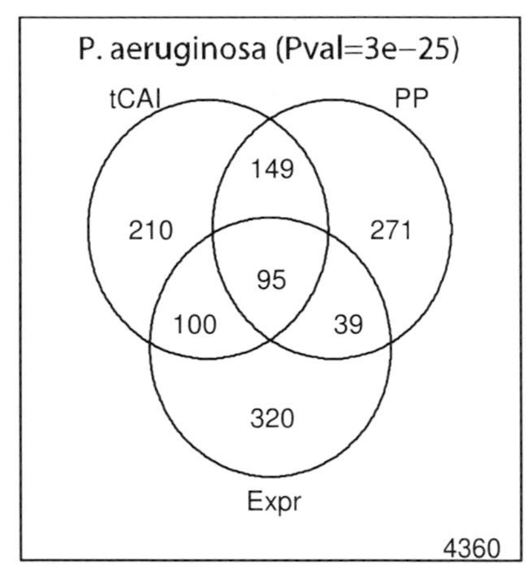

Figure 1.9 Venn diagrams of the 10% most highly expressed genes in *P. aeruginosa* PAO1 as identified by several methods. The tCAI and dCAI circles represent the predictions made by the translational and dominating CAIs, respectively, while PP is the "Position Preference" trinucleotide model and Expr is measured microarray expression values. The figure is modified from Refs. [1,2].

Table 1.2 Correlations between translational and dominating CAI (dCAI).

Organism	Correlation tCAI versus dCAI
P. aeruginosa PAO1	0.42484871
P. entomophila L48	0.52157109
P. fluorescens Pf-5	0.04194573
P. fluorescens PfO-1	0.49853369
P. putida KT2440	0.76612008
P. syringae 1448A	0.77669291
P. syringae B728a	0.65205812
P. syringae DC3000	0.64478432

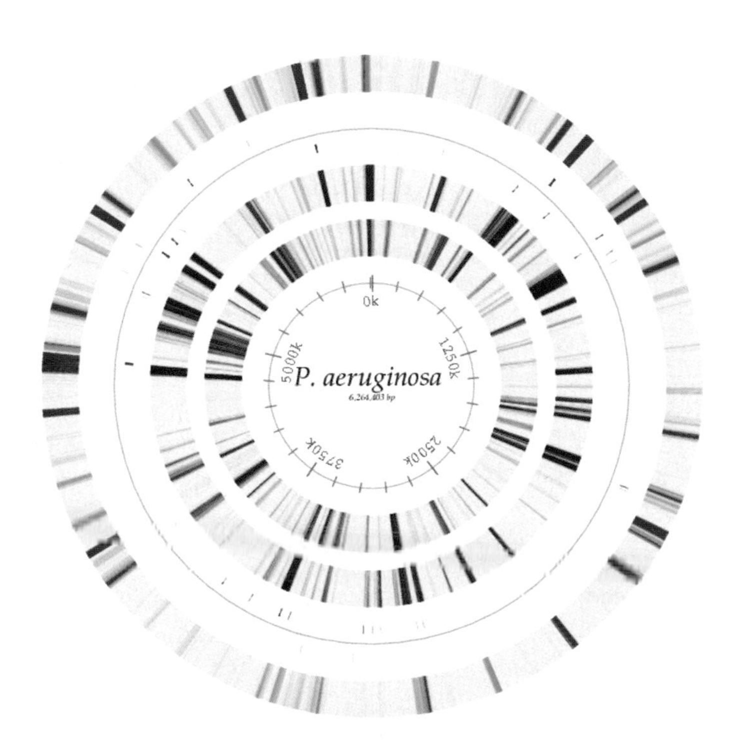

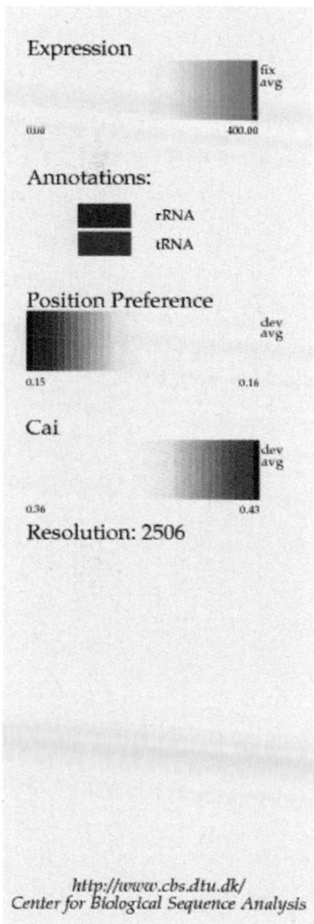

Figure 1.10 Expression atlas of *P. aeruginosa* PAO1. The figure shows a list of measurements relevant for expression in relation to genomic position. From the top (outer ring) and down: the averages of expression values for each gene obtained from microarray experiments; the "Position Preference" of the DNA towards chromatin binding (i.e. a measure of DNA flexibility); estimated values for the tCAI. See text for further explanation of methods.

either towards or away from the nucleosome histone core [42]. By using the absolute values, the position preference value is also a measure of anisotropic DNA flexibility. Consequently, the "Position Preference" measure also describes a more general structural property of DNA, i.e. how easily can it be wrapped around chromatin proteins. As a result, "Position Preference" can be used for prediction of highly expressed genes. This makes sense because regions of DNA that are not condensed into chromatin are more accessible to the RNA polymerase. Although the "Position Preference" measure is less efficient than the tCAI measure in predicting highly expressed genes in *P. aeruginosa* PAO1 (see figure in [2]), chromosomal regions with low "Position Preference" do seem to coincide with regions concentrated in highly expressed genes – in particular, for the regions containing the ribosomal RNAs (Figure 1.10).

1.7 Future Outlook

Only a few years ago to sequence an entire *Pseudomonas* genome in one contiguous piece would literally cost millions of dollars and take more than a year to complete, with an additional year or two to annotate and write up the manuscript. At the time of writing, it is possible to sequence a *Pseudomonas* genome for less than US$10 000, taking less than a day to finish, with an overnight run on the computers to annotate it. To put things in perspective, if the goal of sequencing a human genome for US$1000 or less is achieved, as it likely will within a year or two [56,57], this means to sequence a *Pseudomonas* genome will cost in the range of US$2, i.e. it will take perhaps less time and money than a cup of coffee at a nice restaurant! As the cost of sequencing genomes becomes less expensive, it will become possible to routinely have projects that, instead of obtaining the sequence of just one particular genome from a well-characterized strain, now go for obtaining multiple genomes (perhaps hundreds or even thousands) from entire strain collections. This revolution has potential for a fantastic explosion of data (and perhaps also an explosion of understanding) of our knowledge of *Pseudomonas* genomics. However, there is also the potential for difficult times for many older traditional microbiologists, who have spent the past several decades studying their "favorite gene" in one particular *Pseudomonas* strain and who might not be prepared for so much data – indeed too much data for many microbiologists. This need not be the case, however, since there is now more than ever a need for knowledge of traditional methods, including physiology, in order to put the pieces back together again, in a sense, and understand the whole (very complicated) system.

The challenge in the future will be to develop robust new methods to be able to deal with the enormous amount of valuable sequence information. The beginnings of a new field of study can already be seen – it is now possible to estimate the "pan-genome" and "core genome" of several different *Pseudomonas* genomes, as has been hinted at in this chapter. Perhaps in the future, comparative genomics will move towards comparative "pan-genomics", where one compares the "pan-genome" and "core genome" of species (or other taxonomic groups) against each other, to determine which genes seem to be commonly conserved within groups or ecological niches and which sets of "additional genes" are often found in different environments.

References

1 Willenbrock, H., Friis, C., Friis, A.S. and Ussery, DW. (2006) An environmental signature for 323 microbial genomes based on codon adaptation indices. *Genome Biol*, **7**, R114.

2 Willenbrock, H. and Ussery, D.W. (2007) Prediction of highly expressed genes in microbes based on chromatin accessibility. *BMC Mol Biol*, **8**, 11.

3 Lyczak, J.B., Cannon, C.L. and Pier, G.B. (2000) Establishment of *Pseudomonas aeruginosa* infection: lessons from a versatile opportunist. *Microbes Infect*, **2**, 1051–1060.

4 Stover, C.K., Pham, X.Q., Erwin, A.L., Mizoguchi, S.D., Warrener, P., Hickey, M.J., Brinkman, F.S., Hufnagle, W.O., Kowalik, D.J., Lagrou, M. *et al.* (2000) Complete genome sequence of *Pseudomonas aeruginosa* PAO1, an opportunistic pathogen. *Nature*, **406**, 959–964.

5 Klockgether, J., Würdemann, D., Wiehlmann, L., Binnewies, T.T., Ussery, D.W. and Tümmler, B. (2008) Genome diversity of *Pseudomonas aeruginosa*, in *Pseudomonas: Genetics and Molecular Biology* (ed. P. Cornelis), Horizon Press, Norwich., in press.

6 Brint, J.M. and Ohman, D.E. (1995) Synthesis of multiple exoproducts in *Pseudomonas aeruginosa* is under the control of RhlR-RhlI, another set of regulators in strain PAO1 with homology to the autoinducer-responsive LuxR-LuxI family. *J Bacteriol*, **177**, 7155–7163.

7 Pearson, J.P., Pesci, E.C. and Iglewski, B.H. (1997) Roles of *Pseudomonas aeruginosa las* and *rhl* quorum-sensing systems in control of elastase and rhamnolipid biosynthesis genes. *J Bacteriol*, **179**, 5756–5767.

8 Van Alst, N.E., Picardo, K.F., Iglewski, B.H. and Haidaris, C.G. (2007) Nitrate sensing and metabolism modulate motility, biofilm formation, and virulence in Pseudomonas aeruginosa. *Infect Immun*, **75**, 3780–3790.

9 Furukawa, S., Kuchma, S.L. and O'Toole, GA. (2006) Keeping their options open: acute versus persistent infections. *J Bacteriol*, **188**, 1211–1217.

10 Smith, E.E., Buckley, D.G., Wu, Z., Saenphimmachak, C., Hoffman, L.R., D'Argenio, D.A., Miller, S.I., Ramsey, B.W., Speert, D.P. and Moskowitz, S.M. *et al.* (2006) Genetic adaptation by *Pseudomonas aeruginosa* to the airways of cystic fibrosis patients. *Proc Natl Acad Sci USA*, **103**, 8487–8492.

11 Lee, B., Haagensen, J.A., Ciofu, O., Andersen, J.B., Hoiby, N. and Molin, S. (2005) Heterogeneity of biofilms formed by nonmucoid *Pseudomonas aeruginosa* isolates from patients with cystic fibrosis. *J Clin Microbiol*, **43**, 5247–5255.

12 Jelsbak, L., Johansen, H.K., Frost, A.L., Thogersen, R., Thomsen, L.E., Ciofu, O., Yang, L., Haagensen, J.A., Hoiby, N. and Molin, S. (2007) Molecular epidemiology and dynamics of *pseudomonas aeruginosa* populations in lungs of cystic fibrosis patients. *Infect Immun*, **75**, 2214–2224.

13 Livermore, D.M. (1995) β-Lactamases in laboratory and clinical resistance. *Clin Microbiol Rev*, **8**, 557–584.

14 Poole, K. (2001) Multidrug efflux pumps and antimicrobial resistance in *Pseudomonas aeruginosa* and related

organisms. *J Mol Microbiol Biotechnol*, **3**, 255–264.

15 Livermore, D.M. (2002) Multiple mechanisms of antimicrobial resistance in *Pseudomonas aeruginosa*: our worst nightmare? *Clin Infect Dis.* **34**, 634–640.

16 Franklin, F.C., Bagdasarian, M., Bagdasarian, M.M. and Timmis, K.N. (1981) Molecular and functional analysis of the TOL plasmid pWWO from *Pseudomonas putida* and cloning of genes for the entire regulated aromatic ring meta cleavage pathway. *Proc Natl Acad Sci USA*, **78**, 7458–7462.

17 Regenhardt, D., Heuer, H., Heim, S., Fernandez, D.U., Strompl, C., Moore, E.R. and Timmis, K.N. (2002) Pedigree and taxonomic credentials of *Pseudomonas putida* strain KT2440. *Environ Microbiol*, **4**, 912–915.

18 Dejonghe, W., Boon, N., Seghers, D., Top, E.M. and Verstraete, W. (2001) Bioaugmentation of soils by increasing microbial richness: missing links. *Environ Microbiol*, **3**, 649–657.

19 Timmis, K.N., Steffan, R.J. and Unterman, R. (1994) Designing microorganisms for the treatment of toxic wastes. *Annu Rev Microbiol*, **48**, 525–557.

20 Lehrbach, P.R., Zeyer, J., Reineke, W., Knackmuss, H.J. and Timmis, KN. (1984) Enzyme recruitment*in vitro*: use of cloned genes to extend the range of haloaromatics degraded by *Pseudomonas* sp. strain B13. *J Bacteriol*, **158**, 1025–1032.

21 Nozaki, M., Kagamiyama, H. and Hayaishi, O. (1963) Crystallization and some properties of metapyrocatechase. *Biochem Biophys Res Commun*, **11**, 65–69.

22 Nakazawa, T. (2002) Travels of a *Pseudomonas*, from Japan around the world. *Environ Microbiol*, **4**, 782–786.

23 Federal Register . (1982) Certified Host–Vector Systems. 17197.

24 Nelson, K.E., Weinel, C., Paulsen, I.T., Dodson, R.J., Hilbert, H., Martins dos Santos, V.A., Fouts, D.E., Gill, S.R., Pop, M., Holmes, M. *et al.* (2002) Complete genome sequence and comparative analysis of the metabolically versatile *Pseudomonas putida* KT2440. *Environ Microbiol*, **4**, 799–808.

25 Hirano, S.S. and Upper, C.D. (2000) Bacteria in the leaf ecosystem with emphasis on *Pseudomonas syringae* – a pathogen, ice nucleus, and epiphyte. *Microbiol Mol Biol Rev*, **64**, 624–653.

26 Joardar, V., Lindeberg, M., Jackson, R.W., Selengut, J., Dodson, R., Brinkac, L.M., Daugherty, S.C., Deboy, R., Durkin, A.S., Giglio, M.G. *et al.* (2005) Whole-genome sequence analysis of *Pseudomonas syringae* pv. phaseolicola 1448A reveals divergence among pathovars in genes involved in virulence and transposition. *J Bacteriol*, **187**, 6488–6498.

27 Feil, H., Feil, W.S., Chain, P., Larimer, F., DiBartolo, G., Copeland, A., Lykidis, A., Trong, S., Nolan, M., Goltsman, E. (2005) *et al.* Comparison of the complete genome sequences of *Pseudomonas syringae* pv. syringae B728a and pv. tomato DC3000. *Proc Natl Acad Sci USA*, **102**, 11064–11069.

28 Buell, C.R., Joardar, V., Lindeberg, M., Selengut, J., Paulsen, I.T., Gwinn, M.L., Dodson, R.J., Deboy, R.T., Durkin, A.S., Kolonay, J.F. *et al.* (2003) The complete genome sequence of the *Arabidopsis* and tomato pathogen *Pseudomonas syringae* pv. *tomato* DC3000. *Proc Natl Acad Sci USA*, **100**, 10181–10186.

29 Haas, D. and Defago, G. (2005) Biological control of soil-borne pathogens by fluorescent pseudomonads. *Nat Rev Microbiol*, **3**, 307–319.

30 Lalucat, J., Bennasar, A., Bosch, R., Garcia-Valdes, E. and Palleroni, N.J. (2006) Biology of *Pseudomonas stutzeri*. *Microbiol Mol Biol Rev*, **70**, 510–547.

31 Vodovar, N., Vallenet, D., Cruveiller, S., Rouy, Z., Barbe, V., Acosta, C., Cattolico, L., Jubin, C., Lajus, A., Segurens, B. (2006) *et al.* Complete genome sequence of the entomopathogenic and metabolically versatile soil bacterium *Pseudomonas entomophila*. *Nat Biotechnol*, **24**, 673–679.

32 NCBI Genome Project. http://www.ncbi.nlm.nih.gov/genomes/lproks.cgi.

33 Pathogen Sequencing, Group at the Sanger Institute. ftp://ftp.sanger.ac.uk/pub/pathogens/pae.

34 Larsen, T.S. and Krogh, A. (2003) EasyGene, – a prokaryotic gene finder that ranks ORFs by statistical significance. *BMC Bioinformatics*, **4**, 21.

35 Binnewies, T., Hallin, P.F., Wassenaar, T.M. and Ussery, D.W. (2007) Tools for comparison of bacterial genomes, in *Comparative Genomics and Bioinformatics for the Microbiologist*, Field, D. (Ed.). Horizon Press, Norwich., pp. 00–00.

36 WU-BLAST. http://blast.wustl.edu.

37 Lagesen, K., Hallin, P., Rodland, E.A., Staerfeldt, H.H., Rognes, T. and Ussery, DW. (2007) RNAmmer: consistent and rapid annotation of ribosomal RNA genes. *Nucleic Acids Res*, **35**, 3100–3108.

38 Thompson, J.D., Higgins, D.G. and Gibson, T.J. (1994) CLUSTAL W: improving the sensitivity of progressive multiple sequence alignment through sequence weighting, position-specific gap penalties and weight matrix choice. *Nucleic Acids Res*, **22**, 4673–4680.

39 Gascuel, O. (1997) BIONJ: an improved version of the NJ algorithm based on a simple model of sequence data. *Mol Biol Evol*, **14**, 685–695.

40 Galtier, N., Gouy, M. and Gautier, C. (1996) SEAVIEW and PHYLO_WIN: two graphic tools for sequence alignment and molecular phylogeny. *Comput Appl Biosci*, **12**, 543–548.

41 Hallin, P.F., Binnewies, T.T. and Ussery, D.W. (2004) Genome update: chromosome atlases. *Microbiology*, **150**, 3091–3093.

42 Satchwell, S.C., Drew, H.R. and Travers, A.A. (1986) Sequence periodicities in chicken nucleosome core DNA. *J Mol Biol*, **191**, 659–675.

43 Ornstein, R., Rein, R., Breen, D. and Macelroy R. (1978) An optimized potential function for the calculation of nucleic acid interaction energies, I – base stacking. *Biopolymers*, **17**, 2341–2360.

44 Shpigelman, E.S., Trifonov, E.N. and Bolshoy, A. (1993) CURVATURE: software for the analysis of curved DNA. *Comput Appl Biosci*, **9**, 435–440.

45 Lee, D.G., Urbach, J.M., Wu, G., Liberati, N.T., Feinbaum, R.L., Miyata, S., Diggins, L.T., He, J., Saucier, M., Deziel, E. (2006) *et al.* Genomic analysis reveals that *Pseudomonas aeruginosa* virulence is combinatorial. *Genome Biol*, **7**, R90.

46 He, J., Baldini, R.L., Deziel, E., Saucier, M., Zhang, Q., Liberati, N.T., Lee, D., Urbach, J., Goodman, H.M. and Rahme, LG. (2004) The broad host range pathogen *Pseudomonas aeruginosa* strain PA14 carries two pathogenicity islands harboring plant and animal virulence genes. *Proc Natl Acad Sci USA*, **101**, 2530–2535.

47 Qiu, X., Gurkar, A.U. and Lory, S. (2006) Interstrain transfer of the large pathogenicity island (PAPI-1) of *Pseudomonas aeruginosa*. *Proc Natl Acad Sci USA*, **103**, 19830–19835.

48 Tatusov, R.L., Koonin, E.V. and Lipman D.J. (1997) A genomic perspective on protein families. *Science*, **278**, 631–637.

49 Markowitz, V.M., Korzeniewski, F., Palaniappan, K., Szeto, E., Werner, G., Padki, A., Zhao, X., Dubchak, I., Hugenholtz, P. Anderson, I. (2006) *et al.* The Integrated Microbial Genomes (IMG) system. *Nucleic Acids Res*, **34**, D344–D348

50 Raghava, G.P. and Han, J.H. (2005) Correlation and prediction of gene expression level from amino acid and dipeptide composition of its protein. *BMC Bioinformatics*, **6**, 59.

51 Karlin, S., Barnett, M.J., Campbell, A.M., Fisher, R.F. and Mrazek, J. (2003) Predicting gene expression levels from codon biases in alpha-proteobacterial genomes. *Proc Natl Acad Sci USA*, **100**, 7313–7318.

52 Sharp, P.M. and Li, W.H. (1987) The Codon Adaptation Index – a measure of directional synonymous codon usage bias, and its potential applications. *Nucleic Acids Res*, **15**, 1281–1295.

53 Carbone, A., Kepes, F. and Zinovyev, A. (2005) Codon bias signatures, organization of microorganisms in codon space, and lifestyle. *Mol Biol Evol*, **22**, 547–561.

54 Carbone, A., Zinovyev, A. and Kepes, F. (2003) Codon adaptation index as a measure of dominating codon bias. *Bioinformatics*, **19**, 2005–2015.

55 Willenbrock, H. and Ussery, D.W. (2004) Chromatin architecture and gene expression in *Escherichia coli*. *Genome Biol*, **5**, 252.

56 Service R.F. (2006) Gene sequencing. The race for the $1000 genome. *Science*, **311**, 1544–1546.

57 The Archon X-PRIZE for Genomics. http://www.xprize.org/xprizes/genomics_x_prize.html.

2
Clinical Relevance of *Pseudomonas aeruginosa*: A Master of Adaptation and Survival Strategies

Niels Høiby, Helle Krogh Johansen, Claus Moser, and Oana Ciofu

2.1
Introduction

P. aeruginosa is found in water, including drinking water and sewage, and may survive for days on dry surfaces if it is present in biological secretions such as pus [1]. More interestingly, *P. aeruginosa* is also able to cause a multitude of different infections in man (Table 2.1) [2]. Since the taxonomic work by Stanier *et al.* in 1966 [3] it has been known that the physiology, biochemistry, and resistance to antibiotics and disinfectants explain *P. aeruginosa*'s comprehensive adaptability which allows it to easily survive and multiply in many different habitats in nature and man. Sequencing the genome of the PAO1 strain of *P. aeruginosa* [4] further revealed a large number of regulatory genes, and a large number of genes involved in the catabolism, transport and efflux of organic compounds.

The importance of *P. aeruginosa* as a cause of infections in man is especially due to its resistance to many antibiotics. Concerning parenterally administered antibiotics, wild-type *P. aeruginosa* strains are sensitive to aminoglycosides, some third- and fourth-generation cephalosporins (ceftazidime, cefepime, cefpirone), and ureidopenicillins such as piperacillin with and without tazobactam, carbapenems, monobactams, colistin and some fluoroquinolones (e.g. ciprofloxacin) [5]. However, *P. aeruginosa* strains – even epidemic strains in some hospitals – are frequently found which are resistant to one or more or even all of these antibiotics [6]. The resistance mechanisms may be efflux pumps, β-lactamases, aminoglycoside-modifying enzymes, modified lipopolysaccharide (LPS), and mutations in gyrase and/or topoisomerase genes [5]. The only orally administered antibiotics that are efficient against *P. aeruginosa* are the fluoroquinolones. The clinical consequences of this inherited and acquired antibiotic resistance are that for severe infections, e.g. suspected sepsis in neutropenic patients, the empiric antibiotic therapy should cover *P. aeruginosa* and many clinicians, therefore, use combination therapy of two

Pseudomonas. Model Organism, Pathogen, Cell Factory. Edited by Bernd H.A. Rehm

ISBN: 978-3-527-31914-5

Table 2.1 Range of *P. aeruginosa* infections in man [2].

Organ	Infection	Acute/chronic	Origin	Prevention
Eye	contact lens keratitis	acute	water	hygiene
Skin	folliculitis wound/ulcer infections burn infection	acute	water	hygiene
Ear	external otitis	acute/chronic	water?	hygiene
Nasal sinuses	sinusitis	chronic	water?	?
Urine bladder	urinary tract infection	acute/chronic	?	?
Bones	diabetic osteomyelitis in feet	chronic	water?	hygiene
Lungs/bronchi	ventilator-associated pneumonia	acute	nosocomial humidifiers	hygiene
	endobronchiolitis, cystic fibrosis, bronchiectasis	chronic	nosocomial, environment	hygiene, early aggressive antibiotic eradication therapy of intermittent colonization
Blood	sepsis, neutropenic patients	acute	?	antibiotic prophylaxis

antibiotics with activity against *P. aeruginosa*, e.g. an aminoglycoside and a β-lactam or a carbapenem antibiotic [7]. In intensive care unit patients aminoglycosides with their well-known risk of renal toxicity are often replaced by ciprofloxacin. If the empiric antibiotic therapy does not work due to resistance, the mortality may increase even if the therapy is subsequently changed to effective antibiotics when results of culture and susceptibility testing become available [8–10].

Apart from the environmental reservoirs and antibiotic resistance, which are also found among other species, acute infections caused by *P. aeruginosa* do not reveal the impressive adaptability of these bacteria, whereas this unique property explains the clinical problems characteristic of chronic *P. aeruginosa* infections. Most of our understanding of the ability of *P. aeruginosa* to adapt to different niches in the host during chronic infections comes from studies on the chronic lung infection caused by these bacteria in cystic fibrosis (CF) patients [11] and this topic will therefore be discussed in more detail.

2.2 CF

The consequence of the mutations in the CF transmembrane conductance regulator (*CFTR*) gene is malfunction of the chloride channel in CF patients, which leads to a decreased volume of paraciliary fluid in the lower respiratory tract that in turn leads to impaired mucociliary clearance of inhaled microbes [12]. This impairment of the noninflammatory defense mechanism of the respiratory tract leads to early recruitment of the inflammatory defense mechanisms, e.g. polymorphonuclear leukocytes (PMNs) and antibodies [11,13]. Therefore, from early childhood, CF patients suffer from recurrent and chronic respiratory tract infections characterized by PMN inflammation. In spite of the inflammatory response and intensive antibiotic therapy, however, infections caused by *P. aeruginosa* persist and lead to respiratory failure and lung transplantation or death of the patients [14] (Figure 2.1). Sophisticated adaptive mechanisms of *P. aeruginosa* exist that explain why this pathogen is able to survive and persist for several decades in the respiratory tract of CF patients in spite of the defense mechanisms of the host and intensive antibiotic therapy. As a clinical example we have examined explanted lungs from a 42-year-old CF male with chronic *P. aeruginosa* infection over 28 years. The patient had developed a pronounced antibody response at the beginning of the infection in 1977 and at the same time he began to be treated with regular suppressive antibiotic therapy every 3 months. He had been treated with a total of 114 2-week courses of intravenous tobramycin and anti pseudomonal β-lactam antibiotics, and additionally daily nebulized colistin since 1987. In spite of this massive immunologic and therapeutic attack on his mucoid *P. aeruginosa* strain it continued to multiply and survive in biofilms in his lungs surrounded by activated PMNs which had destroyed his lungs (Figure 2.2). As illustrated by this case, therefore, *P. aeruginosa* is able to survive due to adaptation (i) to the inflammatory defense mechanism, (ii) to the respiratory zone of the lungs, (iii) to the conductive zone of the lungs and (iv) to the antibiotic therapy [15].

2.3 Survival of *P. aeruginosa* by Adaptation to the Inflammatory Defense System

Bronchoalveolar lavage studies in newly diagnosed CF infants have shown that in cases of infection caused by virus or bacteria there are significantly increased numbers of PMNs and alveolar macrophages, whereas this is not the case in noninfected CF infants and normal children [13]. When PMNs try to phagocytose bacteria like *P. aeruginosa* a metabolic burst is produced that leads to liberation of highly reactive oxygen species (ROS) which will kill or induce mutations in the bacteria and damage the surrounding tissue [16]. Such oxygen radical damage has been detected in CF lungs [17]. We have shown that hydrogen peroxide or activated PMNs can induce the same mutation in the *mucA* gene of *P. aeruginosa* PAO1 [18] and we have found mutations in that gene in 93% of mucoid *P. aeruginosa* from Scandinavian CF patients [19], in accordance with other publications [20,21]. The mutation changes the *P. aeruginosa* PAO1 strain to the characteristic mucoid phenotype [18]. Alginate is an oxygen radical scavenger, and protects *P. aeruginosa* against phagocytosis and clearance from the lungs, so that the mucoid phenotype is better protected than the nonmucoid phenotype against the inflammatory defense mechanisms of the host [22], and this is in accordance with our findings that a nonmucoid clinical strain

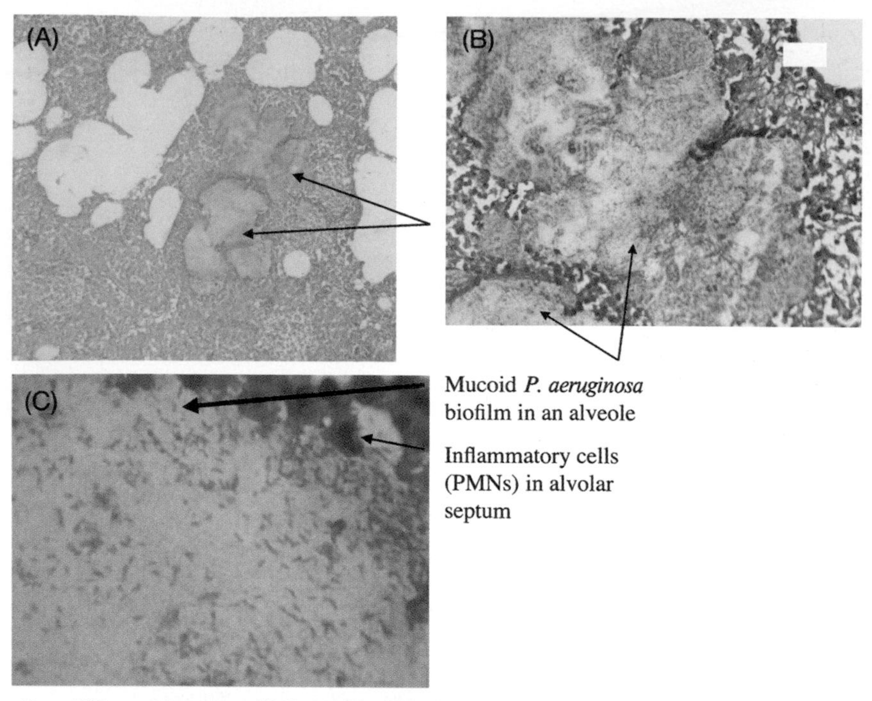

Mature GFP-tagged *P. aeruginosa* biofilm in a flow cell (Søren Molin) similar to the packed aerobic CF alveolar incubation chamber filled with *P. aeruginosa* biofilm

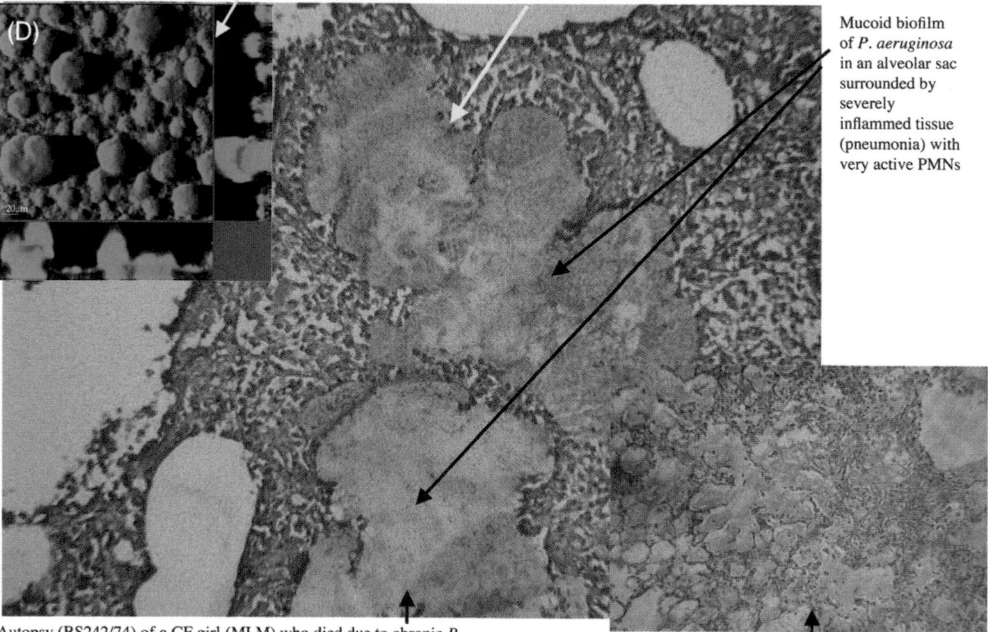

Autopsy (BS242/74) of a CF girl (MLM) who died due to chronic *P. aeruginosa* lung infection. HE stain. · 100

Mucoid *P. aeruginosa* biofilm in alveoli of a CF mouse. HE + Alcian blue stain for alginate. · 40

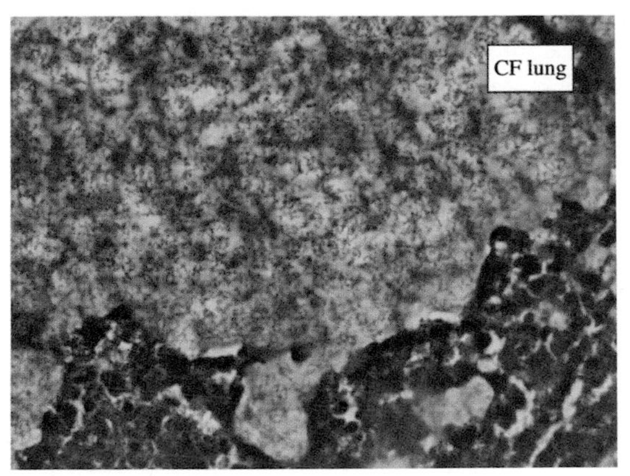

Figure 2.2 Micrograph of an explanted lung from a 41-year-old CF male with chronic *P. aeruginosa* lung infection (mucoid and nonmucoid phenotype) for 28 years. He had 46 precipitating antibodies against *P. aeruginosa*. He had been treated with 114 2-week anti-*P. aeruginosa* treatment courses. Gram stain, magnification ×1000 of a *P. aeruginosa* biofilm surrounded by PMNs.

had higher oxidative damage and mutation frequency than the mucoid isogenic strain [23]. When mucoid phenotypes of *P. aeruginosa* are isolated from sputum of CF patients biofilm growth can be detected by microscopy of Gram-stained smears of sputum from these patients (Figure 2.3) and the identity of the rods has been assured by fluorescence *in situ* hybridization [24]. *In vitro* experiments on biofilms growing in flow cells and animal experiments suggest that resistance of nonmucoid *P. aeruginosa* biofilms to PMNs and the antibiotic tobramycin is dependent on production of quorum sensing (QS)-regulated virulence factors, notably rhamnolipid, and that QS mutants as well as use of QS inhibitors (QSI) such as garlic extract makes the biofilm susceptible to PMN attack and killing as well as eradication by tobramycin [25–28]. Similar results have been obtained in animal experiments by treatment of the *P. aeruginosa* lung infection (mucoid strain) with azithromycin, which also inhibits QS [28].

Figure 2.1 Autopsy of a Danish CF girl who died due to chronic *P. aeruginosa* lung infection. She had 21 precipitating antibodies in serum against *P. aeruginosa*. (A) Hematoxylin & eosin stain ×40. The arrows show mucoid biofilms of *P. aeruginosa* surrounded by pronounced inflammation in the lung tissue. (B) Magnification ×100. (C) Magnification ×1000. The arrows show a mucoid biofilm of *P. aeruginosa* in an alveole (fat arrow) surrounded by severely inflamed tissue (slim arrow, PMNs, pneumonia). (D) Comparison of *P. aeruginosa* growing as an *in vitro* biofilm in a flow cell, growing as a chronic alginate biofilm in a CF mouse model [28] and in a CF patient. Obvious morphological similarities are seen. GFP, green fluorescent protein; HE, hematoxylin & eosin.

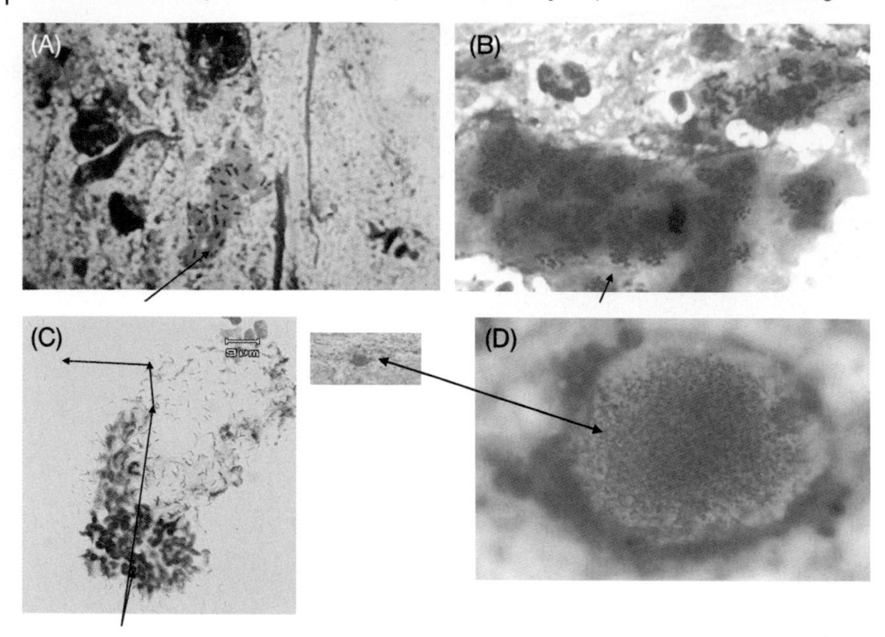

Figure 2.3 Gram-stained sputum from four CF patients. (A and B) *P. aeruginosa* biofilms and PMNs are seen. (C) A *P. aeruginosa* biofilm with two distinct morphologies is seem maybe representing release of nonmucoid revertants from the mucoid biofilm. (D) Sputum from a CF female, 44 years old, growth of mucoid and nonmucoid *P. aeruginosa* in sputum. *P. aeruginosa* since 1970, chronic mucoid *P. aeruginosa* since 1971. She has had 35–44 precipitating antibodies in serum against *P. aeruginosa* for many years. A detached alveolus with a *P. aeruginosa* biofilm is seen (double arrow, magnification ×100 and ×1000).

In conclusion, *P. aeruginosa* adapts to the inflammatory defense system in the respiratory tract of CF patients by forming mucoid biofilms. Possible prophylactic measures could be use of antiinflammatory drugs or antioxidant therapy to prevent ROS-induced mutations in the *mucA* gene or use of macrolides such as azithromycin in doses which inhibit QS, and also alginate synthesis as suggested by the *in vitro* results and animal experiments [28–30]. In addition, QSI treatment of biofilms in the CF lungs may be a promising research subject in the future. Currently, however, the only available therapy is early aggressive eradication therapy of the initial *P. aeruginosa* colonization caused by nonmucoid strains in order to prevent progression to chronic biofilm infection [31], which can be distinguished from intermittent colonization by the antibody response to *P. aeruginosa* (Figure 2.4) and by the characteristic mucoid phenotype (Figure 2.5) [32,33]. This therapy has been used successfully in the Danish CF Center since 1989 and can prevent about 80% of chronic *P. aeruginosa* lung infection in CF patients [34]. Once the chronic biofilm infection is established, the only efficient therapy is chronic suppressive, maintenance therapy with regular anti-pseudomonal antibiotic courses systemically and/or nebulized in order to maintain the lung function for years [35].

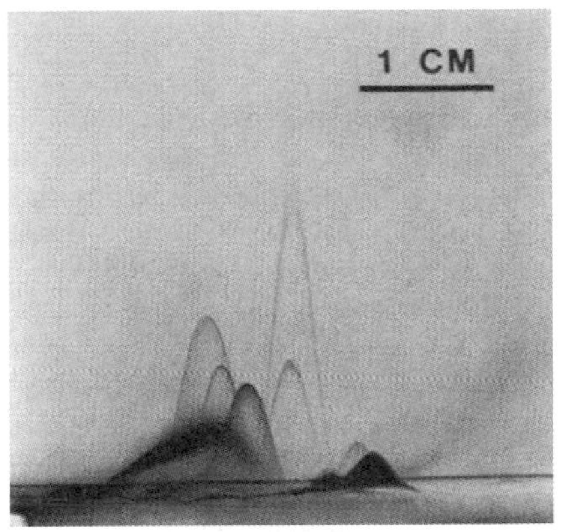

Figure 2.4 Crossed immunoelectrophoresis of *P. aeruginosa* standard antigen against serum from a CF patient with precipitating antibodies against 26 different *P. aeruginosa* antigens. Normal: 0–1 precipitating antibodies; standard antigen: water-soluble antigens from 17 different O-groups obtained by sonication [55].

2.4 Conductive and the Respiratory Zones of the Lungs

The lungs consist of the smaller conductive zone and the larger respiratory zone (Figure 2.6). The respiratory zone (approximately 3000 ml, 95% of the lung volume)

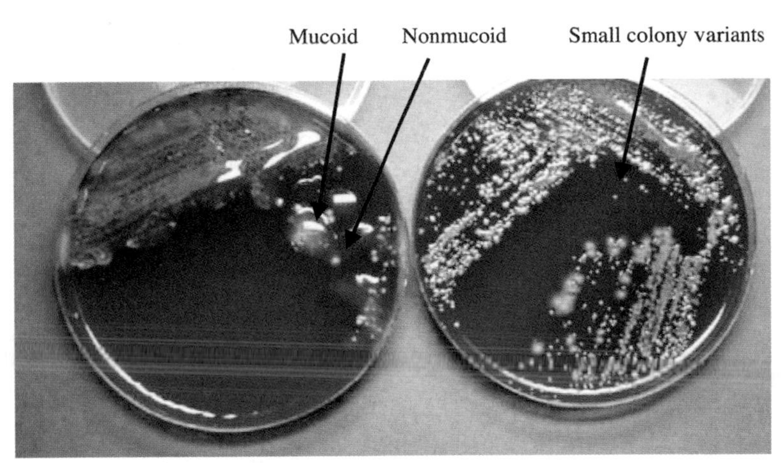

Figure 2.5 Adaptive divergence *in vitro* of *P. aeruginosa* from a CF patient. *P. aeruginosa* CF 57388A (mucoid) after 14 days shaken (left) or nonshaken static (right) aerobic culture in a flask; notice the appearance of three phenotypes: mucoid, nonmucoid and SCVs. Shaken: mucoid colonies dominate (after 1 week there was only mucoid colonies). Static: nonmucoid and SCV colonies dominate (similar results after 1 week). CF 57388A (mucoid) has in *mucA*: at 170 deletion C, frame shift, stop codon 282 and at 293 insertion 105 bp. CF 57388B (nonmucoid) has *additionally* in *algT*: at 147 insertion AGCCCAGGA, frame shift [60].

includes respiratory bronchioles, alveolar ducts and alveolar sacs [36]. This part of the lungs has no cilia, no goblet cells and no submucosal glands, and the defense system consists of alveolar macrophages and defensins, for example. All of the venous blood of the body passes through the capillaries of the alveoles which consist of a nearly continuous sheet of blood, and only a very thin barrier is present between the air and the blood. The smaller conducting zone (approximately 150 ml, 5% of the lung volume) includes trachea, the bronchi and the terminal bronchioles. This part of the lungs has cilia, goblet cells and submucal glands, and has an ordinary arterial blood supply from aorta. The mucus is produced in the respiratory zone and the major defense system consists of the mucociliary escalator [36]. Nebulized tobramycin and colistin and other antibiotics are widely used to treat *P. aeruginosa* lung infection in CF patients. Very high concentrations of these drugs are obtained in the conductive zone (sputum), whereas very little actually reaches the respiratory zone, since the measurable concentration in serum, which reflects the amount in the respiratory zone, is very low [37–42]. On the contrary, when antibiotics are administered intravenously or orally, very low concentrations are found in sputum, but high concentrations in respiratory tissue because the whole dose of, for example, an intravenous bolus of antibiotics is transported directly by the blood to the alveolar capillaries before being distributed to the rest of the body. Since both the respiratory and the conductive zones of the lungs are infected with *P. aeruginosa*, there is pharmacokinetic/pharmacodynamic evidence for using combined systemic and nebulized antibiotics in CF patients [43,44].

2.5 Survival of *P. aeruginosa* by Adaptation to the Respiratory Zone of the Lungs

The condition inside the airspace of the respiratory zone is aerobic (13% O_2, 5% CO_2) if the air supply and the blood supply is normal or if the air supply is normal but the blood supply absent (20% O_2, 0.003% CO_2). If there is mucus plugging and therefore no airflow, then there is a microaerophilic condition (5% O_2, 6% CO_2) [36]. In the case of abscess formation, the condition is anaerobic, and characterized by necrosis of the cells and tissue. *P. aeruginosa* can grow under all these conditions, but the generation time is shorter in an aerobic atmosphere where O_2 is the final electron acceptor due to the much higher ATP yield compared to anaerobic conditions where NO_3^- from, for example, alveolar macrophages or PMNs [45] [iNOS $\rightarrow$ NO + ROS ($O_2 \rightarrow NO_3$)] is the final electron acceptor [46]. Neither macrophages nor PNMs survive for prolonged time in anaerobic niches.

Mucoid *P. aeruginosa* biofilms are located in the respiratory zone of CF lungs according to autopsy findings (Figure 2.1) and a pronounced antibody response against alginate is produced by the patients (Figure 2.4) [47,48]. This is in accordance with the immunological role of alveolar macrophages as antigen-presenting cells in the response to offending pulmonary pathogens. The pronounced PMN-dominated inflammation which surrounds biofilm-containing alveoles leads to tissue damage and a decrease of lung function [49]. Detached biofilm-containing alveoles can sometimes be seen in sputum (Figure 2.3D), and degradation products from elastin and collagen can be detected in the urine [50,51]. The healthy lungs consist of 300 000 000 alveoles [36] and a 1–2% decline of pulmonary function per year implies, therefore, a loss of $3–6 \times 10^6$ alveoles per year = 2700–5400 alveoles per day. The pathogenesis of the tissue damage is immune complex-mediated inflammation promoted by the pronounced antibody response [11] which is characteristic for a T helper 2-polarized immune response [52]. The *P. aeruginosa* biofilm infection is a focal infection, where terminal tissue damage can be detected by high-resolution computed tomography scan in chronically infected CF patients even with normal lung function [53]. Gradually, however, the focal infection spreads to new areas of the lungs and the respiratory function declines in spite of antibiotic therapy. An important observation is, however, that there are many PMNs in the tissue surrounding the mucoid *P. aeruginosa* biofilms, whereas there are no or few PMNs inside the biofilms in accordance with the protective activity of alginate (Figures 2.1 and 2.3). In conclusion, *P. aeruginosa* adapts to the respiratory zone by forming mucoid biofilms which survive for decades in spite of the inflammatory response, whereas the lung tissue is gradually destroyed. Possible prophylactic measures are those which are suggested above to prevent adaptation to the inflammatory response.

2.6
Survival of *P. aeruginosa* by Adaptation to the Conductive Zone of the Lungs

The conductive zone of the adult lungs consists of 16 generations of bronchi from trachea to the terminal bronchioles (Figure 2.6) [36]. Only the four to five proximal generations can be directly observed by bronchoscopy *in vivo*, whereas the distal generations can be indirectly investigated by lavage or directly in lungs from transplanted or succumbed patients. Worlitzsch *et al.* [54] have shown that *P. aeruginosa* in the conductive zone is mainly localized inside sputum in biofilms (microcolonies), whereas very few bacteria are localized at the epithelial surface of the bronchi. Furthermore, they showed that there are anaerobic conditions in sputum. Routine microcopy of Gram-stained smears of sputum from CF patients shows that, in addition to mucoid biofilms, planktonic single *P. aeruginosa* cells are also found [55]. Additionally, cultures of sputum regularly show mucoid, nonmucoid and sometimes also small colony variant (SCV) phenotypes in sputum from each individual CF patient [55–57]. The reason for this characteristic phenotypic variation has been unclear for many years, but the elegant experiments with *P. fluorescens* published by Rainey *et al.* [58] and with *P. aeruginosa* by [59] Wyckoff *et al.* and by my group using *P. aeruginosa* isolates from a CF patient [60] have shown that such phenotypic variation occurs when different niches are present in the growth media. The fundamental driving forces for the phenotypic variation are mutations (or recombinations) caused, for example, by DNA-damaging free radicals and the bacterial SOS response, followed by competition and selection for different niches in a spatially heterogenous environment that, in turn, leads to adaptive divergence [61]. Such different niches occur during stationary batch culture in Erlenmeyer

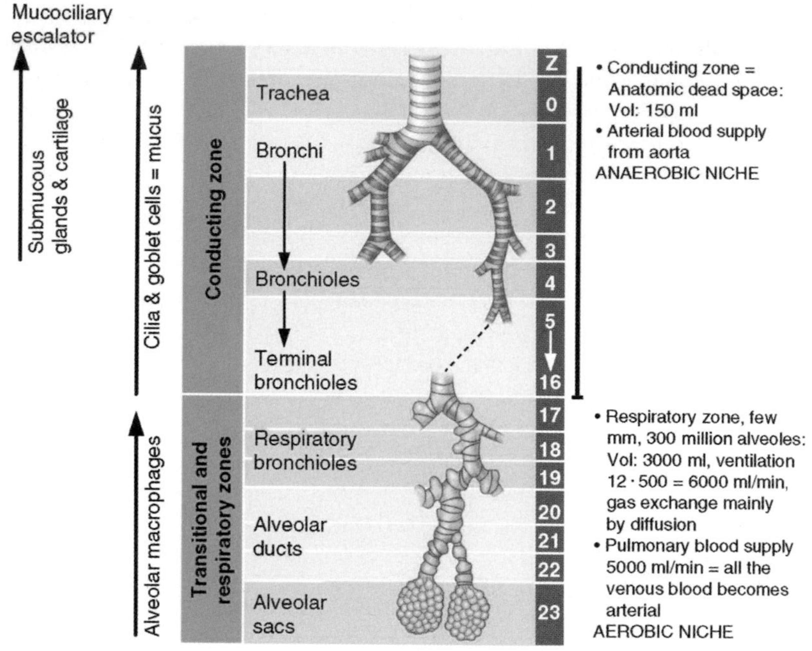

Figure 2.6 The conductive and respiratory zone of the lungs [15].

flasks, where oxygen is plentiful at the surface, whereas it is restricted at the bottom. After 2–3 days of culture this leads to a split-off of new specialist phenotypes of *P. aeruginosa* in addition to the wild phenotype. This does not occur in shaken batch cultures where homogenous conditions exists in the flasks. Similar results have been obtained by my group [29] with a clinical mucoid *P. aeruginosa* CF strain which split-off a nonmucoid phenotype and sometimes also SCVs (Figures 2.2 A and C, and 2.6) which are also characteristic of CF patients [56,57]. The mucoid phenotype has been shown to be unstable during anaerobic conditions and the advantage of the non-mucoid phenotype is thought to be due to its flagella-mediated motility that is negatively regulated by the σ factor *algT*, which also positively regulates the alginate biosynthesis [59]. Nonmucoid *P. aeruginosa* are therefore able to move to the aerobic surface, where the aerobic growth gives the selective advantage of a faster growth rate. As mentioned above, most mucoid *P. aeruginosa* strains from CF patients have a mutation/insertion in *mucA* and this is also the case in the great majority of nonmucoid (revertant) isolates from the same patients [19–21]. Some, but not all, of the nonmucoid revertant phenotypes have mutations/insertions in *algT* which could explain the phenotype [21]. Furthermore, many of the nonmucoid revertants are QS deficient and therefore not virulent [62]. Nonmucoid phenotypes generally do not give raise to a pronounced antibody response in the few patients who only are colonized with these phenotypes [47,63,64]. This is probably due to their location in sputum in the conductive zone far from the key cells of the immune system. Likewise, nonmucoid phenotypes of *P. aeruginosa* have been shown not to be associated with poor prognosis or deteriorating lung function [63], and this is in accordance with their location in sputum in the conductive zone [54] where they probably merely induce increase sputum production and obstruction. According to these results, therefore, mucoid biofilms from the aerobic respiratory zone are transported to the anaerobic sputum in the conductive zone and this is where the nonmucoid phenotypes split-off. Furthermore, the anaerobic conditions of sputum probably means that the abundance of dead PMNs in sputum in the conductive zone have been activated already in the respiratory zone. This is in accordance with the observation that sputum contains an abundance of released DNA, elastase and myeloperoxidase from the dead PMNs [65–67]. In conclusion, *P. aeruginosa* adapts to the conductive zone by splitting off nonmucoid revertants which probably do not contribute much to the tissue damage and therefore do not require special therapeutic or prophylactic attention.

2.7 Survival of *P. aeruginosa* by Adaptation to the Antibiotic Therapy

In spite of intensive antibiotic therapy, chronic *P. aeruginosa* lung infection is rarely if ever eradicated. The main reason is most likely the biofilm way of growth [68–74], but the frequent occurrence of multiply resistant *P. aeruginosa* in CF patients implies that conventional resistance mechanisms probably also play a role [75]. There are many reports of β-lactam-resistant *P. aeruginosa* strains from CF patients [76–79] due to stable or partly stable derepression of chromosomal β-lactamase [80], multiple mutations causing resistance to ciprofloxacin [81], frequent occurrence of resistance to tobramycin [82–85] and also resistance to colistin [74,86,87]. We have shown that biofilm-growing *P. aeruginosa* respond to β-lactam antibiotics by increased production of chromosomal β-lactamase [80,88,89], and that nonmucoid phenotypes are more resistant (higher minimum inhibitory concentration to antibiotics) [75] and have higher oxygen radical damage measured as 8-oxodG/10^6 dG than mucoid phenotypes from the same patients [23] (Table 2.2). An important reason for the occurrence of multiply resistant *P. aeruginosa* strains in CF patients is the high frequency of mutator strains as shown by Oliver *et al.* [90] and confirmed by us [23] (Figure 2.7) Mutator strains are characterized by defects of their DNA repair enzymes and the consequence of this is a much higher frequency of mutations compared to wild-type strains [91]. Such mutator strains of *P. aeruginosa* from CF patients are frequently multiply resistant to antibiotics [23,90–93]. We have shown that such mutator strains generally have increased levels of DNA damage which may be caused by ROS from activated PMNs during the chronic infection [23]. We have also found that no mutator strains were detectable during the first few years of the chronic infection, but after 5 years the prevalence of mutator strains increased with time [23]. By population analysis for mutators, in sputum of CF patients we found a prevalence of 54% [23]. *P. aeruginosa*, therefore, adapts to the intensive antibiotic therapy by producing mucoid biofilms and by becoming mutators. In conclusion, *P. aeruginosa* adapts to the intense antibiotic therapy by forming mucoid biofilms and by

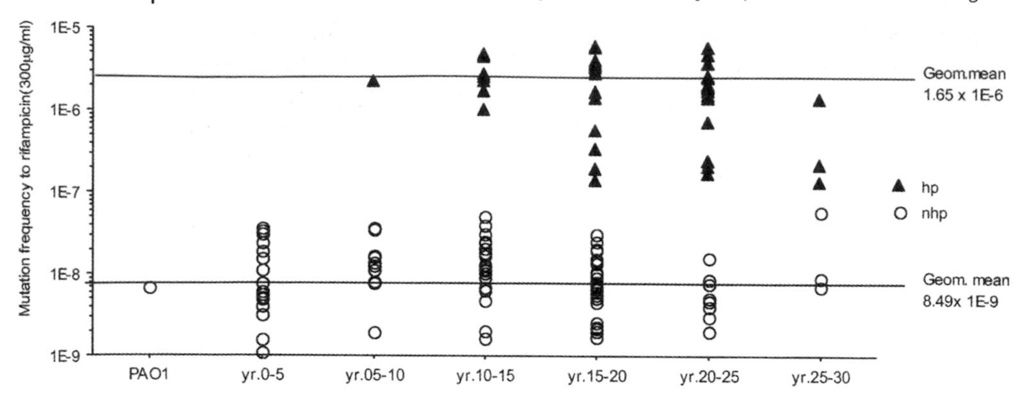

Figure 2.7 Mutation frequency after exposure to rifampicin of 141 *P. aeruginosa* isolates collected in five periods of chronic lung infection (0–5, 5–10, 10–15, 15–20 and 20–25 years) from 11 patients with CF. The results for hypermutable (filled triangles) and nonhypermutable (open circles) *P. aeruginosa* isolates are presented. The mutation frequency after exposure to rifampicin of reference strains PAO1 is shown [23].

Table 2.2 Levels of 8-oxodG/10^6 dG in hypermutable and nonhypermutable *P. aeruginosa* isolates from cystic fibrosis patients[a] [23].

	8-oxodG/10^6 dG	
CF patient no.	**Hypermutable isolates**	**Nonhypermutable isolates**
CF 1	52	39.6
CF 2	62.3	60.1
CF 3	71.9	57.8
CF 4	70.2	40.8
CF 5	105.4	37.3
CF 5		49
CF 6	51.4	ND[b]
CF 6	70.6	
CF 6	90.2	
CF 7	50.3	ND[b]
CF 7	56	
CF 8	117.9[c]	75[d]
CF 9	67.4[d]	58.8[e]

[a] For patients CF 1–CF 7 the 8-oxodG levels were measured by the use of single isolates. These isolates were collected from the population analysis of the sputum samples (cross-sectional study). For the patients CF 8 and CF 9 the levels are means of the levels for several isolates from the longitudinal study.
[b] Not determined.
[c] Mean for five isolates.
[d] Mean for four isolates.
[e] Mean for two isolates.

conventional resistance mechanisms which act synergistically to allow *P. aeruginosa* to survive for decades in spite of the antibiotic therapy, whereas the lung tissue is gradually destroyed. Possible prophylactic measures are those which are suggested above to prevent adaptation to the inflammatory response, but additionally we have shown that induction of neutralizing antibodies against chromosomal β-lactamase of *P. aeruginosa* improves the clinical outcome of antibiotic therapy with β-lactam antibiotics [94–96]. This strategy may therefore be utilized clinically in the future.

2.8 Evolutionary Implications of the Adaptability of *P. aeruginosa*

The *P. aeruginosa* PAO1 strain was sequenced by Stover *et al.* in 2000 [4]. It has 6.2 Mbp and 5570 genes, among which is a high proportion of regulatory, catabolic, transport, efflux and chemotaxis genes, which explains much of its adaptability [4]. When investigating clinical strains from chronic lung infection of CF patients, there are some obvious differences from the PAO1 strain as described above. CF strains are alginate producers, grow in biofilms, split-off nonmucoid and SCV phenotypes, are often mutators, and have accumulated mutations and have become auxotrophs. We have found that clinical CF strains grow much slower than PAO1: the mean generation time is 2- to 3-fold slower than environmental isolates in laboratory media and *in vivo* in the lungs of CF patients the generation time is 100–200 min [97]. When such clinical strains have survived in CF lungs for 30 years, they may have experienced 65000 divisions during which time niche specialists have evolved which are well adapted to survive in CF lungs, but may have less capability to survive elsewhere. Compared to the evolution of our own species, the genetic and phenotypic difference between PAO1 and CF *P. aeruginosa* strains may be comparable to the distance between *Homo erectus* and modern *H. sapiens.*

We have also found that the biofilm-forming ability *in vitro* decreased over time when subsequent nonmucoid isolates of the same clone of *P. aeruginosa* colonizing CF patients were studied for many years [62]. This change was associated with a loss of motility, and with decreased production of virulence factors and QS molecules, and the isolates also became hypermutable [62]. This was not the case with mucoid isolates of the same clone [98]. These findings are in accordance with the results of Smith *et al.* who studied a series of subsequent nonmucoid isolates during 8 years of infection in a CF patient [99]. They found numerous genetic adaptations during the 8-year period. Sequencing of the genome of early and late mucoid and nonmucoid isolates of *P. aeruginosa* from individual CF patients is therefore urgently needed to investigate the evolution of the *P. aeruginosa* genome in CF lungs during decades of chronic biofilm infection in spite of immune response and antibiotic therapy.

References

1 Zimakoff, J., Høiby, N., Rosendal, K. and Guilbert, J.P. (1983) Epidemiology of *Pseudomonas aeruginosa* infection and the role of contamination of the environment

in a cystic fibrosis clinic. *J Hosp Infect*, **4**, 31–40.

2 Campa, M., Bendinelli, M. and Friedman, H. (1993) Pseudomonas aeruginosa *as an Opportunistic Pathogen*, Plenum Press, New York.

3 Stanier, RY., Palleroni, NJ. and Doudoroff, M. (1966) The aerobic pseudomonads: a taxonomic study. *J Gen Microbiol*, **43**, 159–271.

4 Stover, CK., Pham, XQ., Erwin, AL., Mizoguchi, SD., Warrener, P., Hickey, MJ., Brinkman, FSL., Hufnagle, WO., Kowalik, DJ., Lagrou, M., Garber, RL., Goltry, L., Tolentino, E., WestbrockWadman, S., Yuan, Y., Brody, LL., Coulter, SN., Folger, KR., Kas, A., Larbig, K., Lim, R., Smith, K., Spencer, D., Wong, GKS., Wu, Z., Paulsen, IT., Reizer, J., Saier, MH., Hancock, REW., Lory, S. and Olson, MV. (2000) Complete genome sequence of *Pseudomonas aeruginosa* PAO1, an opportunistic pathogen. *Nature*, **406**, 959–964.

5 Ciofu, O. (2003) *Pseudomonas aeruginosa* chromosomal β-lactamase in patients with cystic fibrosis and chronic lung infection – mechanism of antibiotic resistance and target of the humoral immune response. *APMIS*, **111**, 4–47.

6 Obritsch, M., Fish, DN., Maclaren, R. and Jung, R. (2004) National surveillance of antimicrobial resistance in *Pseudomonas aeruginosa* isolates obtained from intensive care unit patients from 1993–2002. *Antimicrob Agents Chemother*, **48**, 4606–4610.

7 Bodey, GP. (2005) Management of febrile neutropenia: current strategies. *Clin Infect Dis*, **40** (Suppl 4), s237–s358.

8 Kang, C.-I., Kim, S.-H., Park, WB., Lee, K.-D., Kim, H.-B., Kim, E.-C., Oh, M. and Choe, K.-W. (2005) Bloodstream infections caused by antibiotic-resistant Gram-negative bacilli: risk factors for mortality and impact of inappropriate initial antimicrobial therapy on outcome. *Antimicrob Agents Chemother*, **49**, 760–766.

9 Russell, JA. (2006) Management of sepsis. *N Engl J Med*, **355**, 1699–1713.

10 Osih, RB., McGregor, JC., Rich, SE., Moore, AC., Furono, JP., Perencevich, EN. and Harris, AD. (2007) Impact of empiric antibiotic therapy outcomes in patients with *Pseudomonas aeruginosa* bacteremia. *Antimicrob Agents Chemother*, **51**, 839–844.

11 Høiby, N., Johansen, H.K., Moser, C., Song, Z.J., Ciofu, O. and Kharazmi, A. (2001) *Pseudomonas aeruginosa* and the biofilm mode of growth. *Microb Infect*, **3**, 1–13.

12 Boucher, R.C. (2004) New concepts of the pathogenesis of cystic fibrosis lung disease. *Eur Respir J*, **23**, 146–158.

13 Armstrong, D.S., Hook, S.M., Jamsen, K.M., Nixon, G.M., Carzino, R., Carlin, J.B., Robertson, C.F. and Gromwood, K. (2005) Lower airway inflammation in infants with cystic fibrosis detected by newborn screening. *Pediatr Pulmonol*, **40**, 500–510.

14 Frederiksen, B., Koch, C. and Høiby, N. (1999) The changing epidemiology of *Pseudomonas aeruginosa* infection in Danish cystic fibrosis patients 1974–1995. *Pediatr Pulmonol*, **28**, 159–166.

15 Høiby, N. (2006) *P aeruginosa* in cystic fibrosis patients resists host defenses, antibiotics. *Microbe*, **1**, 571–577.

16 Miller, R.A. and Britigan, B.E. (1997) Role of oxidants in microbial pathophysiology. *Clin Microbiol Rev*, **10**, 1–18.

17 Hull, J., Vervaart, P., Grimwood, K. and Phelan, P. (1997) Pulmonary oxidative stress response in young children with cystic fibrosis. *Thorax*, **52**, 557–560.

18 Mathee, K., Ciofu, O., Sternberg, C., Lindum, P.W., Campbell, J.I.A., Jensen, P., Johnsen, A.H., Givskov, M., Molin, S., Høiby, N. and Kharazmi, A. (1999) Mucoid conversion of *Pseudomonas aeruginosa* by hydrogen peroxide: a mechanism for virulence activation in the cystic fibrosis lung. *Microbiology*, **145**, 1349–1357.

19 Ciofu, O., Johanneson, M., Hermansen, N.O., Meyer, P. and Høiby, N. (2008) Regulation of the algT operon sequence in mucoid and non-mucoid *P. aeruginosa* isolates from 115 Scandinavian patients

with cystic fibrosis D and in 88 *in vitro* non-mucoid revertants. *Microbiology*, **154 (pt1)**, 103–113.

20 Boucher, J.C., Martinezsalazar, J., Schurr, M.J., Mudd, M.H., Yu, H. and Deretic, V. (1996) Two distinct loci affecting conversion to mucoidy in *Pseudomonas aeruginosa* in cystic fibrosis encode homologs of the serine protease HtrA. *J Bacteriol*, **178**, 511–523.

21 Bragonzi, A., Wiehlmann, L., Klockgether, J., Cramer, N., Worlizsch, D., Döring, G. and Tümmler, B. (2006) Sequence diversity of the mucoid *mucABD* locus in *Pseudomonas aeruginosa* isolates from patients with cystic fibrosis. *Microbiology*, **152**, 3261–3269.

22 Song, Z.J., Wu, H., Ciofu, O., Kong, K.F., Høiby, N., Rygaard, J., Kharazmi, A. and Mathee, K. (2003) *Pseudomonas aeruginosa* alginate is refractory to T_h1 immune, response and impedes host immune clearance in a mouse model of acute lung infection. *J Med Microbiol*, **52**, 731–740.

23 Ciofu, O., Riis, B., Pressler, T., Poulsen, H.E. and Høiby, N. (2005) Occurrence of hypermutable *P aeruginosa* in cystic fibrosis patients is associates with the oxidative stress caused by chronic lung inflammation. *Antimicrob Agents Chemother*, **49**, 2276–2282.

24 Bjarnsholt, T., Jensen, P.Ø., Madsen, K., Andersen, C.B., Pressler, T., Givskov, M. and Høiby, N. (2007) The bacterial burden of the CF lung – uncovering the black box. *J Cystic Fibrosis*, **6**, (Suppl 1), s19.

25 Bjarnsholt, T., Jensen, P.Ø., Rasmussen, T.B., Christophersen, L., Calum, H., Hentzer, M., Hougen, H.P., Rygaard, J., Eberl, L., Høiby, N. and Givskov, M. (2005) Garlic blocks quorum sensing and promotes rapid clearing of pulmonary *Pseudomonas aeruginosa* infections. *Microbiology*, **151**, 3873–3880.

26 Bjarnsholt, T., Jensen, P.-Ø., Burmølle, M., Hentzer, M., Haagensen, J.A.J., Hougen, H.P., Calum, H., Madsen, K.G., Moser, C., Molin, S., Høiby, N. and Givskov, M. (2005) *Pseudomonas aeruginosa* tolerance to tobramycin, hydrogen peroxide and polymorphonuclear leukocytes is quorum-sensing dependent. *Microbiology*, **151**, 373–383.

27 Jensen P.Ø., Bjarnsholt, T., Phipps, R., Rasmussen, T.B., Calum, H., Christoffersen, L., Moser, C., Williams, P., Pressler, T., Givskov, M. and Høiby, N. (2007) Rapid necrotic killing of polymorphonuclear leukocytes is caused by quorum-sensing-controlled production of rhamnolipid by *Pseudomonas aeruginosa*. *Microbiology*, **153**, 1329–1338.

28 Hoffmann, N., Lee, B., Hentzer, M., Rasmussen, T.B., Song, Z., Johansen, H.K., Givskov, M. and Høiby, N. (2007) Azithromycin blocks quorum sensing and alginate polymer formation and increases the sensitivity to serum and stationary growth phase killing of *P aeruginosa* and attenuates chronic *P. aeruginosa* lung infection in $Cftr^{-/-}$ mice. *Antimicrob Agents Chemother*, **51**, 3677–3687.

29 Kobayashi, H. (1995) Airway biofilm disease: its clinical manifestation and therapeutic possibilities of macrolides. *J Infect Chemother*, **1**, 1–15.

30 Tateda, K., Comte, R., Pechere, J.C., Kohler, T., Yamaguchi, K. and VanDelden, C. (2001) Azithromycin inhibits quorum sensing in *Pseudomonas aeruginosa*. *Antimicrob Agents Chemother*, **45**, 1930–1933.

31 Döring, G. and Høiby, N. (2004) Early intervention and prevention of lung disease in cystic fibrosis: a European consensus. *J Cystic Fibrosis*, **3**, 67–91.

32 Proesmans, M., Balinska-Miskiewicz, W., Dupont, L., Bossuyt, X., Verhaegen, J., Høiby, N. and De Boeck, K. (2006) Evaluating the "Leeds criteria" for *Pseudomonas aeruginosa* infection in a cystic fibrosis center. *Eur Respir J*, **27**, 937–943.

33 Pressler, T., Frederiksen, B., Skov, M., Garred, P., Koch, C. and Høiby, N. (2006) Early rise of anti-*Pseudomonas* antibodies and a mucoid phenotype of *Pseudomonas aeruginosa* are risk factors for development of chronic lung infection – a case control study. *J Cystic Fibrosis*, **5**, 9–15.

34 Høiby, N., Frederiksen, B. and Pressler, T. (2005) Eradication of early *Pseudomonas aeruginosa* infection. *J Cystic Fibrosis*, **4**, 49–54.

35 Döring, G., Conway, S.P., Heijerman, H.G.M., Hodson, M.E., Høiby, N., Smyth, A. and Touw, D.J. (2000) Antibiotic therapy against *Pseudomonas aeruginosa* in cystic fibrosis: a European consensus. *Eur Respir J*, **16**, 749–767.

36 Westh, J.B. (2001) *Pulmonary Physiology and Pathophysiology*, Lippincott Williams & Wilkins, Philadelphia, PA.

37 Permin, H., Koch, C., Høiby, N., Christensen, H.O., Møller, A.F. and Møller, S. (1983) Ceftazidime treatment of chronic *Pseudomonas aeruginosa* respiratory tract infection in cystic fibrosis. *J Antimicrob Chemother*, **12** (Suppl A), 313–323.

38 Levy, J., Smith, A.L., Koup, J.R., Williams-Warren, J. and Ramsey, B. (1984) Disposition of tobramycin in patients with cystic fibrosis: a prospective controlled study. *J Pediatr*, **105**, 117–124.

39 Ramsey, B.W., Pepe, M.S., Quan, J.M., Otto, K.L., Montgomery, A.B., WilliamsWarren, J., Vasiljev, K.M., Borowitz, D., Bowman, C.M., Marshall, B.C., Marshall, S. and Smith, A.L. (1999) Intermittent administration of inhaled tobramycin in patients with cystic fibrosis. *N Engl J Med*, **340**, 23–30.

40 Gibson, R.L., Emerson, J., McNamara, S., Burns, L.L., Rosenfeld, M., Yunker, A., Hamblett, N., Accurso, F., Dovey, M., Hiatt, P., Konstan, M.W., Moss, R., RetschBogart, G., Wagener, J., Waltz, D., Wilmott, R., Zeitlin, P.L. and Ramsey, B. (2003) Significant microbiological effect of inhaled tobramycin in young children with cystic fibrosis. *Am J Respir Crit Care Med*, **167**, 841–849.

41 Geller, D.E., Pitlick, W.H., Nardella, P.A., Tracewell, W.G. and Ramsey, B.W. (2002) Pharmacokinetics bioavailability of aerosolized tobramycin in cystic fibrosis. *Chest* **122**, 219–226.

42 Le Brun, P.P.H. (2001) *Optimization of Antibiotic Inhalation Therapy in Cystic Fibrosis. Studies on Nebulized Tobramycin. Development of a Colistin Dry Powder Inhaler System*. University of Groningen.

43 Jensen, T., Pedersen, S.S., Garne, S., Heilmann, C., Høiby, N. and Koch, C. (1987) Colistin inhalation therapy in cystic fibrosis patients with chronic *Pseudomonas aeruginosa* lung infection. *J Antimicrob Chemother*, **19**, 831–838.

44 Valerius, N.H., Koch, C. and Høiby, N. (1991) Prevention of chronic *Pseudomonas aeruginosa* colonisation in cystic fibrosis by early treatment. *Lancet*, **338**, 725–726.

45 Cedergren, J., Follin, P., Forslund, T., Lindmark, M., Sundquist, T. and Skogh, T. (2003) Inducible nitric oxide synthase (NOS II) is constitutive in human neutrophils. *APMIS*, **111**, 963–968.

46 Hassett, D.J., Cuppoletti, J., Trapnell, B., Lymar, S.V., Rowe, J.J., Yoon, S.S., Hilliard, G.M., Parvatiyar, K., Kamani, M.C., Wozniak, D.J., Hwang, S.H., McDermott, T.R. and Ochsner, U.A. (2002) Anaerobic metabolism and quorum sensing by Pseudomonas aeruginosa biofilms in chronically infected cystic fibrosis airways: rethinking antibiotic treatment strategies and drug targets. *Adv Drug Deliv Rev*, **54**, 1425–1443.

47 Pedersen, S.S., Espersen, F., Høiby, N. and Jensen, T. (1990) Immunoglobulin-A and immunoglobulin-G antibody responses to alginates from *Pseudomonas aeruginosa* in patients with cystic fibrosis. *J Clin Microbiol*, **28**, 747–755.

48 Pedersen, S.S., Kharazmi, A., Espersen, F. and Høiby, N. (1990) *Pseudomonas aeruginosa* alginate in cystic fibrosis sputum and the inflammatory response. *Infect Immun*, **58**, 3363–3368.

49 Goldstein, W. and Döring, G. (1986) Lysosomal enzymes from polymorphonuclear leukocytes and proteinase inhibitors in patients with cystic fibrosis. *Am Rev Respir Dis*, **134**, 49–56.

50 Bruce, M.C., Poncz, L., Klinger, J.D., Stern, R.C., Tomashefski, J.F. and Dearborn,

D.G. (1985) Biochemical and pathologic evidence for proteolytic destruction of lung connective tissue in cystic fibrosis. *Am Rev Respir Dis*, **132**, 529–535.

51 Ammitzbøll, T., Pedersen, S.S., Espersen, F. and Schiøler, H. (1988) Excretion of urinary collagen metabolites correlates to severity of pulmonary disease in cystic fibrosis. *Acta Paediatr Scand*, **77**, 842–846.

52 Johansen, H.K. (1996) Potential of preventing *Pseudomonas aeruginosa* lung infections in cystic fibrosis patients: Experimental studies in animals. *APMIS*, **104** (Suppl 63), 5–42.

53 Tiddens, H.A.W.M. (2002) Detecting early structural lung damage in cystic fibrosis. *Pediatr Pulmonol*, **34**, 228–231.

54 Worlitzsch, D., Tarran, R., Ulrich, M., Schwab, U., Cekici, A., Meyer, K.C., Birrer, P., Bellon, G., Berger, J., Weiss, T., Botzenhart, K., Yankaskas, JR., Randell, S., Boucher, R.C. and Döring, G. (2002) Effects of reduced mucus oxygen concentration in airway *Pseudomonas* infections of cystic fibrosis patients. *J Clin Invest*, **109**, 317–325.

55 Høiby, N. (1977) *Pseudomonas aeruginosa* infection in cystic fibrosis. Diagnostic and prognostic significance of *Pseudomonas aeruginosa* precipitins determined by means of crossed immunoelectrophoresis. A survey. *Acta Pathol Microbiol Scand Suppl*, **262(C)**, 3–96.

56 Haussler, S., Tummler, B., Weissbrodt, H., Rohde, M. and Steinmetz, I. (1999) Small-colony variants of *Pseudomonas aeruginosa* in cystic fibrosis. *Clin Infect Dis*, **29**, 621–625.

57 Haussler, S. (2004) Biofilm formation by the small colony variant phenotype of *Pseudomonas aeruginosa*. *Environ Microbiol*, **6**, 546–551.

58 Rainey, P.B. and Travisano, M. (1998) Adaptive radiation in a heterogenous environment. *Nature*, **394**, 69–72.

59 Wyckoff, T.J.O., Thomas, B., Hassett, D.J. and Wozniak, D.J. (2002) Static growth of mucoid *Pseudomonas aeruginosa* selects for non-mucoid variants that have acquired flagellum-dependent motility. *Microbiology*, **148**, 3423–3430.

60 Hoffmann, N., Rasmussen, T.B., Jensen, P.O., Stub, C., Hentzer, M., Molin, S., Ciofu, O., Givskov, M., Johansen, H.K. and Høiby, N. (2005) Novel mouse model of chronic *Pseudomonas aeruginosa* lung infection mimicking cystic fibrosis [Correction]. *Infect Immun*, **73**, 5290.

61 Spiers, A.J., Buckling, A. and Rainey, P.B. (2000) The causes of *Pseudomonas* diversity. *Microbiology*, **146**, 2345–2350.

62 Lee, B., Haagensen, J.A.J., Ciofu, O., Andersen, J.B., Høiby, N. and Molin, S. (2005) Heterogeneity of biofilms formed by non-mucoid *Pseudomonas aeruginosa* isolates from patients with cystic fibrosis. *J Clin Microbiol*, **43**, 5247–5255.

63 Pedersen, S.S., Høiby, N., Espersen, F. and Koch, C. (1992) Role of alginate in infection with mucoid *Pseudomonas aeruginosa* in cystic fibrosis. *Thorax*, **47**, 6–13.

64 Pedersen, S.S., Møller, H., Espersen, F., Sørensen, C.H., Jensen, T. and Høiby, N. (1992) Mucosal immunity to *Pseudomonas aeruginosa* alginate in cystic fibrosis. *APMIS*, **100**, 326–334.

65 Shah, P.L., Scott, S.F., Geddes, D.M. and Hodson, ME. (1995) Two years experience with recombinant human DNase I in the treatment of pulmonary disease in cystic fibrosis. *Respir Med*, **89**, 499–502.

66 Brandt, T., Breitenstein, S., Vonderhardt, H. and Tummler, B. (1995) DNA concentration and length in sputum of patients with cystic fibrosis during inhalation with recombinant human DNase. *Thorax*, **50**, 880–882.

67 Ratjen, F. and Tummler, B. (1999) Comparison of the *in vitro* and *in vivo* response to inhaled DNase in patients with cystic fibrosis. *Thorax*, **54**, 91.

68 Anwar, H., Strap, J.L. and Costerton, J.W. (1992) Establishment of aging biofilms: possible mechanism of bacterial resistance to antimicrobial therapy. *Antimicrob Agents Chemother*, **36**, 1347–1351.

69 Anwar, H., Strap, J.L., Chen, K. and Costerton, J.W. (1992) Dynamic interactions of biofilms of mucoid *Pseudomonas aeruginosa* with tobramycin and piperacillin. *Antimicrob Agents Chemother,* **36**, 1208–1214.

70 Smith, A.L., Fiel, S.B., MayerHamblett, N., Ramsey, B. and Burns, J.L. (2003) Susceptibility testing of *Pseudomonas aeruginosa* isolates and clinical response to parenteral antibiotic administration – lack of association in cystic fibrosis. *Chest,* **123**, 1495–1502.

71 Borriello, G., Werner, E., Roe, F., Kim, A.M., Ehrlich, G.D. and Stewart, P.S. (2004) Oxygen limitation contributes to antibiotic tolerance of *Pseudomonas aeruginosa* in biofilms. *Antimicrob Agents Chemother,* **48**, 2659–2664.

72 Keren, I., Kaldalu, N., Spoering, A., Wang, Y.P. and Lewis, K. (2004) Persister cells and tolerance to antimicrobials. *FEMS Microbiol Lett,* **230**, 13–18.

73 Moskowitz, S.M., Foster, J.M., Emerson, J. and Burns, J.L. (2004) Clinically feasible biofilm susceptibility assay for isolates of *Pseudomonas aeruginosa* from patients with cystic fibrosis. *J Clin Microbiol,* **42**, 1915–1922.

74 Haagensen, J., Klausen, M., Ernst, R.K., Miller, S.I., Folkesson, A., Tolker-Nielsen, T. and Molin, S. (2007) Differentiation and distribution of colistin- and sodium dodecyl sulfate-tolerant cells in *Pseudomonas aeruginosa* biofilms. *J Bacteriol,* **189**, 28–37.

75 Ciofu, O., Fussing, V., Bagge, N., Koch, C. and Hoiby, N. (2001) Characterization of paired mucoid/non-mucoid *Pseudomonas aeruginosa* isolates from Danish cystic fibrosis patients: antibiotic resistance, β-lactamase activity and RiboPrinting. *J Antimicrob Chemother,* **48**, 391–396.

76 Giwercman, B., Lambert, P.A., Rosdahl, V.T., Shand, G.H. and Høiby, N. (1990) Rapid emergence of resistance in *Pseudomonas aeruginosa* in cystic fibrosis patients due to *in vivo* selection of stable partially derepressed β-lactamase producing strains. *J Antimicrob Chemother,* **26**, 247–259.

77 Giwercman, B., Meyer, C., Lambert, P.A., Reinert, C. and Høiby, N. (1992) High-level β-lactamase activity in sputum samples from cystic fibrosis patients during antipseudomonal treatment. *Antimicrob Agents Chemother,* **36**, 71–76.

78 Dibdin, G.H., Assinder, S.J., Nichols, W.W. and Lambert, P.A. (1996) Mathematical model of β-lactam penetration into a biofilm of *Pseudomonas aeruginosa* while undergoing simultaneous inactivation by released β-lactamases. *J Antimicrob Chemother,* **38**, 757–769.

79 Denton, M., Littlewood, J.M., Brownlee, K.G., Conway, S.P. and Todd, N.J. (1996) Spread of β-lactam-resistant *Pseudomonas aeruginosa* in a cystic fibrosis unit. *Lancet,* **348**, 1596–1597.

80 Bagge, N., Ciofu, O., Hentzer, M., Campbell, J.I.A., Givskov, M. and Høiby, N. (2002) Constitutive high expression of chromosomal β-lactamase in *Pseudomonas aeruginosa* caused by a new insertion sequence (IS1669) located in *ampD*. *Antimicrob Agents Chemother,* **46**, 3406–3411.

81 Jalal, S., Ciofu, O., Høiby, N., Gotoh, N. and Wretlind, B. (2000) Molecular mechanisms of fluoroquinolone resistance in *Pseudomonas aeruginosa* isolates from cystic fibrosis patients. *Antimicrob Agents Chemother,* **44**, 710–712.

82 Saiman, L., Mehar, F., Niu, W.W., Neu, H.C., Shaw, K.J., Miller, G. and Prince, A. (1996) Antibiotic susceptibility of multiply resistant *Pseudomonas aeruginosa* isolated from patients with cystic fibrosis, including candidates for transplantation. *Clin Infect Dis,* **23**, 532–537.

83 WestbrockWadman, S., Sherman, D.R., Hickey, M.J., Coulter, S.N., Zhu, Y.Q., Warrener, P., Nguyen, L.Y., Shawar, R.M., Folger, K.R. and Stover, C.K. (1999) Characterization of a *Pseudomonas aeruginosa* efflux pump contributing to aminoglycoside impermeability.

Antimicrob Agents Chemother, **43**, 2975–2983.

84 Burns, J.L., VanDalfsen, J.M., Shawar, R.M., Otto, K.L., Garber, R.L., Quan, J.M., Montgomery, A.B., Albers, G.M., Ramsey, B.W. and Smith, A.L. (1999) Effect of chronic intermittent administration of inhaled tobramycin on respiratory microbial flora in patients with cystic fibrosis. *J Infect Dis*, **179**, 1190–1196.

85 MacLeod, D.L., Nelson, L.E., Shawar, R.M., Lin, B.B., Lockwood, L.G., Dirks, J.E., Miller, G.H., Burns, J.L. and Garber, R.L. (2000) Aminoglycoside-resistance mechanisms for cystic fibrosis *Pseudomonas aeruginosa* isolates are unchanged by long-term, intermittent, inhaled tobramycin treatment. *J Infect Dis*, **181**, 1180–1184.

86 Denton, M., Kerr, K., Mooney, L., Keer, V., Raijgoppaal, A., Browniee, K., Arundel, P. and Conway, S. (2002) Transmission of Colistin-resistant *Pseudomonas aeruginosa* between patients attending a pediatric cystic fibrosis center. *Pediatr Pulmonol*, **34**, 257–261.

87 Moskowitz, S.M., Ernst, R.K. and Miller, S.I. (2004) PmrAB, a two-component regulatory system of *Pseudomonas aeruginosa* that modulates resistance to cationic antimicrobial peptides and addition of aminoarabinose to lipid A. *J Bacteriol*, **186**, 575–579.

88 Bagge, N., Hentzer, M., Andersen, J.B., Ciofu, O., Givskov, M. and Høiby, N. (2004) Dynamics and spatial distribution of β-lactamase expression in *Pseudomonas aeruginosa* biofilms. *Antimicrob Agents Chemother*, **48**, 1168–1174.

89 Bagge, N., Schuster, M., Hentzer, M., Ciofu, O., Givskov, M., Greenberg, E.P. and Høiby, N. (2004) *Pseudomonas aeruginosa* biofilms exposed to imipenem exhibit changes in global gene expression and β-lactamase and alginate production. *Antimicrob Agents Chemother*, **48**, 1175–1187.

90 Oliver, A., Canton, R., Campo, P., Baquero, F. and Blazquez, J. (2000) High frequency of hypermutable *Pseudomonas aeruginosa* in cystic fibrosis lung infection. *Science*, **288**, 1251–1253.

91 Oliver, A., Baquero, F. and Blazquez, J. (2002) The mismatch repair system (*mutS*, *mutL* and *uvrD* genes) in *Pseudomonas aeruginosa*: molecular characterization of naturally occurring mutants. *Mol Microbiol*, **43**, 1641–1650.

92 Macia, M.D., Borrell, N., Perez, J.L. and Oliver, A. (2004) Detection and susceptibility testing of hypermutable *Pseudomonas aeruginosa* strains with the Etest and disk diffusion. *Antimicrob Agents Chemother*, **48**, 2665–2672.

93 Macia, M.D., Blanquer, D., Togores, B., Sauleda, J., Perez, J.L. and Oliver, A. (2005) Hypermutation is a key factor in development of multiple-antimicrobial resistance in *Pseudomonas aeruginosa* strains causing chronic lung infections. *Antimicrob Agents Chemother*, **49**, 3382–3386.

94 Ciofu, O., Petersen, T.D., Jensen, P. and Høiby, N. (1999) Avidity of anti-*P. aeruginosa* antibodies during chronic infection in patients with cystic fibrosis. *Thorax*, **54**, 141–144.

95 Ciofu, O., Bagge, N. and Høiby, N. (2002) Antibodies against β-lactamase can improve ceftazidime treatment of lung infection with β-lactam-resistant *Pseudomonas aeruginosa* in a rat model of chronic lung infection. *APMIS*, **110**, 881–891.

96 Ciofu, O. (2003) *Pseudomonas aeruginosa* chromosomal β-lactamase in patients with cystic fibrosis and chronic lung infection – mechanism of antibiotic resistance and target of the humoral immune response. *APMIS*, **111**, 4–47.

97 Yang, L., Haagensen, J.A.J., Jelsbak, L., Johansen, H.K., Sternberg, C., Høiby, N. and Molin, S. (2007) In situ growth rates and biofilm development of *Pseudomonas aeruginosa* populations in chronic lung infections. *J Bacteriol*, Epub.

98 Lee, B., Ciofu, O., Wu, H., Hoffmann, N., Høiby, N. and Molin, S. (2007) Phenotypic and biofilm forming differences of mucoid and non-mucoid *Pseudomonas aeruginosa* isolates from patients with cystic fibrosis. *J.Clin. Microbiol.*

99 Smith, E.E., Buckley, D.G., Wu, Z.A., Saenphimmacjak, C., Hoffman, L.R., D'Argenio, D.A., Miller, S.I., Ramsey, B.W., Speert, D.P., Moskowwitz, S.M., Burns, J.L., Kaul, R. and Olson, M.V. (2006) Genetic adaptation by *Pseudomonas aeruginosa* to the airways of cystic fibrosis patients. *Proc Natl Acad Sci USA*, **103**, 8487–8492.

3
Adherence of *Pseudomonas aeruginosa*

Randall T. Irvin

3.1
Introduction

Pseudomonas aeruginosa is a widely distributed environmental Gram-negative bacterium that is a significant opportunistic pathogen of immunocompromised or immunosuppressed individuals. *Pseudomonas* infections in these susceptible patients are frequently life threatening due to the virulence of the organism and its high antimicrobial resistance, due both to a high innate resistance level and plasmid- or transposon-encoded antimicrobial resistance factors. A significant portion of the literature concerning *P. aeruginosa* has focused on its role as a human pathogen, but this is an anthropocentric view as the organism can and does infect a wide range of organisms, and survives well in the natural environment without infecting a host. This chapter will focus on *P. aeruginosa* adherence, and while respecting the anthropocentric view due to the significant impact of *Pseudomonas* infections, will attempt to bring a more balanced view on the mechanisms and biological role of adherence that extends beyond the framework of human infection.

P. aeruginosa is an extraordinarily successful organism that colonizes an extensive number of environments (some authors have described it as a ubiquitously distributed organism). *Pseudomonas* may perhaps be best described as an effective opportunist (as opposed to an organism that has adapted to a specific environment) given its diverse physiology and its well-documented ability to survive in physically diverse environments. The success of the organism lies in its flexible physiology, its extensive genetic regulatory networks that allows it to adapt to its environment [1], its ability to form extensive biofilms [2], its ability to control the permeability of its outer membrane [3,4] and its ability to produce an extensive capsule [5]. An additional aspect of the success of *Pseudomonas* is its ability to move from one environment to another and its ability to adhere to multiple surfaces.

Pseudomonas. Model Organism, Pathogen, Cell Factory. Edited by Bernd H.A. Rehm

ISBN: 978-3-527-31914-5

3.2 What is Adherence?

It is obvious that bacterial adherence entails the interaction or binding of bacterial cell(s) to another entity or "receptor" such that the bacteria become bound. Interactions occur with great frequency, occasionally resulting in the formation of a "complex" between two or more interacting components. The question of whether the interaction is the result of a defined process is frequently not readily discernable. Our clothes readily get dirty when we make contact with mud, but few of us would describe getting "dirty" as a specific process, we consider the event to be a random or "nonspecific" interaction. One cannot determine if an interaction is due to a specific process solely by the strength of the interaction – grease on clothes interacts very tightly, but the interaction is not due to a specific process. Bacterial adherence to substrates is considered to be a specific process and not just a nonspecific interaction. Bacterial adherence is dependent upon specific bacterial surface components, termed adhesins, which interact specifically with a substrate or receptor. However, bacteria are intrinsically sticky and nonspecific interaction of bacteria with other materials occurs. Differentiating bacterial adherence from nonspecific binding is accomplished by demonstrating the specificity of the interaction under appropriate conditions. The binding affinity of bacteria (or apparent association constant) for a receptor that the organisms binds to is higher than what is observed for nonspecific interactions. The interaction of bacteria with a substrate involves both specific and nonspecific events, and with high concentrations of bacteria or high ratios of bacteria to available receptors, the number of bacteria that will bind specifically to a receptor may not be greater than those that bind due to nonspecific interactions. Given enough mud, everything eventually gets coated with dirt. Thus, measuring specific bacterial adherence to a given receptor requires utilizing the lowest concentration of bacteria that can be readily monitored in the assay, demonstrating specificity in the interaction with the receptor and demonstrating that the number of bacteria that interacts with the receptor is significantly higher than what occurs due to nonspecific interactions under the same experimental conditions. Experimental conditions have to be carefully structured such that one can actually reproducibly measure the number of bacteria that bind to a receptor while maintaining a relatively low level of nonspecific interactions. The scope of this chapter is such that only specific bacterial interactions with receptors will be discussed.

3.3 Role of Adherence in Infection

Beachey [6] first proposed that the first step of most infections is the adherence of a pathogen to a mucosal surface – an event that is due to specific adhesin–receptor interactions – and suggested that the adhesins would constitute excellent vaccine targets. These concepts have been substantiated in a range of systems and in animal vaccine studies. However, the anticipated emergence of a significant number of antiadhesin-based vaccines or therapeutics has not occurred. The intuitively satisfying antiadhesin concept has, to date, not proven to be an easy vaccine or therapeutic target despite extensive efforts. Why?

The initial studies of Johanson *et al.* [7] showed that Gram-negative oral pharyngeal colonization precedes the onset of clinical infection in hospitalized patients. Woods *et al.* [8–10] then demonstrated that *P. aeruginosa* utilized pili as an adhesin and that *Pseudomonas* adherence to the respiratory epithelial cells of normal healthy humans is substantially lower than that observed for hospitalized patients that become infected in a hospital setting. These initial studies were carried out with buccal epithelial cells (BECs) – a somewhat troubling choice of a cellular target as around 95% of BECs obtained by the gentle scrapping of the interior cheek surface are nonviable when examined by Trypan blue dye exclusion. While BECs are easily obtained there have been questions raised as to their suitability for pathogenesis studies. To answer those questions, a prospective study was carried out to examine the adherence of *P. aeruginosa* to the ciliated tracheal epithelial cells of intensive care unit patients obtained by bronchoscopy (these cells were viable, free of respiratory mucus and generally had functional, beating cilia at the time of the assay). The results, somewhat surprisingly, strongly supported the results of the earlier studies [7] in that bacterial adherence to a susceptible patient's epithelial cell is elevated and closely associated with the development of clinical pneumonia [11]. Further, as was observed for adherence to BECs [12], we were able to demonstrate conclusively that purified pili competitively inhibited bacterial binding to the tracheal epithelial cells [13]. While our assay utilized fractionated tracheal epithelial cells that had been removed from the mucosal epithelial surface, the adherence of the bacteria was specific for the shaft of the cilia and minimal binding was observed on other regions of the cell (e.g. we observed little or no binding to the basolateral cell surface that is not normally exposed in the airway) [14] (Figure 3.1). Our results suggested that the increased ability of *Pseudomonas* to bind to a patient's respiratory epithelial cell is closely associated with a physiological change in the patient's epithelial cell surface (increased adherence occurs at the onset and during clinical pneumonia, but with the clearing of pneumonia the change in the patient's cell surface is reversed and bacterial adherence decreases). The adherence of *Pseudomonas* to a ciliated tracheal epithelial cell does not have an immediate impact on the epithelial cell as the cilia function continues and cell integrity was maintained until the conclusion of our assays. The change in a patient's epithelial cell surface was not specifically associated with ventilator-associated pneumonia (VAP) due to *P. aeruginosa*, but rather VAP in general [11]. While the adherence of *Pseudomonas* to a susceptible patient's respiratory epithelial cells is elevated relative to that observed with healthy individuals, the binding affinity for those cells is still fairly low and relatively few cells are bound per epithelial cell.

3.4 How is Bacterial Adherence Associated with Virulence?

The elevated adherence to a patient's ciliated tracheal epithelial cells is significantly associated with clinical infection [11], and many have interpreted this to indicate the

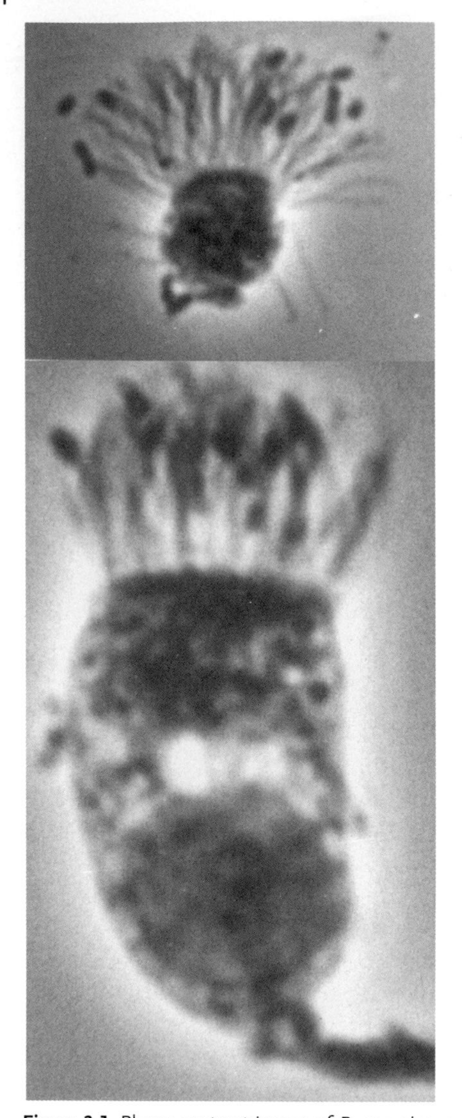

Figure 3.1 Phase contrast image of *P. aeruginosa* cells bound to cilia of (a) human transitional tracheal epithelial cell and (b) human columnar tracheal epithelial cell.

limited ability of *Pseudomonas* to infect healthy individuals and its opportunistic pathogen characteristics is the inability of *Pseudomonas* to bind effectively to normal mucosal surfaces. This interpretation has been supported by a number of *in vitro* and animal infection studies.

Woods *et al.* [12] established that pili played a significant role in mediating adherence to respiratory epithelial cells, but a closer examination of the binding kinetics revealed that there were at least two separate adhesin–receptor interactions that contributed to *Pseudomonas* binding to human respiratory epithelial cells, and that *Pseudomonas* binding attained a dynamic equilibrium and that bound bacteria could be desorbed from the surface [15]. Antipilus antibodies or direct competition with purified pili also could not completely inhibit *Pseudomonas* adherence to BECs [13,16], which further suggested that other nonpilus adhesins played a role in adhesion to mucosal epithelial surfaces. The capsular component, alginate, of certain strains of *Pseudomonas* was subsequently demonstrated to function as an adhesin for BECs, likely by interacting with human cell surface lectins [17]. Alginate was found to mediate low-affinity binding to a fairly high number of potential cell surface receptors (considerably higher than that observed for the pilus adhesin) [17].

We investigated the adhesion characteristics of a number of different clinical isolates and observed that not all *Pseudomonas* strains produced a capsule that mediated binding to respiratory epithelial cells, and that pilus-mediated binding to cells varied considerably in their apparent affinities. In particular, strains PAK and PAO varied approximately 10-fold in their apparent whole-cell binding affinities for BECs (PAO having a much higher affinity than PAK). We subsequently demonstrated that the ratio of the apparent binding affinities of the purified pili of the two strains accurately reflected the outcome of direct binding competition studies for both whole cells and purified pili [18]. We reasoned that a higher-binding-affinity adhesin (pili) would bind more extensively to surface receptors than a low-binding-affinity adhesin at low concentrations of bacteria (i.e. the initial stages of infection where the biological burden of a pathogen is low) even if the total number of potential receptors for the lower affinity adhesin (alginate) was much higher. We further reasoned that the affinity of an adhesin should be correlated to the virulence of a pathogen if adherence to a susceptible host tissue constrains the development of an infection. This has proven to be difficult to rigorously test experimentally, but we have established that PAO is much more virulent than PAK in a mouse infection and that abolishing the adhesin function of pili (while retaining ability to produce pili) but retaining other adhesins compromised the virulence of PAO in the same mouse infection model [19]. Our data suggests, but does not rigorously prove, that the apparent affinity of the pilus directly modulates the virulence of *Pseudomonas*. Studies on *Pseudomonas*-mediated corneal infections also support a critical role for adherence in the initial stages of infection as corneal scratches expose cryptic pilus receptors leading to substantially enhanced susceptibility for infection [20].

The data certainly supports a role for the pilus adhesin in initiating *Pseudomonas* infection and may suggest that other adhesins could play a more significant role at later stages of the infection process. The initial stage of infection would consist of the first colonization of the host, the subsequent development a higher bacterial population and the initial stages of biofilm formation. Lower-affinity adhesins could contribute significantly as the population rises, e.g. for the development of a descending respiratory infection from an extensively colonized endotracheal tube where microcolonies can be shed to descend into the lower respiratory tract, as microcolonies can bind to respiratory epithelial cells as a unit [21] which would increase the probability of an infection being established in the lower airway.

The preceding evidence supports Beachey's original proposal [6] that adhesins should be an excellent therapeutic and vaccine target. However, the dearth of antiadhesin therapeutics and vaccines some two and half decades later argues that either the proposal is (was) flawed or that things are certainly not as straightforward as one would hope. The evidence supporting the role of adhesins as virulence factors is fairly clear, and it is likely that challenges to antiadhesin therapeutics and vaccines lies in the redundancy of adhesins, in the complexity of adherence and the subsequent host interactions, and in our assessment of how critical adherence is for the establishment of an infection. It is also possible that the anthropocentric view of the function of *Pseudomonas* adherence has distorted our understanding of *Pseudomonas* pathogenesis.

3.5
P. aeruginosa Adhesins

P. aeruginosa utilizes a number of different adhesins to mediate attachment to either biotic or abiotic surfaces, including the alginate capsule [17], flagella [22,23], outer membrane proteins [24] and pili [12,13]. The existence of a significant number of alternative adhesins has been generally interpreted to imply that adherence is critical or at least plays a very significant role for the organism and that the redundancy of multiple adhesin–receptor systems represents a failsafe system. Given a human-centered view, this is intuitively satisfying and very logical. However, these nonpilus adhesin–receptor interactions are of low affinity and they collectively account for around 10% of bacterial–host binding events that are observed [19]. The observed kinetics of these alternative adhesin–receptor interactions indicates that these interactions will not occur at significant levels with low bacterial concentrations (such as in the early or initial stages of infection) and thus can only play a role in pathogenesis at high bacterial population levels (such as would be observed during the course of an active clinical infection). *P. aeruginosa* is an opportunistic pathogen which competes successfully in a wide range of environments and it thus seems more likely that these redundant adhesins may actually have nonadhesive functions that are of much higher significance to the organism.

Bacterial surface structures that mediate interactions with the host are generally viewed as adhesins, but that may prove to be a biased view. The capsular alginate of a number of *P. aeruginosa* strains mediates interactions with human mucosal epithelial cells [17]. However, the actual binding event appears to be a host function where the alginate is likely bound by a surface lectin [25]. Under conditions where microcolonies of cells have formed, the alginate polymer can serve to allow for the attachment of an entire microcolony to the mucosal epithelial cell surface [21] which may be of considerable significance during pathogenesis. The primary function of alginate does not appear to reside in its role as an adhesin, particularly given that some strains produce alginate that does not bind to respiratory epithelial cells [17].

The interaction of flagella with the host epithelial surface is of considerable interest. The flagella binds specifically to epithelial surfaces via the flagellin structural protein FliC [22] and through the flagella cap protein FliD [26]. The flagella binds to mucin [27] and asialo-GM_1 [28]. The interaction of flagella with the mucosal epithelial surface results in significant cell signaling events and secretion of a number of proinflammatory cytokines [29]. Flagella also interact with Toll-like receptors, specifically Toll-like receptor (TLR)-5 [29], thus triggering a substantial inflammatory response. *P. aeruginosa* flagella also bind to the epithelial cell sodium channel [30] to alter sodium transport. The interaction of flagella with the mucosal epithelial surface is complex and appears to reflect a sophisticated innate immune system monitoring of the airway. What is particularly interesting is that FliC and PilA have very similar receptor specificities as they both can bind to asialo-GM_1 [28,31,32]. Flagella interaction with mucosal epithelial cells leads to activation of TLR-5 signaling [29], which is likely an intracellular event given the cytochemical localization pattern of TLR-5 [33]. Given that flagella bind effectively to ENaC, it appears likely that ENaC cycling in mucosal epithelial cells [34] will result in the internalization of either free flagella or bacterial cells with the associated flagella for antigen sampling and presentation to the innate immune system (i.e. exposure to TLRs and other innate receptors [30]). *P. aeruginosa* pili have also been reported to activate cellular signaling pathways in airway mucosal epithelial surfaces, and given the similar binding specificity of pili and flagella it seems likely that ENaC cycling may mediate some of the pilus-associated cell signaling events associated with exposure to pili. The role of the type III protein secretion system in directly modulating host responses also complicates the analysis of the interaction of *P. aeruginosa* with the host [35].

3.6 Surface Receptor Requirements of the Pilus Adhesin

Pili specifically recognize the disaccharide GalNAc–Gal [36], but the receptor structural studies revealed somewhat limited requirements for the receptor as explored through a large number of synthetic disaccharide constructs [37]. Indeed, it was surprising that the pilin receptor-binding domain demonstrated no requirement for any hydroxyl group on the disaccharide, but rather the interaction appeared to depend upon the hydrophobic face of the of GalNAc–Gal (Figure 3.2) and was substantially enhanced by increasing the hydrophobicity of the C_8 alkyl linker of the synthetic analogs [37]. As is typical with protein carbohydrate interactions the interaction was observed be of low affinity with methyl-GalNAc–Gal having a IC_{50} of 79 μM for inhibiting PAK pili binding to immobilized asialo-GM_1 [37], which is also consistent with the relatively low affinity of pili for mucosal epithelial cells. The predominance of what appear to be predominantly hydrophobic interactions in the molecular interaction was surprising, as was the increased affinity of carbohydrate analogs where the hydrophobicity of the carbohydrate was increased at either the 3-hydroxyl group of the Gal moiety or increasing the hydrophobicity of normal C_8 linker upon which synthetic carbohydrates are normally synthesized [37]. The structural basis for the increased affinity of pili to more hydrophobic receptor analogs is unclear, but it clearly indicates that the hydrophobic interactions are important in

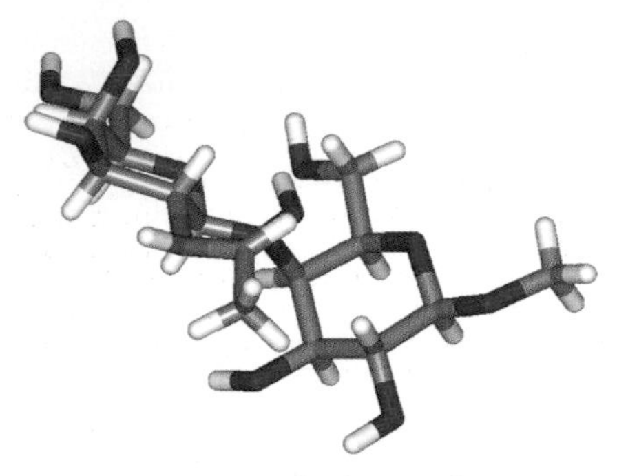

Figure 3.2 Structure of GalNAc–Gal (β-D-*N*-acetyl-galactosamine (1 → 4)-β-D-galactose): carbon atoms are colored grey, oxygen atoms are colored red, nitrogen atoms are colored blue and hydrogen atoms are colored in off white. One face of the GalNAc–Gal (the bottom in this orientation) is essentially hydrophobic, displaying primarily hydrogens and the *N*-acetyl group, while the other face displays primarily hydroxyl groups. GalNAc–Gal represents the minimal carbohydrate moiety required for pilus receptor activity [36].

the adhesin–receptor interaction. While we have defined the minimal receptor requirements of pili to be the GalNAc–Gal disaccharide which is displayed in asialo-GM_1 and asialo-GM_2, we have not ambiguously determined the nature of the mucosal epithelial surface. While a number of reports have highlighted asialo-GM_1 as the presumptive receptor for pili, there is evidence that pili actually utilize other surface structures as cell surface receptors [38,39], and given the rather limited structural requirements for pilus receptor activity it is likely that a number of mucosal surface components can and do function as surface receptors. It is perhaps more appropriate to anticipate that pili will bind to a number of surface components that display a GalNAc–Gal-like structure, i.e. a variety of glycoconjugates. As pili are likely multivalent in terms of the number of receptor-binding domains that are presented at the tip of a pilus, a reasonable amount of variation in the receptor can be readily accommodated with minimal affect on the net ability of the cell to bind to a mucosal surface.

The pilin receptor-binding domains vary considerably in terms of their amino acid sequence, and the C-terminal disulfide loop region has been described as semi-conserved [40] and a brief examination of a even a limited number of pilin receptor-binding domain sequences reveals only limited sequence similarity (Figure 3.3). While the binding kinetics of the pili vary considerably from strain to strain, they all compete for binding sites on respiratory epithelial cells [41] despite the sequence variability in their receptor-binding domains (Figure 3.4). *P. aeruginosa* adherence to human ciliated tracheal epithelial cells varies as patients become more or less susceptible to clinical pneumonia [11] and some have correlated this change to expression levels of asialo-GM_1 in the lung [42]. Binding isotherm studies of *P. aeruginosa* adherence to human tracheal epithelial cells indicated that at least for one patient, the number of receptor sites on the patient's cell surface does not increase when the patient become susceptible to infection, but rather the affinity of cells increases [43]. This observation is consistent with the early studies of Woods *et al.* [8,44]

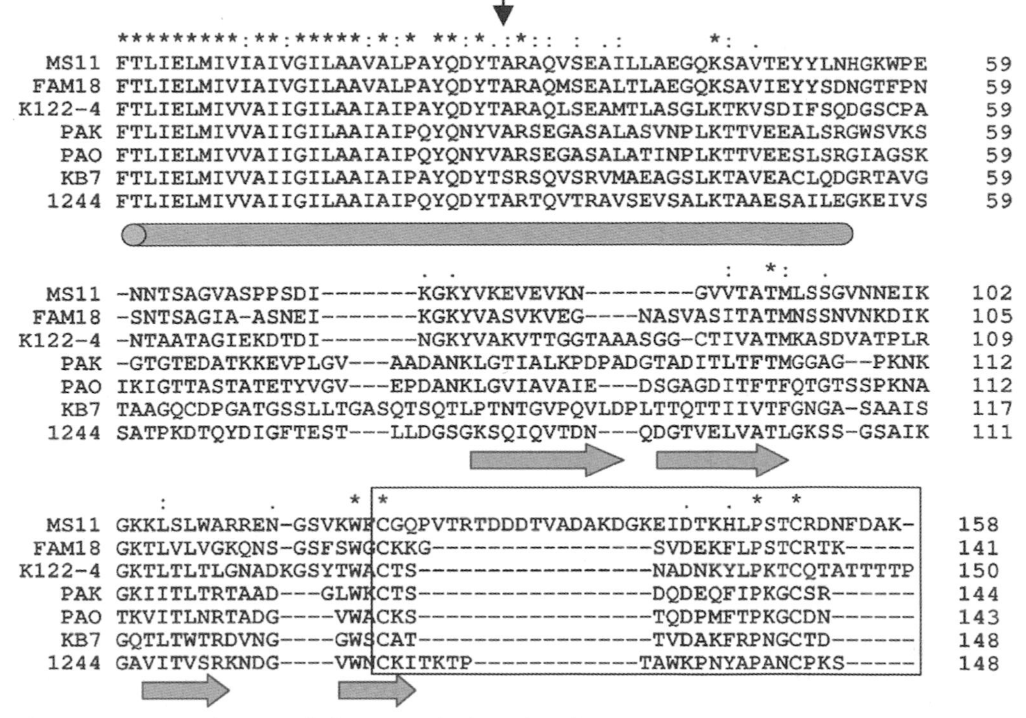

Figure 3.3 Sequence alignment of pilin structural subunits based on secondary structure (α-helix is represented by a cylinder below the sequence alignment and β-structure is represented by arrows below the sequence alignment). The alignment includes two *Neisseria* PilE sequences (MS11 and FAM18), and *P. aeruginosa* PilA for PAK, PAO, K122-4, KB7 and 1244.

who observed that trypsinization of cells resulted in increased binding to respiratory epithelial cells. Our data suggests that changes in the physiology of a patient alter the display or accessibility of surface receptors to pili, which in turn results in increased or decreased ease in binding of *P. aeruginosa* to the airway. The complex and dynamic interaction of the pathogen with the host mucosal surface that occurs following adherence to the host cell results in complex physiological changes, the outcome being colonization and infection of a susceptible host.

3.7
How Does PilA Mediate Attachment to Human Mucosal Surfaces?

P. aeruginosa initially binds to human mucosal epithelial cells via the tip of the pilus [31] (Figure 3.5). An interesting aspect of the tip-specific binding and thus localization of the adhesin is that purified pili cannot aggregate or agglutinate cells

PAO_Scrambled	N-C-P-D-F-D-P-T-K-K-G-M-Q-A-C-T-S
PAO(128144)C129A/C142A_Scrambled	N-**A**-P-D-F-D-P-T-K-K-G-M-Q-A-**A**-T-S
2Pfs(128-144)ox	K-C-T-S-D-Q-D-E-Q-F-I-P-K-G-C-S-K
PAO(128-144)ox	A-C-K-S-T-Q-D-P-M-F-T-P-K-G-C-D-N
KB7(128-144)ox	S-C-A-T-T-V-D-A-K-F-R-P-N-G-C-T-D
PAK(128-144)ox	K-C-T-S-D-Q-D-E-Q-F-I-P-K-G-C-S-R
K122-4(128-150)ox	A-C-T-S-N-A-D-N-K-Y-L-P-K-T-C-Q-T-A-T-T-T-P
CD4	T-C-T-S-T-Q-E-E-M-F-I-P-K-G-C-N-K
P1	N-C-K-I-T-K-T-P-T-A-W-K-P-N-Y-A-P-A-N-C-P-K-S
Cs1 [PAK(128-144)ox T130K/E135P]	K-C-**K**-S-D-Q-D-**P**-Q-F-I-P-K-G-C-S-K
PAK_C129A/C142A_Scrambled	K-**A**-D-D-F-K-Q-G-T-Q-E-K-S-S-**A**-P-I
MS11	FCGQPVTRTDDDTVADAKDGKEIDTKHLPSTCRDNFDAK

Figure 3.4 *P. aeruginosa* PilA receptor C-terminal sequences that contain a functional receptor-binding domain.

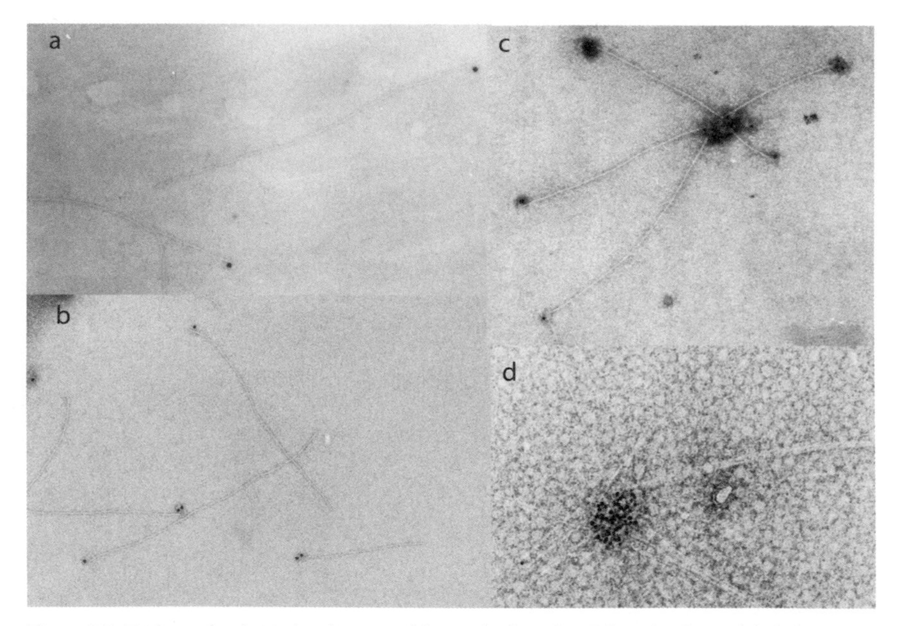

Figure 3.5 Evidence for the tip localization of the receptor-binding domain on the PAK pilus. The monoclonal antibody PK99H binds to only to one end of the pilus (a) , while monoclonal antibody PK3B also only binds to one end of the pilus (b) , which is different from the end that PK99H binds to as labeling with both PK99H and PK3B results in both ends of the pilus being labeled (c) . PK99H inhibits *P. aeruginosa* binding, while PK3B does not. Asialo-GM$_1$ absorbed onto colloidal gold only associates with one end the of the pilus (d) which is necessarily the tip of the pilus structure.

unless they themselves are clumped (an event that generally does not occur under normal experimental conditions). The *P. aeruginosa* type IV pilus is unusual in that there is no evidence of a specific adhesin that is displayed at the tip of the pilus. Proving that there is no tip-specific adhesin is a challenging issue as the proof basically relies on negative evidence to a large extent. However, we have established that the PilA structural protein contains a functional receptor-binding domain that is encoded with the C-terminal disulfide loop region of the protein [45]. The receptor-binding domain is functional as a synthetic peptide, and we have demonstrated that this receptor-binding domain is necessary and sufficient to mediate binding to mucosal epithelial cells. A murine monoclonal antibody, PK99H, has proven to be very useful in the analysis of the localization of the receptor-binding domain, and demonstrating the role of the binding domain in adherence and virulence in animal infection models. PK99H was produced against pili purified from strain 2Pfs and observed to be fairly strain specific, and capable of inhibiting *P. aeruginosa* adherence to respiratory epithelial cells [16]. The antigenic epitope of PK99H was determined through the use of synthetic peptide analogs to reside within the C-terminal disulfide loop region of PilA – a linear epitope consisting of residues 134–140 (DEQFIPK) [46]. Immunocytochemical studies demonstrated that PK99H only bound to the tip of the pilus [31], while PK3B (another antibody generated at the same time as PK99H [16]) bound the basal end of the pilus and did not inhibit bacterial adherence to respiratory epithelial cells [31] or confer protection against *P. aeruginosa* in an animal infection model [41].

Nuclear magnetic resonance (NMR) studies of a number of PilA C-terminal peptides established that the synthetic peptides were structured in solution and that surprisingly the peptide structures were strikingly similar despite extensive sequence variation [47–49]. The peptide structure was derived from the disulfide constraint, a hydrophobic core region and two β-turns [47,50,51]. Interestingly, the structure of the peptides did not change when they were bound to either GalNAc–Gal or a cross-reactive monoclonal antibody [48,52,53]. None of the residues displayed any changes in the environment of their amino acid residue side-chains that could be detected by NMR. This suggested that the peptide structures were similar to that displayed in the native protein.

An extensive structure–activity relationship study of the receptor-binding domains of PAK and KB7 utilizing a series of alanine-mutated peptides demonstrated that only a limited number of amino acid residues were critical for adhesin function [54]. The results were complex and difficult to interpret, but indicated that positionally equivalent residues were critical for adhesin function in both receptor-binding domain sequences, but that one could not substitute a KB7 critical residue for a critical PAK residue or a PAK critical residue for a critical KB7 residue [54]. We formulated a hypothesis that a number of the residues played a critical role in adhesin function by maintaining a specific peptide conformation. The structures of the peptides are somewhat difficult to interpret as one obtains an assembly of potential conformations of the peptide generated from modeling the nuclear Overhauser effect distance constraints obtained from the spectra and they reflect the flexibility of the peptide in solution. The receptor-binding domains all contain proline residues that are observed in both the *trans* and *cis* isomers [49], resulting in two families of conformers that are in dynamic equilibrium [51]. Adhesin function was associated with the *cis* proline isomer conformers [47,54]. The high degree of flexibility of the synthetic receptor-binding domains precludes extensive or detailed analysis of the structure–activity relationship.

3.8
X-ray Crystallographic Structural Studies of the Pilin Structural Protein

Following the initial determination of the *Neisseria* MS11 type IV pilin structure by Parge *et al.* [55], the structures of a number of type IV pilin structural proteins have now been determined and despite the extensive sequence variation they are all remarkably similar (although they have differentiated into type a and b variants with *P. aeruginosa* producing a type IVa pilin [56]) (Figure 3.6). The common structure has been termed the pilin-fold which is an αβ-fold where a long linear α-helix packs against a four-stranded antiparallel β-sheet to form a "lollypop" type of structure where the long N-terminal α-helix extends below a globular head of the protein.

The long N-terminal α-helix is very hydrophobic and is highly conserved in its sequence running from residues 1 to around 54 (Figure 3.3), but the helix has two separate kinks, the first occurring at Ala20 and the second occurring at Pro42 in PAK PilA (Gly42 in *Neisseria* PilE) (Figure 3.7). Arg30 is one of the few charged residues in the N-terminal α-helix, is highly conserved in *P. aeruginosa* PilA and is positioned very close the receptor-binding domain. The N-terminal α-helix is also a highly conserved feature of a number of pilin-like proteins that are either associated with the type II protein secretion system and/or the secretion, processing and assembly of type IV pilins [57]. The N-terminal α-helix appears to play a significant role in the polymerization of pilin into a pilus, with the helices creating a central core of the pilus fiber [55,56,58–60].

A second feature of type IV pilins is that the globular head of the protein consists of a multistranded β-sheet that wraps around the α-helix such that the globular head of the protein forms a pie-shaped wedge that can be assembled around the central core of the α-helices, with the outer surface of the resulting fiber or pilus thus consisting of the antiparallel β-sheet (Figure 3.8). The-wedge shaped globular heads assemble nicely to generate a fiber structure with the α-helices forming a central core structure.

Type IV pilins also have what has been termed a "semiconserved" C-terminal disulfide loop [61,62], which constitutes a receptor-binding domain in *P. aeruginosa* PilA [45,48,60]. While the amino acid sequence of the disulfide loop is not well conserved, structurally there is surprising structural similarity (Figure 3.9) [63]. A significant portion of both the disulfide loops of the *P. aeruginosa* PilA and the *Neisseria* PilE can be superimposed even though *Neisseria* pili has not been reported to bind to asialo-GM_1 (Figure 3.9). Such a strong conservation of structure generally implies a common function. The C-terminal disulfide loop region is quite unusual in that a significant number of main-chain atoms are solvent exposed and are in fact clustered together on the surface of PilA to generate what has been proposed to be a

Figure 3.6 Structure of *Neisseria* PilE of MS11 (PDD file 1AY2) in stick representation (this figure and all subsequent structure figures were prepared with WebLeb ViewerPro version 3.1) [55].

very shallow receptor-binding domain [60]. Generally, atoms of the Cα backbone are not solvent exposed. The structure of the C-terminal region is similar, but not identical to the structures of the synthetic peptides consisting of the same sequence – there is evidence of the two β-turns [60,61], but the last β-strand extends into the disulfide loop region and there is no evidence for any β-structure in the synthetic peptides [47]. Interestingly, there is no evidence of a *cis* configuration of the proline within the loop in the X-ray structures [60,61,64], but there is clear experimental evidence of both *cis* and *trans* proline configurations in dynamic equilibrium in the NMR studies of the synthetic peptides [49,51]. The NMR studies of the K122-4 pilin indicated that the loop has fairly extensive dynamic motion [65] which precluded an accurate determination of the structure of the loop [66], which suggests that the C-terminal disulfide loop region may have a number of conformations *in vivo*. The position of the C-terminal extension beyond the disulfide bridge is unclear as this

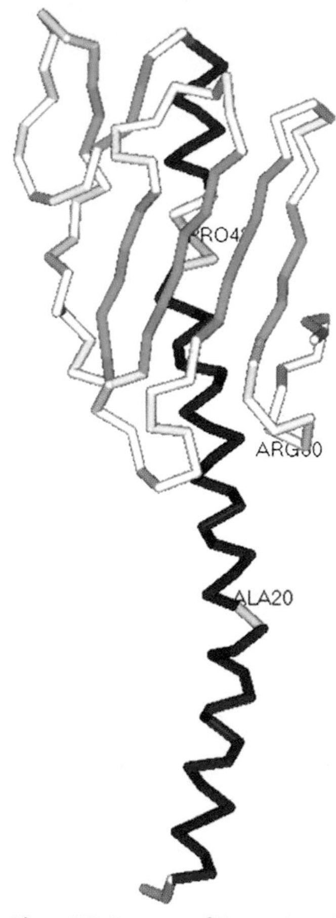

Figure 3.7 Structure of *P. aeruginosa* PilA of strain PAK (this is the full-length PilA structure, PDD file 1OQW), note the structural similarities with *Neisseria* PilE, and also note the two distinct kinks in the N-terminal α-helix that occur at positions Ala20 and Pro42. Also note Arg30 which is in close apposition to the C-terminal disulfide loop as it is one of the few charged residues in the helix.

region has generally been involved in interfacial crystal contacts and thus the current structures for this region are likely suspect [61].

An unambiguous determination of the nature of the pilin receptor-binding domain has not been possible to date, as a structure of a receptor analog complexed to pilin has not yet been solved despite considerable effort. Attempts to generate cocrystals of pilin and a receptor analog have not been successful (the hydrophobic nature of the synthetic carbohydrates and low binding affinity have presented significant challenges) nor have attempts to soak in receptor analogs (the crystals fracture as the binding domain is very close to crystal contacts and there is limited solvent space within the crystal itself). The hypothesis that the receptor-

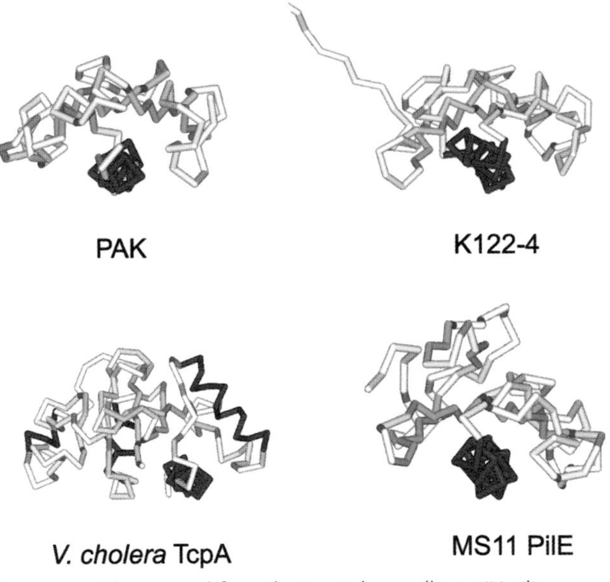

Figure 3.8 When viewed from the "top" down, all type IV pilin structural proteins display a wedge or sector shape that readily packs into a fiber.

binding domain consists primarily of exposed main chain atoms is interesting, and does have reasonable, if somewhat indirect, experimental support. The monoclonal antibody PK99H can recognize bound synthetic receptor-binding domains on human respiratory epithelial cells [45,67] and yet preincubation of pili or peptide with PK99H inhibits binding to mucosal epithelial cells [16,68–70]. These results strongly suggest that the amino acid side-chains of the PK99H epitope (residues 134–140) are not directly involved in binding, but are very close to the receptor-

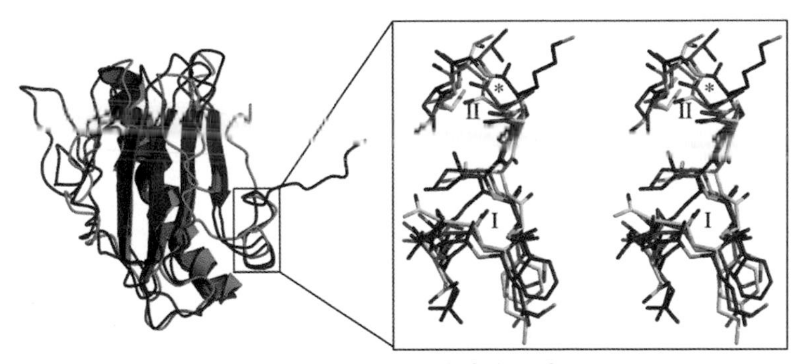

Figure 3.9 Superimposition of the globular head of pilins of PAK, K122-4 and MS11 with the C-terminal disulfide loop region boxed and a stereo image which highlights the structural homology. Reprinted with permission from [61].

binding domain, such that antibody binding to those residues occludes the receptor-binding domain and thus inhibits binding. Only antibodies that recognize the C-terminal disulfide loop region of PilA inhibit pilus mediated binding to respiratory epithelial cells [69].

3.9 Structure of the Pilus Fiber

The *P. aeruginosa* type IV pilus is a deceptively simple structure consisting of an asymmetrical linear fiber composed of a single structural protein. The pilus itself is remarkably stable and resistant to depolymerization or disassembly *in vitro* – to date it has been extremely difficult to disassemble the pilus without using denaturing conditions [58,71] and yet the type IV pilus is readily dissociated or assembled *in vivo* at quite remarkable rates [57,72–76]. The synthesis, export and assembly of the type IV pilus is a very complex tightly coordinated process involving a large number of discrete highly regulated gene products [57]. The synthesis and assembly of the type IV pilus is beyond the scope of this review.

The actual structure of the pilus fiber has largely remained an enigma despite the determination of the molecular structure of the pilin structural subunit and excellent fiber diffraction studies. Parge *et al.* [55] had initially proposed that the central core region of the pilus fiber consisted of bundled N-terminal α-helices which mediated assembly of the fiber and suggested that the resulting pilus (in this case, the *Neisseria* MS11 pilus) would be around 60 Å in diameter. The structural modeling was reasonably consistent with the fiber diffraction studies on purified PAK pili. Folkhard *et al.* [77] established that the pilus fiber is a helical assembly of the pilin structural protein (PilA) such that a long fiber of around 52 Å in diameter with an inner core of 12 Å was formed. A subsequent re-examination of the fiber diffraction data and extensive modeling studies provided strong experimental support for a central core structure consisting of bundled α-helices forming an internal structure that is around 10 Å in diameter, but surprisingly a strong diffraction pattern arising from the globular head of the pilin subunit was not observed [78]. The absence of any diffraction pattern from the globular portion of the protein in the fiber diffraction is perplexing and suggests either disorder within the globular head of the protein or potentially disorder in the packing arrangements of head of the protein within the fiber. As Marvin *et al.* [78] observed, there is little disorder observed in solution in the K122-4 PilA globular head [65,66], which suggests that there may be surprising flexibility in the pilus fiber that introduces disorder in the relative positions of the globular domain. The only clearly defined aspect of the pilus fiber structure is that the helices cluster or bundle together to form a central core around which the pilus globular head is packed.

Based on molecular modeling studies, a variety of pilus fiber models have been proposed that differ in various aspects. The initial pilus model of Parge *et al.* [55] is a simple one-start right-handed helical model where the N-terminal α-helices serve to anchor the pilus into the membrane to an identical model, but where the tip of the

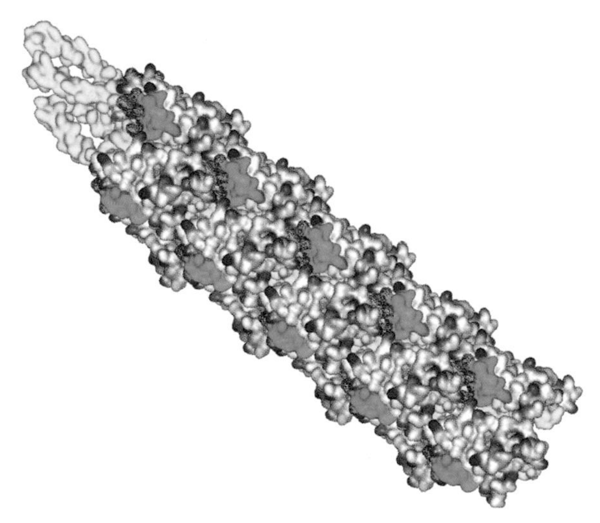

Figure 3.10 The left-handed three-start helix model of the *P. aeruginosa* PAK pilus fiber as proposed by Craig *et al.* [56]. The PDB coordinates were generously provided by Lisa Craig. In this model, the N-terminal α-helix (residues 1–52) is colored yellow, while the C-terminal disulfide loop region (residues 128–144) is colored coded in green and the remainder of the surface is coded according to its electrostatic potential (acidic regions are red, basic regions are blue and hydrophobic or neutral surface regions are colored in off-white).

pilus consists of exposed N-terminal α-helices, to a left-handed one-start helical model [66], to a two-start right-handed or left-handed helix model [56,58]. The closely related type IVb pili appear to consist of a three-start left-handed helical assembly of the pilin structural subunit [56,58,59] for which the crystals of the pilin subunit provided strong evidence for. Recently, Craig *et al.* [59] presented elegant evidence, based on cryo-electron microscopy, which supports a left-handed three-start helix for the *Neisseria* pilus. The type IV pilus thus appears to consist of three filaments that are wound left-handedly to form a linear fiber. A molecular model of the three-start PAK pilus is presented in Figure 3.10. The implications of a fiber of this type is that the pilus will present three independent receptor-binding domains at the tip of the pilus and thus pilus-mediated adherence will benefit by the avidity associated with having three binding domains. Optimal binding of a pilus to a respiratory epithelial cell would require fairly close positioning of receptors to engage more than one receptor-binding domain. A clustering of receptors could readily generate rather specific binding sites for *P. aeruginosa*, such as was observed for ciliated human tracheal epithelial cells [14].

Parge *et al.* [55] proposed that the N-terminal α-helices constitute an oligomerization domain that mediates assembly of the pilus fiber. Truncation of the initial 28 residues of pilin structural proteins results in a truncated monomeric pilin that is well folded [60,66] and whose structure does not differ significantly from a full-length protein [58]. While the central core of the pilus clearly contains bundles of N-terminal

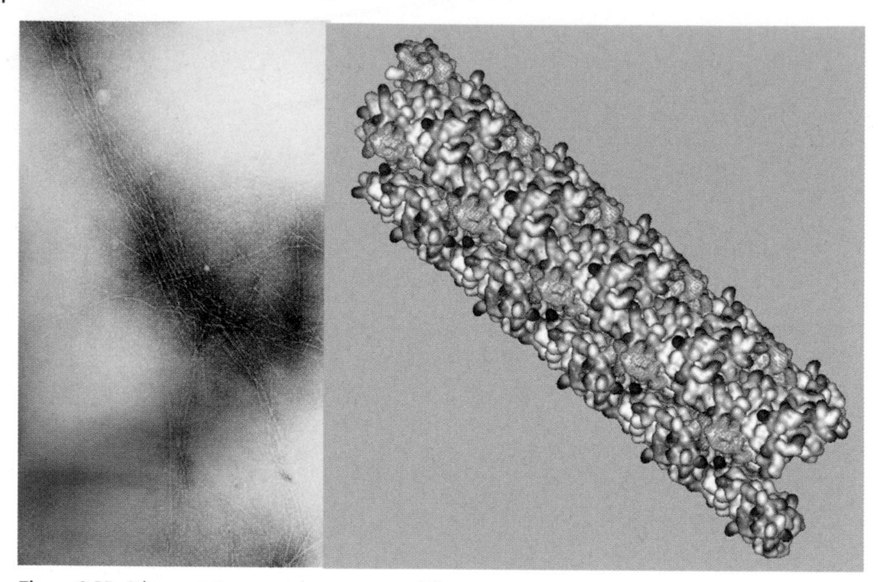

Figure 3.11 Pilin protein nanotubes generated from monomeric truncated K122-4 PilA that self assemble following exposure of the pilin to a C_{11}-alkylthiol. (A) Negative stained electron microscopy image demonstrating the extreme length of 6-nm diameter protein nanotubes. (B) A left-handed three-start molecular model of PAK protein nanotubes.

α-helices, the actual role of the N-terminal α-helices in pilus assembly may reside in the export and trafficking of the pilin subunits to the pilus assembly site rather than in mediating direct formation of the pilus fiber as was originally thought. Truncated monomeric PilA generated from strain K122-4 has been particularly useful in pointing this out, as Δ(1–28)K122-4 PilA when exposed to a hydrophobic molecule such as an alkyl C_{11} thiol undergoes a conformational change and spontaneously assembles into very long pilin protein nanotubes [79] (Figure 3.11). Exposure of the monomeric PilA to an alkyl thiol results in a conformational change in the pilin monomer that leads to self-assembly of nanotubes that reach a length of hundreds of micrometers. The assembled pilin nanotubes display a variety of features associated with a pilus fiber, including pilus-associated antigens and DNA binding [79,80]. The solution and X-ray structures of the K122-4 PilA are very similar, but the orientation and distance between the α-helix and the β-sheet differs between the NMR and X-ray structures [61]. There is an unexpected degree of flexibility in the structure of the PilA globular head [61,65], but it is unclear whether either the NMR or X-ray structures reflect the actual structure of the globular head either in the pilus or the pilin nanotube. The demonstration of a conformational change in PilA that results in altered PilA function raises a number of issues and possibilities that have not been addressed at this time, and suggests that the type IV pilus will continue to offer surprises.

3.10
Structure of the Receptor-Binding Domain and Location on the Pilus

The left-handed three-start helix model of the pilus structure is based on rather minor surface structural features of the pilus detected by examination of cryo-preserved pili preparations [59] – evidence which one might be somewhat hesitant to accept without further experimental support. The three-start helix model for the pilus structure is actually quite useful in that a number of rather specific hypotheses can be formed and tested. The most obvious feature is that the pilus will necessarily display three receptor-binding domains as three discrete filaments that end at the pilus tip (assembly is from the pilus base within the periplasm at or near the surface of the plasma membrane [81,82]) and asialo-GM_1 only binds at the tip [31]. The implications of the three-start helix model of the pilus fiber include (i) that the pilus will display three functional receptor-binding domains at the pilus tip, (ii) that the pilin–pilin interactions that form the small filaments and/or the pilin–pilin interactions created by the intercalation or wrapping together of the filaments into the pilus fiber occlude the receptor-binding domain along the shaft of the pilus fiber, and (iii) that the PK99H antigenic epitope (residues 134–140 of PAK PilA or DEQFIPK) is only available at the tip of the pilus. The left-handed three-start helix model of the pilus fiber proposed by Craig *et al.* [56] is based entirely on structural data, originally through molecular modeling studies, and then subsequently more directly on *Neisseria* pili by the cryo-electron microscopy [59]. Interestingly, the original left-handed three-start helix model proposes that the N-terminal α-helices are positioned at the base of the pilus where they serve to anchor the pilus to the cell envelope [56–59], while we have proposed that the N-terminal α-helices are positioned at the tip of the pilus [60,66].

An inspection of the model of the PAK three-start helix pilus fiber generously provided by Lisa Craig indicates that the PK99H antigenic epitope is occluded at the base and along the shaft of the pilus fiber, but the epitope is fully accessible at the tip of the pilus fiber, as are the protruding N-terminal α-helices (see deep pink cross-hatched surface of the 128–144 green coded surface residues in Figure 3.12). We had previously established that the PK99H epitope is displayed at one end of the pilus and that binding of PK99H to the pili significantly inhibits pilus binding to both mucosal respiratory epithelial cells and asialo-GM_1 [16,31,36] (Figure 3.5). Further, we have established that colloidal gold coated with the receptor analog asialo GM_1 also only binds at one end of the pilus [31] (Figure 3.5). We have established that the pilin receptor domain resides within residues 128–144, and that this PilA domain is necessary and sufficient for pilus-mediated binding to epithelial cells. Inspection of the PAK two-start helix pilus fiber model also indicates that this domain is only fully accessible at what we describe as the tip of the pilus, the region where the N-terminal α-helices are positioned (see Figure 3.12). Thus there is significant independent experimental support for two of the three predictive aspects of the three-start helix model of the pilus fiber.

The third aspect of the three-start helix model is that the pilus is trivalent in terms of the number of displayed receptor-binding domains. This can be experimentally

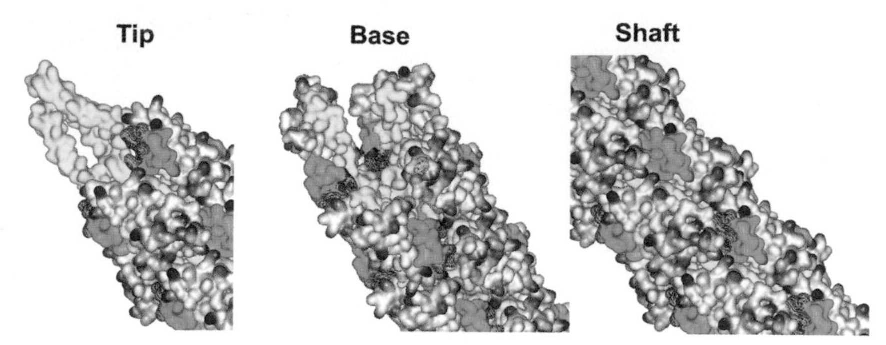

Figure 3.12 Molecular three-start helix model of PAK pilus examining the tip, shaft and base of the pilus to determine that the receptor-binding domain (colored in green) is fully accessible at the tip of the pilus, but occluded along the shaft and at the base. Similarly, the PK99H epitope (residues 134–140, colored in green with red hatching) is only exposed at the tip of the pilus, and is occluded along the shaft of the pilus and at the base of the pilus. Note the hydrophobic N-terminal α-helices (colored yellow) are exposed at the tip of the pilus and through "holes" or "caves" along the pilus shaft.

tested by examining the apparent binding affinity for pili, truncated monomeric pilin and pilin protein nanotubes. Our experimental approach has been to utilize a competitive inhibition of ligand binding (either pili, monomer or nanotubes) using a synthetic receptor-binding domain [the synthetic peptide PAK(128–144)ox]. We chose to utilize K122-4 pilin protein nanotubes [and thus K122-4 pili and Δ(1–28) K122-4 PilA monomeric pilin] in this study as the hydrophobe induced self-assembled nanotubes display many of the same features as native pili with the obvious difference being that they are much longer and do not include the N-terminal α-helices [79]. We determined that K122-4 pili, monomeric pilin and nanotubes bound to human BECs in a concentration-dependent manner (Figure 3.13). The pili bound much more effectively than did either the monomeric pilin or the nanotubes. The concentration of PilA on a molar basis in the pili sample is less than that observed in the monomeric pilin sample and the avidity afforded by a multivalent display of receptor-binding domains could have accounted for the increased binding of the pili relative to the monomeric pilin. However, the valency of the pili and the nanotubes is similar, if not identical, and the pili also bound much more effectively than did the nanotubes. The binding of monomer and nanotubes to the cells was equivalent, even though the valency of the nanotubes is higher. These results strongly suggest that the pilus receptor-binding domain differs from the receptor-binding domain that is displayed in the pilin monomer or the pilin nanotubes. Structurally, the monomeric pilin and the full length pilin do not differ significantly [58] other than the N-terminal α-helix is truncated. A similar involvement of the N-terminal α-helix in binding to steel has not been observed, and peptide competitive inhibition studies have strongly supported the three-start helix fiber model. We can definitely rule out a single start helix in terms of the pilus fiber structure and the three-start model appears likely. The three-start helix fiber model provides structural indication of three independent receptor-binding domains clustered around three protruding hydrophobic α-helices

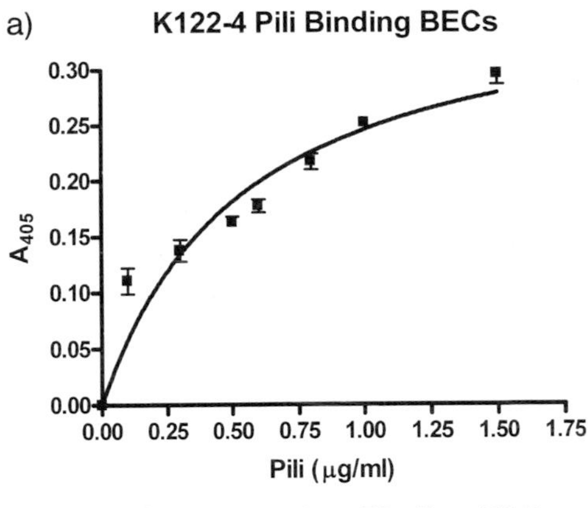

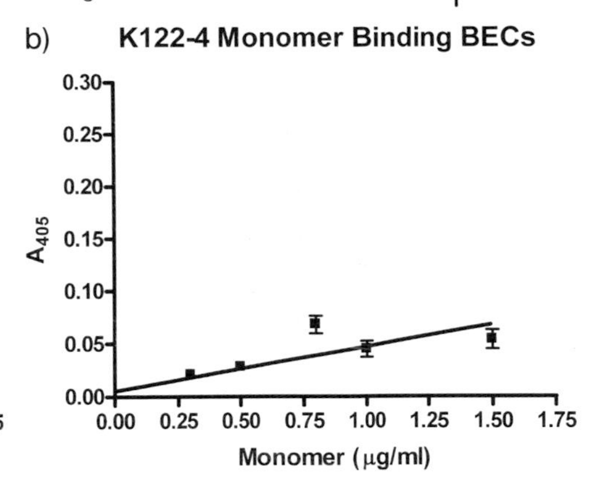

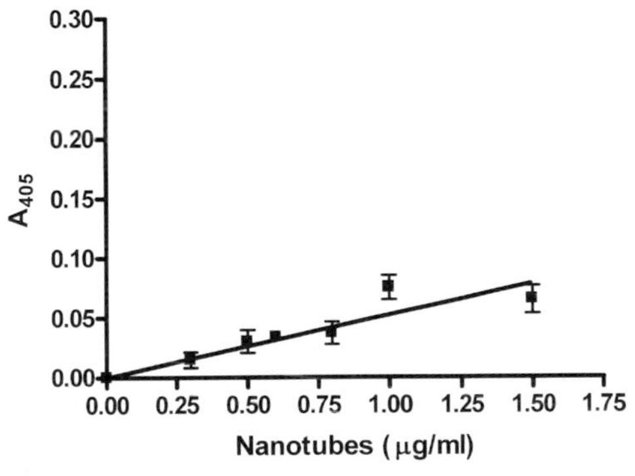

Figure 3.13 Direct binding of K122-4 pili (a), Δ(1–28) K122-4 monomeric PilA (b) and K122-4 nanotubes (c) to immobilized human BECs. Bound pili, monomeric pilin and pilin protein nanotube binding to BECs was probed utilizing a polyclonal anti-K122-4 antiserum in a modified ELISA protocol. Ligand binding was normalized by protein concentration, and thus the molar concentration of PilA in the monomer and nanotubes is equivalent (although there are considerably fewer nanotubes present relative to the number of monomeric pilin subunits given the extreme length of the nanotubes), while the molar concentration of PilA present in the pili sample is around 20% lower than the other samples (due to the higher molecular mass of the full-length PilA present in the pili). There will be much fewer pili relative to monomeric pili, but more pili than nanotubes.

and it is difficult to envision how three binding domains could simultaneously bind to three different receptors without the α-helices penetrating into the host membrane. The structure suggests that pilus binding would (initially at least) occur on an angle with one to two receptor-binding domains actually being able to initially engage separate receptors simultaneously, although the nanotubes would not face a similar constraint. On balance, I believe the experimental data supports the left-handed three-start helix model for the *P. aeruginosa* type IV pilus (manuscript in preparation).

Given the experimental support for the three-start helix model, a close examination of the proposed pilus fiber is quite interesting for at least one additional aspect. When the N-terminal α-helices are colored coded in yellow, it is clear that the shaft of the pilus fiber is proposed to contain holes in the outer surface such that the internal hydrophobic core of the pilus is somewhat solvent exposed (see Figure 3.10). This prospect raises the possibility that the pilus could bind hydrophobic compounds to the pilus through hydrophobic interactions and actually accumulate those molecules into the periplasm of the cell when the pilus retracts. Potentially this could allow the cell to use the pilus to probe the external environment and bring molecules into a compartment where the cell could respond.

3.11
Structural Nature of the Receptor-Binding Domain

A definitive structure for the PilA receptor-binding domain has not been determined as the structure of a bound receptor analog with PilA has not been determined despite extensive efforts. However, the structural studies suggest that a small binding pocket is generated by the conformation of the Cα backbone of the C-terminal disulfide loop region of PilA. If exposed main-chain atoms mediate the direct interaction of PilA with receptors, then the amino acid side-chains of the residues in the receptor-binding domain do not mediate direct contact with a receptor and should therefore be available for interaction with antibodies following binding of the pilus, monomeric pilin or peptides to a surface receptor or receptor analog. PK99H is an ideal antibody to test this as PK99H binds only to the tip of the pilus. We have now tested this hypothesis and observed that PK99H does indeed bind to pili if they are bound to either immobilized asialo-GM_1 or human BECs (Figure 3.14 and van Schaik *et al.*, submitted). These observations confirm our earlier studies where we were able to detect PAK(128–144)ox bound to human ciliated epithelial cells with PK99H by immunofluorescence [45]. Thus it seems likely that the binding domain is indeed formed by exposed main chain Cα backbone atoms, this would be consistent with the results of the structure–activity relationship studies of Wong *et al.* [54] who observed that while positionally equivalent residues of the KB7 and PAK receptor binding were critical for adhesin activity, the different amino acid residues at those positions could not be reciprocally substituted without destroying adhesin function. Studies on developing a consensus sequence anti-receptor-binding domain vaccine have observed that a much broader cross-reactive antibody response can be generated substituting two PAO residues into the PAK receptor-binding domain [83,84]. Recently, the structure of a PAK PilA mutant termed Cs1 (a Thr130Lys/Pro135Glu double mutant in the receptor-binding domain) was determined by X-ray crystallography and the backbone structure of the receptor-binding domain was essentially unaltered from the PAK wild-type (Koa *et al.*, submitted).

The question of whether the entire receptor-binding domain is included within the last 17 residues of the C-terminus is challenging to address as the C-terminal domain is both necessary and sufficient for adhesin function, but competitive inhibition studies of monomeric pilin of either PAK or K122-4 PilA indicate that the monomeric pilin only inhibits 20–40% of the binding of native pili, even at high monomeric pilin

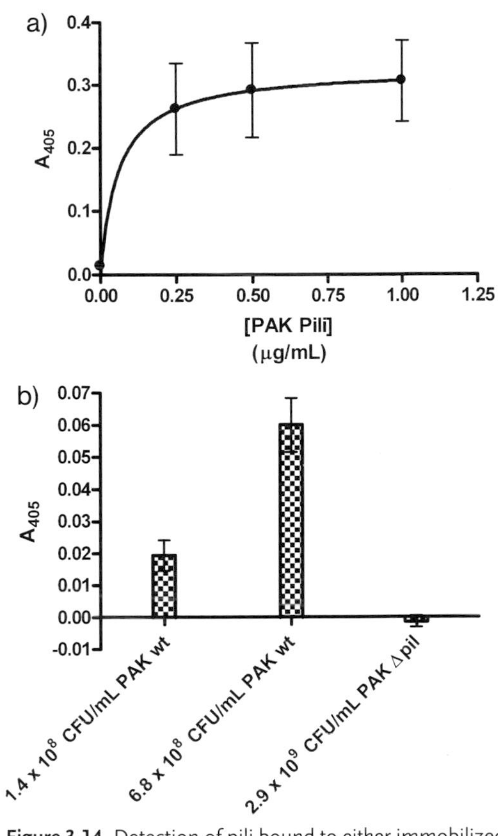

Figure 3.14 Detection of pili bound to either immobilized asialo-GM_1 (a) or pili on whole cells of PAK bound to human BECs (employing a pilin-deficient strain as a negative control) detected with monoclonal antibody PK99H that binds to the pilus tip and inhibits adherence of pili and whole cells if preincubated with the pili or cells before the adherence assay is performed (b).

concentrations [60,61]. This could be due to avidity, but it could also be due to the presence of the hydrophobic N-terminal α-helix which could potentially interact with either the mucosal membrane or the lipid portion of asialo-GM_1. As was noted earlier, there is a fairly highly conserved arginine residue (Arg30) in the N-terminal α-helix that is structurally positioned such that it is contiguous with the proposed receptor-binding domain (Figure 3.15). A role for the hydrophobic helix in mediating binding to receptors would be consistent with our previous studies with synthetic analogs of GalNAc–Gal where derivatives that were more hydrophobic in the C_1 synthetic linker region had much higher receptor affinity than did more hydrophilic linkers [37]. I feel that it is likely that both Arg30 and the N-terminal helix contribute to bacterial adherence to mucosal epithelial cells and likely interact to some degree with the mucosal epithelial surface membrane.

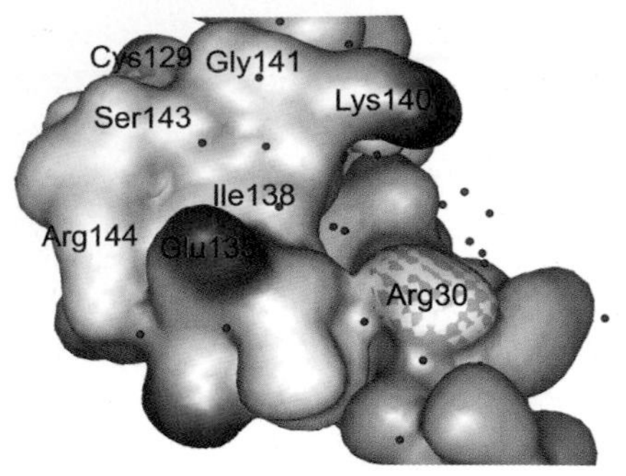

Figure 3.15 Proposed structure of the PAK PilA receptor-binding domain with major residues that surround the binding cleft labeled.

3.12 Twitching Motility

Bradley [85] initially established that *P. aeruginosa* type IV pili are retractable and function to mediate twitching motility. Type IV pili are very dynamic structures that are retractable or extendable at a speed of around 0.5 $\mu m\ s^{-1}$ with a tensional force of around 10 pN without load [76]. A single retracting pilus can generate large forces (above 100 pN) and the pilus structure can withstand up to around 120 pN per pilus (which is equivalent to a stress of 919 $lbf\,in^{-1}$) before the pilus breaks [74,75]. The force generated by retraction of type IV pili powers twitching motility and allows cells to readily move along solid surfaces. The structural strength of the pilus may be due to the central core assembly of hydrophobic α-helices [59], but the extreme length obtained by pilin monomers that assemble into nanotubes suggests that the pilin interfacial subunit interactions could also confer substantial strength to the pilus fiber [79]. As cells generally have more than one pilus, bound cells can resist significant shear forces ensuring that surfaces are firmly colonized and are resistant to release from the surface by current flow. The rate of pilin assembly or retraction is remarkably high; around 1000 pilin structural subunits s^{-1} are added to or removed from the base of the pilus fiber [74], which indicates that each of the three pilin filaments of the fiber are elongated or retracted equivalently at a rate of around 333 monomers s^{-1}. Recent direct biophysical studies on pilus retraction or elongation as a function of force applied to the pilus indicates that pilus retraction or extension is controlled by a force or pilus tension sensor switch [86], which limits the stress that is applied to an individual pilus (below the breaking point of the fiber). The complexity of pilus morphogenesis coupled with the retraction–elongation process that is controlled, in part, by a tension sensor and integrated with the cell's chemotactic response mechanisms [87] indicates that the pilus functions are complex and not yet completely characterized. Pilus expression is highly regulated [40], specifically modulated during quorum sensing and biofilm development [88], and results in considerable movement along a solid surface. Indeed, twitching-based motility allows the organism to travel faster and further along a solid surface than the net movement generated by flagella-based swimming in low viscosity media.

The basis of twitching motility is that the pilus makes contact with a surface and then binds firmly to that surface; the cell then retracts the pilus by disassembling pilin monomers from the base of the fiber (an ATP-dependent process mediated by the pilus associated proteins PilT and PilU [57]) generating a pulling force in the range of 10–120 pN [74–76] that physically pulls the cell along the solid interface. Another pilus is extended from the cell, makes contact with and binds to the surface, and then begins to retract. For net movement, the first pilus must release from the surface, or the cell becomes trapped and stuck to one position on the surface. Thus, the pilus must be designed such that it can bind readily to solid surfaces and resist the retraction of the pilus (i.e. sustain contact with the surface under forces ranging from 10 to 120 pN before detaching from the surface), but the pilus is also designed to be released from the same surface in a coordinated fashion such that net directed motion of the cell occurs.

Twitching motility plays a major role in both pathogenesis and biofilm formation. *P. aeruginosa* cells that bind to mucosal epithelial cells induce a variety of host cell signaling events and physiological responses [73,88–91]. In particular, the pilin-associated proteins PilT and PilU, which are required for twitching motility, appear to play a significant role in mediating cytotoxicity in epithelial cells, independently of the type III protein secretion events [92]. Twitching also plays a role in biofilm development; the classical definition of a biofilm is microbial growth on any substratum which is likely initiated by means of pilus-mediated binding. The biofilm form of growth is now generally thought to be the dominant form of bacterial growth in the natural environment and thus represents a significant role for the organism. Biofilm growth contrasts with another form of growth, i.e. the planktonic mode or free-living single cells associated with classical broth cultures (the basis for the classic bacteriology literature), which is characterized by bacterial replication when unattached to surfaces and not in a complex "community". Type IV pili are essential for the normal development of *P. aeruginosa* biofilms as mutants lacking functional pili are not able to develop past the microcolony stage in static or flow biofilm systems [93,94]. Our recent work suggests that type IV pili are critical for the first step of biofilm formation, i.e. the adherence or binding of the cells to a substrate [95]. Twitching motility itself is also required for normal biofilm development as retraction-deficient cells do mature past an undifferentiated layer of cells [96]

3.13
How Does the Pilus Attach to a Solid Surface?

We recently became interested in how *P. aeruginosa* colonizes surfaces; in our case we were interested how *P. aeruginosa* binds to metal surfaces, particularly stainless steel surfaces. The aggressive colonization of stainless steel surfaces, apart from being of industrial significance, is also of medical relevance. *P. aeruginosa* infections are prevalent in burn units where large stainless steel hydrotherapy units are often used to treat patients with severe burns [97]. Tredget *et al.* [97] have demonstrated a significant decrease in *P. aeruginosa* infection rates in burn units where stainless steel hydrotherapy units were removed. Biofilm formation on stainless steel and other substrata, as a function of physical and chemical modifications, has been widely investigated [98–101]. *P. aeruginosa* biofilm formation on these abiotic surfaces in hospital environments allows for significant exposure of immunocompromised individuals to *P. aeruginosa*, and results in high infection rates and significant morbidity and mortality in the hospital environment. Rough stainless steel surfaces more readily develop biofilms compared with smooth or electropolished steel [99,101–104]. We recently sought to clarify the role of the *P. aeruginosa* type IV pilus in the initial colonization of abiotic surfaces, particularly stainless steel, given the classic genetic evidence that flagella are more likely responsible for the initial stages of *P. aeruginosa* biofilm formation [93]. We demonstrated that the initial binding of *P. aeruginosa* to stainless steel is concentration dependent, exhibits classical saturation kinetics, and that pili clearly play a major role in mediating whole cell binding to stainless steel as (i) pili-deficient strains are unable to adhere, (ii) pili bind to steel in a concentration-dependent, saturable manner, (iii) pili competitively inhibit whole-cell binding in a direct competition assay and (iv) pilus specific antibodies inhibit binding to steel [95]. We demonstrated, through the use of synthetic peptides, that type IV pili bind to the steel surface through the C-terminal disulfide loop region of the pilin structural subunit, and that the previously identified pilin receptor-binding domain functions to mediate binding to both biotic and abiotic surfaces [95]. Strikingly, the synthetic receptor-binding domain PAK(128–144)ox has a very high affinity for steel displaying an apparent K_i of 4.2 and 0.2 nM for inhibition of viable cells and purified pili binding to steel, respectively (Figure 3.16) [95]. As the apparent K_i of PAK(128–144)ox for inhibition of PAK bacteria binding to human respiratory epithelial cells is around 120 000 nM, it would appear that the receptor-binding domain has evolved to mediate high-affinity interactions with abiotic surfaces rather than biotic surfaces. This suggests that *P. aeruginosa* is thus truly an opportunistic pathogen rather than an acute pathogen that has evolved to become an opportunistic or less virulent pathogen. The very high affinity of the receptor-binding domain for steel surfaces was unexpected, particularly given the flexibility of the peptide and the presumptive entropic penalty for associated with binding.

Peptides with marginal affinity (in the order of millimolar) for steel have been identified by phage display methodology and their affinity for steel correlated to their amino acid composition [105]. While investigation into biometalic interfaces has received attention from both industry and academia, there has been minimal progress in the development of high-affinity biological ligands for metal surfaces. The interaction of the *P. aeruginosa* receptor-binding domain with steel is not simply a function of the peptide amino acid composition as two scrambled sequences of the PAO PilA receptor-binding domain (one retaining the disulfide bridge and the other a linear variant where the two cysteine residues are replaced with alanine residues) failed to inhibit binding to steel [95]. We have been able to demonstrate that the binding function of the *P. aeruginosa* pilin receptor-binding domain for abiotic and

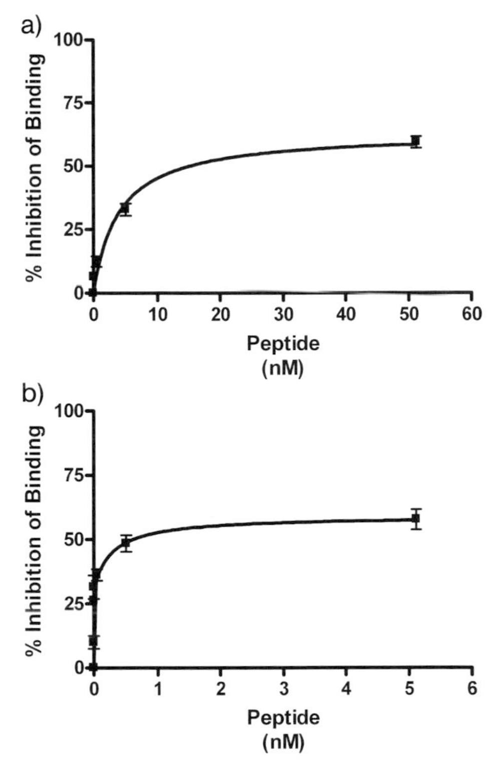

Figure 3.16 Competitive inhibition of PAK whole cells (a) or PAK pili (b) by the synthetic receptor-binding domain PAK(128–144) ox. Note that while it takes considerably more synthetic peptide to inhibit whole-cell binding than pili binding (due to avidity of the whole cells as each cell has more than one pilus) the IC_{50} is still less than 10 nM. Reproduced with permission from Giltner *et al.* [17].

biotic surfaces can be readily differentiated as a single point mutation of the PAO PilA receptor-binding domain [PAO(128–144)oxT130I] failed to inhibit binding to steel, but was much more effective in inhibiting *P. aeruginosa* binding to human respiratory epithelial cells without altering receptor specificity (Figure 3.17) [95]. Attachment of *P. aeruginosa* to abiotic surfaces appears to be highly optimized and thus likely confers a significant advantage for the organism in the natural environment. Bacteria preferentially bind to surfaces (nutrient availability on or near surfaces is higher) and form biofilms until nutrient limitation occurs when the population "emigrates" or moves to a more favorable environment (either by moving along the surface via twitching motility or by detaching from the surface to engage in a planktonic or free-living life form where they swim to find a more favorable environment). A significant question that arises is how the high-affinity interaction of the PilA receptor-binding domain is released from the surface, i.e. the same issue that twitching motility raises.

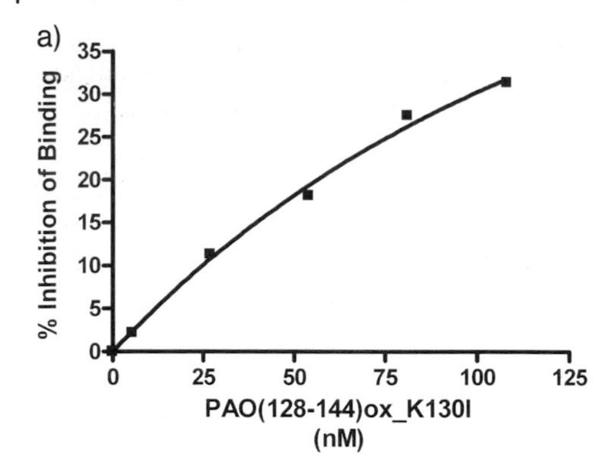

Figure 3.17 Examination of a single point mutation, K130I, in the sequence of the PAO PilA receptor-binding domain region. The change in sequence enhances the peptide's affinity for human BECs (the apparent K_i for the inhibition of bacterial binding to BECs drops from around 1200 to 38 nM) (a), but the same sequence change abolishes the peptide's affinity for stainless steel (b). Reproduced with permission from Giltner *et al.* [17].

3.14 The Monkey-Bar Swing Paradox

A challenging aspect of type IV pilus function arises with the documented extremely high affinity of a relatively flexible 17mer peptide for stainless steel (the entropic penalty for binding is substantial and yet the apparent binding affinity of a monomeric peptide is subnanomolar) in a structure that binds to and releases from the surface during twitching motility [75]. Twitching motility *in situ* is significant (estimated at around 2 mm/h for the bacterium), energetically expensive (around

1000 PilA subunits are assembled or disassembled per second per pilus), highly regulated and allows for chemotactically directed movement on solid surfaces [57]. An analogy to a childhood activity can be made, i.e. a child can swing hand-over-hand on a set of monkey bars to accomplish rapid movement across the bars. A child learning to cross monkey bars (or an unpracticed parent) grabs onto the bar and swings, but if they do not release one hand to reach for and grab the next bar, they get stuck hanging in one position on the bars until they fall off. Similarly, for a bacterium to move across a surface by twitching motility, a pilus must bind to the surface, initiate retraction (to generate cell movement), a second pilus then must extend to and then bind to the surface further away from the cell and start to retract, and the first pilus must then release from the surface for directed cell movement to occur, but if the first pilus does not release then the cell is trapped at that position. The paradox that arises is thus how a very-high-affinity binding event allows for release from the surface in a rapid enough manner to allow for directed motion along the surface. Release of pili from a solid surface is a critical aspect of twitching motility, biofilm formation/development, emigration from biofilms and dissemination from an initial infection site. Based on a comparison of NMR and X-ray structures [61,66], and the ability of monomeric PilA to form pilin nanotubes on exposure to a hydrophobe [63], we suspect that PilA may have multiple conformations and that the answer to the paradox may rest there. However, Chiang and Burrows [96] have reported that retraction-deficient strains of *P. aeruginosa* that are deficient in PilU function can bind to abiotic surfaces, but cannot detach from the surface or form differentiated biofilms. This suggests that pilus release from a surface may be physiologically regulated and controlled.

3.15
Molecular Basis for Receptor-binding Domain Interaction with Steel Surfaces

Biofilm formation on metal surfaces has been reported to occur at grain boundaries [106] and we have established that *P. aeruginosa* initially binds preferentially to grain boundaries on stainless steel surfaces via the type IV pili (Figure 3.18 [95]). Metals are polycrystalline and their grain boundaries, the interfaces between crystals, strongly influence the surface properties of the metal, including the ability of the metal surface to stick to another surface, which in materials science terminology is termed the adhesive force of the surface [107,108]. Typical grain size in grade 304 stainless steel is larger than 100 μm. At a grain boundary, the misorientation between two adjacent metallic crystals or grains results in various types of dislocation or defects [109–111]. Interfacial imperfections at grain boundaries lead to distorted crystal lattices that have interfacial disorder, which render electrons more active or more capable of interacting with other entities outside of the metal surface that in turn leads to larger adhesive force, including frictional forces when under low loads. The surface electron activity of a metal surface can be characterized by the electron work function (EWF), which is the minimum energy required to remove an electron from the Fermi level to a point immediately above the metal surface with no kinetic

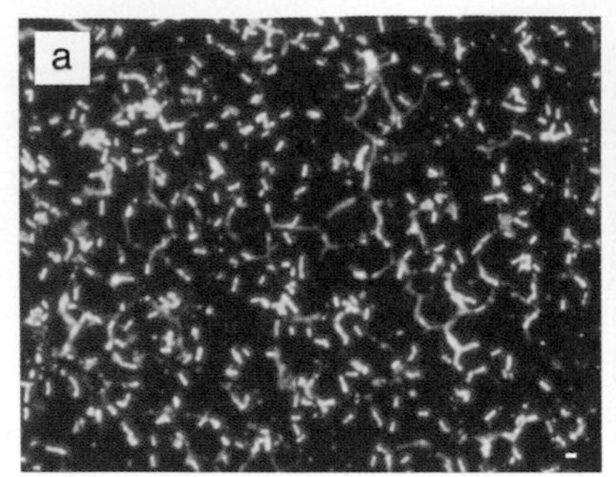

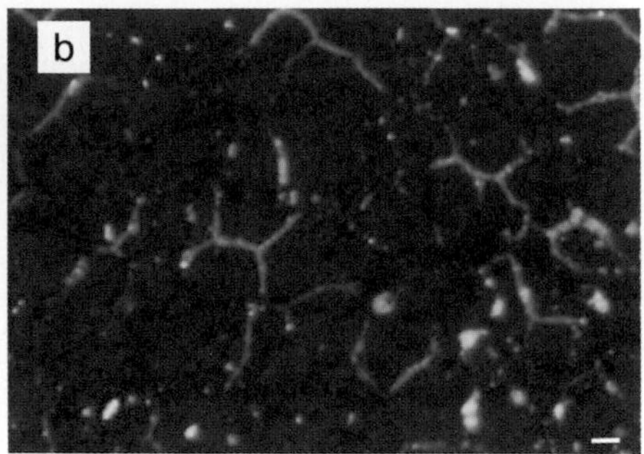

Figure 3.18 *P. aeruginosa* binds preferentially the metal grain boundaries on stainless steel due to interaction of the pili with the steel surface. Epifluorescence micrograph of strain PAK bound to steel (a: the cells are stained with acridine orange and thus appear orange) while strain PAKNP (a pilin-deficient strain) does not bind to the steel surface (b: the green fluorescence is nonspecific interaction of the acridine orange with grain boundaries). Reproduced with permission from Giltner *et al.* [95].

energy [112]. Recent studies have demonstrated that a lower EWF corresponds to a higher adhesive force [104]. Electrons in the vicinity of a grain boundary are more active, corresponding to a lower EWF, which in turn is reflected by a higher adhesive force [113]. The increased electron activity of a grain boundary results in a higher surface free energy for that surface region, and the state of the surface free energy has long been tenuously associated with increased bacterial adherence and biofilm formation [114].

We have used a direct force measurement approach (which has also been termed force mapping) utilizing atomic force microscopy (AFM) in a "dry" (normal building relative humidity, 30–40% relative humidity) air environment in which polished and etched steel are utilized (the polishing and etching removes surface-conditioning films, and surface adsorption of organics from air onto steel is very low). As the synthetic receptor-binding domain is small (roughly rectangular in nature with dimensions of around 15.7×17.5 Å) coupling the peptide directly to an AFM tip could compromise the interaction of the peptide with the steel surface. An approach has been developed to indirectly derivatize AFM tips with the peptide presented on a coiled-coil structure such that the peptide would be maintained around 42.5 Å away from the AFM tip surface and thus allow the receptor-binding domain more freedom to interact with the metal surface. A *de novo* designed heterodimeric coiled-coil system was utilized for this application [115–117], which consists of a 35-residue K-coil which is coupled to the AFM tip and serves to capture a 35-residue E-coil [115] with the receptor-binding domain [PAK(128–144)ox] fused to the C-terminus of the E-coil peptide. The affinity of the coiled-coil interaction is 60 pM [118] and the resulting coiled-coil is an extraordinarily stable structure that is only partially dissociated in solvent at an ambient temperature of 80°C in denaturing conditions [115,116,119]. The construction of the derivatized AFM tip is described in Figure 3.19. We then proceeded to force map the interaction of the receptor-binding

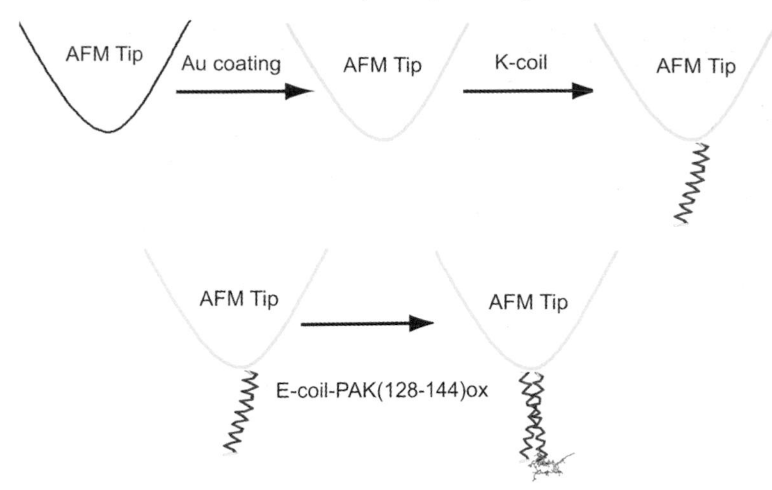

Figure 3.19 Diagram describing the fabrication an AFM tip with a coiled-coil peptide construct where the PAK receptor-binding domain is displayed on the AFM tip. The coiled-coil system has been previously described Tripet *et al.* [115].

domain with the steel surface relative to grain boundary (Figure 3.20) (Yu *et al.*, submitted). The receptor-binding domain thus appears to interact directly with metal grain boundaries – an area of high electron activity. The nature of the forces involved in the molecular interaction is unknown at this point, but clearly the binding event is not dependent upon a strong hydrophobic effect as there was no bulk phase water in

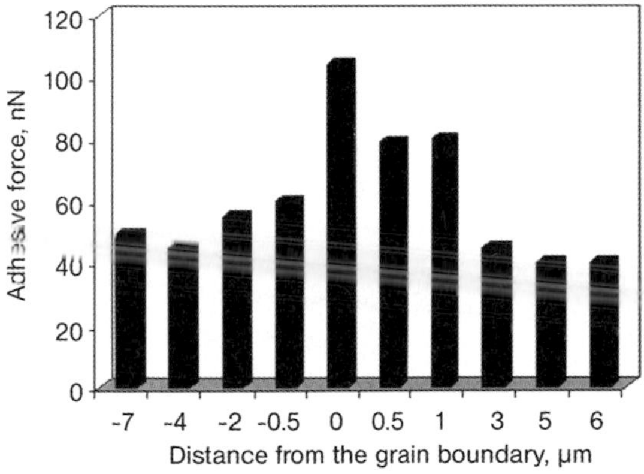

Figure 3.20 Force mapping of the interaction of the PAK receptor-binding domain displayed on a coiled-coil derivatized on an AFM tip. The force required to disrupt the interaction of the peptide-steel interaction from a "dry" polished and etched stainless steel surface as a function of the distance from the grain boundary as determined by optical microscopy using polarized light to identify the grain boundary (Yu *et al.*, submitted).

this study. The strength of this interaction is very high and suggests that the biometallic interaction likely stabilizes the electrons in the grain boundary, possibly like a metal ion coordination interaction.

3.16
Pili as Nanowires for Redox Reactions

A couple of very novel and interesting publications have reported that type IV pili in *Geobacter* [120,121] and *Shewanella* [122] function as nanowires to conduct electrons towards or away from bacterial cells during the oxidation or reduction of growth substrates. While it may seem somewhat unlikely, it is possible that type IV pili have evolved to bind to surfaces and act as nanowires to conduct electrons to or away from the cell. While *P. aeruginosa* pili do not appear to function as nanowires [120], the experimental surface did not utilize pilus-mediated contact with a metal, thus the issue remains somewhat open.

3.17
What is the Most Important Role of Adherence to *P. aeruginosa*

A balanced view of *P. aeruginosa* adherence suggests that binding to solid surfaces and the development of differentiated biofilms is likely the critical function for the organism in the natural environment. It seems likely that type IV pili are the elegant end result of the evolution of a multifunctional organelle that plays a major role in biofilm formation and differentiation, motility, and DNA binding; and potentially may serve as a "nose" (as one of my colleagues has described the potential accumulation of small molecules from the environment by the pilus) and may even potentially serve as means of extracting electrons from metallic surfaces. The role of adherence and the pili in infection is significant – both for the initial colonization of mucosal surfaces and, perhaps more importantly, for the development of biofilms following initial colonization; however, in terms of the overall survival of the organism this is likely a "fortuitous" accident from the perspective of the organism. It is clear that a significant number of aspects of pilus structure and function are poorly understood even with this simple extracellular appendage.

Acknowledgments

I would like to gratefully acknowledge the funding support of the Canadian Institutes for Health Research and the Natural Sciences and Engineering Research Council. I would also like to thank my long-term collaborators, and my former students and post-docs whose efforts have helped illuminate the puzzles of *P. aeruginosa* adherence and type IV pilus structure and function.

References

1 Stover, C.K. *et al.* (2000) Complete genome sequence of *Pseudomonas aeruginosa* PA01, an opportunistic pathogen. *Nature*, **406**, 959–964.

2 Stoodley, P. *et al.* (2002) Biofilms as complex differentiated communities. *Annu Rev Microbiol*, **56**, 187–209.

3 Sugawara, E. *et al.* (2006) *Pseudomonas aeruginosa* porin OprF exists in two different conformations. *J Biol Chem*, **281**, 16220–16229.

4 Hancock, R.E. (1985) The *Pseudomonas aeruginosa* outer membrane permeability barrier and how to overcome it. *Antibiot Chemother*, **36**, 95–102.

5 Ramsey, D.M. and Wozniak, D.J. (2005) Understanding the control of *Pseudomonas aeruginosa* alginate synthesis and the prospects for management of chronic infections in cystic fibrosis. *Mol Microbiol*, **56**, 309–322.

6 Beachey, E.H. (1981) Bacterial adherence: adhesin–receptor interactions mediating the attachment of bacteria to mucosal surface. *J Infect Dis*, **143**, 325–345.

7 Johanson, W.G. Jr. *et al.* (1980) Bacterial adherence to epithelial cells in bacillary colonization of the respiratory tract. *Am Rev Respir Dis*, **121**, 55–63.

8 Woods, D.E. *et al.* (1981) Role of salivary protease activity in adherence of Gram-negative bacilli to mammalian buccal epithelial cells *in vivo*. *J Clin Invest*, **68**, 1435–1440.

9 Woods, D.E. *et al.* (1981) Role of fibronectin in the prevention of adherence of *Pseudomonas aeruginosa* to buccal cells. *J Infect Dis*, **143**, 784–790.

10 Woods, D.E. *et al.* (1983) Factors influencing the adherence of *Pseudomonas aeruginosa* to mammalian buccal epithelial cells. *Rev Infect Dis*, **5** (Suppl 5), S846–S851.

11 Todd, T.R. *et al.* (1989) Augmented bacterial adherence to tracheal epithelial cells is associated with Gram-negative pneumonia in an intensive care unit population. *Am Rev Respir Dis*, **140**, 1585–1589.

12 Woods, D.E. *et al.* (1980) Role of pili in adherence of *Pseudomonas aeruginosa* to mammalian buccal epithelial cells. *Infect Immun*, **29**, 1146–1151.

13 Doig, P. *et al.* (1988) Role of pili in adhesion of *Pseudomonas aeruginosa* to human respiratory epithelial cells. *Infect Immun*, **56**, 1641–1646.

14 Franklin, A.L. *et al.* (1987) Adherence of *Pseudomonas aeruginosa* to cilia of human tracheal epithelial cells. *Infect Immun*, **55**, 1523–1525.

15 McEachran, D.W. and Irvin, R.T. (1985) Adhesion of *Pseudomonas aeruginosa* to human buccal epithelial cells: evidence for two classes of receptors. *Can J Microbiol*, **31**, 563–569.

16 Doig, P. *et al.* (1990) Inhibition of pilus-mediated adhesion of *Pseudomonas aeruginosa* to human buccal epithelial cells by monoclonal antibodies directed against pili. *Infect Immun*, **58**, 124–130.

17 Doig, P. *et al.* (1987) Characterization of the binding of *Pseudomonas aeruginosa* alginate to human epithelial cells. *Infect Immun*, **55**, 1517–1522.

18 Irvin, R.T., Doig, P.C., Sastry, P.A., Heller, B. and Paranchych, W. (1989) Competition for bacterial receptor sites on respiratory epithelial cells by *Pseudomonas aeruginosa* strains of heterologous pilus type: usefulness of kinetic parameters in predicting the outcome. *Microb Ecol Health Dis*, **3**, 39–47.

19 Farinha, M.A. *et al.* (1994) Alteration of the pilin adhesin of *Pseudomonas aeruginosa* PAO results in normal pilus biogenesis but a loss of adherence to human pneumocyte cells and decreased virulence in mice. *Infect Immun*, **62**, 4118–4123.

20 Sato, H. and Okinaga, K. (1987) Role of pili in the adherence of *Pseudomonas*

aeruginosa to mouse epidermal cells. *Infect Immun*, **55**, 1774–1778.

21 Doig, P.C., Tapping, R., Mankinen-Irvin, P. and Irvin, R.T. (1989) Effect of microcolony formation on the adherence of *Pseudomonas aeruginosa* to human buccal epithelial cells. *Microb Ecol Health Dis*, **2**, 203–209.

22 Lillehoj, E.P., Kim, B.T. and Kim, K.C. (2002) Identification of *Pseudomonas aeruginosa* flagellin as an adhesin for Muc1 mucin. *Am J Physiol Lung Cell Mol Physiol*, **282**, L751–L756.

23 Arora, S.K. *et al.* (1998) The *Pseudomonas aeruginosa* flagellar cap protein, FliD, is responsible for mucin adhesion. *Infect Immun*, **66**, 1000–1007.

24 Carnoy, C. *et al.* (1994) *Pseudomonas aeruginosa* outer membrane adhesins for human respiratory mucus glycoproteins. *Infect Immun*, **62**, 1896–1900.

25 Irvin, R.T. and Ceri, H. (1985) Immunochemical examination of the *Pseudomonas aeruginosa* glycocalyx: a monoclonal antibody which recognizes L-guluronic acid residues of alginic acid. *Can J Microbiol*, **31**, 268–275.

26 Scharfman, A. *et al.* (2001) Recognition of Lewis X derivatives present on mucins by flagellar components of *Pseudomonas aeruginosa*. *Infect Immun*, **69**, 5243–5248.

27 Ramphal, R., Arora, S.K. and Ritchings, B.W. (1996) Recognition of mucin by the adhesin–flagellar system of *Pseudomonas aeruginosa*. *Am J Respir Crit Care Med*, **154**, S170–S174.

28 McNamara, N. *et al.* (2001) ATP transduces signals from ASGM1, a glycolipid that functions as a bacterial receptor. *Proc Natl Acad Sci USA*, **98**, 9086–9091.

29 Prince, A. (2006) Flagellar activation of epithelial signaling. *Am J Respir Cell Mol Biol*, **34**, 548–551.

30 Kunzelmann, K. *et al.* (2006) Flagellin of *Pseudomonas aeruginosa* inhibits Na^+ transport in airway epithelia. *FASEB J*, 05–4454fje.

31 Lee, K.K. *et al.* (1994) The binding of *Pseudomonas aeruginosa* pili to glycosphingolipids is a tip-associated event involving the C-terminal region of the structural pilin subunit. *Mol Microbiol*, **11**, 705–713.

32 Hahn, H. *et al.* (1997) Pilin-based anti-*Pseudomonas* vaccines: latest developments and perspectives. *Behring Inst Mitt*, 315–325.

33 Zhang, J. *et al.* (2003) Toll-like receptor 5-mediated corneal epithelial inflammatory responses to *Pseudomonas aeruginosa* flagellin. *Invest Ophthalmol Vis Sci*, **44**, 4247–4254.

34 Butterworth, M.B., Helman, S.I. and Els, W.J. (2001) cAMP-sensitive endocytic trafficking in A6 epithelia. *Am J Physiol Cell Physiol*, **280**, C752–C762.

35 Sato, H. and Frank, D.W. (2004) ExoU is a potent intracellular phospholipase. *Mol Microbiol*, **53**, 1279–1290.

36 Sheth, H.B. *et al.* (1994) The pili of *Pseudomonas aeruginosa* strains PAK and PAO bind specifically to the carbohydrate sequence βGalNAc(1 → 4)βGal found in glycosphingolipids asialo-GM_1 and asialo-GM_2. *Mol Microbiol*, **11**, 715–723.

37 Schweizer, F. *et al.* (1998) Interaction between the pili of *Pseudomonas aeruginosa* PAK and its carbohydrate receptor beta-D-GalNAc(1 → 4)beta-D-Gal analogs. *Can J Microbiol*, **44**, 307–311.

38 Emam, A. *et al.* (2006) Laboratory and clinical *Pseudomonas aeruginosa* strains do not bind glycosphingolipids *in vitro* or during type IV pili-mediated initial host cell attachment. *Microbiology*, **152**, 2789–2799.

39 Doig, P. *et al.* (1989) Human buccal epithelial cell receptors of *Pseudomonas aeruginosa*: identification of glycoproteins with pilus binding activity. *Can J Microbiol*, **35**, 1141–1145.

40 Mattick, J.S., Whitchurch, C.B. and Alm, R.A. (1996) The molecular genetics of type-4 fimbriae in *Pseudomonas aeruginosa* – a review. *Gene*, **179**, 147–155.

41 Sheth, H.B. *et al.* (1995) Development of an anti-adhesive vaccine for *Pseudomonas aeruginosa* targeting the C-terminal region of the pilin structural protein. *Biomed Pept Proteins Nucleic Acids*, **1**, 141–148.

42 Tang, H., Kays, M. and Prince, A. (1995) Role of *Pseudomonas aeruginosa* pili in acute pulmonary infection. *Infect Immun*, **63**, 1278–1285.

43 Irvin, R. (1993) Attachment and colonization and *P. aeruginosa*: role of the surface structures, in *Pseudomonas aeruginosa as an Opportunistic Pathogen*, (eds M. Campa, M. Bendinelli and H. Friedman), Plenum, New York. pp. 19–42.

44 Woods, D.E. *et al.* (1980) Role of adherence in the pathogenesis of *Pseudomonas aeruginosa* lung infection in cystic fibrosis patients. *Infect Immun*, **30**, 694–699.

45 Irvin, R.T. *et al.* (1989) Characterization of the *Pseudomonas aeruginosa* pilus adhesin: confirmation that the pilin structural protein subunit contains a human epithelial cell-binding domain. *Infect Immun*, **57**, 3720–3726.

46 Wong, W.Y. *et al.* (1992) Antigen–antibody interactions: elucidation of the epitope and strain-specificity of a monoclonal antibody directed against the pilin protein adherence binding domain of *Pseudomonas aeruginosa* strain K. *Protein Sci*, **1**, 1308–1318.

47 Campbell, A.P. *et al.* (1995) Comparison of NMR solution structures of the receptor binding domains of *Pseudomonas aeruginosa* pili strains PAO, KB7, and PAK: implications for receptor binding and synthetic vaccine design. *Biochemistry*, **34**, 16255–16268.

48 Campbell, A.P. *et al.* (1997) Interaction of the receptor binding domains of *Pseudomonas aeruginosa* pili strains PAK, PAO, KB7 and P1 to a cross-reactive antibody and receptor analog: implications for synthetic vaccine design. *J Mol Biol*, **267**, 382–402.

49 McInnes, C. *et al.* (1994) Conformational differences between *cis* and *trans* proline isomers of a peptide antigen representing the receptor binding domain of *Pseudomonas aeruginosa* as studied by $^1H - NMR$. *Biopolymers*, **34**, 1221–1230.

50 Campbell, A.P. *et al.* (1997) Solution secondary structure of a bacterially expressed peptide from the receptor binding domain of *Pseudomonas aeruginosa* pili strain PAK: a heteronuclear multidimensional NMR study. *Biochemistry*, **36**, 12791–12801.

51 Campbell, A.P. *et al.* (2000) Backbone dynamics of a bacterially expressed peptide from the receptor binding domain of *Pseudomonas aeruginosa* pilin strain PAK from heteronuclear $^1H - ^{15}N.2trueem$NMR spectroscopy. *J Biomol NMR*, **17**, 239–255.

52 Campbell, A.P *et al.* (2003) Interaction of a peptide from the receptor-binding domain of *Pseudomonas aeruginosa* pili strain PAK with a cross-reactive antibody: changes in backbone dynamics induced by binding. *Biochemistry*, **42**, 11334–11346.

53 Campbell, A.P. *et al.* (2000) Interaction of a bacterially expressed peptide from the receptor binding domain of *Pseudomonas aeruginosa* pili strain PAK with a cross-reactive antibody: conformation of the bound peptide. *Biochemistry*, **39**, 14847–14864.

54 Wong, W.Y. *et al.* (1995) Structure–function analysis of the adherence-binding domain on the pilin of *Pseudomonas aeruginosa* strains PAK and KB7. *Biochemistry*, **34**, 12963–12972.

55 Parge, H.E. *et al.* (1995) Structure of the fibre-forming protein pilin at 2.6 A resolution. *Nature*, **378**, 32–38.

56 Craig, L., Pique, M.E. and Tainer, J.A. (2004) Type IV pilus structure and bacterial pathogenicity. *Nat Rev Microbiol*, **2**, 363–378.

57 Mattick, J.S. (2002) Type IV pili and twitching motility. *Annu Rev Microbiol*, **56**, 289–314.

58 Craig, L. *et al.* (2003) Type IV pilin structure and assembly: X-ray and EM analyses of *Vibrio cholerae* toxin-coregulated pilus and *Pseudomonas aeruginosa* PAK pilin. *Mol Cell*, **11**, 1139–1150.

59 Craig, L. *et al.* (2006) Type IV pilus structure by cryo-electron microscopy and crystallography: implications for pilus assembly and functions. *Mol Cell*, **23**, 651–662.

60 Hazes, B. *et al.* (2000) Crystal structure of *Pseudomonas aeruginosa* PAK pilin suggests a main-chain-dominated mode of receptor binding. *J Mol Biol*, **299**, 1005–1017.

61 Audette, G.F., Irvin, R.T. and Hazes, B. (2004) Crystallographic analysis of the *Pseudomonas aeruginosa* strain K 122-4monomeric pilin reveals a conserved receptor-binding architecture. *Biochemistry*, **43**, 11427–11435.

62 Beard, M.K. *et al.* (1990) Morphogenetic expression of *Moraxella bovis* fimbriae (pili) in *Pseudomonas aeruginosa*. *J Bacteriol*, **172**, 2601–2607.

63 Audette, G.F., van Schaik, E.J., Hazes, B. and Irvin, R.T. (2004) DNA-binding protein nanotubes: learning from nature's nanotech examples. *Nano Lett*, **4**, 1897–1902.

64 Dunlop, K.V., Irvin, R.T. and Hazes, B. (2005) Pros and cons of cryocrystallography: should we also collect a room-temperature data set? *Acta Crystallogr D*, **61**, 80–87.

65 Suh, J.Y. *et al.* (2001) Backbone dynamics of receptor binding and antigenic regions of a *Pseudomonas aeruginosa* pilin monomer. *Biochemistry*, **40**, 3985–3995.

66 Keizer, D.W. *et al.* (2001) Structure of a pilin monomer from *Pseudomonas aeruginosa*: implications for the assembly of pili. *J Biol Chem*, **276**, 24186–24193.

67 Lee, K.K. *et al.* (1995) Use of synthetic peptides in characterization of microbial adhesins. *Methods Enzymol*, **253**, 115–131.

68 Lee, K.K. *et al.* (1996) Anti-adhesin antibodies that recognize a receptor-binding motif (adhesintope) inhibit pilus/fimbrial-mediated adherence of *Pseudomonas aeruginosa* and *Candida albicans* to asialo-GM_1 receptors and human buccal epithelial cell surface receptors. *Can J Microbiol*, **42**, 479–486.

69 Lee, K.K. *et al.* (1989) Mapping the surface regions of *Pseudomonas aeruginosa* PAK pilin: the importance of the C-terminal region for adherence to human buccal epithelial cells. *Mol Microbiol*, **3**, 1493–1499.

70 Yu, L. *et al.* (1996) Use of synthetic peptides to confirm that the *Pseudomonas aeruginosa* PAK pilus adhesin and the *Candida albicans* fimbrial adhesin possess a homologous receptor-binding domain. *Mol Microbiol*, **19**, 1107–1116.

71 Watts, T.H., Kay, C.M. and Paranchych, W. (1982) Dissociation and characterization of pilin isolated from *Pseudomonas aeruginosa* strains PAK and PAO. *Can J Biochem*, **60**, 867–872.

72 Bradley, D.E. (1972) Evidence for the retraction of *Pseudomonas aeruginosa* RNA phage pili. *Biochem Biophys Res Commun*, **47**, 142–149.

73 Merz, A.J., Enns, C.A. and So, M. (1999) Type IV pili of pathogenic *Neisseriae* elicit cortical plaque formation in epithelial cells. *Mol Microbiol*, **32**, 1316–1332.

74 Merz, A.J. and Forest, K.T. (2002) Bacterial surface motility: slime trails, grappling hooks and nozzles. *Curr Biol*, **12**, R297–R303.

75 Merz, A.J., So, M. and Sheetz, M.P. (2000) Pilus retraction powers bacterial twitching motility. *Nature*, **407**, 98–102.

76 Skerker, J.M. and Berg, H.C. (2001) Direct observation of extension and retraction of type IV pili. *Proc Natl Acad Sci USA*, **98**, 6901–6904.

77 Folkhard, W. *et al.* (1981) Structure of polar pili from *Pseudomonas aeruginosa* strains K and O. *J Mol Biol*, **149**, 79–93.

78 Marvin, D.A., Nadassy, K., Welsh, L.C. and Forest, K.T. (2003) Type-4 bacterial pili: molecular models and their simulated diffraction patterns. *Fibre Diffract Rev*, **11**, 87–94.

79 Audette, G.F. *et al.* (2004) DNA-binding protein nanotubes: learning from nature's nanotech examples. *Nano Lett.*, **4**, 1897–1902.

80 van Schaik, E.J. *et al.* (2005) DNA binding: a novel function of *Pseudomonas aeruginosa* type IV pili. *J Bacteriol*, **187**, 1455–1464.

81 Collins, R.F. *et al.* (2001) Analysis of the PilQ secretin from *Neisseria meningitidis* by transmission electron microscopy reveals a dodecameric quaternary structure. *J Bacteriol*, **183**, 3825–3832.

82 Wolfgang, M. *et al.* (2000) Components and dynamics of fiber formation define a ubiquitous biogenesis pathway for bacterial pili. *EMBO J*, **19**, 6408–6418.

83 Cachia, P.J. *et al.* (1998) The use of synthetic peptides in the design of a consensus sequence vaccine for *Pseudomonas aeruginosa*. *J Pept Res*, **52**, 289–299.

84 Cachia, P.J. and Hodges, R.S. (2003) Synthetic peptide vaccine and antibody therapeutic development: prevention and treatment of *Pseudomonas aeruginosa*. *Biopolymers*, **71**, 141–168.

85 Bradley, D.E. (1980) A function of *Pseudomonas aeruginosa* PAO polar pili: twitching motility. *Can J Microbiol*, **26**, 146–154.

86 Maier, B., Koomey, M. and Sheetz, M.P. (2004) A force-dependent switch reverses type IV pilus retraction. *Proc Natl Acad Sci USA*, **101**, 10961–10966.

87 Whitchurch, C.B. *et al.* (2004) Characterization of a complex chemosensory signal transduction system which controls twitching motility in *Pseudomonas aeruginosa*. *Mol Microbiol*, **52**, 873–893.

88 Glessner, A. *et al.* (1999) Roles of *Pseudomonas aeruginosa las* and *rhl* quorum-sensing systems in control of twitching motility. *J Bacteriol*, **181**, 1623–1629.

89 Comolli, J.C. *et al.* (1999) *Pseudomonas aeruginosa* gene products PilT and PilU are required for cytotoxicity *in vitro* and virulence in a mouse model of acute pneumonia. *Infect Immun*, **67**, 3625–3630.

90 Kang, P.J. *et al.* (1997) Identification of *Pseudomonas aeruginosa* genes required for epithelial cell injury. *Mol Microbiol*, **24**, 1249–1262.

91 Taha, M.K. *et al.* (1998) Pilus-mediated adhesion of *Neisseria meningitidis*: the essential role of cell contact-dependent transcriptional upregulation of the PilC1 protein. *Mol Microbiol*, **28**, 1153–1163.

92 Comolli, J.C. *et al.* (1999) Pili binding to asialo-GM_1 on epithelial cells can mediate cytotoxicity or bacterial internalization by *Pseudomonas aeruginosa*. *Infect Immun*, **67**, 3207–3214.

93 O'Toole, G.A. and Kolter, R. (1998) Flagellar and twitching motility are necessary for *Pseudomonas aeruginosa* biofilm development. *Mol Microbiol*, **30**, 295–304.

94 Klausen, M. *et al.* (2003) Involvement of bacterial migration in the development of complex multicellular structures in *Pseudomonas aeruginosa* biofilms. *Mol Microbiol*, **50**, 61–68.

95 Giltner, C.L. *et al.* (2006) The *Pseudomonas aeruginosa* type IV pilin receptor binding domain functions as an adhesin for both biotic and abiotic surfaces. *Mol Microbiol*, **59**, 1083–1096.

96 Chiang, P. and Burrows, L.L. (2003) Biofilm formation by hyperpiliated mutants of *Pseudomonas aeruginosa*. *J Bacteriol*, **185**, 2374–2378.

97 Tredget, E.E. *et al.* (1992) Epidemiology of infections with *Pseudomonas aeruginosa* in burn patients: the role of hydrotherapy. *Clin Infect Dis*, **15**, 941–949.

98 Arnold, J.W. *et al.* (2004) Multiple imaging techniques demonstrate the manipulation of surfaces to reduce bacterial contamination and corrosion. *J Microsc*, **216**, 215–221.

99 Balazs, D.J. *et al.* (2004) Inhibition of bacterial adhesion on PVC endotracheal tubes by RF-oxygen glow discharge, sodium hydroxide and silver nitrate treatments. *Biomaterials*, **25**, 2139–2151.

100 Groessner-Schreiber, B. *et al.* (2004) Do different implant surfaces exposed in the oral cavity of humans show different biofilm compositions and activities? *Eur J Oral Sci*, **112**, 516–522.

101 Lomander, A. *et al.* (2004) Evaluation of chlorines' impact on biofilms on scratched stainless steel surfaces. *Bioresour Technol*, **94**, 275–283.

102 Characklis, W.G. and Marshall, K.C. (1990) *Biofilms*, Wiley, New York.

103 Vanhaecke, E. *et al.* (1990) Kinetics of *Pseudomonas aeruginosa* adhesion to 304 and 316-L stainless steel: role of cell surface hydrophobicity. *Appl Environ Microbiol*, **56**, 788–795.

104 Bagge, D. *et al.* (2001) *Shewanella putrefaciens* adhesion and biofilm formation on food processing surfaces. *Appl Environ Microbiol*, **67**, 2319–2325.

105 Zuo, R., Ornek, D. and Wood, T.K. (2005) Aluminum- and mild steel-binding peptides from phage display. *Appl Microbiol Biotechnol*, **68**, 505–509.

106 Sreekumari, K.R., Nandakumar, K. and Kikuchi, Y. (2001) Bacterial attachment to stainless steel welds: significance of substratum microstructure. *Biofouling*, **17**, 303–316.

107 Mclean, D. (1957) *Grain Boundaries in Metals*, Clarendon Press, Oxford.

108 Chawick, G. and Smith, D. (1976) *Grain Boundary Structure and Properties*, Academic Press, London.

109 Fionova, L. and Artemyev, A. (1993) *Grain Boundaries in Metals and Semiconductors*, Editions De Physique, Les Ulis.

110 Song, S. (1990) *in situ* high-resolution electron microscopy observation of grain-boundary migration through ledge motion in an Al–Mg alloy. *Phil Mag Lett*, **79**, 511–517.

111 Chan, S. *et al.* (1998) High resolution transmission electron microscopy of $Ba_{1-x}K_xBO_3$ superconductor–insulator –superconductor grain boundary tunnel junctions. *J Mater Res*, **13**, 1774–1779.

112 Ashcroft, N. and Mermin, N.D. (1976) *Solid State Physics*, Holt Rinehart & Winston, New York.

113 Wang, X. and Li, D. (2003) Mechanical, electrochemical and tribological properties of nano-crystalline surface of 304 stainless steel. *Wear*, **255**, 836–845.

114 Fletcher, M. (1987) How do bacteria attach to solid surfaces? *Microbiol Sci*, **5**, 133–136.

115 Tripet, B. *et al.* (1996) Engineering a *de novo*-designed coiled-coil heterodimerization domain off the rapid detection, purification and characterization of recombinantly expressed peptides and proteins. *Protein Eng*, **9**, 1029–1042.

116 Chao, H. *et al.* (1998) Use of a heterodimeric coiled-coil system for biosensor application and affinity purification. *J Chromatogr B*, **715**, 307–329.

117 Chao, H. *et al.* (1996) Kinetic study on the formation of a *de novo* designed heterodimeric coiled-coil: use of surface plasmon resonance to monitor the association and dissociation of polypeptide-chains. *Biochemistry*, **35**, 12175–12185.

118 De Crescenzo, G. *et al.* (2003) Real-time monitoring of the interactions of two-stranded *de novo* designed coiled-coils: effect of chain length on the kinetic and thermodynamic constants of binding. *Biochemistry*, **42**, 1754–1763.

119 Litowski, J.R. and Hodges, R.S. (2001) Designing heterodimeric two-stranded alpha-helical coiled-coils: the effect of chain length on protein folding, stability and specificity. *J Pept Res*, **58**, 477–492.

120 Reguera, G. *et al.* (2005) Extracellular electron transfer via microbial nanowires. *Nature*, **435**, 1098–1101.

121 Reguera, G. *et al.* (2006) Biofilm and nanowire production leads to increased current in *Geobacter sulfurreducens* fuel cells. *Appl Environ Microbiol*, **72**, 7345–7348.

122 Gorby, Y.A. *et al.* (2006) Electrically conductive bacterial nanowires produced by *Shewanella oneidensis* strain MR-1 and other microorganisms. *Proc Natl Acad Sci USA*, **103**, 11358–11363.

4
Flagella and Pili of *Pseudomonas aeruginosa*

Jeevan Jyot and Reuben Ramphal

4.1
Introduction

Motility is one of the most impressive features of microbial life and requires a large percent of cellular energy. Its roles include increased efficiency of nutrient acquisition, evasion of toxic substances, ability to translocate to preferred hosts and access to optimal colonization sites within them, as well as dispersal in the environment.

Movement in aqueous environments by swimming or along surfaces by using different modes of translocation has been classified into several distinct types [1]. *Pseudomonas aeruginosa* exhibits three types of motility – flagellum-mediated swimming, flagellum and type IV pilus-mediated swarming [2,3], and type IV pilus-mediated twitching [4–6]. Swarming motility can be distinguished from swimming motility in that swarming is required to move across a hydrated, viscous semisolid surface, while swimming allows movement through a relatively low-viscosity liquid environment. The direction of movement is biased by chemotactic responses to chemical stimuli [7–9]. This chapter focuses on the two motility organelles of *P. aeruginosa*, i.e. the flagellum and the pilus, their structure, regulation and role in pathogenesis.

4.2
Flagellum of *P. aeruginosa*

P. aeruginosa has a single, unsheathed polar flagellum similar to that of *Caulobacter crescentus* and *Vibrio parahaemolyticus*, but unlike the polar, sheathed flagellum observed in *Helicobacter* and the peritrichous flagella seen in *Escherichia coli* and *Salmonella enterica* serovar Typhimurium. However, swarmer cells of *P. aeruginosa* appear to possess on average two polar flagella and no lateral flagella [2,10].

At the genetic level, the expression of flagellar genes in *P. aeruginosa* is tightly regulated, involving transcriptional activators and multiple σ factors as in other enteric bacteria. Flagellar structural genes are usually conserved across the genomes

Pseudomonas. Model Organism, Pathogen, Cell Factory. Edited by Bernd H.A. Rehm

ISBN: 978-3-527-31914-5

of flagellated bacteria, but differences in regulatory genes are not uncommon [11]. *P. aeruginosa* has a four-tiered flagellar hierarchy, but how this flagellar hierarchy differs from that in other organisms will not be discussed in this chapter. Readers are directed to some excellent reviews on the subject [11,12].

A role for flagella in virulence has been demonstrated for numerous pathogenic bacteria including *P. aeruginosa, Campylobacter jejuni, Proteus mirabilis, V. cholerae* and *H. pylori* [13,14]. In *P. aeruginosa,* the early importance of flagella lay in the fact that the heat-labile antigens (H antigen) of flagella could be used for typing of *Pseudomonas* strains [15], but the clinical significance of the flagellum emerged after loss of virulence was associated with the absence of a flagellum in an isogenic mutant of *P. aeruginosa* in the burned-mouse model [16]. Passive transfer of antiflagellar serum increased protection and opsonophagocytosis against *P. aeruginosa* infection [17–19]. These studies led to increased interest in the *P. aeruginosa* flagellum, providing a deeper insight into its genetic regulation, and its role in inflammation and disease, all which are discussed in detail in the following sections.

4.2.1
Structure of the *P. aeruginosa* Flagellum

Wild-type *P. aeruginosa* have typical straight swimming behavior interrupted by occasional tumbling. The majority of these bacteria have a velocity of 30–50 $\mu m\,s^{-1}$ ($35.4 \pm 13.6\,\mu m\,s^{-1}$) [20]. The flagellum of *P. aeruginosa* (Figure 4.1A) is about 2.5 wave units (1 wave unit = 1.8 μm) [21]. The flagellum has three physical sections – the basal body, the hook and the filament (Figure 4.1B). The basal body is embedded within the cell surface and based on homology is composed of three rings – the cytoplasmic membrane supramembrane (MS) ring, the peptidoglycan (P) ring and the outer membrane lipopolysaccharide (L) ring – all encircling a rod that traverses the periplasm. The hook is a flexible universal joint between the basal body and the filament, and is surface exposed. The filament is made up of polymerized flagellin monomers capped by the flagellar cap, FliD which is a mucin adhesin [22].

4.2.2
Chromosomal Organization of the Flagellar Genes of *P. aeruginosa*

The genome sequences of *P. aeruginosa* strains PAO1 and PA14 are available at www.pseudomonas.com. In addition, genome sequences of *P. aeruginosa* strains C3719 and 2192 are complete and annotated, and those of PA7 and PACS2 are complete and being assembled. Studies in *P. aeruginosa* or homology to proteins of known function in other flagellar systems [12] led to the identification and annotation of the genes involved in the regulation of flagellar biogenesis in *P. aeruginosa*.

P. aeruginosa strains are classified into a or b types based on sequence variation in *fliC*, the flagellin gene [23]. Two distinct antigenic types of FliD in *P. aeruginosa*, named A and B types [24], are coinherited with a- or b-type flagellins, respectively [22,24].

The genomes of the sequenced strains PAO1, PA14 (www.pseudomonas.com) and C3719 (Broad institute) encode type b flagellin. However, the majority of clinical

(A)

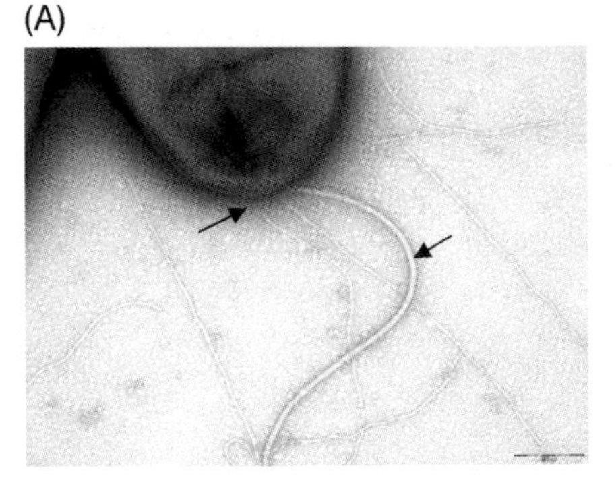

(B)

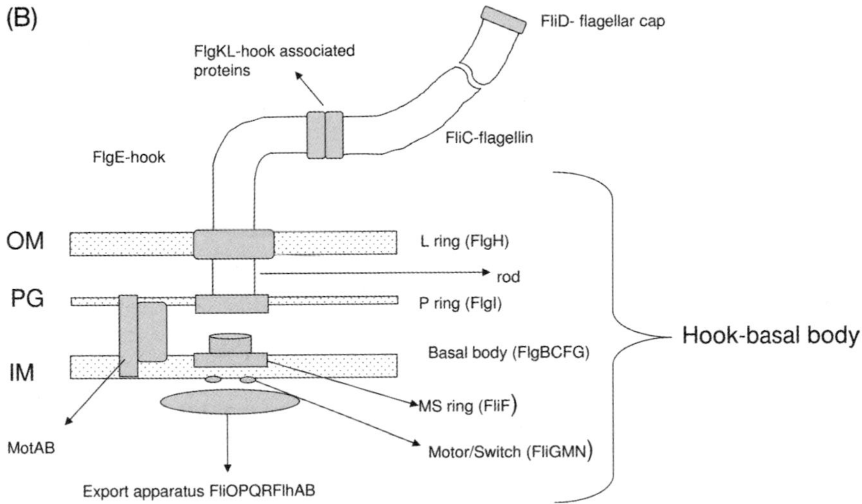

Figure 4.1 (A) Electron microscopic image of *P. aeruginosa* strain PAK showing its single polar flagella and multiple pili (pointed with arrows). (B) Schematic diagram of the flagellar structure in *P. aeruginosa* showing the flagellar proteins and their locations.

isolates surveyed express type a flagellin [25,26], and include strain PAK, and sequenced strains CS2 (University of Washington Genome Center), PA7 (TIGR) and 2192 (Broad institute). For ease and clarity, we shall refer to the flagellar organization in *P. aeruginosa* PAK as the representative a-type strain and PAO as the representative b-type strain instead of referring to all strains.

There are 41 genes encoding structural/assembly or regulatory components of the flagellar organelle in the sequenced PAO1 genome (Table 4.1). These operons are located in three regions of the chromosome (Table 4.1). Most of the structural genes coding for the basal body rod, rings, hook, filament, cap and basal body (*flgBCDE, flgFGHIJKL, fliCfleL, fliDS* and *fliEFGHIJ*) are clustered in region I of the PAO1 genome. The genes encoding the transcriptional regulator FleQ and two-component system FleSR are also located within this region. Region II contains genes encoding the hook length regulator, switch, export apparatus, flagellar placement determinant,

Table 4.1 Annotation of flagellar biogenesis genes.

Chromosomal organization[a]	PA number[b]	Gene name[c]	Protein/product name/function[d]
Region I (1164275–1197833)	1077	*flgB*	flagellar basal-body rod protein
	1078	*flgC*	flagellar basal-body rod protein
	1079	*flgD*	flagellar basal-body rod modification protein
	1080	*flgE*	flagellar hook protein
	1081	*flgF*	flagellar basal-body rod protein
	1082	*flgG*	flagellar basal-body rod protein
	1083	*flgH*	flagellar L-ring protein precursor
	1084	*flgI*	flagellar P-ring protein precursor
	1085	*flgJ*	flagellar protein
	1086	*flgK*	flagellar hook-associated protein 1
	1087	*flgL*	flagellar hook-associated protein type 3
	1092	*fliC*	flagellin type B
	1093	*fleL/flag*	hypothetical protein
	1094	*fliD*	flagellar capping protein FliD
	1095	*fliS*	hypothetical protein/flagellin-specific chaperone
	1096	*fleP*	type IV pili length control
	1097	*fleQ*	σ^{54} transcriptional regulator
	1098	*fleS*	two-component sensor kinase protein
	1099	*fleR*	two-component response regulator
	1100	*fliE*	flagellar hook–basal body complex protein
	1101	*fliF*	flagella M-ring outer membrane protein precursor
	1102	*fliG*	flagellar motor switch protein
	1103	*fliH*	probable flagellar assembly protein
	1104	*fliI*	flagellum-specific ATP synthase
	1105	*fliJ*	flagellar biosynthesis chaperone
Region II (1570496–1585538)	1441	*fliK*	flagellar hook-length control protein
	1442	*fliL*	flagellar basal body-associated protein
	1443	*fliM*	flagellar motor switch protein
	1444	*fliN*	flagellar motor switch protein
	1445	*fliO*	flagellar biogenesis protein/export pathway
	1446	*flip*	flagellar biogenesis protein/export pathway
	1447	*fliQ*	flagellar biogenesis protein/export pathway
	1448	*fliR*	flagellar biogenesis protein/export pathway
	1449	*flhB*	flagellar biogenesis protein/export pathway
	1452	*flhA*	flagellar biogenesis protein/export pathway
	1453	*flhF*	polar flagellar placement
	1454	*fleN*	flagellar number regulator

Table 4.1 *(Continued)*

Chromosomal organization[a]	PA number[b]	Gene name[c]	Protein/product name/function[d]
	1455	*fliA*	σ^{28} factor
Region III (3761961–3763652)	3350	*flgA*	flagellar basal body P-ring biosynthesis protein
	3351	*flgM*	negative regulator of flagellin synthesis (anti-σ^{28} factor)
	3352	*flgN*	flagellar biosynthesis/type III secretory pathway chaperone
Flagellar Motor genes			
	4954	*motA*	flagellar motor protein
	4953	*motB*	flagellar motor protein
	1460	*motC*	flagellar motor component
	1461	*motD*	flagellar motor protein
	3353	*ycgR*	hypothetical protein – may be involved in the flagellar motor function
	3526	*motY*	probable outer membrane protein precursor
Chemotaxis genes	1456	*cheY*	two-component response regulator
	1457	*cheZ*	chemotaxis protein CheZ
	1458	*cheA*	probable two-component sensor
	1459	*cheB*	probable methyltransferase
	1464	*chew*	probable purine-binding chemotaxis protein
	3349	*cheV*	probable chemotaxis protein methyltransferase
	3348	*cheR*	probable chemotaxis protein
Extra gene in PAK		*fliS′*; exists downstream of *fliS*	unknown

[a]Flagellar genes are organized in three regions on the chromosome. The regions I–III are marked along with the PAO1 genome numbers.
[b]Annotation based on information from www.pseudomonas.com.
[c]The flagellar genes in regions I-III and their operon arrangements are indicated by continuity.
[d]Protein/product name/function based on studies in *P. aeruginosa* or based on homology to proteins of known function found in other flagellar systems.

flagellar number regulator and alternative σ factor FliA (*fliK*, *fliLMNOPQRflhB*, *flhA*, *flhFfleN* and *fliA*) (Table 4.1). Region III consists of genes coding for the flagellar export apparatus, anti-σ factor (*flgA* and *flgMN*) (Table 4.1).

P. aeruginosa has two sets of the flagellar motor genes, now annotated as *motAB* (PA1460–PA1461) and *motCD* (PA4954–PA4953), both of which contribute to motility (Table 4.1). The *motCD* genes lie in region II, whereas *motAB* is located at a separate chromosomal locus. An additional protein MotY (PA3526) found at a separate locus has been found to associate with MotAB. Another protein YcgR (PA3353) located in region III has been found to play a role in flagellar motor function and is a new member of the flagellar regulon (www.pseudomonas.com).

Region II also includes genes encoding motor and chemotaxis proteins CheYZ, CheAB and CheW, and region III includes additional chemotaxis regulatory proteins CheVR (Table 4.1).

4.2.3
Transcriptional Hierarchy of the Flagellar Genes

P. aeruginosa utilizes a four-tiered hierarchy (Figure 4.2) involving RpoN, FleQ, FleS, FleR, FliA and FlgM regulatory proteins to control and coordinate the transcription of the flagellar regulon.

FleQ is the master regulator of the flagellar regulon in *P. aeruginosa*. It directly or indirectly regulates the expression of the majority of flagellar gene promoters (Figure 4.2) with the exception of *fliA*. Thus, *fleQ* and *fliA* are grouped in class I. Although these promoters are primarily regulated by factors outside of the flagellar regulon [27], they are independently controlled. σ^{70} is involved in the transcription of *fleQ* and *fleQ* is repressed by Vfr, the *P. aeruginosa* homolog of *E. coli* CRP [27]. The molecular mechanism that controls the expression of *fliA* remains unknown. The transcription of *fliA* appears to be constitutive and not dependent on σ^{54}, and there is no σ^{70}-binding site in its promoter. Further studies to determine dependence on *rpoD* or any other σ factor are warranted.

The class II flagellar genes require FleQ and RpoN (σ^{54}) for transcriptional activation. The *fliDSS′fleP* operon shows dependence on FleQ in the reporter assays [12,22], but microarray data suggests that the operon is both FleR- and FleQ-dependent [12]. The class II genes encode structural components of the basal

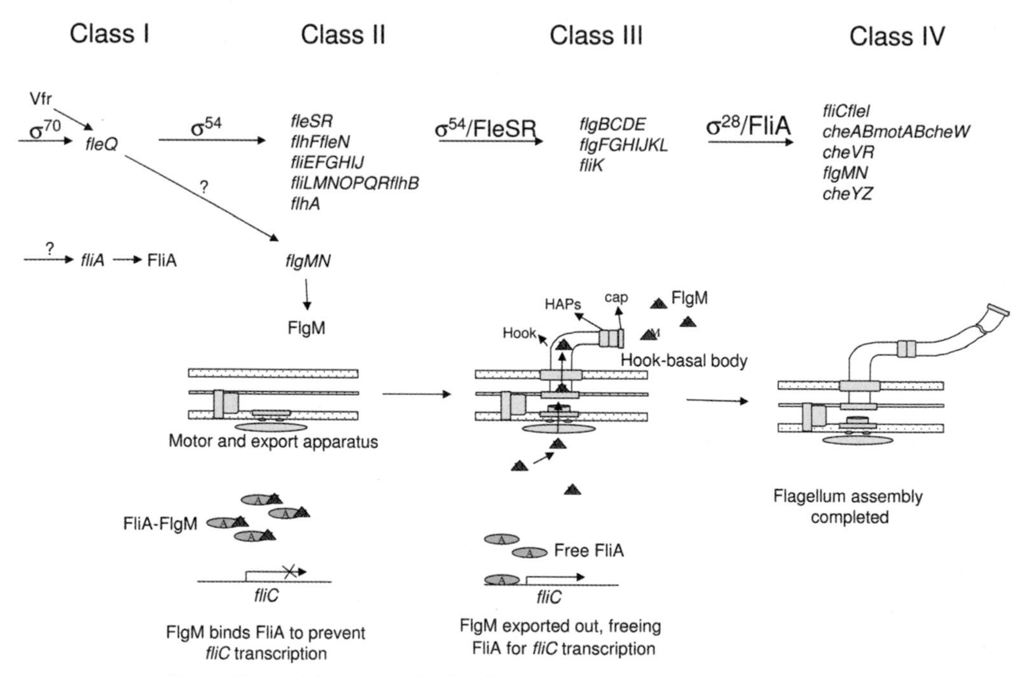

Figure 4.2 Model proposed for flagellar assembly showing the transcriptional hierarchy of the flagellar genes (class I–IV) in *P. aeruginosa*.

body, MS ring, P ring, motor, switch, flagellar export apparatus and filament cap. Regulatory proteins expressed from class II genes include FlhF, FleN, FleS and FleR [12]. In *P. aeruginosa* the loss of FlhF results in defective swimming and swarming, and a Δ*flhF* mutant assembles lateral flagella and swims at decreased velocity [10]. The FlhF protein localizes to the flagellar pole, but the signal required for its localization remains unknown [10].

FleN, the antiactivator of FleQ, downregulates FleQ activity through direct interactions and thereby plays a crucial role in maintaining a single flagellum [28,29]. FleS and FleR comprise a two-component system in which FleS is a sensor kinase for the response regulator FleR [30]. In *P. aeruginosa,* the signal sensed FleS is unknown but the phosphorylation-dependent activation of FleR by its cognate sensor kinase FleS is necessary for the transcriptional progression from class II to class III promoters and serves as an additional check point in the flagellar biogenesis of this organism.

The promoters of the operons *flgBCDE, flgFGHIJKL* and *fliK* are grouped under class III as their expression required RpoN, FleQ and FleR [12]. The expression of the *fleSR* promoter also requires RpoN as do all other class II promoters. In addition to its indirect effect, RpoN is also required for the expression of class III promoters as evidenced by the identification of σ^{54}-binding sites in each promoter element [12]. Class III genes encode the basal body rod, L ring, hook, hook–cap scaffold and hook–filament junctional proteins [12].

The class IV genes require the gene products of the preceding classes as well as FliA. The *cheABmotABcheW, cheVR, fliCfleL* and *flgMN* operons showed FliA dependence, and are grouped under class IV genes. Unlike the other class IV gene, *fliC,* which loses most of its promoter activity in the *rpoN, fleQ, fleR* and *fliA* mutants, the *flgMN* promoter retains partial transcriptional activity in all the above mutant backgrounds [31].

4.2.4
Model Proposed for Flagellar Assembly in *P. aeruginosa*

In free-swimming *P. aeruginosa,* FleQ is the master regulator of flagellar biogenesis. The lack of a FleQ-binding consensus in the class II gene promoters [32] suggests that FleQ probably has variable affinity for the different promoters it controls. This may represent a mechanism by which this single transcription factor could control the timing and level of expression of various components required at different stages in the flagellar assembly process.

The model for the assembly of the single polar flagellum in *P. aeruginosa* presented in Figure 4.2 is based on the transcriptional and posttranslational regulation of the various flagellar genes. Initiation of the assembly is most likely to begin with FlhF determining the polar placement site for the new flagellum. The other class II gene products comprising the MS ring, switch, basal body and export apparatus are subsequently assembled in the maturing flagellum. Therefore, even though *fliD* is a class II gene, its product is required after the class III-encoded hook and hook–filament junction proteins are assembled. Meanwhile, simultaneous synthesis of FleN, another class II gene product, helps maintain the monoflagellate status by inhibiting FleQ activity and downregulating further synthesis of most of the structural components. The FleSR two-component system subsequently activates the class III genes coding for proteins needed for the completion of the hook–basal body structure (Figure 4.2). This allows secretion of the anti-σ factor FlgM, and consequent transcription of FliA-dependent genes coding for flagellin (FliC) and some chemotaxis proteins (Figure 4.2). A gene *fleL*, which is cotranscribed with *fliC*, is implicated in the filament length control and maintains the flagellar filament length through an unknown mechanism. Some sequence heterogeneity in the flagellar genes *flgK*, *flgL*, *fliC*, *fleL*, *fliD* and *fliS* found in region I flanked by *flgJ* and *fleQ* has been reported between PAK and PAO1 [25]. This region contains a polymorphic gene *fleP*, which in other *Pseudomonas* strains influences stability of the mature flagellar filament. Type a strains possess an additional gene in the *fliDS* operon, designated *fliS′* (because of its sequence similarity to *fliS*), which is located between *fliS* and *fleP* (Table 4.1) [22]. The role of *fliS′* in flagellar biogenesis is unknown.

4.2.5
Environmental/Nonflagellar Regulators of Flagellar Expression

An *E. coli* CRP homolog, Vfr, is capable of repressing *fleQ* transcription by binding to its consensus sequence in the *fleQ* promoter. This is the first evidence that a nonflagellar regulator (Vfr) affects the topmost regulator (FleQ) in the flagellar biogenesis hierarchy by downregulating its transcription [27].

Another regulator reported to inhibit flagellum biosynthesis in mucoid, nonmotile *P. aeruginosa* cystic fibrosis (CF) isolates is AlgT, which by promoting expression of AmrZ, represses *fleQ*. Since *fleQ* directly or indirectly controls the expression of almost all flagellar genes, its repression ultimately leads to the loss of flagellum biosynthesis [33].

Another novel membrane-localized regulator, MorA, controls the timing of flagellar development, and affects motility, chemotaxis and biofilm formation in *P. putida*. However, unlike the motility of *P. putida*, the motility of the *P. aeruginosa* mutants was unaffected [34].

Another protein, SadB, is required for an early step in biofilm formation by the opportunistic pathogen *P. aeruginosa*. SadB inversely regulates biofilm formation and swarming motility via its ability both to modulate flagellar reversals in a viscosity-dependent fashion and to influence the production of the Pel exopolysaccharide. SadB is required to properly modulate flagellar reversal rates via chemotaxis cluster IV (CheIV cluster) [35]. However, in *P. aeruginosa*, the role of flagella in biofilm formation still remains an open question as a recent study of biofilm development showed that flagella are not involved in the attachment and that initial microcolony formation occurs by clonal growth [36].

More recently it has been observed that the type III secretion system (T3SS) affects flagellar gene expression [37], e.g. in a strain lacking flagella, T3SS gene expression, effector secretion and cytotoxicity were increased. Inversely, it was revealed that flagellar gene expression and motility was decreased in a strain overproducing ExsA, the T3SS master regulator [37]. Thus, flagellar gene expression may be affected by other systems, some of which may be involved in virulence.

When *P. aeruginosa* is grown in a purulent mucus environment as is the case in CF patients, it downregulates synthesis of its flagellin [38,39] The shut-off of flagellin synthesis in response to mucus is rapid, and independent of quorum sensing and the known regulatory networks controlling the hierarchical expression of flagellar genes [38]. Later studies revealed the flagellin repression in mucus to be a result of the action of neutrophil elastase, a protease, on the flagellar hook [40].

4.2.6
Posttranslational Modification of Flagellin

4.2.6.1 Flagellar Glycosylation Islands (GIs) in *P. aeruginosa*

All type a strains carry a GI, a cluster of genes demonstrated to be involved in the glycosylation of the type a flagellin, leading to a flagellin molecular weight in the range of 45–52 kDa [25]. In strain PAK, the 16-kb genomic island is composed of 14 genes, *orfA* to *orfN*, arranged in several putative operons and located between *flgL* and *fliC* (Table 4.2).

Table 4.2 GI genes in type a strains that lies between *flgL* and *fliC* genes.

Gene name	Homolog/organism	Function of homologs
orfA	VioA/*E. coli*	synthesis of dTDP-4-amino-4,6 dideoxyglucose (dTDP-viosamine) (nucleotide sugar transaminase)
orfB	putative ACP/*C. jejuni*	carrier of the growing fatty acid chain in fatty acid biosynthesis
orfC	FABH, 3-oxoacyl-ACP synthase/*Streptomyces coelicolor*	catalyzes the condensation reaction of fatty acid synthesis by the addition of an acyl acceptor of two carbons from malonyl-ACP
orfD	FABG, 3-oxoacyl-ACP reductase/*Chlamydia pneumonia*	first reduction step in the fatty acid biosynthesis pathway
orfE	serine O-acetyltransferase/*Bacillus* subtilis	involved in methylation of Nod factors (NoeI)
orfF	putative aromatic ring hydroxylating dioxygenase α subunit/*Sphingomonas* sp.	involved in napthalene catabolic pathway
orfG	putative acetyl transferase/*Legionella pneumophilia*	involved in lipopolysaccharide biosynthesis
orfH	nodulation protein NoeI/*Sinorhizobium fredii*	involved in Nod factor biosynthesis
orfI	no homolog	
orfJ	2-deoxy-manno-octulosonate cytidyl transferase/*E. coli*	activates KDO for incorporation into lipopolysaccharide in Gram-negative bacteria

(Continued)

Table 4.2 (*Continued*)

Gene name	Homolog/organism	Function of homologs
orfK	no homolog	
orfL	3-demethylubiquinone-9 3-methyltransferase/Neisseria meningitidis	involved in ubiquinone biosynthesis pathway
orfM	cmtG, 4-hydroxy-2-oxovalerate aldolase/P. putida	involved in *p*-isopropylbenzoate (*p*-cumate) catabolic pathway
fgtA	O-antigen biosynthesis protein RfbC/Myxococcus xanthus	O-antigen biosynthesis

GenBank accession number for the PAK glycosylation island genes is AF332547.

Inactivation of either *orfA* or *orfN* (now annotated as *fgtA* for flagellar glycosyl transferase) [41] abolishes flagellin glycosylation. *OrfA* appears to exist in an operon with *orfB* and *orfC*. Based on sequence homologies *orfB* encodes a putative acyl carrier protein (ACP) and *orfC* encodes FabH, 3-oxoacyl-ACP synthase (Table 4.2). The first gene of the next operon *orfDEF* bears homology to *fabG*, encoding 3-oxoacyl-ACP reductase. While FabH along with its ACP catalyzes the condensation reaction of fatty acid biosynthesis, FabG catalyzes the first reduction step in the fatty acid elongation cycle (Table 4.2). In *P. syringae orf3* is required for acyl homoserine lactone synthesis [42]. The exact role of each of these genes in glycosylation and other cellular functions in *P. aeruginosa* remains to be elucidated.

In the type b strain PAO1, with an invariant 53-kDa flagellin [43], there exist only four genes, PA1088, PA1089, PA1090 and PA1091, in the same chromosomal location as in a-type strains. PA1091 (*fgtA*) is homologous, but not identical, to the corresponding gene in a-type strains and probably is a glycosyltransferase that attaches a deoxyhexose to the protein. One of the other genes, PA1090 also appears to play a role in glycosylation, encoding a nucleotidyltransferase. Another gene, PA1089, may be involved in attaching a phosphate group to the glycan chain. Strains PA14 and CF3719 have the same four genes, PA1088–PA1091, with almost complete sequence identity. Polymerase chain reaction (PCR) analysis of 12 other b-type strains suggests that all b-type strains possess this island [41].

4.2.6.2 Polymorphism of the *P. aeruginosa* a-type GI

In the a-type strains, differences in the observed molecular masses of flagellin suggest differences in the levels or nature of glycosylation of the flagellin molecule exist [25]. Analysis of strains lacking the complete GI, by microarray, PCR and sequencing, led to an identification of an abbreviated version of the GI (short island) in which *orfD*, *E* and *H* are polymorphic and *orfI*, *J*, *K*, *L* and *M* are absent [44].

Further two distinct flagellin subtypes were noticed, designated A1 and A2, with A2 flagellins having a short deletion in the central region. Long GIs associate with A1-type strains, whereas strains carrying the short island associated with both A1- and A2-type flagellins [44]. These polymorphisms indicate that *P. aeruginosa* probably has the capacity to further diversify the antigenicity of this surface protein by the use of its GIs. The origin of these short islands is also unknown and evidence for transposition is also lacking.

4.2.7 Role of Flagella in Inflammation

Toll-like receptors (TLRs) are key components of the immune system that detect microbial infection and trigger antimicrobial host defense responses. TLR5 is a sensor for monomeric flagellin, known to be a virulence factor [45]. Flagellin released from *P. aeruginosa* triggers airway epithelial TLR5 signaling NF-κB, and causing production and release of proinflammatory cytokines that recruit neutrophils to the infected region. This response has been termed hyperinflammatory because a large number of neutrophils accumulate damaging the CF lung tissue [46–49].

The exact region of the *P. aeruginosa* flagellin molecule interacting with TLR5 has been determined [50,51]. In addition, it has been demonstrated that interleukin-8 signaling from the flagella purified from the two glycosylation-defective mutant strains is reduced up to 50% compared to the levels for their respective wild-type strains [51]. Whether flagellin glycan moieties would aid in the binding of flagellin to TLR5 or glycans have some signaling activity of their own through another TLR or other cellular receptor is not known.

While flagellin binding to TLR5 triggers inflammation, *P. aeruginosa* downregulates flagellin transcription when it is grown in purulent mucus from patients with CF and non-CF bronchiectasis [38]. This response possibly abrogates the potent inflammatory response mediated by the interaction of flagellin with TLR5. Neutrophil elastase released from neutrophils in mucus proteolytically cleaves the flagellar hook, resulting in FlgM accumulation within the cell and causing repression of flagellin synthesis [40]. It is speculated that the cyclical bouts of inflammation observed in CF patients may result from flagellin synthesis and its repression, caused by the absence or presence of neutrophils at the site of infection [40].

4.2.8 Role of Flagellum in Pathogenesis

In an era of increasing drug resistance, immunotherapy is a desirable treatment against *P. aeruginosa* infections. Earlier studies showed two murine monoclonal antibodies to *P. aeruginosa* type a and b flagella prepared by conventional hybridoma methodology provided specific and significant prophylactic and therapeutic protection against lethal challenge with *P. aeruginosa* strains in a mouse burn wound sepsis model [52]. Human monoclonal antibodies also strongly protected burned mice challenged parenterally with *P. aeruginosa* bearing b-type flagella from death [53]. Clear additive effects in terms of mouse survival were observed with a combination of either carbapenem or aminoglycoside antibiotics with an antiflagellar monoclonal antibody, suggesting that an antiflagellar monoclonal antibody would be effective against systemic infection in combination with some kinds of antibiotics [54].

Immunotherapy with N-terminal of flagellin anti-*N*′-fla-b IgG, given either as prophylaxis or therapeutically, effectively reduced mortality and morbidity, and improved wound healing in a severely *P. aeruginosa*-infected murine burn model [55]. The N′-terminal domain of *P. aeruginosa* flagellin harbors critically important bioactive domains and an antibody-targeted, neutralization approach directed at this region could provide a novel therapeutic strategy to combat *P. aeruginosa* infection [56].

A chemically mutagenized nonflagellated strain of *P. aeruginosa* was observed to be less virulent than its motile counterpart in the burned-mouse model of infection [16]. Nonmotile mutants proliferated in the wound, but the characteristic bacteremia and systemic invasion were markedly absent, whereas wild-type organisms revealed a characteristic, rapidly systemic infection in burned mice [57]. In addition, a neonatal mouse pneumonia model also showed reduced virulence of a genetically characterized nonflagellated mutant of *P. aeruginosa*. Histopathological studies demonstrated that the *fliC* mutants caused very focal inflammation and that the organisms did not spread through the lungs as seen in infection due to wild-type strain PAK [58].

The contributions of flagellar motility, flagellin structure and its glycosylation in *P. aeruginosa* have been tested in a burned-mouse model of infection. A higher LD_{50} was obtained for PAKΔC (non-motile), PAK*motABCD* (paralyzed flagella) and PAK*rfbC* (non-glycosylated) mutants as compared to the wild-type PAK strain [59]. Recently the role of flagellin versus motility was examined in a lung model of acute infection using different *P. aeruginosa* mutants [60]. Absence of motility does not significantly alter the LD_{50}, whereas the production of excess amounts of flagellin lowers it and results in early death. Further, it was noted that the absence of the TLR5 ligand, flagellin, results in slower clearance of this organism from the lungs and a delay in the time of death [60].

Thus, flagella appear to play different roles in lung infections. They may lead to death when present in excess and presumably in the appropriate amount they stimulate a protective innate immune response.

4.3
Pili of *P. aeruginosa*

Rapid colonization of new surfaces under conditions of high nutrient availability occurs via a flagellum-independent form of bacterial translocation known as twitching motility. It occurs on wet surfaces, and is important for host colonization and other forms of complex colonial behavior, including the formation of biofilms [61,62]. Twitching motility in *P. aeruginosa* is mediated by type IV pili, forming flat, spreading colonies with a characteristic rough appearance and a small peripheral twitching zone consisting of a thin layer of cells [63,64]. It has been reported that pili in conjunction with flagella are required for swarming motility [2,3]. However, mutations in pilus synthesis genes in *P. aeruginosa* PA14 had no effect on swarming motility [2,65], whereas a *P. aeruginosa* PAO1 *pilA* mutant (*pilA* encodes pilin) and mutants in the other pilus synthesis genes were defective in swarming motility [3,66]. Thus, the role of pili may be strain dependent.

Apart from acting as organelles of motility, these pili are also adhesins, mediating the adherence to eukaryotic cell surfaces [67–69] and probably to abiotic surfaces as well [64]. They are also essential for the normal development of *P. aeruginosa* biofilms as mutants lacking functional pili are not able to develop past the microcolony stage in static or flow biofilm systems [61,36]. Pili have additional roles in *P. aeruginosa*, functioning as receptors for the binding and entry of bacteriophages [4]. In the following section, in addition to reviewing the structure, regulation and assembly of type IV pili, we will briefly discuss the chaperone–usher pathways and the *flp–tad–rcp* locus in *P. aeruginosa*.

4.3.1
Structure of *P. aeruginosa* Pilus

The *P. aeruginosa* type IV pili are flexible surface filaments about 6 nm in diameter [70,71], several micrometers in length (Figure 4.1A) [72], and very strong in tension due to the hydrophobic and ionic bonds between subunits. Pili are polymers of a single gene product, called PilA or pilin, which is a protein derived from a precursor by cleavage of its six N-terminal residues and the *N*-methylation of the newly revealed phenylalanine at this terminus [73]. Crystal structures of *P. aeruginosa* PAK type IV pilin monomer and pilus fiber as well as a fiber model of *P. aeruginosa* PAK pilin are known [74]. Electron microscopic imaging of PAK pili reveal long, straight, unbundled filaments with a diameter of 50–60 Å [72], while X-ray fiber diffraction analyses of PAK pili and the closely related PAO pili indicate hollow cylinders with a 52-Å outer diameter, a 12-Å inner diameter and a roughly 31-Å diameter ring of hydrophobic residues [75] composed of five subunits per turn packed into a conserved helical structure [76,77].

Pili appear to bind via their tip to specific receptors on mammalian epithelial cells and other cell types [67]. In *P. aeruginosa* pilin a region of 17 semiconserved amino acid residues, located within residues 128–144 of the C-terminal region of PilA, which is otherwise buried within the filament, is exposed at the tip of the pilus and binds to the βGalNAc(1→4)βGal moieties of the glycosphingolipids asialo-GM_1 and asialo-GM_2 on epithelial cells [71,78,79].

4.3.2
Pilus/Fimbrial Genes of *P. aeruginosa*

The biogenesis and function of type IV pili in *P. aeruginosa* are controlled by more than 40 genes, including proteins involved in the structure, regulation of pilus assembly and twitching motility (Table 4.3) [71,80,81]. These genes are expressed from several unlinked gene clusters throughout the *P. aeruginosa* genome (Table 4.3). A number of genes required for pilus assembly are homologous to genes involved in type II protein secretion and competence for DNA uptake, suggesting that these systems share a common architecture.

Table 4.3 Annotation of the pilus/fimbrial biogenesis genes in *P. aeruginosa.*

PA number	Gene	Product name
PA0395	*pilT*	twitching motility protein
PA0396	*pilU*	twitching motility protein
PA0408	*pilG*	twitching motility protein
PA0409	*pilH*	twitching motility protein
PA0410	*pilI*	twitching motility protein
PA0411	*pilJ*	twitching motility protein
PA0412	*pilK*	methyltransferase
PA1822	*fimL*	hypothetical protein
PA2960	*pilZ*	type 4 fimbrial biogenesis protein
PA3115	*fimV*	motility protein
PA3805	*pilF*	type 4 fimbrial biogenesis protein
PA4525	*pilA*	type 4 fimbrial precursor
PA4526	*pilB*	type 4 fimbrial biogenesis protein
PA4527	*pilC*	still frameshift type 4 fimbrial biogenesis protein
PA4528	*pilD*	type 4 pre-pilin peptidase
PA4546	*pilS*	two-component sensor
PA4547	*pilR*	two-component response regulator
PA4549	*fimT*	type 4 fimbrial biogenesis protein
PA4550	*fimU*	type 4 fimbrial biogenesis protein
PA4551	*pilV*	type 4 fimbrial biogenesis protein
PA4552	*pilW*	type 4 fimbrial biogenesis protein
PA4553	*pilX*	type 4 fimbrial biogenesis protein
PA4554	*pilY1*	type 4 fimbrial biogenesis protein
PA4555	*pilY2*	type 4 fimbrial biogenesis protein
PA4556	*pile*	type 4 fimbrial biogenesis protein
PA4959	*fimX*	FimX (phosphodiesterase activity)
PA5040	*pilQ*	type 4 fimbrial biogenesis outer membrane protein PilQ precursor
PA5041	*pilP*	type 4 fimbrial biogenesis protein
PA5042	*pilO*	type 4 fimbrial biogenesis protein
PA5043	*pilN*	type 4 fimbrial biogenesis protein
PA5044	*pilM*	type 4 fimbrial biogenesis protein
PA5262	*fimS/algZ*	alginate biosynthesis protein AlgZ/FimS

Annotation of the pilus/fimbrial biogenesis genes in *P. aeruginosa* is based on information at www.pseudomonas.com.

P. aeruginosa has 26 *pil* genes from *pilA* to *pilZ* and has five *fim* genes (Table 4.3). The functions of the major proteins are mentioned below, but the rest are listed in Table 4.3. The cleavage and methylation of the *pilA* encoded pre-pilin is carried out by a single membrane-located bifunctional enzyme which is encoded by *pilD/xcpA*, which is located in the same cluster as the *pilA* gene, but in the opposite transcriptional orientation [73,82,83]. The PilB ATPase powers the extrusion of the pilus, while the PilT ATPase is responsible for pilus retraction [6,71]. The exact function of a third ATPase, PilU (a paralog of PilT), is unclear [84]. Mutation of any of the three ATPases results in the loss of twitching motility [6,84,85]. Inactivation of the *pilA* or *pilB* gene results in a lack of pilus formation, while mutations in *pilU* or *pilT* lead to overproduction of pili and a lack of motility on solid surfaces [86]. The *pilM* gene encodes a 38-kDa protein, exists as a multimeric complex and its deletion has a dominant-negative effect [87].

PilQ is a member of the general secretory pathway secretin superfamily, members of which translocate a variety of macromolecules across the outer membrane. PilQ forms a dodecameric doughnut-shaped complex whose internal cavity diameter closely matches that of the diameter of the corresponding pilus in *P. aeruginosa* [88,89], implicating a role of PilQ in pilus extrusion.

Pili assembly and twitching motility in *P. aeruginosa* also requires the fimbrial proteins FimV [90], FimX [91] and FimL [92]. A cluster of genes encoding a number of minor pilin-like proteins (PilE, PilV, PilW, PilX, FimT and FimU) is also required for pili assembly, twitching motility and infection by pilus-specific phage [93,94].

4.3.3 Regulation of Pilus Assembly and Twitching Motility

Regulatory proteins involved in type IV pili assembly include the sensor kinase/response regulator pairs PilS/PilR and FimS/AlgR, Vfr [70], and FimL, AlgZ and the Chp system.

PilS/PilR, a two-component sensor–regulator pair, [95,96] along with RpoN is involved in transcription of *pilA* in *P. aeruginosa* [97]. PilS a transmembrane sensor protein located at the pole of the cell, which when stimulated by an unknown environmental signal(s) [98] activates PilR through kinase activity. PilR then activates transcription of *pilA*. PilA expression in *P. aeruginosa* may be autoregulated [99], which might be expected if cells are to control the size of the pool of pilins in the cell membrane.

FimS/AlgR, an atypical sensor–regulator pair, controls twitching motility in *P. aeruginosa* [100]. FimS and AlgR do not affect PilA expression, and so must be regulating some other aspect of the system, as both *fimS* and *algR* mutants lack extracellular pili [100].

AlgZ (AmrZ), a ribbon-helix-helix DNA-binding protein, is also reported to control twitching motility and biogenesis of type IV pili [101]. The *P. aeruginosa* AlgZ/AmrZ is required for proper assembly of surface-exposed type IV pili, but not for expression of the major subunit *pilA*. The twitching motility genes under AlgR control are in the *fimTU–pilVWXY1Y2E* operon [102]. Both *amrZ* and *algR* mutants exhibit similar twitching phenotypes [103], but how they regulate twitching is not known. Involvement of these two alginate synthesis genes in twitching motility is interesting and could lend further insight into the infection process of *P. aeruginosa* in CF patients.

Vfr is known to control twitching motility in *P. aeruginosa* [70]. Another regulator FimL has been identified that affects twitching motility at least in part through modulation of Vfr production [92]. FimL affects the regulation of type IV pilus assembly and function rather than production. While both *fimL* and *vfr* mutants show reduced levels of surface-assembled pili compared with wild-type, the defect is more severe in *fimL* mutants – an observation which supports the notion that FimL might also be controlling additional gene products necessary for functional type IV pili.

chpABCDE, a chemosensory phosphotransfer signal transduction system found downstream of *pilK*, in *P. aeruginosa* also controls twitching motility [104,105]. This system operates via methyl-accepting chemotaxis proteins (MCPs), which can induce autophosphorylation of a central histidine kinase (CheA) which in turn phosphorylates a response regulator (CheY). CheW, CheR, CheB and CheZ act as adaptors between the MCPs and CheA or are involved in modulating the methylation state of the MCPs and in the re-setting (dephosphorylation) of CheY. The *chp* gene cluster in *P. aeruginosa* contains three CheY-like response receiver domains – two encoded by adjacent genes (*pilG* and *pilH*,) and another at the C terminus of the CheA homolog (*pilL*/*chpA*) [5,104,106]. Mutations affecting PilG (CheY), PilI (CheW) and PilJ (MCP) are all defective in twitching motility and extracellular pili, whereas those affecting PilH (the second CheY) show an aberrant pattern of twitching motility [5,106].

4.3.4 Assembly of Type IV Pili

Type IVa pili are assembled with the help of a complex molecular machine, the core of which consists of a traffic NTPase, a polytopic inner membrane protein and an outer membrane channel or secretin. A pre-pilin peptidase, PilD, cleaves the pilin signal sequence. The pilin monomer is embedded in the inner membrane bilayer with its hydrophilic head in the periplasm. PilC and minor pilins (PilE, V, W and X, and FimU) form a platform within the cytoplasmic membrane through which pilins and pseudopilins are driven to the periplasmic side of the cytoplasmic membrane by energy provided by the traffic NTPase PilB. The pilus is extruded through the outer membrane via a pore composed of multimeric PilQ, [82,95,96,104,107,108] stabilized by the lipoprotein PilP.

After extension is completed, and possibly following a signal from the pilus tip, retraction commences, driven by PilT. The PilT motor is drawn as a hexameric ATPase, a member of the AAA family of motor proteins [109]. PilT is extracted in the membrane fraction, and extends into the periplasmic space between the inner and outer membrane. As a result of its hexameric geometry and the location and structure of its nucleotide binding site, PilT may be homologous to the β-subunit of F_1 ATPase [110]. The assembly/disassembly process must occur from the base, as the hydrophobic α-helical core of the type IV pilus is too tightly packed to allow macromolecular transport through the pilus [77,89].

4.3.5 Pilin Classification

4.3.5.1 Type IV a and b Pilins

The type IV pilins are divided into two subclasses, type IVa (found in a wide range of Gram-negative bacteria) and IVb (restricted to human enteric bacteria, e.g. toxin coregulated pilus from *V. cholerae*), on the basis of differences in amino acid sequence and length. The *P. aeruginosa* type IV pili are members of the type IVa family, which is characterized by a number of features including the size of the pre-pilin [72]. The type

IVa pilins have a shorter leader sequence (five or six amino acids), a shorter mature sequence (average length of around 150 amino acids) and the *N*-methylated N-terminal residue is phenylalanine for the type IVa pilins [72].

4.3.5.2 Group I–V Pilins

The pilins of *P. aeruginosa* can also be divided into five distinct phylogenetic groups (designated I–V), based on amino acid sequence and the presence of unique accessory genes immediately downstream of *pilA* [111]. Only pilins belonging to groups I and IV are known to be glycosylated. Group II pilin genes (strains PAO1, PA103 and PAK) were immediately upstream of a $tRNA^{Thr}$ gene [97,112,113], while group I pilin genes contained *tfpO*/*pilO*, between *pilA* and $tRNA^{Thr}$. TfpO is a glycosyltransferase involved in transfer of the O-antigen unit to the C-terminal Ser residue of PilA. In group III strains a gene *tfpY* homologous to FimB, a pilin accessory protein of unknown function from *Dichelobacter nodosus*, exists downstream of the *pilA* gene [114]. In group IV strains two genes are present downstream of *pilA*: *tfpW*, which encodes a large hydrophobic hypothetical transmembrane protein, and *tfpX*, which encodes a putative pilin accessory protein most similar to TfpZ and to PilB and FimB of *Eikenella corrodens* and *D. nodosus*, respectively. Type V type strains contain a *tfpZ* gene which encodes a putative pilin accessory protein homologous (31% identity, 53% similarity) to the PilB protein of *E. corrodens* strain VA1, respectively downstream of the *pilA* gene [115].

4.3.6 Pseudopilins

There are multiple sets of homologies between the assembly proteins PilB, C, D, N and Q, and the five minor pilins (PilE, V, W and X, and FimU) to components of the type II secretory machinery (Xcp secreton consisting of genes *xcpP–Z* and *xcpA*), the pseudopilins XcpT–X [116,117]. The homology between type IV pilins and the pseudopilins is restricted, however, to the N-terminal 30 hydrophobic amino acids that interact to enable the pilus to assemble [118]. Type IV pilins and pseudopilins are processed and *N*-methylated at their N-terminal ends by the same pre-pilin peptidase XcpA/PilD [96,117,119,120]. XcpT but not the four other pseudopilins (XcpU–X) can be assembled into single filaments about 6 nm thick, which stick together under certain circumstances to form a multifibrillar structure rather than tight bundles, called the pseudopilus [121]. In *P. aeruginosa*, where both the secreton and type IV piliation pathways coexist, PilA, the most abundant pilin, can be cross-linked to the pseudopilins and is required for efficient secretion of exoenzymes [122]. Thus, the secreton and the piliation machinery are intimately related and might have overlapping functions.

4.3.7 Chaperone–Usher Pathways

In *P. aeruginosa*, three gene clusters (*cupA*, *cupB* and *cupC*) encoding chaperone–usher pathway components have been identified in the genome sequence of the

PAO1 strain [123]. The *cupA*, *cupB* and *cupC* gene clusters each encode an usher, a chaperone and at least one fimbrial subunit [124]. There are five *cupA* (*A1–A5*), six *cupB* (*B1–B6*) and three *cupC* (*C1–C3*) genes. The chaperone proteins bind to the fimbrial subunits and the usher protein, which forms a pore in the outer membrane. This usher protein allows the pilin subunits to be finally released from the chaperone and assembled into fibrils while crossing the outer membrane [125].

The *cup* gene cluster is poorly expressed under laboratory conditions [125] and is regulated by MvaT [125,126]. Another regulator, a two-component regulatory system, i.e. the RocS1 (the sensor of the Roc1 [regulation of cup 1] system)–RocR–RocA1 system, has recently been identified as controlling *cupB* and *cupC* gene cluster expression [127]. The CupC1-containing fimbriae are peritrichously distributed in strain PAO1 at the surface of the bacterial cell envelope in a RocS1 overproducing strain [128].

4.3.8
flp–tad–rcp Gene Cluster

The Flp pilus has all the features characteristic of type IVb pre-pilins, and requires for its assembly a subset of components called Tad and Rcp. Homologs of the *flp–tad–rcp* genes reported to support tight adherence and biofilm formation [129,130] were identified in *P. aeruginosa* PAO1. PA4297–PA4305 encode Tad–Rcp homologous proteins (the *tad* locus is also named "widespread colonization island" [131]) and PA4306 was identified as an *flp* (fimbrial low-molecular-weight protein)-like gene [132]. Flp pilus-dependent adherence was, however, only observed in a mutant strain deficient in the production of type IVa pili and flagella, which can mediate attachment [133], underscoring the complexity of adherence, colonization and biofilm formation.

FppA, the Flp pre-pilin peptidase, appears to be specific for Flp pre-pilins and PilD cannot substitute for FppA function [133]. Moreover, overproduction of FppA in a *pilD* mutant does not allow restoration of type IVa pilus-associated phenotypes such as twitching motility. To observe Flp pili an overexpression of only the *flp* gene is required with basal expression of the *tad–rcp* gene cluster [133]. No Flp pili can be seen at the surface of *tadA*, traffic NTPase (PA4302) and *rcpA*, secretin (PA4304) mutants. Further, two open reading frames encode PilC homologs, *tadB* (PA4301) and *tadC* (PA4300). The Flp assembly machinery seems to require an additional subset of components, *rcpC* (PA4305), *tadZ* (PA4303), *tadD* (PA4296) and *tadG* (PA4297). A two-component system adjacent to the *tad–rcp* locus was reported as PprA and PprB [134], but its role in the expression of the *flp–tad–rcp* operon is not known.

The redundancy of the chaperone–usher systems and the presence of the *flp–tad–rcp* genes in the *P. aeruginosa* genome, combined with a complex regulatory network tightly controlling their expression, highlight the diversity of organelles that the bacterium can expose at its surface while encountering different environments or supports. However, particular traits of each fimbrial structure may determine a specific function for the colonization of an ecological niche and may confer tissue-specific adhesive properties [135].

4.3.9
Posttranslational Modifications of Pilin

A number of posttranslational modifications of pilin have been observed. Group I pilin from strain 1244 has been shown to be posttranslationally modified by *O*-glycosylation on the C-terminal serine residue [136]. It has a trisaccharide substituent containing xylopyranosyl and furanosyl residues, but glycosylation has not been observed in other strains including the widely studied PAK and PAO strains [137]. TfpO is the glycosyltransferase responsible for attachment of the O-antigen repeating unit to pilin [138]. It is the only glycosylation factor required that is not part of either the pilin or the O-antigen pathways. The importance of type IV pili in the biology and virulence of *P. aeruginosa* has made them attractive targets for vaccine development [139,140]. Saiman et al. raised cross-reactive monoclonal antibodies that recognized some pilin variants and reduced binding to eukaryotic cells [141]. *P. aeruginosa* strains producing glycosylated pili are commonly found among clinical isolates and particularly among those strains isolated from sputum. Pilin glycosylation increases colonization as determined by the mouse acute pneumonia model. Competition index analysis using a mouse respiratory model comparing strains 1244 and 1244G7 indicated that the pilin glycan is a significant virulence factor and may aid in the establishment of infection [142].

A novel glycan modification on the pilins of the group IV strain Pa5196 has been identified. Group IV pilins continued to be modified in a lipopolysaccharide (*wbpM*) mutant of Pa5196, showing that, unlike group I strains, the pilins of group IV are not modified with the O-antigen unit of the background strain. Instead, the pilin glycan was determined to be an unusual homo-oligomer of α-1,5-linked D-arabinofuranose (D-Ara*f*). This sugar is uncommon in prokaryotes, occurring mainly in the cell wall arabinogalactan and lipoarabinomannan (LAM) polymers of mycobacteria, including *Mycobacterium tuberculosis* and *M. leprae* [143].

4.3.10
Role of Pili in Pathogenesis

Despite the role of pili in adhesion to cells, biofilm formation and their ability to aid in bacterial motility, the role of type IV pili in the pathogenesis of infection is still not settled. The following observations point towards a role of pilus in disease. Pilin-deficient (*pilA*) mutants have decreased ability to damage epithelial cells, and have reduced cytotoxicity toward A549 and HeLa cells [86], and vaccination with purified pili can protect against infection with serologically related strains [144,145]. Intratracheally, but not subcutaneously, pili protein-immunized mice showed significant improvement of survival after intratracheal challenge with the *P. aeruginosa* PAO1 strain [146]. *P. aeruginosa pilT* mutants are not infective in corneal tissue [147] and exhibit reduced cytotoxicity to epithelial cells in culture [86]. A recent screen for *P. aeruginosa* mutants that are impaired in virulence in *Drosophila* showed that all strains that were strongly impaired in fly killing also lacked twitching motility, with the majority of mutations occurring in the *pilGHIJKL–chpABCDE* chemosensory gene cluster, although this appeared to be a consequence of *chp* control of other virulence factors as other (apparently) nontwitching variants had normal virulence [148]. In an infant mouse model of pulmonary infection the piliated strains were more often associated with severe diffuse pneumonias, while the nonpiliated organisms resulted in less severe, focal pneumonias, although these differences did not achieve statistical significance [149].

4.4
Conclusions

The flagellum and pili of *P. aeruginosa* are important for conferring motility and in aiding the colonization of diverse ecological niches by *P. aeruginosa in vitro*. They may play important roles in host–pathogen interactions serving as adhesins to host surfaces, forming macrocolonies and eliciting host innate immune responses. This chapter has summarized what is known about the flagellar and pilus systems in *P. aeruginosa*.

References

1 Henrichsen, J. (1972) *J Bacteriol Rev*, **36**, 478–503.
2 Rashid, M.H. and Kornberg, A. (2000) *Proc Natl Acad Sci USA*, **97**, 4885–4890.
3 Kohler, T., Curty, L.K., Barja, F., van Delden, C. and Pechere, J.C. (2000) *J Bacteriol*, **182**, 5990–5996.
4 Bradley, D.E. (1980) *Can J Microbiol*, **26**, 146–154.
5 Darzins, A. (1994) *Mol Microbiol*, **11**, 137–153.
6 Whitchurch, C.B., Hobbs, M., Livingston, S.P., Krishnapillai, V. and Mattick, J.S. (1991) *Gene*, **101**, 33–44.
7 Craven, R. and Montie, T.C. (1985) *J Bacteriol*, **164**, 544–549.
8 Masduki, A., Nakamura, J., Ohga, T., Umezaki, R., Kato, J. and Ohtake, H. (1995) *J Bacteriol*, **177**, 948–952.
9 Taguchi, K., Fukutomi, H., Kuroda, A., Kato, J. and Ohtake, H. (1997) *Microbiology*, **143**, 3223–3229.
10 Murray, T.S. and Kazmierczak, B.I. (2006) *J Bacteriol*, **188**, 6995–7004.
11 Aldridge, P. and Hughes, K.T. (2002) *Curr Opin Microbiol*, **5**, 160–165.
12 Dasgupta, N., Wolfgang, M.C., Goodman, A.L., Arora, S.K., Jyot, J., Lory, S. and Ramphal, R. (2003) *Mol Microbiol*, **50**, 809–824.
13 Moens, S. and Vanderleyden, J. (1996) *Crit Rev Microbiol*, **22**, 67–100.
14 Ottemann, K.M. and Miller, J.F. 1997, *Mol Microbiol*, **24**, 1109–1117.
15 Pitt, T.L. (1981) *J Med Microbiol*, **14**, 261–270.
16 Montie, T.C., Doyle-Huntzinger, D., Craven, R.C. and Holder, I.A. (1982) *Infect Immun*, **38**, 1296–1298.
17 Drake, D. and Montie, T.C. (1987) *Can J Microbiol*, **33**, 755–763.
18 Anderson, T.R. and Montie, T.C. (1987) *Infect Immun*, **55**, 3204.
19 Anderson, T.R. and Montie, T.C. (1989) *Can J Microbiol*, **35**, 890–894.
20 Doyle, T.B., Hawkins, A.C. and McCarter, L.L. (2004) *J Bacteriol*, **186**, 634.
21 Suzuki, T. and Iino, T. (1980) *J Bacteriol*, **143**, 1471–1479.
22 Arora, S.K., Ritchings, B.W., Almira, E.C., Lory, S. and Ramphal, R. (1998) *Infect Immun*, **66**, 1000–1007.

23 Spangenberg, C., Heuer, T., Burger, C. and Tummler, B. (1996) *FEBS Lett*, **396**, 213–217.

24 Arora, S.K., Dasgupta, N., Lory, S. and Ramphal, R. (2000) *Infect Immun*, **68**, 1474–1479.

25 Arora, S.K., Bangera, M., Lory, S. and Ramphal, R. (2001) *Proc Natl Acad Sci USA*, **98**, 9342–9347.

26 Mahenthiralingam, E., Campbell, M.E. and Speert, D.P. (1994) *Infect Immun*, **62**, 596–605.

27 Dasgupta, N., Ferrell, E.P., Kanack, K.J., West, S.E. and Ramphal, R. (2002) *J Bacteriol*, **184**, 5240–5250.

28 Dasgupta, N., Arora, S.K. and Ramphal, R. (2000) *J Bacteriol*, **182**, 357–364.

29 Dasgupta, N. and Ramphal, R. (2001) *J Bacteriol*, **183**, 6636–6644.

30 Ritchings, B.W., Almira, E.C., Lory, S. and Ramphal, R. (1995) *Infect Immun*, **63**, 4868–4876.

31 Frisk, A., Jyot, J., Arora, S.K. and Ramphal, R. (2002) *J Bacteriol*, **184**, 1514–1521.

32 Jyot, J., Dasgupta, N. and Ramphal, R. (2002) *J Bacteriol*, **184**, 5251–5260.

33 Tart, A.H., Blanks, M.J. and Wozniak, D.J. (2006) *J Bacteriol*, **188**, 6483–6489.

34 Choy, W.K., Zhou, L., Syn, C.K., Zhang, L.H. and Swarup, S. (2004) *J Bacteriol*, **186**, 7221–7228.

35 Caiazza, N.C., Merritt, J.H., Brothers, K.M. and O'Toole, G.A. (2007) *J Bacteriol*, **189**, 3603–3612.

36 Klausen, M., Heydorn, A., Ragas, P., Lambertsen, L., Aaes-Jorgensen, A., Molin, S. and Tolker-Nielsen, T. (2003) *Mol Microbiol*, **48**, 1511–1524.

37 Soscia, C., Hachani, A., Bernadac, A., Filloux, A. and Bleves, S. (2007) *J Bacteriol*, **189**, 3124–3132.

38 Wolfgang, M.C., Jyot, J., Goodman, A.L., Ramphal, R. and Lory, S. (2004) *Proc Natl Acad Sci USA*, **101**, 6664–6668.

39 Palmer, K.L., Mashburn, L.M., Singh, P.K. and Whiteley, M. (2005) *J Bacteriol*, **187**, 5267–5277.

40 Jyot, J., Sonawane, A., Wu, W. and Ramphal, R. (2007) *Mol Microbiol*, **63**, 1026–1038.

41 Verma, A., Schirm, M., Arora, S.K., Thibault, P., Logan, S.M. and Ramphal, R. (2006) *J Bacteriol*, **188**, 4395–4403.

42 Taguchi, F., Takeuchi, K., Katoh, E., Murata, K., Suzuki, T., Marutani, M., Kawasaki, T., Eguchi, M., Katoh, S., Kaku, H., Yasuda, C., Inagaki, Y., Toyoda, K., Shiraishi, T. and Ichinose, Y. (2006) *Cell Microbiol*, **8**, 923–938.

43 Brimer, C.D. and Montie, T.C. (1998) *J Bacteriol*, **178**, 3209–3217.

44 Arora, S.K., Wolfgang, M.C., Lory, S. and Ramphal, R. (2004) *J Bacteriol*, **186**, 2115–2122.

45 Hayashi, F., Smith, K.D., Ozinsky, A., Hawn, T.R., Yi, E.C., Goodlett, D.R., Eng, J.K., Akira, S., Underhill, D.M. and Aderem, A. (2001) *Nature*, **410**, 1099–1103.

46 Hybiske, K., Ichikawa, J.K., Huang, V., Lory, S.J. and Machen, T.E. (2004) *Cell Microbiol*, **6**, 49–63.

47 Honko, A.N. and Mizel, S.B. (2004) *Infect Immun*, **72**, 6676–6679.

48 Honko, A.N. and Mizel, S.B. (2005) *Immunol Res*, **33**, 83–101.

49 Feuillet, V., Medjane, S., Mondor, I., Demaria, O., Pagni, P.P., Galan, J.E., Flavell, R.A. and Alexopoulou, L. (2006) *Proc Natl Acad Sci USA*, **15** (103), 12487–12492.

50 Jacchieri, S.G., Torquato, R. and Brentani, R.R. (2003) *J Bacteriol*, **185**, 4243–4247.

51 Verma, A., Arora, S.K., Kuravi, S.K. and Ramphal, R. (2005) *Infect Immun*, **73**, 8237–8246.

52 Rosok, M.J., Stebbins, M.R., Connelly, K., Lostrom, M.E. and Siadak, A.W. (1990) *Infect Immun*, **58**, 3819–3828.

53 Ochi, H., Ohtsuka, H., Yokota, S., Uezumi, I., Terashima, M., Irie, K. and Noguchi, H. (1991) *Infect Immun*, **59**, 550–554.

54 Uezumi, I., Terashima, M., Kohzuki, T., Kato, M., Irie, K., Ochi, H. and Noguchi, H. (1992) *Antimicrob Agents Chemother*, **36**, 1290–1295.

55 Barnea, Y., Carmeli, Y., Gur, E., Kuzmenko, B., Gat, A., Neville, L.F., Eren, R., Dagan, S. and Navon-Venezia, S. (2006) *Plast Reconstr Surg*, **117**, 2284–2291.

56 Neville, L.F., Barnea, Y., Hammer-Munz, O., Gur, E., Kuzmenko, B., Kahel-Raifer, H., Eren, R., Elkeles, A., Murthy, K.G., Szabo, C., Salzman, A.L., Dagan, S., Carmeli, Y. and Navon-Venezia, S. (2005) *Int J Mol Med*, **16**, 165–171.

57 Drake, D. and Montie, T.C. (1988) *J Gen Microbiol*, **134**, 43–52.

58 Feldman, M., Bryan, R., Rajan, S., Scheffler, L., Brunnert, S., Tang, H. and Prince, A. (1998) *Infect Immun*, **66**, 43–51.

59 Arora, S.K., Neely, A.N., Blair, B., Lory, S. and Ramphal, R. (2005) *Infect Immun*, **73**, 4395–4398.

60 Balloy, V., Verma, A., Kuravi, S., Si-Tahar, M., Chignard, M. and Ramphal, R. (2007) *J Infect Dis*, **196**, 289–296.

61 O'Toole, G.A. and Kolter, R. (1998) *Mol Microbiol*, **30**, 295–304.

62 Watson, A.A., Mattick, J.S. and Alm, R.A. (1996) *Gene*, **175**, 143–150.

63 Mattick, J.S., Whitchurch, C.B. and Alm, R.A. (1996) *Gene*, **179**, 147–155.

64 Semmler, A.B., Whitchurch, C.B. and Mattick, J.S. (1999) *Microbiology*, **145**, 2863–2873.

65 Toutain, C.M., Zegans, M.E.T. and O'Toole, G.A. (2005) *J Bacteriol*, **187**, 771–777.

66 Overhage, J., Lewenza, S., Marr, A.K. and Hancock, R.E. (2007) *J Bacteriol*, **189**, 2164–2169.

67 Hahn, H.P. (1997) *Gene*, **192**, 99–108.

68 Woods, D.E., Straus, D.C., Johanson, W.G., Jr. Berry, V.K. and Bass, J.A. (1980) *Infect Immun*, 1146–1151.

69 Ramphal, R., Sadoff, J.C., Pyle, M. and Silipigni, J.D. (1984) *Infect Immun*, **44**, 38–40.

70 Beatson, S.A., Whitchurch, C.B., Sargent, J.L., Levesque, R.C. and Mattick, J.S. (2002) *J Bacteriol*, **184**, 3605–3613.

71 Mattick, J.S. (2002) *Annu Rev Microbiol*, **56**, 289–314.

72 Craig, L., Pique, M.E. and Tainer, J.A. (2004) *Nat Rev Microbiol*, **2**, 363–378.

73 Strom, M.S., Nunn, D.N. and Lory, S. (1993) *Proc Natl Acad Sci USA*, **90**, 2404–2408.

74 Hazes, B., Sastry, P.A., Hayakawa, K., Read, R.J. and Irvin, R.T. (2000) *J Mol Biol*, **299**, 1005–1017.

75 Folkhard, W., Marvin, D.A., Watts, T.H. and Paranchych, W. (1981) *J Mol Biol*, **149**, 79–93.

76 Watts, T.H., Kay, C.M. and Paranchych, W. (1983) *Biochemistry*, **22**, 3640–3646.

77 Forest, K.T. and Tainer, J.A. (1997) *Gene*, **192**, 165–169.

78 Lee, K.K., Sheth, H.B., Wong, W.Y., Sherburne, R., Paranchych, W., Hodges, R.S., Lingwood, C.A., Krivan, H. and Irvin, R.T. (1994) *Mol Microbiol*, **11**, 705–713.

79 Sheth, H.B., Lee, K.K., Wong, W.Y., Srivastava, G., Hindsgaul, O., Hodges, R.S., Paranchych, W. and Irvin, R.T. (1994) *Mol Microbiol*, **11**, 715–723.

80 Jacobs, M.A., Alwood, A., Thaipisuttikul, I., Spencer, D., Haugen, E., Ernst, S., Will, O., Kaul, R., Raymond, C., Levy, R., Chun-Rong, L., Guenthner, D., Bovee, D., Olson, M.V. and Manoil, C. (2003) *Proc Natl Acad Sci USA*, **100**, 14339–14344.

81 Whitchurch, C.B., Leech, A.J., Young, M.D., Kennedy, D., Sargent, J.L., Bertrand, J.J., Semmler, A.B., Mellick, A.S., Martin, P.R., Alm, R.A., Hobbs, M., Beatson, S.A., Huang, B., Nguyen, L., Commolli, J.C., Engel, J.N., Darzins, A. and Mattick, J.S. (2004) *Mol Microbiol*, **52**, 873–893.

82 Nunn, D., Bergman, S. and Lory, S. (1990) *J Bacteriol*, **172**, 2911–2919.

83 Strom, M.S., Nunn, D. and Lory, S. (1991) *J Bacteriol*, **173**, 1175–1180.

84 Whitchurch, C.B. and Mattick, J.S. (1994) *Mol Microbiol*, 1079–1091.

85 Turner, L.R., Lara, J.C., Nunn, D.N. and Lory, S. (1993) *J Bacteriol*, **175**, 4962–4969.

86 Comolli, J.C., Hauser, A.R., Waite, L., Whitchurch, C.B., Mattick, J.S. and Engel, J.N. (1999) *Infect Immun*, **67**, 3625–3630.

87 Martin, P.R., Watson, A.A., McCaul, T.F. and Mattick, J.S. (1995) *Mol Microbiol*, **16**, 497–508.

88 Bitter, W., Koster, M., Latijnhouwers, M. and de Cock, H. (1998) *J. Tommassen, Mol Microbiol*, **27**, 209–219.

89 Keizer, D.W., Slupsky, C.M., Kalisiak, M., Campbell, A.P., Crump, M.P., Sastry, P.A., Hazes, B., Irvin, R.T. and Sykes, B.D. (2001) *J Biol Chem*, **276**, 24186–24193.

90 Semmler, A.B., Whitchurch, C.B., Leech, A.J. and Mattick, J.S. (2000) *Microbiology*, **146**, 1321–1332.

91 Huang, B., Whitchurch, C.B. and Mattick, J.S. (2003) *J Bacteriol*, **185**, 7068–7076.

92 Whitchurch, C.B., Beatson, S.A., Comolli, J.C., Jakobsen, T., Sargent, J.L., Bertrand, J.J., West, J., Klausen, M., Waite, L.L., Kang, P.J., Tolker-Nielsen, T., Mattick, J.S. and Engel, J.N. (2005) *Mol Microbiol*, **55**, 1357–1378.

93 Alm, R.A., Hallinan, J.P., Watson, A.A. and Mattick, J.S. (1996) *Mol Microbiol*, **22**, 161–173.

94 Alm, R.A. and Mattick, J.S. (1995) *Mol Microbiol*, **16**, 485–496.

95 Hobbs, M., Collie, E.S., Free, P.D., Livingston, S.P. and Mattick, J.S. (1993) *Mol Microbiol*, **7**, 669–682.

96 Strom, M.S. and Lory, S. (1993) *Annu Rev Microbiol*, **47**, 565–596.

97 Johnson, K., Parker, M.L. and Lory, S. (1986) *J Biol Chem*, **261**, 15703–15708.

98 Boyd, J.M. (2000) *Mol Microbiol*, **36**, 153–162.

99 Elleman, T.C. and Peterson, J.E. (1987) *Mol Microbiol*, **1**, 377–380.

100 Whitchurch, C.B., Alm, R.A. and Mattick, J.S. (1996) *Proc Natl Acad Sci USA*, **93**, 9839–9843.

101 Baynham, P.J., Ramsey, D.M., Gvozdyev, B.V., Cordonnier, E.M. and Wozniak, D.J. (2006) *J Bacteriol*, **188**, 132–140.

102 Lizewski, S.E., Schurr, J.R., Jackson, D.W., Frisk, A., Carterson, A.J. and Schurr, M.J. (2004) *J Bacteriol*, **186**, 5672–5684.

103 Whitchurch, C.B., Erova, T.E., Emery, J.A., Sargent, J.L., Harris, J.M., Semmler, A.B.T., Young, M.D., Mattick, J.S. and Wozniak, D.J. (2002) *J Bacteriol*, **184**, 4544–4554.

104 Alm, R.A. and Mattick, J.S. (1997) *Gene*, **192**, 89–98.

105 Wall, D. and Kaiser, D. (1999) *Mol Microbiol*, **32**, 1–10.

106 Darzins, A. (1993) *J Bacteriol*, **175**, 5934–5944.

107 Lory, S. and Strom, M.S. (1997) *Gene*, **92**, 117–121.

108 Martin, P.R., Hobbs, M., Free, P.D., Jeske, Y. and Mattick, J.S. (1993) *Mol Microbiol*, **9**, 857–868.

109 Vale, R.D. and Milligan, R.A. (2000) *Science*, **288**, 88–95.

110 Abrahams, J.P., Leslie, A.G., Lutter, R. and Walker, J.E. (1994) *Nature*, **370**, 621–628.

111 Kus, J.V., Tullis, E., Cvitkovitch, D.G. and Burrows, L.L. (2004) *Microbiology*, **150**, 1315–1326.

112 Pasloske, B.L., Finlay, B.B. and Paranchych, W. (1985) *FEBS Lett*, **183**, 408–412.

113 Sastry, P.A., Finlay, B.B., Pasloske, B.L., Paranchych, W., Pearlstone, J.R. and Smillie, L.B. (1985) *J Bacteriol*, **164**, 571–577.

114 Kennan, R.M., Dhungyel, O.P., Whittington, R.J., Egerton, J.R. and Rood, J.I. (2001) *J Bacteriol*, **183**, 4451–4458.

115 Villar, M.T., Helber, J.T., Hood, B., Schaefer, M.R. and Hirschberg, R.L. (1999) *J Bacteriol*, **181**, 4154–4160.

116 Bleves, S., Voulhoux, R., Michel, G., Lazdunski, A., Tommassen, J. and Filloux, A. (1998) *Mol Microbiol*, **27**, 31–40.

117 Pugsley, A.P. (1993) *Microbiol Rev*, **57**, 50–108.

118 Parge, H.E., Forest, K.T., Hickey, M.J., Christensen, D.A., Getzoff, E.D. and Tainer, J.A. (1995) *Nature*, **378**, 32–38.

119 Nunn, D.N. and Lory, S. (1991) *Proc Natl Acad Sci USA*, **88**, 3281–3285.

120 Nunn, D.N. and Lory, S. (1992) *Proc Natl Acad Sci USA*, **89**, 47–51.

121 Durand, E., Michel, G., Voulhoux, R., Kurner, J., Bernadac, A. and Filloux, A. (2005) *J Biol Chem*, **280**, 31378–31389.

122 Lu, H.M., Motley, S.T. and Lory, S. (1997) *Mol Microbiol*, **25**, 247–259.

123 Vallet, I., Olson, J.W., Lory, S., Lazdunski, A. and Filloux, A. (2001) *Proc Natl Acad Sci USA*, **98**, 6911–6916.

124 Filloux, A., de Bentzmann, S., Aurouze, M., Lazdunski, A. and Vallet, I. (2004) Fimbrial genes in *Pseudomonas aeruginosa* and *Pseudomonas putida*, (ed. J-.L Ramos) in *Pseudomonas*, Kluwer, New York. pp. 721–748.

125 Vallet, I., Diggle, S.P., Stacey, R.E., Cámara, M., Ventre, I., Lory, S., Lazdunski, A., Williams, P. and Filloux, A. (2004) *J Bacteriol*, **186**, 2880–2890.

126 Vallet-Gely, I., Donovan, K.E., Fang, R., Joung, J.K. and Dove, S.L. (2005) *Proc Natl Acad Sci USA* **102**, 11082–11087.

127 Kulasekara, H.D., Ventre, I., Kulasekara, B.R., Lazdunski, A., Filloux, A. and Lory, S. (2005) *Mol Microbiol*, **55**, 368–380.

128 Ruer, S., Stender, S., Filloux, A. and de Bentzmann, S. (2007) *J Bacteriol*, **189**, 3547–3555.

129 Inoue, T., Shingak, R., Sogawa, N., Sogawa, C.A., Asaumi, J., Kokeguchi, S. and Fukui, K. (2003) *Microbiol Immunol*, **47**, 877–881.

130 Kachlany, S.C., Planet, P.J., DeSalle, R., Fine, D.H. and Figurski, D.H. (2001) *Trends Microbiol*, **9**, 429–437.

131 Tomich, M., Planet, P.J. and Figurski, D.H. (2007) *Nat Rev Microbiol*, **5**, 363–375.

132 Kachlany, S.C., Planet, P.J., Desalle, R., Fine, D.H., Figurski, D.H. and Kaplan, J.B. (2001) *Mol Microbiol*, **40**, 542–554.

133 de Bentzmann, S., Aurouze, M., Ball, G. and Filloux, A. (2006) *J Bacteriol*, **188**, 4851–4860.

134 Wang, Y., Ha, U., Zeng, L. and Jin, S. (2003) *Antimicrob Agents Chemother*, **47**, 95–101.

135 Marklund, B.I., Tennent, J.M., Garcia, E., Hamers, A., Baga, M., Lindberg, F., Gaastra, W. and Normark, S. (1992) *Mol Microbiol*, **6**, 2225–2242.

136 Comer, J.E., Marshall, M.A., Blanch, V.J., Deal, C.D. and Castric, P. (2002) *Infect Immun*, **70**, 2837–2845.

137 Castric, P., Cassels, F.J. and Carlson, R.W. (2001) *J Biol Chem*, **276**, 26479–26485.

138 Castric, P. (1995) *Microbiology*, **141**, 1247–1254.

139 Hertle, R., Mrsny, R. and Fitzgerald, D.J. (2001) *Infect Immun*, **69**, 6962–6969.

140 Sheth, H.B., Glasier, L.M., Ellert, N.W., Cachia, P., Kohn, W., Lee, K.K., Paranchych, W., Hodges, R.S. and Irvin, R.T. (1995) *Biomed Pept Proteins Nucleic Acids*, **1**, 141–148.

141 Saiman, L., Sadoff, J. and Prince, A. (1989) *Infect Immun*, **57**, 2764–2770.

142 Smedley, J.G., 3rd Jewell, E., Roguskie, J., Horzempa, J., Syboldt, A., Stolz, D.B. and Castric, P. (2005) *Infect Immun*, **73**, 7922–7931.

143 Voisin, S., Kus, J.V., Houliston, S., St-Michael, F., Watson, D., Cvitkovitch, D., Kelly, G.J., Brisson, J.R. and Burrows, L.L. (2007) *J Bacteriol*, **189**, 151–159.

144 Egerton, J.R., Cox, P.T., Anderson, B.J., Kristo, C., Norman, M. and Mattick, J.S. (1987) *Vet Microbiol*, **14**, 393–409.

145 Tennent, J.M. and Mattick, J.S. (1994) Type 4 fimbriae, *in Fimbriae, Aspects of Adhesion, Genetics, Biogenesis and Vaccines*, (ed. P. Klemm), CRC Press, Boca Raton, FL. pp. 127–146.

146 Ohama, M., Hiramatsu, K., Miyajima, Y., Kishi, K., Nasu, M. and Kadota, J. (2006) *FEMS Immunol Med Microbiol*, **47**, 107–115.

147 Zolfaghar, I., Evans, D.J. and Fleiszig, S.M. (2003) *Infect Immun*, 5389–5393.

148 D'Argenio, D.A., Gallagher, L.A., Berg, C.A. and Manoil, C. (2001) *J Bacteriol*, **183**, 1466–1471.

149 Tang, H., Kays, M. and Prince, A. (1995) *Infect Immun*, **63**, 1278–1285.

5
Pseudomonas Motility and Chemotaxis

Junichi Kato

5.1
Introduction

Motile bacteria have the ability to sense changes in the concentration of chemicals in environments and respond to them by altering their pattern of motility. This behavioral response is called chemotaxis. The pseudomonads also show chemotactic responses to various chemical compounds, including amino acids, organic acids, sugars, aromatic compounds and inorganic ions. Bacterial chemotaxis has been intensively studied in *Escherichia coli* and *Salmonella enterica* serovar Typhimurium. Bacterial chemotaxis has been thought to provide a primitive model for animal sensory systems [1], and most studies have been focused on the genetics and biochemistry of the chemotaxis system in the enteric bacteria. Since the pseudomonads include important plant and animal pathogens, and potential agents of geochemical cycles, biocontrol and bioremediation, and chemotaxis is thought to play an important role in microbe–host and microbe–substrate interactions, ecological aspects of chemotaxis have been intensively investigated in *Pseudomonas* species. I first review chemotaxis assay methods in Section 5.2, and then describe ecological aspects of chemotaxis and molecular biology of the chemotaxis system in *Pseudomonas* in Sections 5.3 and 5.4, respectively.

5.2
Chemotaxis Assay Methods

Several methods have been used for chemotaxis assays of *Pseudomonas* species. A widely used method for measuring chemotactic responses is the capillary assay technique developed by Adler [2]. In this procedure, glass capillary tubes containing medium plus a test chemical or medium alone are immersed in cell suspensions in medium without the test chemical, and, after a suitable interval, the contents of the capillary tubes are plated out and colonies are counted (Figure 5.1A). When the test

Pseudomonas. Model Organism, Pathogen, Cell Factory. Edited by Bernd H.A. Rehm

ISBN: 978-3-527-31914-5

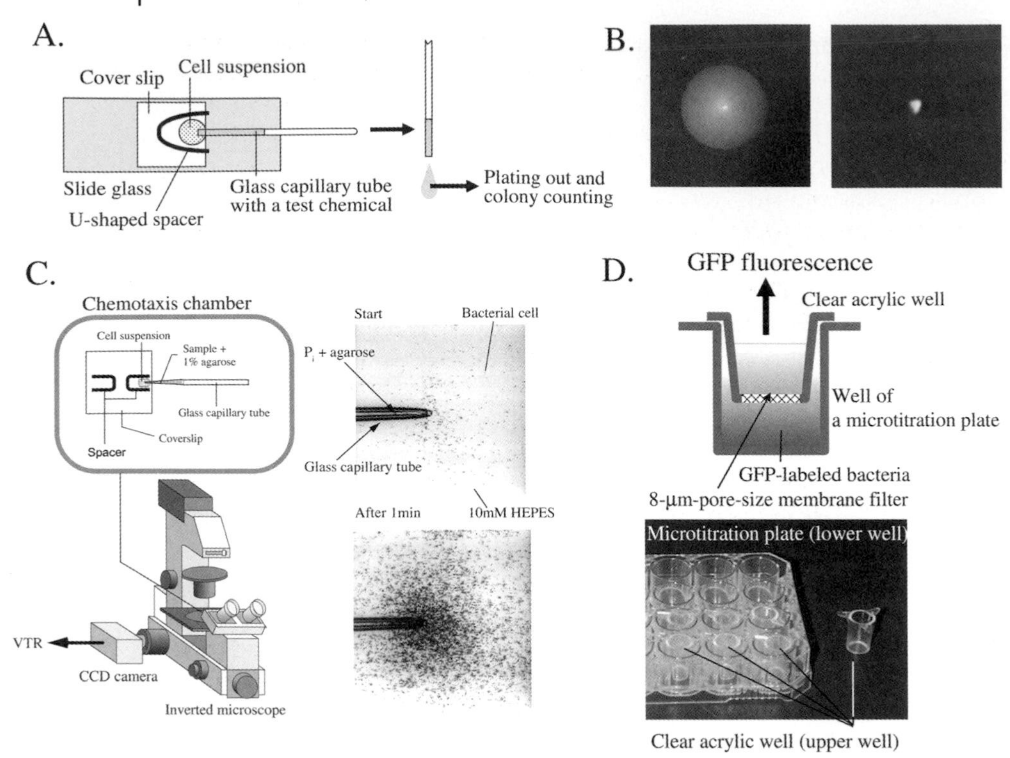

Figure 5.1 Chemotaxis assay methods. (A) The capillary assay. The capillary containing a test chemical is inserted into cell suspension. After a suitable interval, the number of bacterial cells in the capillary is measured by plating out and colony counting. The strength of chemotactic responses is evaluated by the number of bacterial cells in the capillary. (B) The semisolid agar plate assay. Bacterial cells are inoculated on the semisolid growth medium plates. After incubation, chemotactic cells form large swarm rings (left), while nonchemotactic or nonmotile cells form dense and small colonies (right). (C) The computer-assisted capillary assay. Microscopic images of chemotactic responses are videotaped by using an inverted phase contrast microscope equipped with a CCD camera. The strength of chemotactic responses is determined by the number of bacteria per videotape frame. (D) The microtitration plate assay. The upper well is an acrylic well with a bottom of an 8-µm pore membrane filter and the lower well is a well of a 24-well microtitration plate. GFP-labeled bacteria move to the upper well in response to oxygen gradient. GFP fluorescence intensity in the upper well is monitored to measure cell concentrations in the upper well.

chemical is an attractant, the motile bacteria move into the capillary and the concentration of bacteria within the capillary is much higher than that within the control capillary. On the other hand, when the test chemical is a repellent, the motile bacteria swim away from the capillary containing the test chemical and the concentration of bacteria within the capillary is lower than that within the control capillary. The capillary assay technique can quantitatively evaluate bacterial chemotaxis.

The semisolid agar plate method [3] is also widely used for chemotaxis assays. In this procedure, plates containing semisolid growth media with 0.3% agar are inoculated with the bacterial cells to be tested. Bacterial growth forms concentration gradients of growth substrates. Chemotactic cells move outwards as an expanding ring of cells (a swarming ring) in response to self-created substrate gradients (Figure 5.1B). Nonchemotactic or nonmotile cells form dense and small colonies on the semisolid agar plate, but not swarming rings. Since this method can evaluate chemotaxis of a number of cells on one plate, it has been used for isolation of nonchemotactic mutants [4]. It is a simple and easy method; however, it is difficult to apply the semisolid agar plate method to measuring chemotaxis to nonutilizable chemicals.

Although the capillary assay technique provides quantitative information on chemotaxis by using a simple setup, it is time-consuming and the results tend to show great variability from trial to trial. To overcome the demerits of the capillary assay technique, Nikata et al. developed a computer-assisted capillary method [5]. In this method, a capillary containing a known concentration of an attractant and 1% agarose is inserted into cell suspension in a small chamber, similar to that described by Adler [2], mounted on the stage of an inverted phase contrast microscope (Figure 5.1C). Bacterial cells swim toward the orifice of the capillary in response to the attractant diffusing from the capillary. Microscopic images of the bacterial movement are videotaped and the video images are analyzed to count the number of bacteria accumulating toward the orifice of the capillary by digital image processing. The strength of chemotactic response is determined by the number of bacteria per videotape frame. This method can assess bacterial response to an attractant within a few minutes and provide reproducible results. This method can be used for measuring negative chemotaxis, i.e. bacterial movement away from a repellent [6].

The microtitration plate reader method was developed for measuring aerotaxis (the movement of a cell toward oxygen) [7]. In this method, aerotactic responses of bacterial cells are assessed in chemotaxis well chambers (Figure 5.1D). A 1-ml clear acrylic well is used as an upper well. The bottom of the upper well is sealed by a polycarbonate filter with a pore size of 8 μm. The lower well is a well of a 24-well microtitration plate and the upper well is placed in the lower well. For each assay, cell suspension containing green fluorescent protein (GFP)-labeled bacteria and buffer are added to the lower and the upper wells, respectively. Bacterial cells that migrate to the upper well in response to an oxygen gradient are automatically detected by measuring GFP fluorescence intensity using a fluorescence spectrometer. By using this method, more than 20 samples can be assessed automatically and simultaneously.

The agarose plug assay is a simple semiquantitative method and can evaluate chemotactic responses in a relatively short time [8]. This method has been used for measuring positive and negative chemotaxis to organic pollutants with low water solubility, including aromatic hydrocarbons and volatile chlorinated aliphatic hydrocarbons [9–11]. Chemotactic responses to nonutilizable chemicals can be measured by this method.

To investigate an effect of chemotaxis on microbe–plant interactions, it is necessary to measure bacterial chemotactic responses in soils. Some plant-associated *Pseudomonas* were examined for their chemotactic responses in soils [12–15]. In these methods, chemotactic responses are assessed by enumeration of bacterial cells in soil samples.

5.3
Ecological Aspects of Chemotaxis

5.3.1
Host–Microbe Interactions

P. syringae is an agriculturally important foliar phytopathogen, causing chlorosis and necrotic lesions on leaves. This pathogen does not penetrate the plant cell directly – it invades plant tissues through natural openings or wounds on the plant surface [16,17]. In contrast, certain soil-borne fluorescent pseudomonads, including *P. fluorescens* and *P. putida*, exhibit beneficial effects on plants such as the promotion of plant health and development, and biological control of soil-borne diseases [18–23]. Colonization of the plant root system is often necessary before they exert their beneficial effects on their plant hosts [24,25]. These plant-deleterious and plant-beneficial pseudomonads were found to exhibit chemotactic responses toward root and seed exudates, leaf surface water, and organic compounds prevalent in them [12–14,22,26]. Recently, it was reported that ethylene, a plant hormone, also serves as a chemoattractant for *P. fluorescens, P. putida* and *P. syringae* [27]. Therefore, it is reasonable to think that chemotaxis plays an important role in bacterial colonization and infection by the plant-associated pseudomonads, and the significance of motility and chemotaxis has been examined using isogenic motile, nonmotile and nonchemotactic strains. A nonmotile mutant of the bean pathogen *P. syringae* pv. *phaseolicola*, which possesses paralyzed flagella, showed decreased infectivity toward bean leaves compared with that of its motile revertant when bacterial cells were externally inoculated to the leaves [28]. Flagellar motility also affects epiphytic fitness of *P. syringae* on the phylloplane. Loss of motility reduced the ability of a normally motile *P. syringae* strain to establish a large stable population on the bean leaf surface and to compete on leaves [29]. It is probably because the nonmotile mutant less effectively reaches the nutritionally favorable microsites on leaves, due to the lack of motility and chemotaxis. Motility and chemotaxis are a major trait for competitive plant root colonization by the plant growth-promoting *P. fluorescens* strains. Both nonmotile mutants and motile, but nonchemotactic, mutants of good competitive root-colonizing *P. fluorescens* strains were outcompeted in competitive root colonization assays by their wild-type parents [30,31]. Conflicting results were obtained about an effect of motility and chemotaxis of *P. fluorescens* on root colonization itself. It was reported that there was no significant difference in root colonization between nonmotile or nonchemotactic mutants and their wild-type parents when inoculated alone [31–33], while de Weger et al. demonstrated that nonmotile mutants were impaired in the ability to colonize growing potato roots [30]. The differences may be caused by different assays and organisms.

Well-known *P. aeruginosa* is highly invasive in compromised hosts, causing rapid bacteremia. Motility is required for its invasive infection. Drake and Montie reported that nonmotile mutants of virulent *P. aeruginosa*, which were nonflagellated or possessed a paralyzed flagellar, were much less virulent in the burned-mouse model than the wild-type parent [34]. Although colonization in the burn wound by the nonmotile mutants was comparable with that by the wild-type strain, the mutants did not cause the rapid bacteremia and systemic invasion.

A role of bacterial chemotaxis in prey–predator relationships has been investigated. Mitchel et al. isolated two marine *Pseudomonas* strains as predators active against phycomycete *Pythium debaryanum* and diatom *Skeletonema costatum* [35]. These strains not only lysed *P. debaryanum* and *S. costatum* cells, but also showed attractive responses to exudates from the fungus and diatom cells. They found that the threshold concentrations of major components in exudates eliciting chemotactic responses are very high [36]. Therefore, they discussed that bacterial chemotaxis probably assists the predator bacteria in maintaining their proximity to the prey's surface once they have arrived there.

5.3.2 Nitrogen Cycle

Denitrifying bacteria are able to maintain relatively high populations in certain environments and the majority of the predominant denitrifiers are *Pseudomonas* species. Kennedy and Lawless hypothesized that chemotaxis toward nitrate would contribute to the predominance of *Pseudomonas* as nitrifiers, and examined nitrifying *Pseudomonas* species including *P. aeruginosa*, *P. fluorescens* and *P. stetzeri* for their ability to respond to nitrate and nitrite [37]. These *Pseudomonas* species showed strong positive chemotaxis to nitrate and nitrite, regardless of whether the cells were grown on nitrate or nitrite, or whether they were grown aerobically or anaerobically. They also carried out the modified capillary assays to investigate any chemotactic response of the indigenous soil microflora to nitrate and found that the majority of soil bacteria accumulating to nitrate were *Pseudomonas*. The loss of a flagellar motility reduced the fitness of a normally chemotactic nitrifying *P. fluorescens* strain in aerobic and anaerobic soils. From these results, they concluded that chemotaxis may be one of traits by which denitrifiers successfully compete for available nitrate and nitrite.

5.3.3 Bioremediation and Chemotaxis Toward Environmental Pollutants

Pseudomonads are extremely versatile with regard to nutritional properties. Since certain species of pseudomonads are capable of degrading many environmental pollutants, pseudomonads are important agents of bioremediation. Chemotaxis of pollutant-degrading bacteria is a trait important for bioremediation, especially *in situ* bioremediation. The migration of pollutant degraders to pollutants is expected to speed the biodegradation process because it should bring the cells into contact with pollutants [38]. Therefore, there has been much research on chemotaxis of pseudomonads toward environmental pollutants. Benzoate-degrading *P. putida* PRS2000 is attracted by aromatic acids including benzoate, *p*-hydroxybenzoate, toluates, salicylate and chlorobenzoates [39,40]. The chemotaxis toward aromatic acids is induced by β-ketoadipate, an intermediate in metabolic pathways of aromatic acids. Naphthalene-degrading *P. putida* G7 chemotactically responds to naphthalene when it is grown on naphthalene or salicylate [41]. *P. putida* G7 is also attracted by biphenyl although it cannot utilize biphenyl as a growth substrate. Marx and Aitken demonstrated that wild-type *P. putida* G7 showed higher naphthalene degradation rates in the heterogeneous aqueous system than its nonmotile and nonchemotactic mutants [42]. This result supports the idea that bacterial chemotaxis plays an important role in bioremediation. *P. putida* F1 and *P. stutzeri* OX1 are toluene-oxidizing bacteria that utilize toluene for growth through the toluene dioxygenase and monooxygenase pathways, respectively. They also have the ability to cooxidatively degrade trichloroethylene, a major environmental pollutant. They are attracted by not only growth substrate toluene, but also nonutilizable chloroethylenes, including tetrachloroethylene, trichloroethylene and three dichloroethylene isomers [9,10]. The chemotactic response by *P. putida* F1 is induced by toluene. The toluene dioxygenase genes are not required for the chemotactic response to toluene and chloroethylenes in *P. putida* F1; however, the *todS* and *todT* genes encoding two-component regulatory proteins for the toluene dioxygenase operon are essential for induction of toluene and trichloroethylene chemotaxis [9]. In *P. stutzeri* OX1, a deletion of the *touA* gene encoding toluene monooxygenase resulted in loss of the chemotaxis toward tetrachloroethylene [10]. Moreover, when the toluene monooxygenase operon (*touABCDEF*) from *P. stutzeri* OX1 was introduced to *P. putida* PaW340, which does not respond to *o*-xylene, the transformant showed a chemotactic response to *o*-xylene, suggesting that functional toluene monooxygenase is required for chemotaxis toward these environmental pollutants in *P. stutzeri* OX1. In contrast to *P. putida* F1 and *P. stutzeri* OX1, wild-type *P. aeruginosa* PAO1 is repelled by chloroethylenes [7]. Conversely, *P. aeruginosa* PAO1 mutant lacking three major chemosensory protein genes for trichloroethylene shows an attractive response to trichloroethylene [11]. Thus, *P. aeruginosa* PAO1 possesses distinct chemosensory proteins for positive and negative chemotaxis to trichloroethylene.

5.4
Molecular Biology of Chemotaxis in *Pseudomonas*

5.4.1
Molecular Mechanism of Bacterial Chemotaxis

The molecular mechanism of bacterial chemotaxis has been characterized in detail using *E. coli* [43]. In *E. coli*, the chemotaxis system consists of methyl-accepting chemotaxis proteins (MCPs), six cytoplasmic chemotaxis proteins (Che proteins) and flagella (Figure 5.2). CheA and CheY are a histidine protein kinase and a response regulator of a two-component regulatory system, respectively. CheA phosphorylates itself at a specific histidine residue and the phosphoryl group is transferred to CheY. Phosphorylated CheY is an active form and interacts directly with the flagellar motor switch to control the direction of flagellar rotation. CheZ is involved in dephosphorylation of phosphorylated CheY. MCPs are transmembrane chemosensory proteins for environmental stimuli. MCPs, together with CheW,

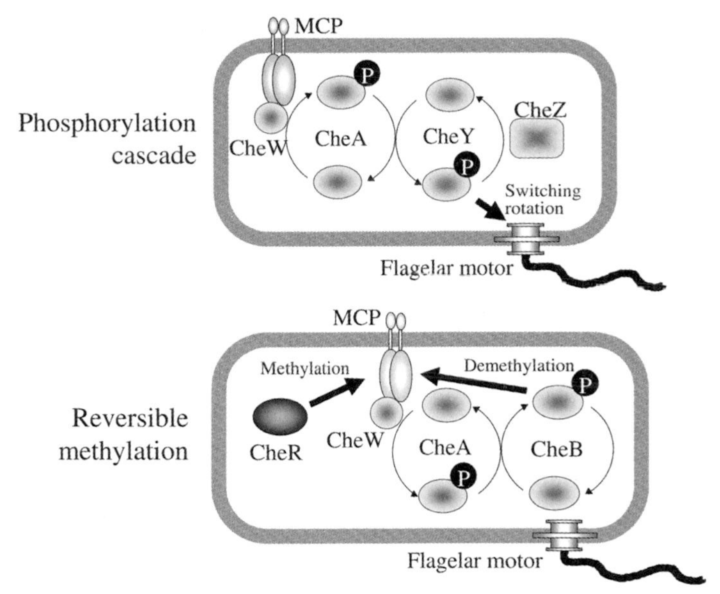

Figure 5.2 Molecular mechanism of chemotaxis in *E. coli*.

modulate the autophospholyration activity of CheA in response to temporal changes in stimulus intensity. MCPs are reversibly methylated at several glutamate residues by methyltransferase CheR and methylesterase CheB. CheB is another response regulator and receives the phosphoryl group from phosphorylated CheA. Phosphorylation of CheB increases its methylesterase activity and, thus, the level of methylation of MCPs is controlled in response to environmental stimuli. This reversible methylation of MCPs is required for temporal sensing of chemical gradients. The chemotaxis machinery allows the organism to move toward attractants and away from repellents by a biased random walk.

The genome sequences have been fully determined in several *Pseudomonas* species. Sequence analysis reveals that these *Pseudomonas* species possess genes encoding MCPs and homologs of six Che proteins. A striking feature of the genomes of *Pseudomonas* species is that the number of putative chemotaxis genes is quite large (Figure 5.3). The best characterized of the chemotaxis systems in pseudomonads is that of *P. aeruginosa* PAO1. From the investigations, it is thought that the chemotaxis machinery of *P. aeruginosa* is basically similar to that of *E. coli*, although *P. aeruginosa* possesses a large number of chemotaxis genes [44,45].

5.4.2
MCPs in *Pseudomonas*

The structure of a typical MCP is shown in Figure 5.4. MCPs are membrane-spanning homodimers. Typical structural features of MCPs are as follows: a positively charged N terminus followed by a hydrophobic membrane-spanning region, a hydrophilic periplasmic domain, a second hydrophobic membrane-spanning region

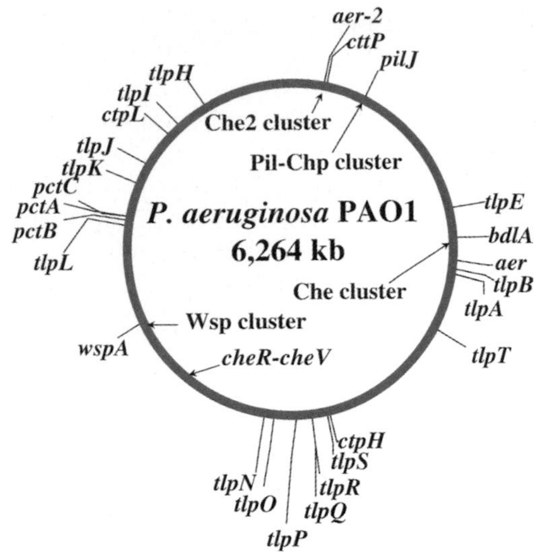

Figure 5.3 Chemotaxis genes on the genome of *P. aeruginosa* PAO1. *P. aeruginosa* PAO1 is predicted to possess 26 putative MCP genes and more than 20 *che* genes in five distinct clusters. The putative MCP genes are shown outside the circle, while the *che* gene clusters are shown inside the circle.

and a hydrophilic cytoplasmic domain. Chemotactic ligands bind to periplasmic domains of MCPs and their binding initiates chemotaxis signaling. The diverse ligand specificities among MCPs reflect amino acid sequence diversities of periplasmic domains of MCPs. In the C-terminal cytoplasmic domain, there are two reversible

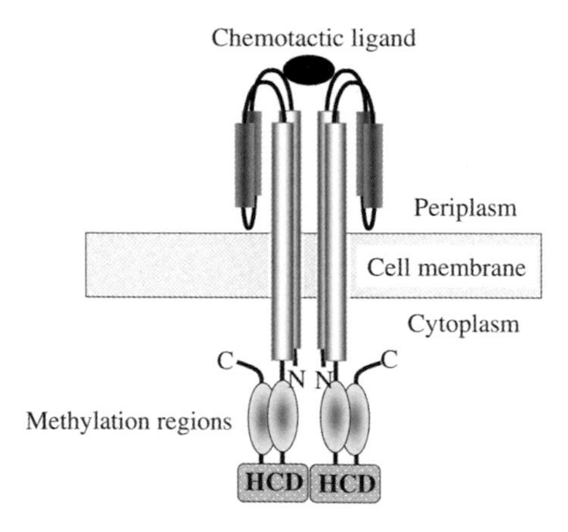

Amino acid sequence of the HCD of *E. coli* Tsr

IISVIDGIAFQTNILALNAAVEAARAGEQGRGFAVVAGEVRNLA

Figure 5.4 The structure of typical MCPs and the HCD sequence of the *E. coli* MCP, Tsr.

methylation regions. A 44-amino-acid highly conserved domain (HCD) is located between two methylation regions. MCPs from phylogenetically diverse bacteria have been shown to possess the HCD [46], which is important for the interaction of MCPs with CheW and CheA [47]. Therefore, it is quite easy to identify putative MCP genes from the genomic sequences. Blastp analysis using the HCD amino acid sequence as a probe predicts that *P. aeruginosa* PAO1, *P. putida* KT2440, *P. fluorescens* Pf-5 and *P. syringae* pv. *syringae* B728a possess 26, 26, 42 and 49 putative MCP genes, respectively, while there are only five MCP genes on the *E. coli* genome, suggesting that pseudomonads have the potential to respond to a wide variety of chemical stimuli. However, a limited number of MCPs have been characterized in pseudomonads (Table 5.1), most of which are MCPs from *P. aeruginosa* PAO1.

Amino acids are strong chemoattractants for many *Pseudomonas* species. Since amino acids are major components in seed and root exudates, it is possible that chemotaxis to amino acids plays an important role in microbe–plant interactions. PctA, PctB and PctC from *P. aeruginosa* have been identified and characterized as MCPs for amino acids [48,49]. There are high degrees of similarity in the periplasmic domains among PctA, PctB and PctC. The *pctA*, *pctB* and *pctC* genes are clustered on the *P. aeruginosa* PAO1 genome (Figure 5.5). Genetic analysis revealed that PctA, PctB and PctC detect 18, seven and two naturally occurring amino acids, respectively. Blastp analysis found that *Pseudomonas* species other than *P. aeruginosa* PAO1 also possess one to three PctA homologs. Probably, these PctA homologs function as MCPs for amino acids in *Pseudomonas* species.

P. aeruginosa PAO1 is attracted by the plant hormone ethylene [27]. TlpQ of *P. aeruginosa* PAO1 was identified as a MCP for ethylene. TlpQ homologs are found in the genomes of plant-associated *Pseudomonas* such as *P. syringae*, *P. fluorescens* and *P. putida*, and it was experimentally demonstrated that several strains of these *Pseudomonas* species are attracted by ethylene. Since essentially all plant tissues produce this hormone, chemotactic responses to ethylene are also interesting with regard to microbe–plant interactions.

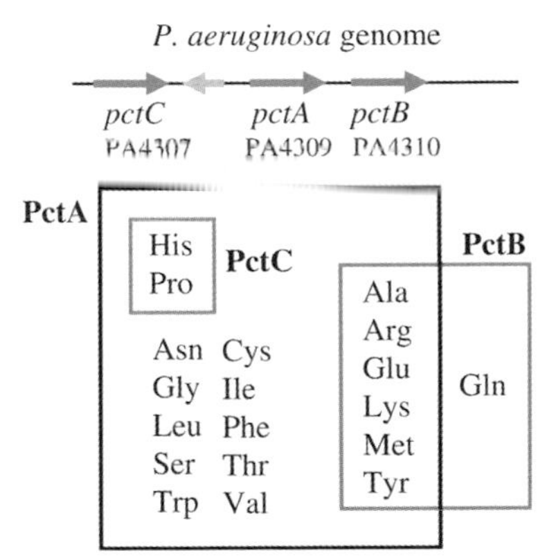

Figure 5.5 The *pctA pctB pctC* region of the *P. aeruginosa* PAO1 genome and amino acids detected by PctA, PctB and PctC [49].

Table 5.1 Characterized MCPs and chemosensory protein.

Protein[a]	Accession no.[b]	Bacterial strain	Chemotactic response[c]	Relevant characteristics	Reference
MCP					
PctA (PA4309)	AAG07697	*P. aeruginosa* PAO1	A	detection of 18 commonly occurring L-amino acids	[48,49]
			R	detection of trichloroethylene, chloroform, methylthiocyanate	[50]
PctB (PA4310)	AAG07698	*P. aeruginosa* PAO1	A	detection of six commonly occurring L-amino acids	[49]
			R	detection of trichloroethylene, chloroform, methylthiocyanate	[50]
PctC (PA4307)	AAG07695	*P. aeruginosa* PAO1	A	detection of two commonly occurring L-amino acids	[49]
			R	detection of trichloroethylene, chloroform, methylthiocyanate	[50]
CtpH (PA2561)	AAG05949	*P. aeruginosa* PAO1	A	detection of high concentrations of inorganic phosphate	[51]
CtpL (PA4844)	AAG08229	*P. aeruginosa* PAO1	A	detection of low concentrations of inorganic phosphate	[51]

Aer (PA1561)	AAG04950	*P. aeruginosa* PAO1	A	aerotaxis transducer	[45]
Aer2 (PA0176)	AAG03566	*P. aeruginosa* PAO1	A	aerotaxis transducer	[45]
CttP (PA0180)	AAG03570	*P. aeruginosa* PAO1	A	detection of trichloroethylene, tetrachloroethylene, dichloroethylenes	[11]
			–	required for biofilm maturation	[52]
TlpQ (PA2654)	AAG06402	*P. aeruginosa* PAO1	A	detection of ethylene	[27]
PilJ (PA0411)	AAG03800	*P. aeruginosa* PAO1	–	required for pilus synthesis	[53]
BdlA (PA1423)	AAG04812	*P. aeruginosa* PAO1	–	involved in biofilm dispersion	[54]
Aer	AAD22405	*P. putida* PRS2000	A	aerotaxis transducer	[55]
NahY	AAD13223	*P. putida* G7	A	detection of naphthalene, encoded on the catabolic plasmid NAH7	[56]
Other protein					
PcaK	AAA85137	*P. putida* PRS2000	A	detection of 4-hydroxybenzoate; permease for 4-hydroxybenzoate and protocatechuate	[57]

[a] Gene ID numbers used in the *P. aeruginosa* sequencing project (http://www.pseudomonas.com) are indicated in parentheses.
[b] Accession numbers in the DDBJ/EMBL/NCBI databases.
[c] A, attractive response; R, repelled response; –, characteristics unrelated to chemotaxis.

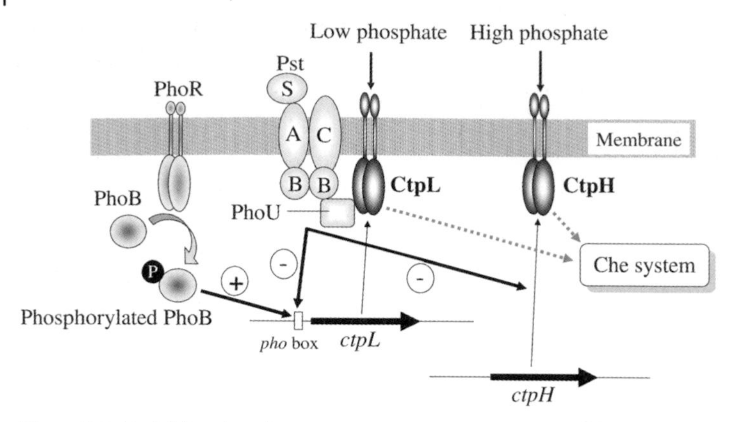

Figure 5.6 Model for phosphate taxis in *P. aeruginosa* PAO1 [51]. CtpL and CtpH are MCPs for chemotaxis to low and high concentrations of phosphate, respectively. Under conditions of phosphate limitation, PhoB is phosphorylated by PhoR. Phosphorylated PhoB activates transcription of the *ctpL* gene. In phosphate-sufficient conditions, the Pst complex and PhoU negatively regulate expression of CtpL and CtpH at transcriptional and posttranscriptional levels, respectively.

P. aeruginosa PAO1 exhibits a positive chemotactic response to inorganic phosphate [58]. The chemotactic response is induced by phosphate limitation. *P. aeruginosa* PAO1 possesses two MCPs for phosphate, CtpH and CtpL, which are functional at different concentrations of phosphate [51] (Figure 5.6). CtpL serves as the major chemoreceptor for phosphate at low concentrations, while CtpH is required for exhibiting phosphate taxis at high concentrations of phosphate. There is no significant similarity between the putative periplasmic domains of CtpH and CtpL. The induction mechanism of phosphate taxis in *P. aeruginosa* is complicated. In *P. aeruginosa,* phosphate limitation elicits the synthesis of several proteins such as alkaline phosphatase, phosphate-specific transport (Pst) complex, a hemolytic phospholipase C and a nonhemolytic phospholipase C [59–61]. These gene promoters are positively regulated by a PhoB/PhoR two-component regulatory system [62], whereas the Pst complex, together with PhoU, negatively regulates the phosphate regulon in *P. aeruginosa* [61,63,64]. CtpH expression is not dependent on the PhoB/PhoR proteins [51]. The Pst complex and PhoU are likely to exert a negative control on CtpH expression at the posttranscriptional level. The *ctpL* gene is a member of the phosphate regulon and PhoB/PhoR are essential for its transcription. A putative PhoB-binding sequence (*pho* box) exists in the *ctpL* promoter region. The *ctpL* gene is constitutively transcribed in *pst* and *phoU* mutants of *P. aeruginosa* PAO1; however, the mutant strains fail to show a chemotactic response toward low concentrations of phosphate, suggesting that the Pst complex and PhoU are required for the phosphate detection by CtpL.

Aerotaxis by *P. aeruginosa* PAO1 was detected by the microtitration plate reader method [7]. Aer and Aer-2 are MCPs for aerotaxis in *P. aeruginosa* PAO1 [45]. The aerotaxis MCP, Aer, was first identified in *E. coli* [65,66]. The *P. aeruginosa* Aer protein is 45% identical to the *E. coli* Aer protein. Like *E. coli, P. aeruginosa* Aer contains a PAS [an acronym of the *Drosophila* period clock protein (PER), vertebrate aryl hydrocarbon receptor nuclear translocator (ARNT) and *Drosophila* single-minded protein (SIM)] motif in the N-terminal domain. The PAS motif have been found in many proteins that sense redox potential, cellular energy levels, and light [67]. In the case of *E. coli* Aer, the PAS motif comprises a binding pocket for flavin adenine dinucleotide (FAD) and it is postulated that this FAD functions as a redox-sensing moiety [66]. There is only one hydrophobic region in *P. aeruginosa* Aer, which may serve to anchor Aer to the cytoplasmic side of the inner membrane. In the C-terminal domain of Aer-2, there are the HCD and two potential methylation regions. Although the C-terminal domain of Aer-2 has typical structural features of MCPs, there is no hydrophobic region in the N-terminal domain. Aer-2 also possesses the PAS motif in the N-terminal domain. Both Aer and Aer-2 may also sense altered redox conditions in the cell. No significant homology other than the PAS motif is detected in the N-terminal domains between Aer and Aer-2. The aerotactic responses of *P. aeruginosa* cells are induced during the transition from the exponential to stationary growth phase. The *aer* and *aer-2* genes are transcriptionally regulated by different regulatory proteins (Figure 5.7). The anaerobic transcriptional regulator ANR [68] is involved in the regulation of *aer* expression [69]. The *aer-2* gene is not under the control of ANR. The *aer* promoter contains two ANR boxes which are the consensus sequence shared by the ANR-dependent promoters [70]. Both ANR boxes are essential for the expression of the *aer* gene. The *aer-2* gene is cotranscribed with *cheY2, cheA2* and *cheW2* [71,72] (Figure 5.8). This transcription is dependent on alternative σ factor RpoS. There is a potential RpoS –10 region with six of the seven bases being identical to the consensus sequence [73]. The transcription of *aer* is independent on RpoS.

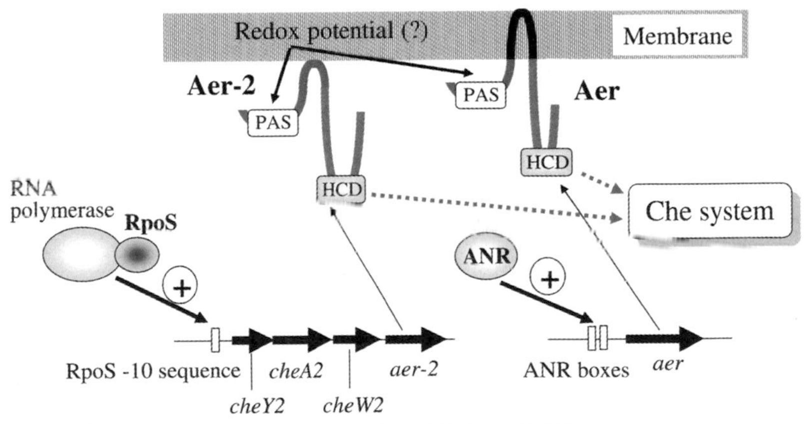

Figure 5.7 Model for aerotaxis in *P. aeruginosa* PAO1 [45,69,72]. *P. aeruginosa* PAO1 possesses two aerotaxis MCPs, Aer and Aer-2. Expression of Aer and Aer-2 is induced during the transition from exponential to stationary growth phase. Alternative σ factor RpoS is involved in induction of Aer-2 expression, while the anaerobic regulator ANR transcriptionally regulates the *aer* gene.

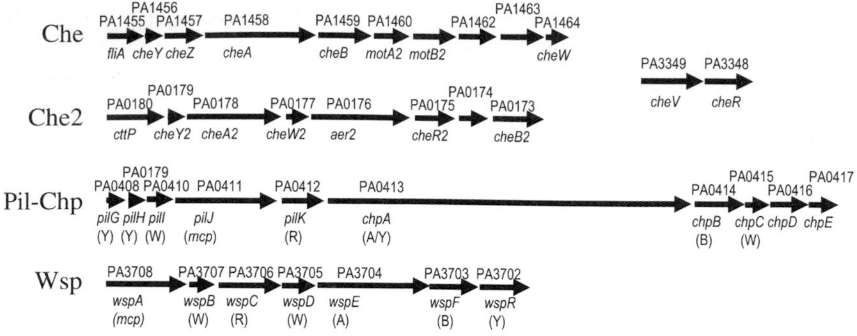

Figure 5.8 Genetic organization of *che* gene clusters of *P. aeruginosa* PAO1. The locations and orientations of individual open reading frames are shown by horizontal arrows. Gene ID numbers in the *P. aeruginosa* genome sequencing project are indicated above horizontal arrows. A, B, R, W and Y in parentheses denote homologs of *cheA*, *cheB*, *cheR*, *cheW* and *cheY*, respectively.

PctA, PctB and PctC, the MCPs for amino acids, also serve as MCPs for negative chemotaxis to trichloroethylene [50]. The *pctA pctB pctC* triple mutant of *P. aeruginosa* PAO1 shows a decreased repelled response to a high concentration (4 mM) of trichloroethylene in the agarose plug assay. Wild-type *P. aeruginosa* PAO1 is weakly repelled by a low concentration (1 mM) of trichloroethylene. However, interestingly, the triple mutant exhibits an attractive response to 1 mM trichloroethylene [11]. Genetic analysis demonstrated that CttP is responsible for this positive chemotaxis to trichloroethylene. When a plasmid containing the *cttP* gene is introduced to the *pctA pctB pctC* triple mutant, the transformant shows an attractive response even to 4 mM trichloroethylene. *P. putida* F1 cells grown in the presence of toluene are attracted by trichloroethylene, while cells grown in the absence of toluene are not chemotactic to it [9]. However, *P. putida* F1 containing the *P. aeruginosa cttP* gene exhibits positive chemotaxis to trichloroethylene even when it is grown in the absence of toluene. CttP detects tetrachloroethylene and three dichloroethylene isomers as well as trichloroethylene, but not ethylene. CttP is predicted to possess only one hydrophobic sequence. Although the N-terminal domain of CttP is identical to that of the *P. aeruginosa* PA14 CttP homolog (Paer03004584), it has no significant similarity to any other known proteins.

Naphthalene-degrading *P. putida* G7 is attracted by naphthalene. This bacterium harbors the NAH7 catabolic plasmid on which the upper and lower naphthalene degradation pathway genes are located. A NAH7-cured derivative of *P. putida* G7 is not chemotactic to naphthalene [41]. Nucleotide sequencing of NAH7 found the MCP-encoding *nahY* gene downstream of the lower pathway [56]. *P. putida* G7 *nahY* mutant is able to grow on naphthalene, but is not chemotactic to it, suggesting that NahY is a MCP for naphthalene chemotaxis. However, the NAH7-cured derivative containing the *nahY* gene but not the naphthalene degradation pathway genes does not show a chemotactic response to naphthalene. Therefore, it is likely that NahY detects a metabolite of naphthalene rather than naphthalene itself. The *nahY* gene is cotranscribed with the lower pathway genes as an operon, suggesting that chemotaxis to naphthalene may play a role in naphthalene degradation.

PcaK from *P. putida* PRS2000 is involved in an attractive response to *p*-hydroxybenzoate [57]. Interestingly, PcaK does not have the structural motifs of MCPs. PcaK belongs to the major facilitator superfamily of transporters [74], and functions as a permease of *p*-hydroxybenzoate and protocatechuate [75]. Ditty and Harwood constructed a series of mutant *pcaK* genes to investigate the relationship between the transport and chemotaxis functions of PcaK [76,77]. They examined cells expressing mutant PcaK proteins for *p*-hydroxybenzoate transport and chemotaxis, and found that mutations that decreased transport of *p*-hydroxybenzoate caused decreased chemotactic responses to *p*-hydroxybenzoate, demonstrating that *p*-hydroxybenzoate transport is required for PcaK to mediate *p*-hydroxybenzoate chemotaxis. Whether PcaK generates chemotaxis signals and directly communicates with the cytoplasmic chemotaxis signal transduction system remains to be elucidated.

PilJ and BdlA from *P. aeruginosa* exhibit typical structural features of MCPs; however, they are involved in biological functions other than chemotaxis, i.e. pilus biosynthesis [53] and biofilm dispersion [54]. CttP, the MCP for positive chemotaxis to chloroethylenes, is also involved in biofilm maturation [52]. The more MCPs in pseudomonads are characterized, the more MCPs involved in functions other than chemotaxis may be found.

5.4.3
Che Proteins in *Pseudomonas*

E. coli possesses only one *che* gene cluster containing six *che* genes, whereas sequence analysis of the complete genomes of *Pseudomonas* species reveals that they possess more than 20 *che* genes in several distinct clusters. Five *che* gene clusters of *P. aeruginosa* PAO1 are shown in Figure 5.8. Among these clusters, the Che cluster and the *cheR* gene were demonstrated to be essential for chemotaxis in *P. aeruginosa* [4,44]. The Che cluster contains *cheY, cheZ, cheA, cheB* and *cheW*. Deletion of each one of the *che* genes in the Che cluster and the *cheR* gene results in a nonchemotactic phenotype in *P. aeruginosa* PAO1. Mutation of these *che* genes affects the frequency of swimming direction reversal. The *cheA, cheR, cheW* and *cheY* mutants of *P. aeruginosa* PAO1 rarely reverse swimming directions, while the *cheB* and *cheZ* mutants change swimming direction much more frequently than wild-type *P. aeruginosa* PAO1. These behaviors of mutants are similar to those of *E. coli che* gene mutants. It was demonstrated that the *cheA* homologs are also essential for chemotaxis in *P. putida* PRS2000 and *P. fluorescens* WCS365 [31,78].

The gene products of *cheA2, cheB2, cheR2* and *cheW2* in the Che2 cluster show higher similarities to their *E. coli* counterparts than the orthologous Che cluster chemotaxis proteins do. Overexpression of *cheA2, cheB2* and *cheW2* inhibits *E. coli* chemotaxis, suggesting that these proteins can compete with *E. coli* chemotaxis proteins [79]. Overexpression of *cheB2* complements the *cheB* mutation in *P. aeruginosa* PAO1. However, deletion of the Che2 cluster genes has little effects on chemotaxis in *P. aeruginosa* PAO1. Hong et al. reported that a Che2 cluster deletion mutant normally responded to peptone in the computer-assisted capillary assay [45]. Ferrández et al. demonstrated that *cheA2* and *cheW2* mutants formed swarm rings on semisolid LB plates comparable to that of the wild-type strain, whereas a *cheB2* mutant formed a relatively smaller swarm ring [79]. Thus, chemotaxis proteins in the Che cluster and CheR usually play the major role in chemotaxis, and the Che2 cluster gene products may function in chemotaxis under certain conditions.

Deletion of the Wsp and the Pil–Chp clusters has no apparent effects on chemotactic responses in *P. aeruginosa* PAO1 [45]. Instead, they were shown to be involved in functions other than chemotaxis. The Pil–Chp cluster is involved in type IV pilus biosynthesis [53] and regulation of twitching motility [80]. The Wsp cluster is involved in biofilm formation [81]. The *wspR* gene of the Wsp cluster encodes an enzyme catalyzing synthesis of cyclic diguanylate (c-diGMP) from GTP and the gene product of *wspF* negatively regulates the activity of WspR [81]. A deletion of *wspF* results in increased intracellular levels of c-diGMP which leads to enhancement of biofilm formation. Conversely, biofilm formation is inhibited when a c-diGMP-degrading enzyme is expressed in wild-type *P. aeruginosa* PAO1. BdlA, one of the MCPs in *P. aeruginosa*, was shown to be involved in biofilm formation through modulating intracellular levels of c-diGMP [54]. Therefore, there may be communication between BdlA and gene products of the Wsp cluster.

5.4.4
Polar Localization of MCPs and Che Proteins

In *E. coli*, MCP, CheA and CheW form a stable ternary complex that localizes at cell poles [82,83]. Polar localization of MCPs and Che proteins is not specific to *E. coli*, but has been observed in many motile bacteria including *Rhodobacter sphaeroides*, *Spirochaeta aurantia*, *Proteus mirabilis*, *Caulobacter cresentus* and *Bacillus subtilis* [84–87]. *Pseudomonas* is not exceptional. Subcellular localization of MCPs and Che proteins has been investigated in *P. aeruginosa* PAO1 by immunofluorescence microscopy or using fluorescent protein-tagged proteins [88,89]. Figure 5.9 shows *P. aeruginosa* PAO1 containing the *pctA–gfp* fusion gene (C. S. Hong and J. Kato, unpublished data). PctA exhibits a typical structural feature, i.e. two transmembrane domains in the N-terminal region. PctA–GFP fusion proteins mainly localize to the cell pole in *P. aeruginosa* (Figure 5.9). MCPs that possess only one transmembrane domain [CttP (PA0180, also referred to as McpA) and Aer] also show polar localization ([88] and C. S. Hong and J. Kato, unpublished data). Polar localization of Aer–GFP is shown in Figure 5.9. Aer-2 (PA0176, also referred to as McpB), BdlA (PA1473) and McpS (PA1930) of *P. aeruginosa* PAO1 lack transmembrane domains, and are predicted to be soluble MCPs. However, Aer-2 and McpS were demonstrated to localize to the cell poles [88,89]. Güvener et al. demonstrated that CheA, CheY and CttP colocalized to the cell poles. Since CheW was required for their localization, they suggested that MCP, CheA, CheY and CheW form signal transduction complexes that localize to the cell poles in *P. aeruginosa* [88], as is the case with *E. coli*. *P. aeruginosa* has a single polar flagellum. The MCP–Che protein complex is mainly at the cell pole where the flagellum is positioned, because CheA and CheY usually

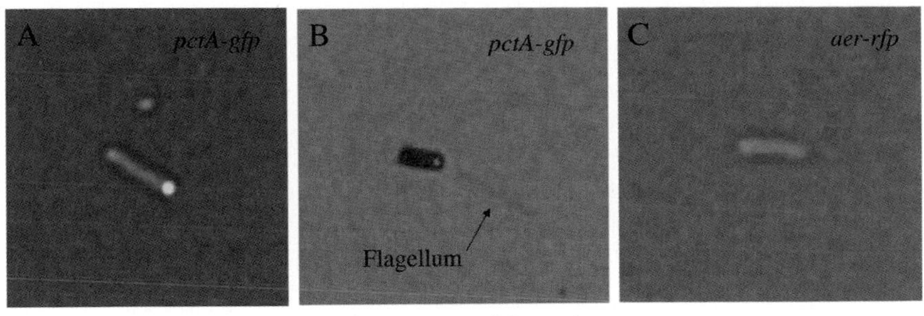

Figure 5.9 Subcellular localization of PctA–GFP and Aer–red fluorescent protein (RFP) by fluorescence microscopy in *P. aeruginosa* PAO1. (A) *P. aeruginosa* PAO1 harboring *pctA–gfp*. (B) *P. aeruginosa* PAO1 harboring *pctA–gfp*. The cell was subject to flagella staining. (C) *P. aeruginosa* PAO1 harboring *aer–rfp*.

localize to the old poles of dividing cells [88]. It is confirmed by the flagella staining of *P. aeruginosa* PAO1 expressing the PctA–GFP fusion protein (C. S. Hong and J. Kato, unpublished data) (Figure 5.9). CheY2 encoded by the Che2 cluster localizes to the cell poles [88]. CheY2 colocalizes with Aer-2, which is also encoded by the Che2 cluster, but not with CheA and CheY. Polar localization of CheY2 is dependent on CheW2 and Aer-2, but not on Che proteins of the Che cluster. Thus, the Che and Che2 cluster proteins seem to form distinct signal transduction complexes in the *P. aeruginosa* cell.

Whether the Wsp and Pil–Chp cluster proteins also form signal transduction complexes, and if they form complexes, whether there are relationships between complex formation and their functions (biofilm dispersion, pilus biosynthesis and twitching motility) are very interesting questions, but these remain to be investigated.

5.5
Pseudomonas as Model Microorganisms for Chemotaxis Research

It is suggested that most microorganisms in heterogeneous environments are motile [90]. Although a large number of genes are required for motility and chemotaxis, these microorganisms have retained motility and chemotaxis under natural selection for very long periods of time. There is also a theoretical research showing that chemotaxis can allow increased population density in unmixed environments, while random motility may lead to decreased population density [91]. Thus, it is reasonable to think that chemotaxis provides a selective advantage for bacteria to survive in unmixed heterogeneous environments. The genomes of many motile bacteria have been completely sequenced, and their sequence analysis reveals that quite a few environmental bacteria possess a number of *che* and MCP genes. Therefore, *Pseudomonas* species that have complex chemotaxis systems with several sets of *che* gene clusters and a number of MCP genes are much better model microorganisms for investigation of ecological aspects of chemotaxis in environmental bacteria than the enteric bacteria of which the chemotaxis system is relatively simple.

References

1 Adler, J., Hazelbauer, G.L. and Dahl, M.M. (1977) *J Bacteriol*, **115**, 824–847.
2 Adler, J. (1972) *J Gen Microbiol*, **74**, 77–91.
3 Armstrong, J.B., Adler, J. and Dahl, M.M. (1967) *J Bacteriol*, **93**, 390–398.
4 Masduki, A., Nakamura, J., Ohga, T., Umezaki, R., Kato, J. and Ohtake, H. (1995) *J Bacteriol*, **177**, 948–952.
5 Nikata, T., Sumida, K., Kato, J. and Ohtake, H. (1992) *Appl Environ Microbiol*, **58**, 2250–2254.
6 Ohga, T., Masduki, A., Kato, J. and Ohtake, H. (1993) *FEMS Microbiol Lett*, **113**, 63–66.
7 Shitashiro, M., Kato, J., Fukumura, T., Kuroda, A., Ikeda, T., Takiguchi, N. and Ohtake, H. (2003) *J Biotechnol*, **101**, 11–18.
8 Yu, H.S. and Alam, M. (1997) *FEMS Microbiol Lett*, **156**, 265–269.
9 Parales, R.E., Ditty, J.L. and Harwood, C.S. (2000) *Appl Environ Microbiol*, **66**, 4098–4104.
10 Vardar, G., Barbieri, P. and Wood, T.K. (2005) *Appl Microbiol Biotechnol*, **66**, 696–701.
11 Kim, H.-E., Shitashiro, M., Kuroda, A., Takiguchi, N., Ohtake, H. and Kato, J. (2006) *J Bacteriol*, **188**, 6700–6702.
12 Arora, D.K., Filonow, A.B. and Lockwood, J.L. (1983) *Can J Microbiol*, **29**, 1104–1109.
13 Scher, F.M., Kloepper, J.W. and Singleton, C.A. (1985) *Can J Microbiol*, **31**, 570–573.
14 Lim, W.C. and Lockwood, J.L. (1988) *Can J Microbiol*, **34**, 196–199.
15 Arora, D.K. and Gupta, S. (1993) *Can J Microbiol*, **39**, 922–931.
16 Getz, S., Fullbright, D.W. and Stephens, C.T. (1983) *Phytopathology*, **73**, 39–43.
17 Schneider, R.W. and Grogan, R.G. (1977) *Phytopathology*, **67**, 898–902.
18 Klopper, J.W., Leong, J., Teintze, M. and Schroth, M.N. (1980) *Nature*, **286**, 885–886.
19 Kloepper, J.W. and Schroth, M.N. (1981) *Phytopathology*, **71**, 1020–1024.
20 Scher, F.M. and Baker, R. (1982) *Phytopathology*, **72**, 1567–1573.
21 Geels, F.P. and Schippers, B. (1983) *Phytopathol Z*, **108**, 207–214.
22 Grewal, S.I.S. and Rainey, P.B. (1991) *J Gen Microbiol*, **137**, 2765–2768.
23 Bolwerk, A., Lagopodi, A.L., Wijfjes, A.H.M., Lamers, G.E.M., Chin-A-Woeng, T.F.C., Lugtenberg, B.J.J. and Bloemberg, G.V. (2003) *Mol Plant Microbe Interact*, **16**, 983–993.
24 Bull, C.T., Weller, D.M. and Thomashow, L.S. (1991) *Phytopathology*, **81**, 954–959.
25 Lugtenberg, B.J.J., De Weger, L.A. and Bennett, J.W.B. (1991) *Curr Opin Biotechnol*, **2**, 457–464.
26 Cuppels, D.A. (1988) *Appl Environ Microbiol*, **54**, 629–632.
27 Kim, H.-E., Kuroda, A., Takiguchi, N. and Kato, J. (2007) *Microbes Environ*, **22**, 186–189.
28 Panopoulos, N.J. and Schroth, M.N. (1974) *Phytopathology*, **64**, 1389–1397.
29 Haefele, D.M. and Lindow, S.E. (1987) *Appl Environ Microbiol*, **53**, 2528–2533.
30 de Weger, L.A., van der Vlugt, C.I., Wijfjes, A., Bakker, P.A., Schippers, B. and Lugtenberg, B.J.J. (1987) *J Bacteriol*, **169**, 2769–2773.
31 de Weert, S., Vemeiren, H., Mulders, I.H.M., Kuiper, I., Hendrickx, N., Bloemberg, G.V., Vanderleyden, J., De Mot, R. and Lugtenberg, B.J.J. (2002) *Mol Plant Microbe Interact*, **15**, 1173–1180.
32 Howie, W.J., Cook, R.J. and Weller, D.M. (1987) *Phytopathology*, **77**, 286–292.
33 Scher, F.M., Kloepper, J.W., Singleton, C., Zaleska, I. and Laliberte, M. (1988) *Phytopathology*, **78**, 1055–1059.
34 Drake, D. and Montie, T.C. (1988) *J Gen Microbiol*, **134**, 43–52.
35 Chet, I., Fogel, S. and Mitchel, R. (1971) *J Bacteriol*, **106**, 863–867.
36 Bell, W. and Mitchel, R. (1972) *Biol Bull*, **143**, 265–277.
37 Kennedy, M.J. and Lawless, J.G. (1985) *Appl Environ Microbiol*, **49**, 109–114.

38 Parales, R.E. and Harwood, C.S. (2002) *Curr Opin Microbiol*, **5**, 266–273.

39 Harwood, C.S., Rivelli, M. and Ornston, L.N. (1984) *J Bacteriol*, **160**, 622–628.

40 Harwood, C.S., Parales, R.E. and Dispensa, M. (1990) *Appl Environ Microbiol*, **56**, 1501–1503.

41 Grimm, A.C. and Harwood, C.S. (1997) *Appl Environ Microbiol*, **6**, 4111–4115.

42 Marx, R.B. and Aitken, M.D. (2000) *Environ Sci Technol*, **34**, 3379–3383.

43 Stock, J.B. and Surette, M.G. (1996) (eds. F.C. Neidhardt, R. Curtiss III, C.A. Gross, J.L. Ingraham, E.C.C. Lin, K.B. Low, B. Magasanik, W. Reznikoff, M. Riley, M. Schaechter and H.E. Umbarger), in Escherichia coli *and* Salmonella: *Cellular and Molecular Biology*, American Society for Microbiology, Washington, DC, pp. 1103–1129.

44 Kato, J., Nakamura, T., Kuroda, A. and Ohtake, H. (1999) *Biosci Biotechnol Biochem*, **63**, 151–161.

45 Hong, C.S., Shitashiro, M., Kuroda, A., Ikeda, T., Takiguchi, N., Ohtake, H. and Kato, J. (2004) *FEMS Microbiol Lett*, **231**, 247–252.

46 Zhulin, I.B. (2001) *Adv Microb Physiol*, **45**, 157–198.

47 Liu, J. and Parkinson, J.S. (1991) *J Bacteriol*, **173**, 4941–4951.

48 Kuroda, A., Kumano, T., Taguchi, K., Nikata, T., Kato, J. and Ohtake, H. (1995) *J Bacteriol*, **177**, 7019–7025.

49 Taguchi, K., Fukutomi, H., Kuroda, A., Kato, J. and Ohtake, H. (1997) *Microbiology*, **143**, 3223–3229.

50 Shitashiro, M., Tanaka, H., Hong, C.S., Kuroda, A., Takiguchi, N., Ohtake, H. and Kato, J. (2005) *J Biosci Bioeng*, **99**, 396–402.

51 Wu, H., Kato, J., Kuroda, A., Ikeda, T., Takiguchi, N. and Ohtake, H. (2000) *J Bacteriol*, **182**, 3400–3404.

52 Southey-Pillig, C.J., Davies, D.G. and Sauer, K. (2005) *J Bacteriol*, **187**, 8114–8126.

53 Darzins, A. (1994) *Mol Microbiol*, **11**, 137–153.

54 Morgan, R., Kohn, S., Hwang, S.H., Hassett, D.J. and Sauer, K. (2006) *J Bacteriol*, **188**, 7335–7343.

55 Nichols, N.N. and Harwood, C.S. (2000) *FEMS Microbiol Lett*, **182**, 177–183.

56 Grimm, A.C. and Harwood, C.S. (1999) *J Bacteriol*, **181**, 3310–3316.

57 Harwood, C.S., Nichols, N.N., Kim, M.-K., Ditty, J.L. and Parales, R.E. (1994) *J Bacteriol*, **176**, 6479–6488.

58 Kato, J., Ito, A., Nikata, T. and Ohtake, H. (1992) *J Bacteriol*, **174**, 5149–5151.

59 Hancock, R.E.W., Poole, K. and Benz, R. (1982) *J Bacteriol*, **150**, 730–738.

60 Ostroff, R.M. and Vasil, M.L. (1987) *J Bacteriol*, **169**, 4597–4601.

61 Poole, K. and Hancock, R.E.W. (1984) *Eur J Biochem*, **144**, 607–612.

62 Siehnel, R.J., Worobec, E.A. and Hancock, R.E.W. (1988) *Mol Microbiol*, **2**, 347–352.

63 Kato, J., Sakai, Y., Nikata, T. and Ohtake, H. (1994) *J Bacteriol*, **176**, 5874–5877.

64 Nikata, T., Sakai, Y., Shibata, K., Kato, J., Kuroda, A. and Ohtake, H. (1996) *Mol Gen Genet*, **250**, 692–698.

65 Rebbapragada, A., Johnson, M.S., Harding, G.P., Zuccarelli, A.J., Fletcher, H.M., Zhulin, I.B. and Taylor, B.L. (1997) *Proc Natl Acad Sci USA*, **94**, 10541–10546.

66 Bibikov, S.I., Biran, R., Rudd, K.E. and Parkinson, J.S. (1997) *J Bacteriol*, **179**, 4075–4079.

67 Taylor, B.L. and Zhulin, I.B. (1999) *Microbiol Mol Biol Rev*, **63**, 479–506.

68 Galimand, M., Gamper, M., Zimmermann, A. and Haas, D. (1991) *J Bacteriol*, **173**, 1598–1606.

69 Hong, C.S., Kuroda, A., Ikeda, T., Takiguchi, N., Ohtake, H. and Kato, J. (2004) *J Biosci Bioeng*, **97**, 184–190.

70 Haas, D., Gamper, M. and Zimmermann, A. (1992) *Pseudomonas: Molecular Biology and Biotechnology*, American Society for Microbiology, Washington, DC. pp. 177–187.

71 Ferrández, A., Hawkins, A.C., Summerfield, D.T. and Harwood, C.S. (2002) *J Bacteriol*, **184**, 4374–4383.

72 Hong, C.S., Kuroda, A., Takiguchi, N., Ohtake, H. and Kato, J. (2005) *J Bacteriol*, **187**, 1533–1535.

73 Espinosa-Urgel, M., Chamizo, C. and Tormo, A. (1996) *Mol Microbiol*, **21**, 657–659.

74 Pao, S.S., Paulsen, I.T. and Saier, M.H. Jr, (1998) *Microbiol Mol Biol Rev*, **62**, 1–34.

75 Nichols, N.N. and Harwood, C.S. (1997) *J Bacteriol*, **179**, 5056–5061.

76 Ditty, J.L. and Harwood, C.S. (1999) *J Bacteriol*, **181**, 5068–5074.

77 Ditty, J.L. and Harwood, C.S. (2002) *J Bacteriol*, **184**, 1444–1448.

78 Ditty, J.L., Grimm, A.C. and Harwood, C.S. (1998) *FEMS Microbiol Lett*, 267–273.

79 Ferrández, A., Hawkins, A.C., Summerfield, D.T. and Harwood, C.S. (2002) *J Bacteriol*, **184**, 4374–4383.

80 Whitchurch, C.B., Leech, A.J., Young, M.D., Kennedy, D., Sargent, J.L., Bertrand, J.J., Semmler, A.B., Mellick, A.S., Martin, P.R., Alm, R.A., Hobbs, M., Beatson, S.A., Huang, B., Nguyen, L., Commolli, J.C., Engel, J.N., Darzins, A. and Mattick, J.S. (2004) *Mol Microbiol*, **52**, 873–893.

81 Hickman, J.W., Tifrea, D.F. and Harwood, C.S. (2005) *Proc Natl Acad Sci USA*, **102**, 14422–14427.

82 Maddock, J.R. and Shapiro, L. (1993) *Science*, **259**, 1717–1723.

83 Sourjik, V. and Berg, H.C. (2002) *Proc Natl Acad Sci USA*, **99**, 123–127.

84 Harrison, D.M., Skidmore, J., Armitage, J.P. and Maddock, J.R. (1999) *Mol Microbiol*, **31**, 885–892.

85 Alley, M.R.K., Maddock, J.R. and Shapiro, L. (1992) *Genes Dev*, **6**, 825–836.

86 Kirby, J.R., Niewold, T.B., Maloy, S. and Ordal, G.W. (2000) *Mol Microbiol*, **35**, 44–57.

87 Gestwicki, J.E., Lamanna, A.C., Harshey, R.M., McCarter, L.L., Kiesling, L.L. and Adler, J. (2000) *J Bacteriol*, **182**, 6499–6502.

88 Güvener, Z.T., Tifrea, D.F. and Harwood, C.S. (2006) *Mol Microbiol*, **61**, 106–118.

89 Bardy, S.L. and Maddock, J.R. (2005) *J Bacteriol*, **187**, 7840–7844.

90 Fenchel, T. (2002) *Science*, **296**, 1068–1071.

91 Lauffenburger, D. (1982) *Biophys J*, **40**, 209–219.

6
Iron Transport and Signaling in Pseudomonads

Francesco Imperi, Karla A. Mettrick, Matt Shirley, Federica Tiburzi, Richard C. Draper, Paolo Visca and Iain L. Lamont

6.1
Introduction

Cells need iron for a variety of metabolic functions. In most biological systems iron is present in two positively charged forms, the ferrous [Fe(II)] and the ferric [Fe(III)] ions, both of which take part in many essential redox reactions contributing to respiration, generation and scavenging of reactive oxygen species, and synthesis of nucleosides and amino acids. Given the high concentration of total intracellular iron (around 10^{-4} M), bacterial cells need to acquire considerable amounts of this metal from the extracellular milieu [1].

Despite its abundance in the Earth's crust, iron has extremely low bioavailability due to its peculiar solution chemistry. In aqueous aerobic environments at neutral pH iron exists mostly as the oxidized Fe(III) species which is endowed with a very low solubility due to its tendency to precipitate as oxy-hydroxy polymer ($K_s \sim 10^{-38}$ M), a property that can be increased by high pH and coprecipitating ions like phosphate [1]. Such a situation is likely to occur in abiotic habitats like aquatic systems and soil. In biological fluids of mammals and birds, iron is tightly sequestered by iron-transport proteins (e.g. transferrin, lactoferrin and ovotransferrin) which have high affinity for Fe(III) ($K_f \sim 10^{-20}$ M^{-1}), while intracellular iron [mostly present as Fe(II)] is either engaged in the prosthetic group of iron proteins or stored in the ferritin core [2–4]. Therefore, iron is poorly bioavailable to most microorganisms, whether they are harmless environmental organisms or infectious pathogens.

Microorganisms have evolved a multiplicity of strategies to acquire iron from environmental sources, to cope with low bioavailability of this metal as well as competition from other microorganisms sharing the same ecological niche. As a general rule, the higher the environmental versatility of a given microbial species, the wider the array of strategies it can use to acquire iron. Not surprisingly, pseudomonads have a wide range of iron acquisition mechanisms.

Pseudomonas. Model Organism, Pathogen, Cell Factory. Edited by Bernd H.A. Rehm

ISBN: 978-3-527-31914-5

Secretion of siderophores (low-molecular-weight iron chelators) represents the most common strategy for active iron transport into bacteria (including pseudomonads) and fungi, and will be the focus of this chapter. The binding constants of siderophores for Fe(III) are extremely high, being typically between 10^{25} M^{-1} and 10^{50} M^{-1}, implying that these compounds can effectively scavenge Fe(III) from a variety of complexes found in natural environments [5,6]. To enable transport of iron in the cytoplasm, bacteria capture iron-loaded (ferri) siderophores at the cell surface and transport them (or iron) into the cytosol. In Gram-negative bacteria this typically requires a combination of protein partners: (i) an outer membrane receptor which specifically binds the ferri-siderophore on the outer membrane and transfers it into the periplasm; (ii) a protein complex containing TonB that transduces energy from the proton motive force into transport-proficient structural changes of the receptor; (iii) a binding protein located in the periplasm that transfers the siderophore-bound iron to a cytoplasmic membrane-associated transporter; and (iv) an ABC transporter composed of a protein channel in the cytoplasmic membrane coupled with a cytoplasmic ATPase that enables ferri-siderophore internalization at the expense of ATP hydrolysis in the cytoplasm [7].

Iron acquisition is finely controlled, since both iron deficiency and iron overload can be highly noxious to cells. Too high intracellular concentration of free Fe(II) can result in the production of hydroxyl radicals through Fenton reactions and these may cause oxidation of various macromolecules [8]. Therefore, bacteria must sense intracellular iron levels and induce or repress active iron transport in response to iron deficiency or proficiency, respectively. Such iron homeostasis requires fine regulation of iron uptake systems which is achieved through the activity of negative and positive regulatory proteins, controlling different stages of gene expression [9]. *Pseudomonas* is a paradigm of multifactorial mechanisms of gene regulation. At least three levels of regulation are involved in controlling iron uptake: (i) intracellular Fe (II) is on top of the hierarchy, acting in combination with the ferric uptake regulator (Fur) protein as a repressor of secondary regulators, (ii) secondary regulators behave as activators of iron uptake genes, and (iii) extracellular iron carriers (endogenous or exogenous siderophores) act as coactivators for the expression of genes involved in their uptake, by either direct or indirect mechanisms (reviewed in Ref. [10,11]).

Fluorescent *Pseudomonas* spp. constitute a distinct group within the rRNA homology group I of the *Pseudomonas* genus [12], and include species endowed with pathogenicity traits, and plant growth-promoting and bioremediation activities. The characteristic fluorescence of these species is due to production of yellow–green fluorescent siderophores called pyoverdines (sometimes termed pseudobactins in rhizosphere species). Pyoverdines are the primary siderophores of fluorescent *Pseudomonas* spp. and are required for a number of biological functions, including plant growth promotion, biocontrol of plant pests, animal pathogenicity and bioremediation [13–16]. Apart from pyoverdines, fluorescent pseudomonads can produce secondary siderophores or acquire iron through a number of exogenous iron carriers. This iron uptake versatility provides fluorescent pseudomonads with a broad ecological fitness, since it enables them to use the most convenient iron source.

Recent studies have shown that, at least *in vitro*, iron acquisition by pseudomonads is also linked to biofilm formation. Subinhibitory concentrations of human lactoferrin, an iron-chelating protein normally present in mucous secretions, inhibit biofilm formation by *P. aeruginosa* PAO1 [17] and it was proposed that lactoferrin-induced iron deficiency stimulates twitching motility – a form of surface motility that is inconsistent with the formation of microcolonies and biofilms. It was subsequently shown that, in iron-poor media, a functional pyoverdine system is necessary for biofilm maturation *in vitro* [18]. Exogenous iron chelates such as ferric citrate, ferrioxamine and ferric chloride could substitute for pyoverdine and restore biofilm formation. These findings emphasize the importance of iron transport in growth and development of pseudomonads.

In this chapter we provide an overview of the many iron transport strategies developed by fluorescent *Pseudomonas* spp. and will address the role of siderophores not only as iron carriers, but also as signaling molecules, with special regard to the type species *P. aeruginosa*.

6.2 Siderophores Used by Pseudomonads

6.2.1 Chemical Diversity of Siderophores

A microbial metabolic product can be classified as a siderophore if it can chelate iron, participates in active transport of iron across the cell membrane(s) and if its biosynthesis is regulated by the intracellular iron concentration [19]. Siderophores need to be chemically stable to be effective. Currently, more than 500 siderophores that are produced by bacteria and fungi have been chemically characterized [20], with up to 150 of these synthesized by fungi [21].

Siderophores have great chemical diversity, although most have a peptide backbone (often cyclic) with several nonprotein amino acid residues. Most siderophores form hexadentate octahedral complexes with Fe (III) that are very thermodynamically stable. A minority bind iron in a bidentate manner, but these compounds (often the precursor of hexadentate binding siderophores) have much lower affinities for iron [5].

Siderophores are generally classified on the basis of the coordinating groups that chelate the Fe(III) ion [5,20,22]. The most common coordinating groups are catecholates, hydroxamates and carboxylates. A minority of siderophores have chemically distinct Fe(III) ion-binding groups including salicylic acid, oxazoline or thiazoline nitrogen [19]. Some siderophores (including pyoverdines) are classified as "mixed ligand", having coordinating groups that fall into chemically different classes.

The catecholate class of siderophores are generally produced by bacteria with the best-described example being enterobactin (Figure 6.1) that is produced by *Escherichia coli* and Enterobacterial cells. They have relatively low affinity for iron at low pH, but form stable tris-catecholate Fe(III) complex in less acidic conditions. Generally,

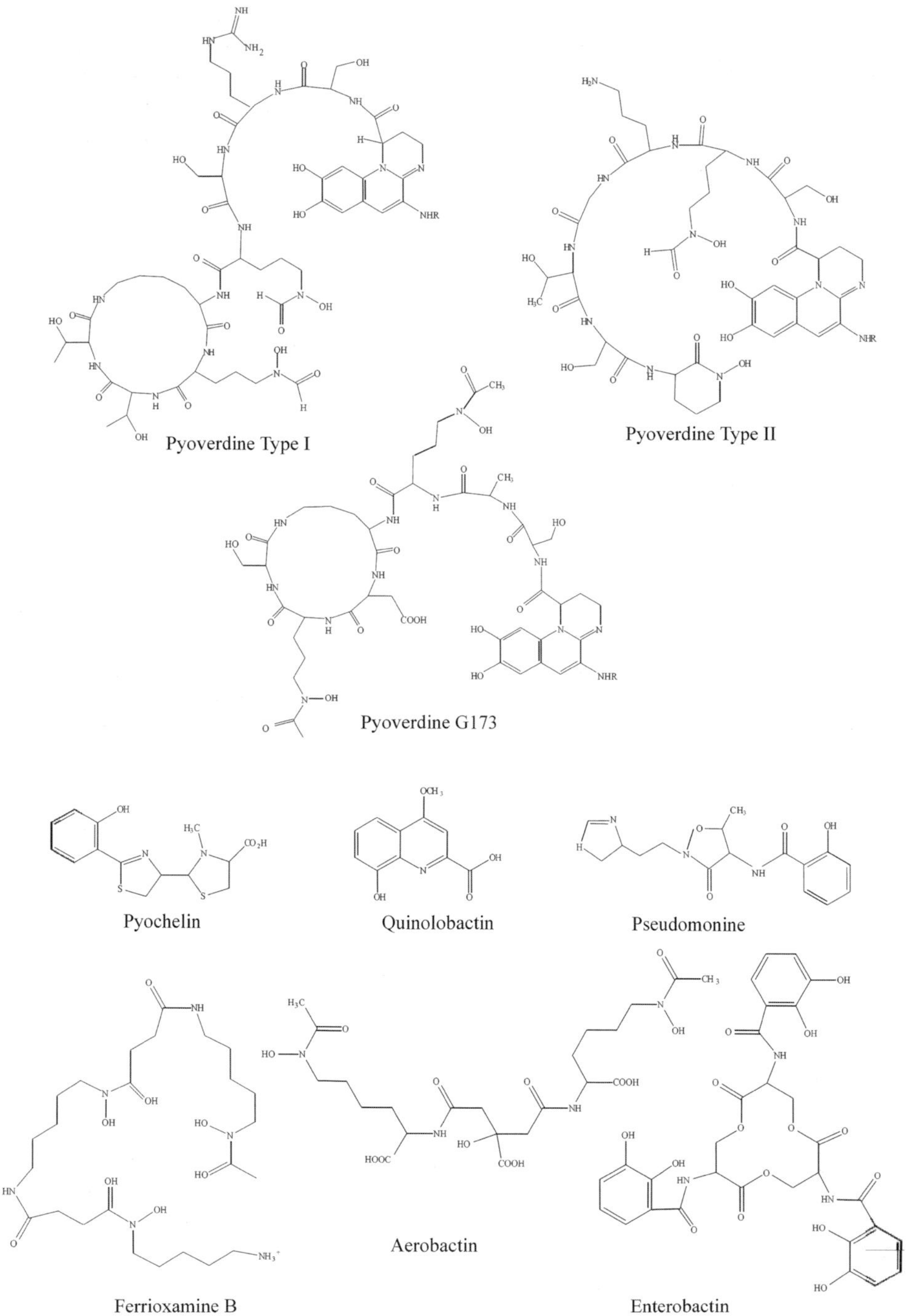

Pyoverdine Type I

Pyoverdine Type II

Pyoverdine G173

Pyochelin

Quinolobactin

Pseudomonine

Aerobactin

Ferrioxamine B

Enterobactin

Figure 6.1 Examples of siderophores that can be utilized by pseudomonads (adapted from [10]). The structures of various siderophores are shown. See text for references and producer species.

they are formed from three units of 2,3-dihydroxybenzoic acid (occasionally salicylic acid) linked by amino acid and/or amine alkane residues forming either a cyclic or a linear molecular backbone. The hydroxamate siderophores such as desferrioxamine and aerobactin (Figure 6.1) are more frequently produced by fungi and Gram-positive filament-forming bacteria (*Streptomycetes*), and form four main groups consisting of rhodotorulic acid, coprogens, ferrichromes and fusarinines [23]. The hydroxamate chelating group is commonly derived from the amino acid ornithine that has been modified through hydroxylation and carbonylation. The carboxylate class of siderophores is the smallest of the groups and polycarboxylates are produced by the zygomycete fungi.

Pseudomonas species secrete siderophores, but can also use a wide range of siderophores made by other organisms. The best characterized of these are described below.

6.2.2 Endogenous Siderophores of Pseudomonads

6.2.2.1 Pyoverdines

Pyoverdines (Figure 6.1) are a heterogenous class of fluorescent siderophores that define fluorescent pseudomonads and are considered to be the primary siderophores of these bacteria [24]. They were first characterized as siderophores in 1978 [25] and the structure of a ferri-pyoverdine complex was first determined in 1981 [26]. Different strains and species of *Pseudomonas* secrete chemically diverse pyoverdines, and over 50 are now known [27]. Pyoverdine from *P. aeruginosa* is required for virulence in animal models of disease [28,29] and pyoverdines from rhizosphere pseudomonads are implicated in biocontrol of pathogens (reviewed in Ref. [13]).

Pyoverdines are mixed catecholate–hydroxamate siderophores. They are highly water soluble with an affinity for iron of around $10^{32}\,M^{-1}$ (stoichiometry 1 : 1) [30]. Structurally, pyoverdines consist of three components – a dihydroxyquinoline chromophore, an acyl side-chain (either dicarboxylic acid or its monoamide) attached to the chromophore and a linear or cyclic peptide chain. The chromophore provides two catecholate iron-coordinating ligands, and is also responsible for the characteristic fluorescence and color pigmentation of pyoverdines and fluorescent pseudomonads [30]; fluorescence is quenched by chelation of iron [31]. The acyl moiety does not contribute to iron chelation or transport and does not have any known biological function. The peptide component is linked via an amide bond to one of the carboxyl groups of the chromophore and provides four Fe(III)-coordinating groups that are hydroxmate or in some cases hydroxyl and/or carbonyl groups. It is between six and twelve amino acids long, with its composition being dependent on the producer strain [27]. It is responsible for the specificity of recognition by the cognate receptor and the characteristic profile of pyoverdines that can be used by different strains provides the basis for a bacterial typing scheme termed siderotyping [24]. The amino acids are frequently nonproteinaceous and can include acylated or cyclic forms of N^{δ}-hydroxyornithine, diaminobutyrate, β-hydroxyaspartate, β-hydroxyhistidine and D-amino acids (see Figure 6.1).

6.2.2.2 Pyochelin

Pyochelin (Figure 6.1) is produced by strains of *P. aeruginosa* and *P. fluorescens*. It is poorly water soluble and although classified as a siderophore, it has a relatively low affinity for Fe(III) ($2.4 \times 10^5\,M^{-1}$) [32]. However, this affinity was found whilst the siderophore was in ethanol and it may exhibit higher affinity in aqueous solution. Iron free pyochelin exhibits a yellow–green fluorescence under ultraviolet light and is unstable to light [32]. Once iron is bound, it is more stable, soluble and no longer fluorescent.

Pyochelin has an unusual structure for a siderophore as it does not have catecholate or hydroxamate groups, but instead is a phenolate and binds Fe(III) via the carboxylate group and the phenolic OH group [33,34]. It has two isomeric states, although only one of these isomers (pyochelin I) is able to bind Fe(III) [35]. The stoichiometry of binding of Fe(III) and other transition metals has not been resolved although recent studies suggest a 1 : 1 ratio [33,36].

Pyochelin is considered to play only a minor role during pathogenicity of *P. aeruginosa* due to its relatively low affinity for Fe(III) and the lack of virulence of pyochelin mutants in animal model systems [28,29]. In addition to iron acquisition it can produce cell-damaging active oxygen species especially in the presence of pyocyanin, a secondary metabolite secreted by *P. aeruginosa* [37,38]. The role of pyochelin in the biology of other pseudomonads has not been investigated.

6.2.2.3 Other Endogenous Siderophores

As well as pyoverdines and pyochelin, *P. fluorescens* strains are able to produce a number of different siderophores including quinolobactin (Figure 6.1), pseudomonine and azotobactins [27,39,40]. Quinolobactin has shown evidence of anti-*Pythium* activity in addition to its iron-chelating ability [41]. Pseudomonine is a novel isoxazolidone that has similarities to pyochelin and is a salicylic acid-based siderophore with a histamine moiety [40]. Azotobactins are closely related to pyoverdines, but they differ in the structure of the dihydroxyquinoline chromophore group (see Ref. [27] for details).

Strains of *P. putida* and *P. stutzeri* produce pyridine-2,6-dithiocarboxylic acid (PDTC). This compound was not essential for growth in an iron-deplete medium [42], but nonetheless there is reason to think that it is a siderophore [43]. PDTC contributes to the chelation and uptake of zinc [44] as well as other transition metals [45]. It also has activity in dechlorination reactions [46], and a role for PDTC in the defence against toxicity from various metals and metalloids has been proposed [47,48]. *P. stutzeri* can synthesize multiple siderophores including desferriferrioxamine E (also known as nocardamin) [49]. *P. corrugata* produces the siderophore corrugatin [50].

Salicylic acid is implicated as a siderophore for *P. fluorescens* CHAO, a strain that produces pyoverdine, but not pyochelin [51]. The low affinity of salicylate for Fe(III) raises the question, is it a siderophore in its own right or just a byproduct of the biosynthesis of other siderophores? [3]. However, evidence for it being more than an intermediate of pyochelin synthesis has been seen for both *P. aeruginosa* and *P. fluorescens* [52].

6.2.3 Exogenous Siderophores Utilized by Pseudomonads

As well as the siderophores that they produce themselves, pseudomonads are able to take up siderophores of both bacterial and fungal origin; indeed, *P. fragi* does not appear to produce any of its own siderophores, but survives through the scavenging of others [53]. These include heterologous pyoverdines and other siderophores produced by other genera. The number of siderophores that can be used reflects the importance of iron acquisition to the bacteria as a vital nutrient. Some examples are listed in Table 6.1 and described below. Utilization of other siderophores has been best studied for *P. aeruginosa* and discussion will be limited to this species; however, the number of genes that can contribute to siderophore uptake in other species (see below) means that they are likely to be able to use a similarly wide range of siderophores.

6.2.3.1 Enterobactin

Enterobactin (Figure 6.1) is produced by most genera of the enteric bacteria, including virtually all strains of *E. coli* [3]. It has the highest affinity for iron of all the known siderophores ($10^{52}\,M^{-1}$) although the iron-free form is chemically labile compared to other siderophores and is poorly soluble As well as enterobactin itself, *P.*

Table 6.1 Siderophores that can be used by *P. aeruginosa*.

Siderophore	Reference
Endogenous	
pyoverdine	[179]
pyochelin	[180]
Bacterial	
enterobactin	[181]
aerobactin	[57]
rhiozobactin 1021	[57]
schizokinen	[57]
cepabactin	[51]
Fungal	
desferrioxamine B	[54]
desferrioxamine E	[51]
deferrichrysin	[51]
copragen	[51]
ferrichrome	[51,54]
desferriferrirubin	[51]
Naturally occurring	
myo-inositol hexakisphosphate	[183]
citrate	[180]
salicylic acid	[51]
Synthetic	
nitrilotriacetic acid	[51]

aeruginosa is able to use as iron sources its precursor dihydroxybenzoic acid and its breakdown product dihydroxybenzoylserine, although the affinities of these compounds for Fe(III) are lower than that of enterobactin.

6.2.3.2 Desferrioxamines

These linear trihydroxamate siderophores (e.g. ferrioxamine B, Figure 6.1) consist of three peptide-linked modified ornithine residues and have an affinity for iron of 10^{31} M^{-1}. They are produced by *Streptomyces* and *Nocardia*. Desferrioxamine is used therapeutically for the binding of excess blood iron in the treatment of thalassemia.

6.2.3.3 Ferrichromes

These are cyclic trihydroxamate siderophores that consist of three modified ornithine residues and three glycines, and are produced by many fungal species. Ferrichromes show variation in either the sequence of amino acids in the peptide ring or in the hydroxamic acid residues although the tripeptide Orn–Orn–Orn is invariant [23]. Ferrichrome is taken up by *P. aeruginosa* through the FiuA receptor and there is speculation that this receptor may also be able to recognize desferrioxamine [54].

6.2.3.4 Aerobactin

This hydroxamate siderophore (Figure 6.1) is commonly produced by *Klebsiella pneumoniae, Aerobacter aerogenes* and other bacteria. It is generally considered an exogenous siderophore of *Pseudomonas,* although it is produced by a *Pseudomonad* of marine origin [55]. Aerobactin has an affinity for Fe(III) of 10^{23} M^{-1}, but this is with fully deprotonated ligands and at nonphysiological pH [56]. It is transported by the receptor ChtA in *P. aeruginosa*. This receptor can also transport the chemically similar siderophores rhiozobactin 1021 and schizokinen [57].

6.2.3.5 Citrate

Citrate is a naturally occurring metabolite that is present in the environment as well as in human cells and serum, and could provide a means for bacteria to obtain iron as ferric citrate during infection. It has a relatively low affinity for iron (10^{16} M^{-1}) and is effective only at relatively high concentrations [58].

6.3 Siderophore Synthesis

6.3.1 Nonribosomal Peptide Synthesis

Although many siderophores contain amino acid residues, the nonproteinogenic nature of many of these indicated that they are not ribosomally synthesized. This hypothesis has been validated by a large amount of experimentation carried out in particular in the Walsh laboratory (reviewed in Ref. [59]). Nonribosomal peptide synthesis occurs via a multiple-carrier thiotemplate mechanism on large enzymes called nonribosomal peptide synthetases (NRPSs) (reviewed in Ref. [60]). NRPSs are organized into sets of iterative catalytic units called modules, with each module responsible for the incorporation of a specific amino acid or other building block into the product. Different modules can be part of the same or different polypeptides. The organization and order of modules encoded in the genome usually correspond to the amino acid sequence of the peptide product. This general correspondence is indicated as *colinearity* rule.

The minimal module required for a single monomer addition consists of: (i) an adenylation domain (A), (ii) a thiolation domain (T) to which a phosphopantetheine cofactor is covalently attached and (iii) a condensation domain (C) (Figure 6.2A). The A domain is involved in selection of the substrate and its subsequent activation as an aminoacyl adenylated intermediate, which is then covalently attached to the enzyme via a thioester bond with the phosphopantetheine arm of the T domain. The C domain catalyzes peptide bond formation between the carboxyl group of the nascent peptide and the amino acid carried by the flanking module, allowing extension of the growing chain (Figure 6.2A). The module involved in the incorporation of the last amino acid usually includes a thioesterase domain that is responsible for the hydrolysis of the thioester bond between the assembled peptide and the phosphopantetheine cofactor, releasing the final product.

Additional enzymatic activities including epimerization to generate D-amino acids, methylation of substrates and peptide cyclization contribute to the chemical diversity of residues present in nonribosomally synthesized peptides (including siderophores). The enzymatic activities that catalyze these can be integral parts of a NRPS module or auxiliary enzymes that act in *trans*. These activities are often involved in providing nascent peptides with functional groups and/or structural constrains essential for their biological functions.

6.3.2
Pyoverdine Biosynthetic Pathways

Pyoverdine synthesis has been studied most intensively in the *P. aeruginosa* strain PAO1. The structure of this pyoverdine (pyoverdine Type I) is shown in Figure 6.1. A number of pyoverdine synthesis genes (*pvd* genes) were identified by traditional genetic methods; the availability of the *P. aeruginosa* PAO1 genome sequence made it possible to identify more (and probably all) of the *pvd* genes in this strain [61,62]. Comparative genome analysis led to the identification of multiple *pvd* genes in other *P. aeruginosa* strains [63] and in other *Pseudomonas* species [64–68].

The likely roles of different enzymes in pyoverdine synthesis are shown in Figure 6.2(B). The four largest *pvd* genes in *P. aeruginosa* strain PAO1, *pvdL, pvdI, pvdJ* and *pvdD*, encode NRPS enzymes [64,69,70]. The predicted substrates for PvdI, PvdJ and PvdD correspond to residues present in the pyoverdine peptide chain, while PvdL is predicted to condensate the amino acid precursors for the chromophore moiety (Figure 6.2B). The order of amino acids in pyoverdine implies the transfer of intermediates from PvdL to PvdI and PvdJ to PvdD. PvdL incorporates L-Glu, D-Tyr and L-2,4-diaminobutyrate (L-Dab) [69]. L-Dab and D-Tyr are condensed to a tetrahydropyrimidine ring, which is then modified to form the dihydroxyquinoline

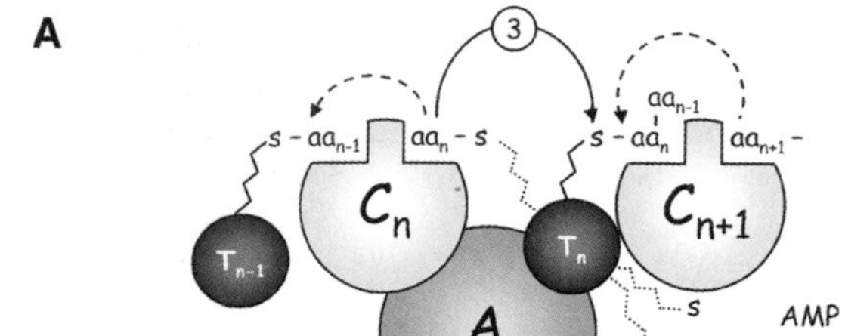
A
Cn
Cn+1
An
AMP
ATP
PP

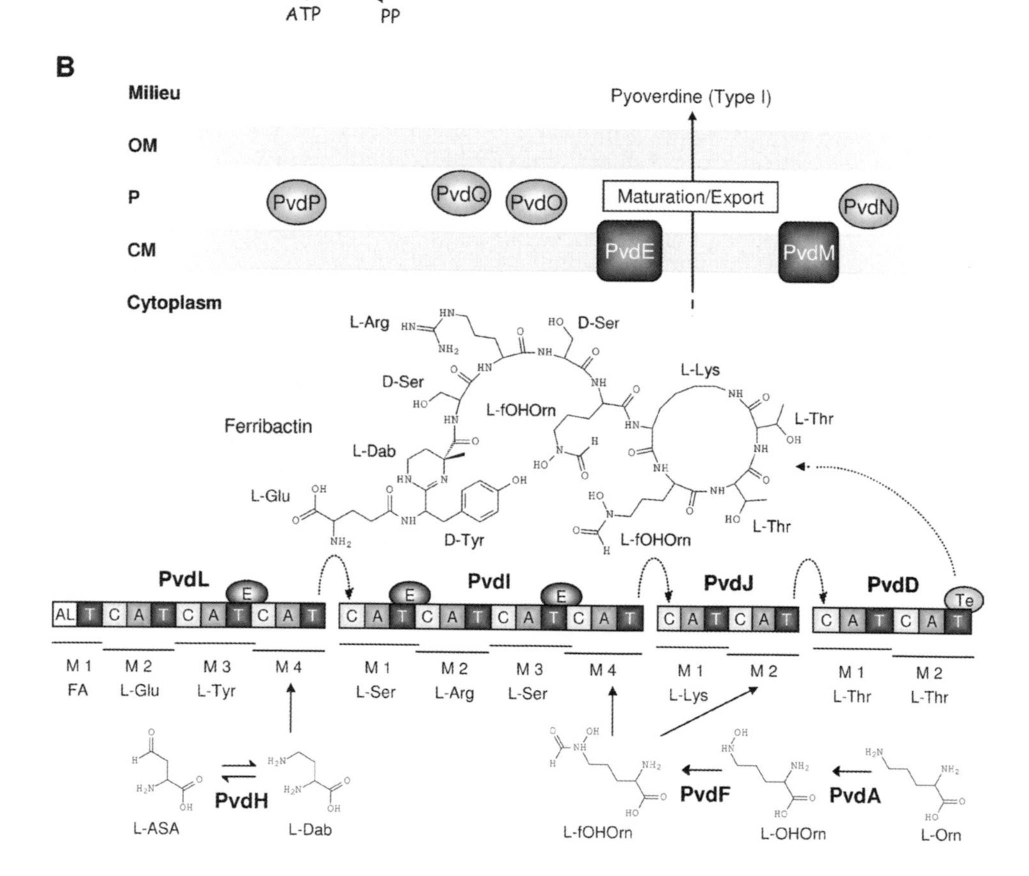
B
Milieu
OM
P
CM
Cytoplasm
Pyoverdine (Type I)
Maturation/Export
PvdP
PvdQ
PvdO
PvdN
PvdE
PvdM
Ferribactin
L-Arg
D-Ser
D-Ser
L-fOHOrn
L-Lys
L-Thr
L-Dab
L-Glu
D-Tyr
L-fOHOrn
L-Thr
PvdL
PvdI
PvdJ
PvdD
M 1 FA
M 2 L-Glu
M 3 L-Tyr
M 4
M 1 L-Ser
M 2 L-Arg
M 3 L-Ser
M 4
M 1 L-Lys
M 2
M 1 L-Thr
M 2 L-Thr
PvdH
L-ASA
L-Dab
PvdF
PvdA
L-fOHOrn
L-OHOrn
L-Orn

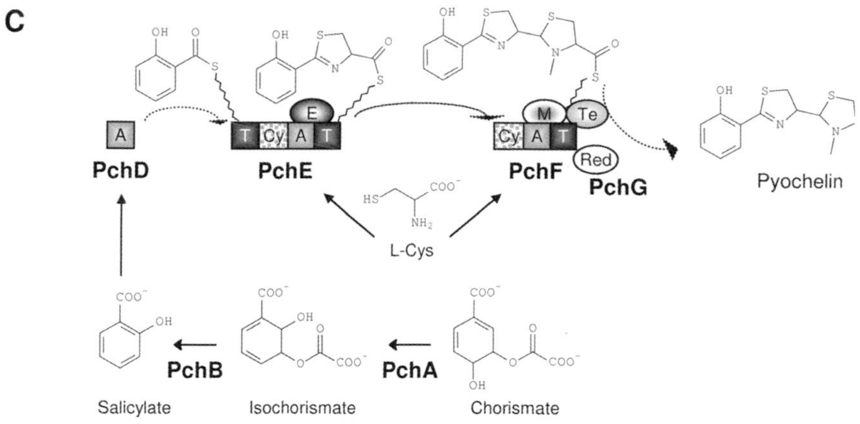

Figure 6.2 Biosynthesis of siderophores in *P. aeruginosa*. (A) Biochemistry of microbial NRPSs. Amino acid incorporation by a minimal NRPS module through substrate activation by consumption of ATP (reaction 1), covalent binding of activated amino acid to the phosphopantetheine arm (reaction 2) and peptide bond formation between amino acids carried by flanking modules (reaction 3). Domain abbreviations: A, adenylation; C, condensation; T, thiolation. Modified from Ref. [14]. (B) Type I pyoverdine biosynthetic pathway in *P. aeruginosa* PAO1. Reactions catalyzed by precursor-generating enzymes (PvdH, PvdA and PvdF) are shown in the lower part (ASA, aspartate β-semialdehyde; fOHOrn, N^5-formyl-N^5-hydroxyornithine; OHOrn, N^5-hydroxyornithine). The NRPS enzymes (PvdL, PvdI, PvdJ and PvdD) are dissected into modular domains: AL, acyl-CoA ligase domain; A, adenylation domain; T, thiolation domain; C, condensation domain. Circles indicate auxiliary domains: E, epimerization domain; Te, thioesterase domain. Substrates recognized by each module (M) are indicated (FA, fatty acid; Dab, 2,4-diaminobutyrate). The combined activity of pyoverdine NRPSs results in the generation of the pyoverdine precursor ferribactin. The presumptive cytoplasmic membrane (CM) and periplasmic (P) localization of uncharacterized proteins that are required for pyoverdine maturation and/or export are shown (OM, outer membrane). Modified from Ref. [14]. (C) Pyochelin assembly line in *P. aeruginosa*. Domain abbreviations are the same as in (B) with the addition of Cy and M for cyclization and methylation domains, respectively. PchG is a thiazolinyl reductase (Red) which acts *in trans* on the precursor carried by the T domain of PchF. Reactions catalyzed by salicylate-generating enzymes (PchA and PchB) are shown in the lower part.

chromophore. The L-Glu residue can be transformed by a still uncharacterized reaction to α-ketoglutaric acid, succinamide, malamide or hydroxylated to free acid, generating pyoverdine isoforms with different acyl side-chains [27].

Biochemical analyses have identified PvdH as the enzyme that generates L-Dab for incorporation into the pyoverdine precursor [71], and PvdA and PvdF as the enzymes that catalyze synthesis of formylhydroxyornithine [72,73] (Figure 6.2B). Two epimerization domains in PvdI result in the presence of D-serine in Type I pyoverdine.

PvdL is atypical among pyoverdine NRPSs because its initial module (PvdL-M1) contains an unusual domain highly similar to acyl-CoA ligases (ALs) (Figure 6.2B). By analogy with the function of similar domains in polyketide and nonribosomal peptide synthesis, the presence of an AL domain suggests that pyoverdine synthesis begins with an acylation event which might deliver a still undetermined fatty acid to the L-Glu carried by the second module of PvdL. However, a fatty acid moiety has never been detected in any of the resolved pyoverdine structures [27]. The identification of the periplasmic acylase PvdQ among essential genes for pyoverdine production [61,62] is suggestive of a mechanism for acyl group removal before release of pyoverdine into the milieu. Intriguingly, PvdQ also hydrolyzes the amide bond linking the acyl group to the lactone ring of several long-chain *N*-acyl-homoserine lactones, including *N*-(3-oxododecanoyl)-L-homoserine lactone which acts as the major quorum sensing (QS) signal molecule in *P. aeruginosa* [74,75]. Although PvdQ most likely uses a pyoverdine intermediate as physiological substrate, its periplasmic location, substrate specificity and regulation by iron have led to the proposal that it may also behave as a QS quencher under iron-limiting conditions [75].

Biochemical events leading to maturation of the pyoverdine chromophore have not been fully characterized. Pyoverdine-related compounds including ferribactin (a nonfluorescent molecule differing from pyoverdine only by the presence of the tripeptide L-Glu–D-Tyr–L-Dab instead of the mature chromophore), dihydropyoverdine (an unsaturated non-fluorescent form of pyoverdine) can coexist with pyoverdine in culture supernatants of fluorescent *Pseudomonas* and may represent precursors [76]. This suggests that synthesis of the peptide backbone may precede maturation of the chromophore. *In vitro*, the D-Tyr–L-Dab moiety of ferribactin can be converted by a polyphenoloxidase enzyme to a mixture of mature pyoverdine chromophore and its dihydro form [77]. The same reaction was obtained with cell-free extracts of iron-deficient *P. aeruginosa*, but not with extracts from cells unable to express *pvd* genes, indicating that this reaction is associated with pyoverdine synthesis. It has also been proposed that maturation of the chromophore takes place in the periplasm and involves redox reactions via one or more hemoproteins [78,79]. Some proteins essential for pyoverdine synthesis are predicted to be periplasmic [80] and this has been confirmed for the putative class V aminotransferase PvdN [81]. A further set of four genes, the *pvc* genes, has also been implicated in chromophore formation [82], but their role remains unclear as *pvc* mutants retain the ability to make pyoverdine under certain conditions and homologs of these genes are absent from the genome of other fluorescent *Pseudomonas* spp. [69,78].

Recent genomic studies provided a biochemical basis for pyoverdine diversity in different *Pseudomonas* species and strains. In a whole-genome diversity study [83] the *pvd* locus was found as the most divergent alignable locus in the genome of multiple *P. aeruginosa* strains, reflecting the strain-specific diversity of pyoverdine structure. Among genes encoding pyoverdine NRPSs, *pvdL* is well conserved in fluorescent *Pseudomonas* spp., consistent with its role in biosynthesis of the invariant chromophore [63,64]. Genes encoding NRPSs responsible for specificity of the peptide chain vary greatly between species and also between *P. aeruginosa* strains, providing a strong basis for the wide variety of pyoverdine structures. Other genes can also contribute to further increase the biochemical diversity between pyoverdines. One of these, encoding an OHOrn acetylase ($PvdY_{II}$), is found only in strains of *P. aeruginosa* producing type II pyoverdine and is thought to contribute to formation of the cyclic hydroxyornithine at the C-terminus of the peptide moiety [84].

6.3.3
Synthesis of other *Pseudomonas* Siderophores

The second siderophore produced by almost all *P. aeruginosa* strains and by strains of *P. fluorescens* is pyochelin (Figure 6.1). Pyochelin is made by condensation of one molecule of salicylate with two Cys residues through a thiotemplate nonribosomal mechanism on a four-protein, 12-domain assembly line (Figure 6.2C) (reviewed in Ref. [59]). The genes for pyochelin synthesis are arranged in two operons, *pchDCBA* and *pchEFGHI*. Pyochelin synthesis starts with the formation of salicylate from chorismate by the joint activity of the isochorismate synthase PchA and the isochorismate pyruvate lyase PchB [85,86]. Activation of salicylate is catalyzed by the adenylating enzyme PchD [87], while the NRPS PchE is responsible for the adenylation of the first Cys residue. Substrate condensation, epimerization and cyclization reactions, catalyzed by specific PchE domains, generate the enzyme-bound intermediate hydroxyphenyl-thiazoline (HPT). Then, the NRPS PchF adenylates the second Cys and generates the second thiazoline ring of pyochelin by condensation of Cys residue with the HPT precursor. This thiazoline ring is subsequently methylated and reduced by the methylation domain of PchF and the reductase PchG, respectively [88]. In both PchE and PchF, C domains are replaced by cyclization variants, plausibly involved in heterocycling the two Cys residues while condensation occurs (Figure 6.2C). Finally, the mature siderophore is released by the C-terminal thioesterase domain of PchF. A second thioesterase (PchC) which acts *in trans* as a proofreading enzyme seems to be involved in optimizing pyochelin biosynthesis by removing wrongly charged molecules from the T domains of PchE and PchF [89]. The salicylate moiety is also present in pseudomonine (Figure 6.1) that is produced by different *P. fluorescens* strains [40,90], but the enzymatic pathway responsible for its biosynthesis remains unclear.

Some *P. putida* and *P. stutzeri* strains produce PDTC. A 25-kbp fragment containing 17 open reading frames (ORFs) has been implicated in PDTC synthesis in *P. stutzeri* [42], although eight of these ORFs seem to be sufficient to confer carbon tetrachloride transformation capability and by implication PDTC synthesis on other pseudomonads [91].

Quinolobactin (Figure 6.1) is secreted by *P. fluorescens* ATCC 17400. The quinolobactin gene cluster consists of two divergently orientated operons of eight and four ORFs, designated *qsbABCDEFGH* and *qsbIJKL*, respectively [92]. It has been proposed that quinolobactin synthesis starts with the formation of xanthurenic acid from tryptophan by the activity of the *qsbF*, *qsbH*, *qsbG* and *qsbB* gene products. Subsequent methylation, sulfurylation and reduction of xanthurenic acid would give 8-hydroxy-4-methoxy-2-quinoline thiocarboxylic acid (thioquinolobactin), a compound which rapidly hydrolyzes to quinolobactin in culture-supernatant [92]. If this proposal is correct, quinolobactin synthesis does not require a NRPS enzyme, distinguishing it from other siderophore synthesis pathways.

Although good progress has been made in understanding the biochemical pathways of siderophore synthesis, the mechanisms by which siderophores are then secreted from *Pseudomonas* cells are not understood. An early report implicated a efflux-like system in secretion of pyoverdine by *P. aeruginosa* [93], but there have been no further reports on this system or on mechanisms of secretion of other siderophores by *Pseudomonas* species.

6.4 Ferri-Siderophore Transport

6.4.1 Overview

Acquisition of iron by Gram-negative bacteria requires active transport across both the outer membrane and the cytoplasmic membrane. Siderophore-bound iron is transported across the outer membrane via high-affinity receptor proteins. Energy for transport is provided by the TonB complex (see below) and the receptors belong to a class of proteins known as TonB-dependent receptors. The receptors effectively act as energy-dependent gated channels. They have a siderophore-binding site on the extracellular side of the channel. Following binding, the ferri-siderophore is transported through the channel into the periplasm of the cell in an energy-dependent manner. From the periplasm, the ferri-siderophore (or released iron) is bound by a periplasmic binding protein and delivered to an ATP-dependent permease located in the cytoplasmic membrane. The iron is transported across the cytoplasmic membrane, via the permease, into the cytoplasm of the cell with energy provided by ATP hydrolysis.

6.4.2 Outer Membrane Ferri-Siderophore Receptors

Outer membrane ferri-siderophore receptors generally have a very high degree of specificity for the cognate siderophore and the ability of a strain to use a specific siderophore is usually dependent on the presence of an appropriate receptor. Gram-negative bacteria often have numerous outer membrane receptor proteins for transporting iron, probably reflecting the importance of iron as an essential element. For example, analysis of the *P. aeruginosa* PAO1 genome has identified 35 predicted TonB-dependent receptors, many of which have a role in iron uptake (Table 6.2). In addition to receptors for the endogenous siderophores ferri-pyoverdine and ferri-pyochelin, PAO1 has numerous receptors for transport of exogenous siderophores, as well as receptors for heme and other iron chelates. Similarly, the presence of multiple TonB-dependent receptors for transport of both endogenous and exogenous siderophores has been reported for several other *Pseudomonas* species (reviewed in Ref. [94]). For some siderophores there appears to be a redundancy of receptors. For example, in *P. aeruginosa*, apart from the main ferri-pyoverdine receptor FpvA, there is a second ferri-pyoverdine receptor, FpvB [95]. Two receptors for ferri-enterobactin have also been identified in *P. aeruginosa* [96]. Genome

Table 6.2 TonB-dependent iron transport proteins of known function in *P. aeruginosa* PAO1.

Locus	Receptor	Description
PA2398	FpvA	ferri-pyoverdine receptor
PA4168	FpvB	second ferri-pyoverdine receptor
PA4221	FptA	ferri-pyochelin receptor
PA3901	FecA	ferric dicitrate receptor
PA2688	PfeA	ferric enterobactin receptor
PA3408	HasR	heme uptake receptor
PA4710	PhuR	heme/hemoglobin uptake receptor
PA2466	FoxA	ferrioxamine B utilization
PA0470	FiuA	ferrichrome utilization
PA0931	PirA	second ferric enterobactin receptor

sequence analysis suggests that receptor redundancy may be a general phenomenon amongst pseudomonads [94].

The three-dimensional structures of two receptors from *P. aeruginosa* have been determined by X-ray crystallography: the ferri-pyoverdine receptor FpvA [97,98] and the ferri-pyochelin receptor FptA [99] (Figure 6.3). These receptors share a common fold with the only previously determined structures for TonB-dependent outer membrane receptor proteins for ferrichrome (FhuA), enterobactin (FepA) and ferric citrate (FecA) as well as the vitamin B12 transporter BtuB, all of which are from *E. coli*. All six proteins are composed of a transmembrane β-barrel domain made up of 22 antiparallel β-strands which form a channel in the outer membrane and a globular domain (the "plug" or "cork") which folds into the barrel from the periplasmic side and occludes the channel of the barrel. On the periplasmic side of the barrel the β-strands are connected by short turns, whereas on the extracellular side of the barrel the strands are connected by loops of varying length, some of which are quite long. A siderophore-binding pocket is formed near the extracellular side of the receptor with residues of both the plug domain and the β-barrel domain (including the extracellular loops) contributing to the pocket. The binding pocket differs between receptors (according to the chemical properties of the specific siderophore) and thus confers the high degree of receptor selectivity [97], enabling uptake of specific ferri-siderophores while maintaining the outer membrane barrier against nonselective entry of other compounds into the periplasm.

A further shared feature of the TonB-dependent receptors is the so-called TonB box. This is a region of five to seven residues, usually near the N-terminus of the receptor, which interacts directly with the TonB protein in the periplasm. The TonB box is essential for ferri-siderophore transport, although the sequence is not strictly conserved – some mutations in the region can be tolerated, indicating that the interaction between TonB and the TonB box is dependent on the conformation of the TonB box rather than the specific sequence. Structural studies of receptor:TonB complexes suggest that the TonB box may form an interprotein β-sheet with TonB that may mediate a mechanical pulling force [100,101]. Some receptors (e.g. FpvA of *P. aeruginosa* and PupB of *P. putida*) have an additional domain (signaling domain) at

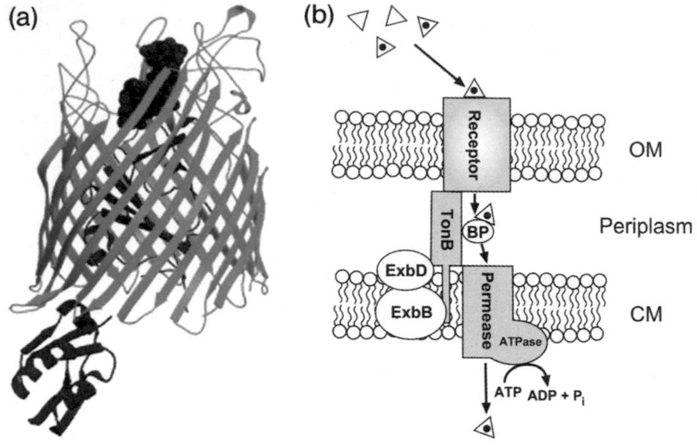

Figure 6.3 Mechanism of ferri-siderophore transport. (A) Structure of the ferri-pyoverdine receptor FpvA with bound ferri-pyoverdine. The ferri-pyoverdine receptor FpvA is composed of a transmembrane β-barrel domain made up of 22 antiparallel β-strands (green), a plug domain (red) and a signaling domain (blue). Bound ferri-pyoverdine is shown in ball-and-stick configuration. The ferri-pyochelin receptor FptA, the only other ferri-siderophore receptor from *Pseudomonas* for which the structure has been determined, shares a common overall structure except that FptA lacks a signaling domain. Reproduced from [98]. (B) General mechanism of iron uptake in pseudomonads. Iron (solid circle) is bound to a siderophore (triangle) in the extracellular environment. The ferri-siderophore binds to a specific receptor protein in the outer membrane (OM) and is transported through the receptor to the periplasm. Energy for transport through the outer membrane receptor is provided by the TonB complex (TonB–ExbB–ExbD). Once in the periplasm, the ferri-siderophore is bound by a periplasmic binding protein (BP), which delivers the ferri-siderophore to a permease located in the cytoplasmic membrane (CM). The ferri-siderophore is then transported through the permease into the cytoplasm of the cell using energy provided by ATP hydrolysis. There are variations to this general mechanism, depending on the system involved, as described in the text.

their N-terminal region that extends into the periplasm and is part of a signal transduction pathway that controls gene expression (Section 6.6).

6.4.3
TonB Complex

Ferri-siderophores are imported against a concentration gradient and so energy is required. Due to the porous nature of the outer membrane, there is no energy gradient across the outer membrane of Gram-negative bacteria. Thus, energy for transport of ferri-siderophores across the outer membrane into the periplasm must be transduced from the cytoplasmic membrane. This process is performed by the TonB protein, which transduces energy [the proton motive force (pmf)] from the cytoplasmic membrane to the receptor protein. In *E. coli*, the TonB protein is found in association with two cytoplasmic membrane-bound proteins, ExbB and ExbD, both of which are required for full function of TonB [102]. The TonB complex has been identified in many Gram-negative bacterial species and is believed to be a conserved mechanism for energy transduction to outer membrane receptor proteins. Exactly how the complex transduces energy to outer membrane receptors is not fully understood. There is strong evidence that an energized form of the TonB protein interacts directly with the receptor protein to transduce the energy generated at the cytoplasmic membrane. It is thought that TonB either undergoes a conformational change that allows it to extend across the periplasm to interact with the receptor or, alternatively, the energized TonB protein is released from the cytoplasmic membrane (and ExbBD) and shuttles across the periplasm [103].

The first *tonB* gene identified in *Pseudomonas* was *tonB* of *P. putida* WCS358 [104]. In *P. aeruginosa* there are three TonB homologs (TonB1, TonB2 and TonB3). TonB1 was identified by complementation of a *tonB* mutation in *P. putida* WCS358 [105]. *tonB1* mutants are defective for siderophore-mediated iron acquisition, are unable to grow in iron-limited medium, have reduced ability to grow *in vivo* and have reduced virulence in animal models [106]. TonB1 is somewhat unusual in that it has an N-terminal extension of approximately 80 amino acids relative to other TonB protein and this extension is required for TonB1 activity [107]. The *tonB2* gene was identified in *P. aeruginosa* by its association with *exbB* and *exbD* homologs [108]. However, none of these three genes are required by *P. aeruginosa* for growth in iron-restricted medium. On the other hand, *tonB1 tonB2* double mutants grow less well under iron limitation than *tonB1* mutants, indicating that TonB2 can at least partially complement TonB1 in its role in iron acquisition [108]. The *tonB3* gene is required for twitching motility and assembly of extracellular pili [109]. It is not known whether or not TonB3 has a role in iron acquisition. Multiple *tonB* systems have also been identified in the genomes of other *Pseudomonas* species.

6.4.4 Transport of Ferri-Siderophores into the Periplasm

Following binding of ferri-siderophores by the cell surface receptors, the ferri-siderophore is imported through the receptor in a TonB-dependent process. The mechanisms involved have been best studied in *E. coli* (reviewed in Ref. [110]). They are not fully understood, but must involve at least partial movement of the plug domain that otherwise blocks the channel. In pseudomonads, the mechanism of transport of ferri-siderophores into the periplasm has only been studied in significant depth for the ferri-pyoverdine receptor FpvA. For this protein, there is evidence that the iron-free siderophore (apo-pyoverdine) is bound in the normal state of the receptor [111]. Binding of apo-pyoverdine occurs at a binding pocket lined with aromatic residues [97]. Apo-pyoverdine is displaced by ferri-pyoverdine in a TonB-dependent manner [111]. An energized conformation of TonB interacts with ferri-pyoverdine-loaded receptor and transduces the energy to the receptor to allow ferri-pyoverdine transport. It has not yet been established whether the plug is displaced from the channel or if it remains in the channel, but undergoes sufficient conformational changes to allow passage of the ferri-siderophore. Once in the periplasm the iron is thought to be released from the siderophore, although the mechanism of this has yet to be resolved. The apo-pyoverdine is recycled to the extracellular environment where it can rebind to the receptor [112].

6.4.5
Iron Transport Across the Cytoplasmic Membrane

A model system for transport of iron across the cytoplasmic membrane is the FhuCDB system of *E. coli*. This system is composed of a periplasmic binding protein (FhuD) and a cytoplasmic membrane permease consisting of an integral membrane protein (FhuB) and an ATP hydrolase (FhuC) that provides the energy for transport [113]. Following transport of ferrichrome across the outer membrane via FhuA, the ferri-siderophore is bound by FhuD, a periplasmic binding protein. Following binding, FhuD delivers the ligand to the cytoplasmic membrane-associated permease FhuBC for transport across the cytoplasmic membrane. There is some evidence that TonB may act as a scaffold for FhuD, connecting the transport between the outer membrane and cytoplasmic membrane [114]. Ferrichrome is then transported across the cytoplasmic membrane by FhuBC in an ATP-dependent manner. The FhuBCD system is able to transport several different hydroxamate ferri-siderophores including ferrichrome, aerobactin, ferrioxamine B, coprogen, schizokinen and other such permeases are also able to transport a variety of substrates [113]. This is in marked contrast to the high specificity of most outer membrane ferri-siderophore receptors.

Analogous systems for transport of ferri-siderophores into the cytoplasm have been found in several other bacteria [113]. Additionally, many bacteria contain transporters that import ferric or ferrous ions from the periplasm into the cytoplasm, so that in principle iron ions can be removed from siderophores in the periplasm and the ions transported into the cytoplasm without cotransport of siderophore. In *Pseudomonas*, cytoplasmic membrane permeases that have been described for ferri-siderophore import are FptX, a permease involved in ferri-pyochelin transport in *P. aeruginosa* [115,116], and PdtE, a permease involved in PDTC uptake in *P. putida* and *P. stutzeri* [117]. The specificities of these permeases have not been investigated nor has a permease associated with pyoverdine-mediated iron acquisition been reported.

6.5
Regulation of Siderophore Synthesis and Transport

6.5.1
Fur Protein as Master Repressor

Iron acquisition by all living cells needs to be finely regulated to prevent the toxic effect due to overloading [118]. In fluorescent pseudomonads, the control of siderophore-dependent iron transport provides a paradigmatic example of a hierarchical and multifactorial mechanism of gene regulation.

The Fur repressor acts on top of this hierarchy. Fur is a protein that requires Fe(II) as a cofactor for DNA binding. When the level of intracellular iron reaches a certain threshold, Fur–Fe(II) complexes bind operator sites (the Fur box) within promoters of target genes thereby blocking their transcription [119]. The *fur* gene of

P. aeruginosa expresses a single 15.2-kDa polypeptide that dimerizes to form the active Fur protein.

The crystal structure of *P. aeruginosa* Fur (PA-Fur) dimer in complex with Zn(II) has been determined [120]. Each monomer is composed of two domains: an N-terminal domain implicated in DNA binding and a C-terminal domain responsible for homodimerization. Two metal coordination sites were identified in each monomer. These are a putative structural Zn(II)-binding site forming a bridge between the dimerization domain and the DNA-binding domain, and a regulatory binding site located in the dimerization domain that can exchange Zn(II) with Fe(II). The molecular basis for DNA recognition by Fur is not fully resolved. DNase I footprinting and modeling data of PA-Fur–DNA complexes indicate that at least two dimers are required to bind a single operator that is no less than 27 bp in length although, in some cases, more dimers are needed to cause full repression [119,120]. This model could in part explain the high amount of PA-Fur, estimated at 5000–10 000 molecules per cell [121]. The tendency of Fur to multimerize on target promoters could also account for the need to have such a high physiological amount of the repressor. A 3- to 4-fold reduction in the amount of PA-Fur causes both derepression of iron repressible genes and increased susceptibility to oxidative damage, suggesting a secondary role in iron scavenging for PA-Fur [121,122].

Global analysis of iron-dependent gene expression has detected 205 iron-regulated genes in *P. aeruginosa* (expression of 118 was induced by iron starvation and 87 were repressed) [61]. This number represents a significant proportion (4%) of *P. aeruginosa* genes, indicating that iron acts as a global regulator of gene expression in this bacterium. A significant number of genes were induced under iron-replete conditions, including genes involved in iron storage, defense from oxidative stress and intermediary metabolism. In particular, the bacterioferritin gene *bfrB* shows an iron-inducible expression that is clearly Fur dependent, suggesting that Fur in pseudomonads, as for a number of other organisms, also acts in positive regulation of gene expression [123].

Fur-dependent positive regulation of gene expression involves two Fur-repressed genes: *prrF1* and *prrF2*. These genes are arranged in tandem and in iron-starved cells give rise to two small RNAs that negatively control the expression of multiple genes at the posttranscriptional level. Under iron-replete conditions Fur represses expression of PrrF1 and PrrF2, relieving the inhibition of expression of their target genes. These small RNAs apparently do not influence expression of *bfrB*, suggesting the existence of other regulators besides Fur and PrrF RNAs [123].

Transcriptome analysis revealed that the majority of the genes induced under iron starvation are directly implicated in iron acquisition and some contribute to *P. aeruginosa* virulence [61]. A few genes are involved in cellular functions other than iron metabolism encoding, for example, alternative isoforms of iron-containing enzymes, and thus provide essential cellular functions in the absence of their iron-cofactored counterparts. Only 27 of the 78 genes displaying the highest fold-change during iron starvation contain a Fur box element in their promoter region, suggesting that, at a lower hierarchical level, additional regulators govern the indirect iron control of a large number of genes. Moreover, transcripts of several genes encoding

TonB-dependent receptors for heterologous siderophores contain Fur boxes in their promoters, but were not detected in the transcriptome analysis. This reflects the fact that expression of many iron-uptake systems in pseudomonads is specifically induced by the cognate iron source, through a variety of regulatory mechanisms. One widespread regulatory mechanism involves specific σ factors belonging to the family of extracytoplasmic function (ECF) σ factors [11].

6.5.2
ECF σ Factors

The ECF family of σ factors constitutes a group of small (typically 20–30 kDa) environmentally responsive transcription factors of the σ^{70} family (reviewed in Ref. [124]). They are found in a wide range of bacterial species and control a very wide range of functions, such as response to heat, osmotic and oxidative stresses, virulence, motility, transport of metal ions, and synthesis of alginate or carotenoids [125]. All known ECF σ factors coordinate transcriptional responses to extracellular signals. Their expression is controlled by other transcriptional regulators. Their activity can be posttranslationally modulated by cognate anti-σ factor proteins, which are typically membrane localized [124–126].

Genome sequence of several *Pseudomonas* species has revealed the presence of a large number of ECF σ factors, ranging from the 10 in *P. syringae*, through the 19 in *P. aeruginosa* and *P. putida*, to a predicted 28 in *P. fluorescens* [11,127,128]. Most of these ECF σ factors are required for the expression of siderophore synthesis and uptake genes, and for this reason they have been classified as iron starvation (IS) σs. Most IS σ genes are cotranscribed with the cognate anti-σ genes and in iron-replete cells, expression is repressed by Fur. The cognate anti-σ factors modulate the activities of these σs at the posttranslational level, through a signal transduction cascade activated by the appropriate siderophore or iron chelate [11,129].

The most extensively characterized IS σ in *Pseudomonas* is the PvdS protein of *P. aeruginosa*; close homologs (PfrI, PbrA and PsbS) have similar functions in other fluorescent pseudomonads. PvdS is required for pyoverdine synthesis. It is a 187-aminoacid protein that forms stable complexes with the core RNA polymerase (RNAP) and requires the presence of core RNAP to bind the promoters of pyoverdine biosynthetic genes, including *pvdA*, *pvdE/F* and *pvdD* [130,131]. PvdS was shown to cause core RNAP enzyme to initiate transcription from *pvdD* and *pvdE/F* promoters *in vitro* [131], consistent with its function as a σ factor. Despite the small size of ECF σ factors, mutational studies revealed similar mechanisms of promoter recognition between PvdS and RpoD, involving regions 2.4 and 4.2, which in RpoD recognize the −10 and −35 hexamers [132]. A typical signature of all PvdS-dependent promoters is represented by the conserved sequence TAAAT, known as the IS box, located approximately at position −33 relative to the transcription start point. This sequence is also present in the promoters of genes regulated by the PvdS homologs PfrI, PbrA and PsbS [133]. The PvdS-dependent promoters also have, at a distance of 16 or 17 nucleotides downstream of the IS box, the consensus sequence CGT that is required for promoter activity.

The *pvdS* gene is transcribed from a typical RpoD-dependent promoter that is overlapped by a single Fur box. Consequently, transcription of *pvdS* (and, hence, all PvdS-dependent genes) is repressed under iron-sufficient conditions, while repression is relieved in a *fur*-defective background [134,135].

Microarray-based transcriptional profiling identified 26 transcriptional units whose expression is reduced in a *pvdS* mutant [61]. These 26 units included the known PvdS-regulated genes and previously unrecognized pyoverdine biosynthesis genes. Other genes were also observed to be under PvdS control, including genes predicted to encode a protein with high identity to a hemolysin protein family and extracellular proteins of unknown function, and the *mvfR* gene that encoded the transcriptional regulator MvfR, which is involved in the synthesis of multiple QS-regulated virulence factors [61,136].

Several lines of evidence suggest that other regulatory factors in addition to PvdS or its homologs could be implicated in the positive control of *pvd* genes [130]. In *P. putida* WCS358 expression of pseudobactin genes requires the positive regulator PfrA [137]. PfrA is 58% identical to and interchangeable with the product of the *P. aeruginosa* PAO1 *algQ* gene that encodes a positive transcriptional regulator involved in the biosynthesis of alginate, polyphosphate and several extracellular virulence factors. AlgQ is required for optimal expression of *pvd* genes, acting as an anti-σ factor for RpoD and so enabling core RNAP recruitment by PvdS and transcription initiation at *pvd* promoters [138].

The other IS σs (in *P. aeruginosa* and in other species) are very likely to direct expression of other genes involved in iron acquisition. This has been demonstrated for FpvI that is required for expression of the gene encoding the FpvA receptor in *P. aeruginosa* [139,140], for PupI that is required for expression of the gene encoding the PupB receptor in *P. putida* [141], and for FoxI and FiuI that are required for synthesis of ferrioxamine and ferrichrome receptors in *P. aeruginosa* [54].

As well as being required for pyoverdine synthesis PvdS is involved in optimal expression of *P. aeruginosa* virulence determinants, such as the PrpL and AprA proteases and the cytotoxic exotoxin A [142–144]. PvdS is expressed in lung infections associated with cystic fibrosis (CF) and is required for optimal virulence of *P. aeruginosa* in a rabbit model of experimental endocarditis [145,146]. These data clearly indicate that the scope of PvdS regulation is broader than just regulation of the biosynthesis of pyoverdine.

6.5.3
QS and Expression of Iron Transport Genes

Various lines of evidence suggest that QS is involved in controlling the response of pseudomonads to iron limitation. A mutant in the N-(3-oxododecanoyl-L-homoserine) lactone synthase LasI showed reduced pyoverdine production and pyoverdine synthetase genes *pvdI* and *pvdL* were induced by the two *N*-acyl-homoserine lactones produced by *P. aeruginosa* [147,148]. Initial microarray analyses [149–151] failed to detect a link between QS and iron acquisition. This may have been because of the use of planktonic cultures. Transcriptome analysis in a mutant lacking the QS-regulator mutant VqsR indicated cross-talk between the QS network and the iron regulon [152]. This mutant showed a strong downregulation in the expression of pyoverdine and pyochelin genes. An overlap of QS-regulated functions and the iron regulon in *P. aeruginosa* was also detected using proteomics [153]. The QS regulon of sessile *P. aeruginosa* cells was recently mapped by transcriptomics, and many QS and iron-coregulated genes were identified [154]. Hence, it is plausible that QS affects the response to iron limitation when cells grow in the biofilm mode [152].

Two different groups have independently shown that 2-heptyl-3-hydroxy-4(1*H*)-quinolone (PQS), a chelator acting as the third QS signal molecule of *P. aeruginosa*, regulates expression of genes involved in iron scavenging, including several *pvd* genes [155,156]. Moreover, a new function for PQS in trapping Fe(III) at the cell surface and facilitating siderophore-mediated iron transport has been proposed [156]. Intriguingly, expression of the PQS regulator MvfR is influenced by the PvdS regulon [61], implying that PvdS could enhance PQS production through MvfR during iron starvation conditions.

6.6
Introduction to Signaling

The ability to rapidly adapt to a changing environment is essential to the survival of most organisms. The production of siderophores and their receptors should be tightly regulated to make the best use of the resources available. Siderophores have been shown to function as autoinducers, regulating the production of their cognate receptors as well as their own biosynthesis. Here, we describe the different siderophore-mediated signaling pathways to illustrate the molecular mechanisms involved and highlight the prevalence of these type of systems in the genomes of pseudomonads.

6.6.1
Pyoverdine-Mediated Signaling

Pyoverdine acts as an inducer of the expression of the ferri-pyoverdine receptor, FpvA, and of *prd* genes. This involves a signal transduction pathway spanning the outer membrane and cytoplasmic membrane that controls the activity of two IS σ factors PvdS and FpvI (Figure 6.4A). In this pathway, the activities of PvdS and FpvI are inhibited by an anti-σ factor FpvR that extends from the cytoplasm (where it interacts with FpvI and PvdS) into the periplasm. Binding of pyoverdine to FpvA affects interactions between FpvA and FpvR in a way that relieves the inhibition of activity of PvdS and FpvI, resulting in gene expression. The role of PvdS and FpvI in the expression of *pvd* and *fpvA* genes is described above (Section 6.5), and here we will focus on the roles of FpvA and FpvR in signaling.

6.6.1.1 FpvA and Pyoverdine

FpvA was implicated in pyoverdine production by the observation that bacteria lacking FpvA produced much less pyoverdine than wild-type bacteria and have lower

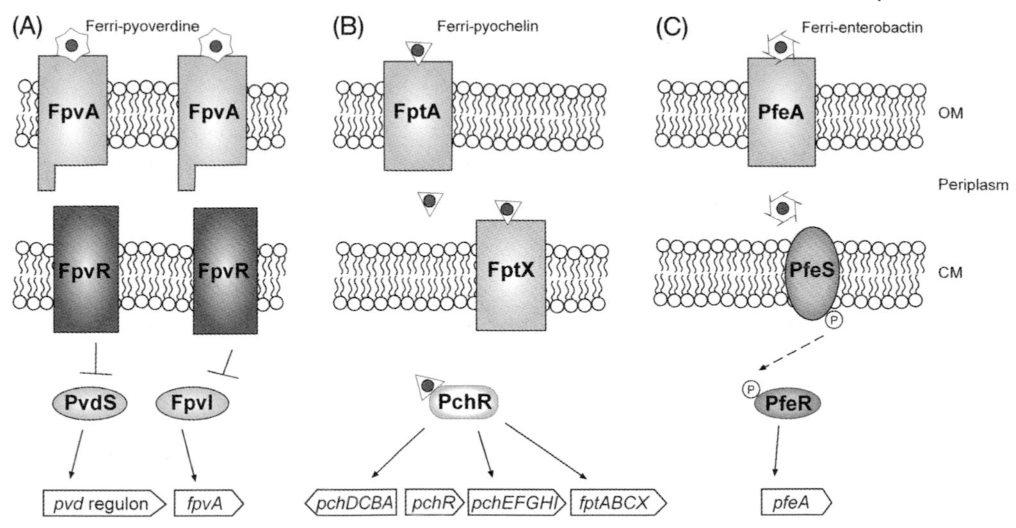

Figure 6.4 Mechanisms for regulation of siderophore synthesis and transport in *P. aeruginosa*. (A) Pyoverdine. The binding of (ferri)pyoverdine at the outer membrane (OM) initiates signal transfer across the periplasm from the periplasmic signaling domain of FpvA to the anti-σ factor FpvR. This deactivates FpvR, and allows the σ factor proteins PvdS and FpvI to promote the expression of *pvd* genes and *fpvA*. (B) Ferri-pyochelin is recognized and transported across the outer membrane and cytoplasmic membrane (CM) by FptA and FptX, respectively. In the cytoplasm pyochelin causes the regulator PchR to bind to the promoter regions of the *pch* and *fpt* operons and promote their expression. (C) Transport of ferri-enterobactin across the outer membrane is mediated by PfeA. Detection of ferri-enterobactin in the periplasm by the sensor domain of PfeS causes the transmitter domain of this protein to autophosphorylate. Transfer of this phosphate group (P) to the receiver domain of PfeR results in the expression of *pfeA*. For further information, see text.

expression of a *prd* gene [157]. FpvA mutants also exhibit less expression from the *fpvA* promoter, suggesting a role for FpvA in *fpvA* expression [139,140]. Pyoverdine-deficient mutants also show reduced expression of both *fpvA* and *prd* genes, although in each case expression can be restored by addition of pyoverdine [139,158]. They accumulate less FpvA in their outer membrane than wild-type bacteria, but this deficit can be restored by the addition of exogenous pyoverdine [139,159]. These data indicate that FpvA could regulate its own expression and that of pyoverdine in response to binding of pyoverdine. FpvA can bind both ferri-pyoverdine and apo-pyoverdine (Section 6.4), and it is not known which form of pyoverdine is active in controlling gene expression. A key part of FpvA that is involved in gene regulation is a signaling domain that is located in the periplasm and is required for induction of gene expression, but not for uptake of ferri-pyoverdine [157,160], indicating that it interacts with FpvR to control its anti-σ activity.

6.6.1.2 Transfer of Signal: Role of FpvR

FpvR is a regulatory protein that spans the cytoplasmic membrane. In the absence of pyoverdine FpvR negatively regulates the activity of PvdS and FpvI, inhibiting expression of *pvd* and *fpvA* genes [139,140,158,161].

Overexpression of *fpvR* reduces PvdS-dependent transcription of *pvd* genes and mutating *fpvR* in a pyoverdine-deficient strain causes restoration of *pvd* expression under conditions where PvdS activity would otherwise be minimal [158]. These data are consistent with FpvR being an anti-σ factor that binds to PvdS to inhibit its activity. FpvR differs from most other anti-σ factors in that it controls the activity of two σ factors, acting on FpvI in the same manner as PvdS and thus inhibiting *fpvA* expression [139,140]. Direct interaction between the N-terminal (cytoplasmic) region of FpvR with PvdS and FpvI has been demonstrated *in vivo*, and site-specific mutations that compromise the FpvR–FpvI interaction also affect interactions between FpvR and PvdS. Residues of PvdS/FpvI involved in binding FpvR appear to be in regions important for core RNAP binding, suggesting a competitive mechanism of anti-σ factor activity [161].

Suppression of PvdS and FpvI activity is overcome by a combination of FpvA and (ferri)pyoverdine, implying an interaction between FpvA and FpvR in the periplasm that affects FpvR–σ factor interactions in the cytoplasm. As described above, FpvA belongs to the subset of TonB-dependent receptors that possess an extended domain at their N-terminus that is implicated in surface signaling (reviewed in Ref. [129]) and is likely to interact with FpvR. The structure of this signaling domain is comprised of two α-helices sandwiched between two β-sheets. This is structurally very similar to the structure of the signaling domain in the *E. coli* ferric citrate receptor FecA, suggesting that the mechanism of signal transduction is likely to be conserved between these receptors [98]. However, exactly how signal is transferred from FpvA to FpvR remains unclear.

Direct interactions between FpvA and FpvR have not been investigated. However, FecA has been shown to directly interact with the anti-σ factor (and FpvR homolog) FecR *in vivo* [162]. Mutations compromising the FecA/FecR interaction cluster around a conserved leucine motif and reduce ferric citrate-mediated induction. It is likely that there are corresponding interactions between FpvA and FpvR to control activity of FpvR in response to the absence/presence of pyoverdine.

The pyoverdine signaling pathway is relatively unusual in that it controls expression not only of FpvA, but also of siderophore synthesis genes and also virulence factors [158]. Signaling pathways of this sort more typically are known to only control ferri-siderophore receptor synthesis in response to the presence of a siderophore. This is the case for genes involved in pyoverdine (pseudobactin) uptake by *P. putida* [141], and for uptake of ferrichrome and ferrioxamine by *P. aeruginosa* [54].

6.6.2
Pyochelin: An Alternative Signaling Mechanism

The second siderophore of *P. aeruginosa*, pyochelin, also acts as a signaling molecule. Pyochelin, in its iron-loaded state, is an inducer of both pyochelin production/biosynthesis and expression of the ferri-pyochelin receptor, FptA [163–166]. Pyochelin-deficient mutants have much lower expression of *fptA* and of pyochelin synthesis genes than wild-type bacteria, and gene expression is restored by the addition of exogenous pyochelin, implicating pyochelin in FptA production.

As with the pyoverdine system described above, siderophore-induced signaling is dependent on the siderophore receptor, in this case FptA [116]. However, the mechanism of induction is quite different, being dependent on the AraC-type regulator PchR [163] (Figure 6.4B). Strains lacking PchR are deficient in the production of both pyochelin and FptA, and fail to express the *fptA* and pyochelin synthesis genes [163,164].

PchR binds to a 32-bp DNA sequence, termed the PchR box, that is located within the promoter regions of *pch* operons [165]. Pyochelin and iron are both required for this interaction, confirming that ferri-pyochelin itself acts as the molecular effector to drive transcription of *pch* genes [165]. The implication of this finding is that ferri-pyochelin is transported to the cytoplasm where it interacts with PchR. Consistent with this interpretation, FptX, an inner membrane permease required for pyochelin-mediated growth [115] is also required for pyochelin-induced transcription of *pch* and *fpt* genes [116].

PchR functions as a repressor as well as an activator of gene expression. Decreased *fptA* expression observed in a pyochelin-deficient strain can be restored by mutating *pchR*. This suggests that PchR is an activator of *pch* and *fptA* expression in the presence of pyochelin, and a repressor in the absence of pyochelin [164].

6.6.3
Enterobactin

The heterologous siderophore enterobactin is a catecholate siderophore produced by *E. coli* and other enterobacteria (Section 6.2). *P. aeruginosa* can utilize enterobactin for iron acquisition via the ferri-enterobactin receptor, PfeA [167]. The expression of *pfeA* is regulated negatively by iron and positively by enterobactin through the *pfeR* and *pfeS* gene products that comprise response regulator and sensor (histidine kinase) components of a typical two-component system [167,168] (Figure 6.4C). Two-component systems are a common regulatory mechanism for enabling bacteria to respond to environmental stimuli, with a large number of them being encoded by *Pseudomonas* genomes.

The importance of PfeR in expression of the PfeA receptor was shown using *pfeA::lacZ* transcriptional fusions. Mutants lacking functional PfeR not only show reduced transcription from the *pfeA* promoter, but can no longer exhibit enterobactin-mediated induction of *pfeA* expression [169]. PfeR has been shown to bind to the promoter region of *pfeA*, suggesting that PfeR, when activated by PfeS in response to enterobactin, functions to drive *pfeA* expression directly [169].

Surprisingly, PfeA is not required for enterobactin-mediated induction of *pfeA*, suggesting the presence of an additional mechanism for import of enterobactin [10,143]. The discovery of PirA, a second receptor for enterobactin and other synthetic chatecholate analogs (Section 6.4), validated this suggestion. PirA is predicted to be regulated by a second two-component system composed of the PirS and PirR genes [10,143]. Why *P. aeruginosa* has two enterobactin receptors is not clear. It is possible that while *pfeA* encodes a high-affinity receptor specific for enterobactin, *pirA* may have a primary ligand that is not yet known [96].

6.6.4
Multiplicity of Heterologous Siderophore Signaling Systems in Pseudomonads

Pseudomonads can utilize many heterologous siderophores for growth (Section 6.2). This is explained at least in part by the large number of TonB-dependent receptors that are predicted or confirmed as being encoded by their genomes [94], e.g. 35 in the case of *P. aeruginosa*. In *P. aeruginosa*, about one-third of these receptors have signaling domains of the sort described above for FpvA; nine are linked to homologs of FpvR and FpvI [11,15], suggesting the involvement of surface signaling systems of the pyoverdine type (Figure 6.4A) in their regulation. Thus far, however, only FpvA (pyoverdine), FiuA (ferrichrome) and FoxA (ferrioxamine) [54] have confirmed ligands that have been shown to activating signaling cascades. Mechanisms controlling expression of receptors that lack signaling domains may well use different regulatory mechanisms such as AraC-type regulators or two-component systems of the type shown in Figure 6.4. Expression of two receptors for heme uptake, HasR and PhuR, is directly under the control of the Fur protein [170]. However, iron limitation alone is not sufficient to cause detectable expression of many uncharacterized siderophore receptor genes [61,171], suggesting there are additional requirements for induction. Determining the ligands, inducing conditions and signaling pathways involved in the regulation of many of these uncharacterized receptors remains an exciting future challenge.

6.7
Concluding Remarks and Future Perspectives

The large number of iron acquisition systems in pseudomonads makes it very clear that obtaining iron is a major biological challenge for these bacteria. The wealth of studies from many research groups that are summarized above mean that many of the processes involved are quite well understood; indeed, fluorescent pseudomonads are amongst the best characterized bacteria in this regard, and serve as a valuable model for other species and genera. Nonetheless, many important questions have yet to be answered. At the pathway level, a complete pathway has been described for biosynthesis of pyochelin (Figure 6.2C), but for other siderophores understanding of the biosynthetic pathways is incomplete or nonexistent. The mechanisms involved in export of siderophores by pseudomonads are also not understood. Similarly, the role of cell surface receptor proteins in uptake of ferri-siderophores has been well documented and proteins enabling import into the periplasm have been identified in some cases, but it remains to be determined how iron is released from ferri-siderophores.

Regulation of siderophore synthesis and uptake is also well understood, at least in general terms, although the exact molecular mechanisms involved in the signal transduction processes (Figure 6.4A) have yet to be determined. The mechanisms controlling synthesis of many of the receptor proteins identified by genome sequence analysis are also unknown. It seems likely that at least some of these will have parallels with known regulatory mechanisms (Figure 6.4), but it may well be that in some cases other mechanisms are involved.

Broader questions, and an area that is less well understood, relate to biological aspects of siderophores. Iron and pyoverdines are primary environmental stimuli for surface motility of fluorescent pseudomonads, since it has been shown that iron deficiency influences both twitching and swarming behavior in these organisms [184]. The importance of iron acquisition in biofilm development [18] as well as the observation that siderophore–metal bonds are formed during the initial stages of attachment of *P. aeruginosa* to both biological and metal surfaces [172,173] indicate that iron uptake is a key factor in biofilm formation, but the reasons for this are not yet clear. It is possible that this relates to the interactions that are being uncovered between iron uptake and QS (Section 6.5), as QS is also a critical factor in biofilm development, but whether this is the case will require further study.

"Siderophore ecology" is also an area that is currently not well understood. Most pseudomonads live predominantly in the rhizosphere – an ecologically complex environment. The large number of ferri-siderophore receptors that are potentially encoded by *Pseudomonas* genomes testify to the large numbers of siderophores that are potentially available, as well as the need to acquire them. A better understanding of the range of ferri-siderophores that can be used by different strains, accompanied by understanding of the range and number of coexisting species (including unculturable microorganisms), will greatly enhance understanding of the soil ecology of pseudomonads. The actual contribution of siderophores, primarily pyoverdines and salicylate (the pyochelin precursor), in plant growth stimulation and/or suppression of deleterious microorganisms of the rhizosphere remains unclear. Further investigations are needed to understand under which natural conditions and to what extent *Pseudomonas* siderophores act as biofertilizers (in plant growth-promoting species) or, conversely, as virulence factors (in phytopathogenic species). For fluorescent pseudomonads, the chemical diversity of pyoverdines has been ascribed to the need to acquire iron, while reducing the likelihood of secreted siderophores being "stolen" by other species. In addition, however, ferri-siderophore receptors can be receptors for bacteriophage and bacteriocins. The diversity of pyoverdines may also reflect evolutionary pressure for diversification of receptors, to reduce susceptibility to phages and bacteriocins [63,174,175]. Fluorescent pseudomonads are able to make other siderophores such as pyochelin or quinolobactin in addition to a pyoverdine and the significance of this is also currently unknown. It may be that different siderophores are important in different environments, e.g. some strains of *P. aeruginosa* isolated from chronically infected patients with CF are unable to synthesize pyoverdine [176] and expression of pyochelin genes is upregulated in response to sputum from these patients [177] so it may be that pyochelin is a more important siderophore in this environment. Lastly, the possibility of applying siderophores from *Pseudomonas* in xenobiotic degradation deserves further investigation. Both pyoverdine and pyochelin can decompose organotin pesticides [16,178] and PDTC was shown to dechlorinate CCl_4 in the presence of reducing agents [46], and the potential of these reactions in bioremediation processes should be further exploited.

Acknowledgments

We apologize to authors whose research we could not refer to because of reasons of space. The work of F.I. and F.T. in P.V. lab was supported by a grant from the Italian Ministry of University and Research (PRIN 2006) and the Italian Cystic Fibrosis Foundation. Work in the Lamont laboratory is supported by grants from the Marsden Fund, administered by the Royal Society of New Zealand, and the Australian National Health and Medical Research Council.

References

1 Crichton, R.R. and Boelart, J.R. (2001) *Inorganic Biochemistry of Iron Metabolism: From Molecular Mechanisms to Clinical Consequences*, Wiley, New York.

2 Bullen, J.J., Rogers, H.J., Spalding, P.B. and Ward, C.G. (2006) Natural resistance, iron and infection: a challenge for clinical medicine. *J Med Microbiol*, **55**, 251–258.

3 Ratledge, C. and Dover, L.G. (2000) Iron metabolism in pathogenic bacteria. *Annu Rev Microbiol*, **54**, 881–941.

4 Smith, J.L. (2004) The physiological role of ferritin-like compounds in bacteria. *Crit Rev Microbiol*, **30**, 173–185.

5 Winkelmann, G. (2002) Microbial siderophore-mediated transport. *Biochem Soc Trans*, **30**, 691–696.

6 Wandersman, C. and Delepelaire, P. (2004) Bacterial iron sources: from siderophores to hemophores. *Annu Rev Microbiol*, **58**, 611–647.

7 Faraldo-Gomez, J.D. and Sansom, M.S. (2003) Acquisition of siderophores in gram-negative bacteria. *Nat Rev Mol Cell Biol*, **4**, 105–116.

8 Touati, D. (2000) Iron and oxidative stress in bacteria. *Arch Biochem Biophys*, **373**, 1–6.

9 Braun, V. and Killmann, H. (1999) Bacterial solutions to the iron-supply problem. *Trends Biochem Sci*, **24**, 104–109.

10 Poole, K. and McKay, G.A. (2003) Iron acquisition and its control in *Pseudomonas aeruginosa*: many roads lead to Rome. *Front Biosci*, **8**, d661–d686.

11 Visca, P., Leoni, L., Wilson, M.J. and Lamont, I.L. (2002) Iron transport and regulation, cell signaling and genomics: lessons from *Escherichia coli* and *Pseudomonas*. *Mol Microbiol*, **45**, 1177–1190.

12 Palleroni, N.J. (2003) Prokaryote taxonomy of the 20th century and the impact of studies on the genus *Pseudomonas*: a personal view. *Microbiology*, **149**, 1–7.

13 Haas, D. and Defago, G. (2005) Biological control of soil-borne pathogens by fluorescent pseudomonads. *Nat Rev Microbiol*, **3**, 307–319.

14 Visca, P., Imperi, F. and Lamont, I.L. (2007) Pyoverdine siderophores: from biogenesis to biosignificance. *Trends Microbiol*, **15**, 22–30.

15 Visca, P. (2004) Iron regulation and siderophore signaling in virulence by *Pseudomonas aeruginosa*, in *Pseudomonas*, (eds J.L. Ramos) Kluwer/Plenum, New York, pp. 69–123.

16 Inoue, H. *et al.* (2003) Tin–carbon cleavage of organotin compounds by pyoverdine from *Pseudomonas chlororaphis*. *Appl Environ Microbiol*, **69**, 878–883.

17 Singh, P.K., Parsek, M.R., Greenberg, E.P. and Welsh, M.J. (2002) A component of innate immunity prevents bacterial biofilm development. *Nature*, **417**, 552–555.

18 Banin, E., Vasil, M.L. and Greenberg, E.P. (2005) Iron and *Pseudomonas aeruginosa* biofilm formation. *Proc Natl Acad Sci USA*, **102**, 11076–11081.

19 Matzanke, B.F. (2005) Iron transport: siderophores, in *Encyclopedia of Inorganic Chemistry*, 2nd edn (ed. R.B. King), Wiley, New York, vol. IV pp. 2619–2646.

20 Drechsel, H. and Winkelmann, G. (1997) Iron chelation and siderophores, in *Transition Metals in Microbial Metabolism*, (eds G. Winkelmann and C.J. Carrano) Harwood Academic, Amsterdam, pp. 1–49.

21 Winkelmann, G. (2007) Ecology of siderophores with special reference to the fungi. *Biometals*, **20**, 379–392.

22 Guerinot, M.L. (1994) Microbial iron transport. *Annu Rev Microbiol*, **48**, 743–772.

23 Haas, H. (2003) Molecular genetics of fungal siderophore biosynthesis and uptake: the role of siderophores in iron uptake and storage. *Appl Microbiol Biotechnol*, **62**, 316–330.

24 Meyer, J.M. (2000) Pyoverdines: pigments, siderophores and potential taxonomic markers of fluorescent *Pseudomonas* species. *Arch Microbiol*, **174**, 135–142.

25 Meyer, J.-M. and Hornspreger, J. (1978) Role of pyoverdine$_{Pf}$, the iron binding fluorescent pigment of *Pseudomonas fluorescens* iron transport. *J Gen Microbiol*, **107**, 329–331.

26 Teintze, M. *et al.* (1981) Structure of ferric pseudobactin, a siderophore from a plant growth promoting *Pseudomonas*. *Biochemistry*, **20**, 6446–6457.

27 Budzikiewicz, H. (2004) Siderophores of the Pseudomonadaceae *sensu stricto* (fluorescent and non-fluorescent *Pseudomonas* spp.). *Fortschr Chem Org Naturst*, **87**, 81–237.

28 Meyer, J.M. *et al.* (1996) Pyoverdin is essential for virulence of *Pseudomonas aeruginosa*. *Infect Immun*, **64**, 518–523.

29 Takase, H., Nitanai, H., Hoshino, K. and Otani, T. (2000) Impact of siderophore production on *Pseudomonas aeruginosa* infections in immunocompromised mice. *Infect Immun*, **68**, 1834–1839.

30 Meyer, J.-M. and Abdallah, M.A. (1978) The fluorescent pigment of *Pseudomonas fluorescens*: biosynthesis, purification and physicochemical properties. *J Gen Microbiol*, **107**, 319–328.

31 Yoder, M.F. and Kisaalita, W.S. (2006) Fluorescence of pyoverdin in response to iron and other common well water metals. *J Environ Sci Health A*, **41**, 369–380.

32 Cox, C.D. and Graham, R. (1979) Isolation of an iron-binding compound from *Pseudomonas aeruginosa*. *J Bacteriol*, **137**, 357–364.

33 Hayen, H. and Volmer, D.A. (2006) Different iron-chelating properties of pyochelin diastereoisomers revealed by LC/MS. *Anal Bioanal Chem*, **385**, 606–611.

34 Ankenbauer, R.G. *et al.* (1988) Synthesis and biological activity of pyochelin, a siderophore of *Pseudomonas aeruginosa*. *J Bacteriol*, **170**, 5344–5351.

35 Schlegel, K., Taraz, K. and Budzikiewicz, H. (2004) The stereoisomers of pyochelin, a siderophore of *Pseudomonas aeruginosa*. *Biometals*, **17**, 409–414.

36 Tseng, C.F. *et al.* (2006) Bacterial siderophores: the solution stoichiometry and coordination of the Fe(III) complexes of pyochelin and related compounds. *J Biol Inorg Chem*, **11**, 419–432.

37 Britigan, B.E., Rasmussen, G.T. and Cox, C.D. (1997) Augmentation of oxidant injury to human pulmonary epithelial cells by the *Pseudomonas aeruginosa* siderophore pyochelin. *Infect Immun*, **65**, 1071–1076.

38 Coffman, T.J., Cox, C.D., Edeker, B.L. and Britigan, B.E. (1990) Possible role of bacterial siderophores in inflamation. *J Clin Invest*, **86**, 1030–1037.

39 Mossialos, D. *et al.* (2000) Quinolobactin, a new siderophore of *Pseudomonas fluorescens* ATCC 17400, the production of which is repressed by the cognate pyoverdine. *Appl Environ Microbiol*, **66**, 487–492.

40 Anthoni, U. *et al.* (1995) Pseudomonine, an isoxazolidone with siderophoric activity from *Pseudomonas* fluorescens

AH2 isolated from Lake Victorian Nile perch. *J Nat Prod*, **58**, 1786–1789.

41 Matthijs, S. *et al.* (2007) Thioquinolobactin, a *Pseudomonas* siderophore with antifungal and anti-*Pythium* activity. *Environ Microbiol*, **9**, 425–434.

42 Lewis, T.A. *et al.* (2000) A *Pseudomonas stutzeri* gene cluster encoding the biosynthesis of the CCl_4-dechlorination agent pyridine-2,6-bis(thiocarboxylic acid). *Environ Microbiol*, **2**, 407–416.

43 Lewis, T.A. *et al.* (2004) Physiological and molecular genetic evaluation of the dechlorination agent, pyridine-2,6-bis (monothiocarboxylic acid) (PDTC) as a secondary siderophore of *Pseudomonas*. *Environ Microbiol*, **6**, 159–169.

44 Leach, L.H., Morris, J.C. and Lewis, T.A. (2007) The role of the siderophore pyridine-2,6-bis (thiocarboxylic acid) (PDTC) in zinc utilization by *Pseudomonas putida* DSM 3601. *Biometals*, **20**, 717–726.

45 Cortese, M.S. *et al.* (2002) Metal chelating properties of pyridine-2,6-bis (thiocarboxylic acid) produced by *Pseudomonas* spp., the biological activities of the formed complexes. *Biometals*, **15**, 103–120.

46 Lee, C.H., Lewis, T.A., Paszczynski, A. and Crawford, R.L. (1999) Identification of an extracellular agent [correction of catalyst] of carbon tetrachloride dehalogenation from *Pseudomonas stutzeri* strain KC as pyridine-2,6-bis (thiocarboxylate). *Biochem Biophys Res Commun*, **261**, 562–566.

47 Zawadzka, A.M., Crawford, R.L. and Paszczynski, A.J. (2007) Pyridine-2,6-bis (thiocarboxylic acid) produced by *Pseudomonas stutzeri* KC reduces chromium(VI) and precipitates mercury, cadmium, lead and arsenic. *Biometals*, **20**, 145–158.

48 Zawadzka, A.M., Crawford, R.L. and Paszczynski, A.J. (2006) Pyridine-2,6-bis (thiocarboxylic acid) produced by *Pseudomonas stutzeri* KC reduces and precipitates selenium and tellurium oxyanions. *Appl Environ Microbiol*, **72**, 3119–3129.

49 Meyer, J.M. and Abdallah, M.A. (1980) The siderochromes of non-fluorescent pseudomonads: production of nocardamine by *Pseudomonas stutzeri*. *J Gen Microbiol*, **118**, 125–129.

50 Risse, D., Risse, D., Beiderbeck, H., Taraz, K., Budzikiewicz, H., Gustine, D. *et al.* (1998) Corrugatin, a lipopeptide siderophore from *Pseudomonas corrugata*. *Z Naturforsch*, **53**, 295–304.

51 Meyer, J.M., Azelvandre, P. and Georges, C. (1992) Iron metabolism in *Pseudomonas*: salicylic acid, a siderophore of *Pseudomonas fluorescens* CHAO. *Biofactors*, **4**, 23–27.

52 Visca, P., Ciervo, A., Sanfilippo, V. and Orsi, N. (1993) Iron-regulated salicylate synthesis by *Pseudomonas* spp. *J Gen Microbiol*, **139**, 1995–2001.

53 Champoner-Verges, M.C., Stintzi, A. and Meyer, J.-M. (1996) Acquisition of iron by the non-siderophore producing *Pseudomonas fragi*. *Microbiology*, **142**, 1191–1199.

54 Llamas, M.A. *et al.* (2006) The heterologous siderophores ferrioxamine B and ferrichrome activate signaling pathways in *Pseudomonas aeruginosa*. *J Bacteriol*, **188**, 1882–1891.

55 Buyer, J.S., de Lorenzo, V. and Neilands, J.B. (1991) Production of the siderophore aerobactin by a halophilic pseudomonad. *Appl Environ Microbiol*, **57**, 2246–2250.

56 Bagg, A. and Neilands, J.B. (1987) Molecular mechanism of regulation of siderophore-mediated iron assimilation. *Microbiol Rev*, **51**, 509–518.

57 Cuiv, P.O., Clarke, P. and O'Connell, M. (2006) Identification and characterization of an iron-regulated gene, *chtA*, required for the utilization of the xenosiderophores aerobactin, rhizobactin 1021 and schizokinen by *Pseudomonas aeruginosa*. *Microbiology*, **152**, 945–954.

58 Wooldridge, K.G. and Williams, P.H. (1993) Iron uptake mechanisms of pathogenic bacteria. *FEMS Microbiol Rev*, **1**, 325–348.

59 Crosa, J.H. and Walsh, C.T. (2002) Genetics and assembly line enzymology of siderophore biosynthesis in bacteria. *Microbiol Mol Biol Rev*, **66**, 223–249.

60 Grunewald, J. and Marahiel, M.A. (2006) Chemoenzymatic and template-directed synthesis of bioactive macrocyclic peptides. *Microbiol Mol Biol Rev*, **70**, 121–146.

61 Ochsner, U.A., Wilderman, P.J., Vasil, A.I. and Vasil, M.L. (2002) GeneChip® expression analysis of the iron starvation response in *Pseudomonas aeruginosa*: identification of novel pyoverdine biosynthesis genes. *Mol Microbiol*, **45**, 1277–1287.

62 Lamont, I.L. and Martin, L.W. (2003) Identification and characterization of novel pyoverdine synthesis genes in *Pseudomonas aeruginosa*. *Microbiology*, **149**, 833–842.

63 Smith, E.E. *et al.* (2005) Evidence for diversifying selection at the pyoverdine locus of *Pseudomonas aeruginosa*. *J Bacteriol*, **187**, 2138–2147.

64 Ravel, J. and Cornelis, P. (2003) Genomics of pyoverdine-mediated iron uptake in pseudomonads. *Trends Microbiol*, **11**, 195–200.

65 Paulsen, I.T. *et al.* (2005) Complete genome sequence of the plant commensal *Pseudomonas fluorescens* Pf-5. *Nat Biotechnol*, **23**, 873–878.

66 Nelson, K.E. *et al.* (2002) Complete genome sequence and comparative analysis of the metabolically versatile *Pseudomonas putida* KT2440. *Environ Microbiol*, **4**, 799–808.

67 Vodovar, N. *et al.* (2006) Complete genome sequence of the entomopathogenic and metabolically versatile soil bacterium *Pseudomonas entomophila*. *Nat Biotechnol*, **24**, 673–679.

68 Feil, H. *et al.* (2005) Comparison of the complete genome sequences of *Pseudomonas syringae* pv. *syringae* B728a and pv. *tomato* DC3000. *Proc Natl Acad Sci USA*, **102**, 11064–11069.

69 Mossialos, D. *et al.* (2002) Identification of new, conserved, non-ribosomal peptide synthetases from fluorescent pseudomonads involved in the biosynthesis of the siderophore pyoverdine. *Mol Microbiol*, **45**, 1673–1685.

70 Ackerley, D.F., Caradoc-Davies, T.T. and Lamont, I.L. (2003) Substrate specificity of the nonribosomal peptide synthetase PvdD from *Pseudomonas aeruginosa*. *J Bacteriol*, **185**, 2848–2855.

71 Vandenende, C.S., Vlasschaert, M. and Seah, S.Y. (2004) Functional characterization of an aminotransferase required for pyoverdine siderophore biosynthesis in *Pseudomonas aeruginosa* PAO1. *J Bacteriol*, **186**, 5596–5602.

72 Visca, P., Ciervo, A. and Orsi, N. (1994) Cloning and nucleotide sequence of the *pvdA* gene encoding the pyoverdine biosynthetic enzyme L-ornithine N^5-oxygenase in *Pseudomonas aeruginosa*. *J Bacteriol*, **176**, 1128–1140.

73 McMorran, B.J., Kumara, H.M.C.S., Sullivan, K. and Lamont, I.L. (2001) Involvement of a transformylase enzyme in siderophore synthesis in *Pseudomonas aeruginosa*. *Microbiology*, **147**, 1517–1524.

74 Huang, J.J., Han, J.I., Zhang, L.H. and Leadbetter, J.R. (2003) Utilization of acyl-homoserine lactone quorum signals for growth by a soil pseudomonad and *Pseudomonas aeruginosa* PAO1. *Appl Environ Microbiol*, **69**, 5941–5949.

75 Sio, C.F. *et al.* (2006) Quorum quenching by an *N*-acyl-homoserine lactone acylase from *Pseudomonas aeruginosa* PAO1. *Infect Immun*, **74**, 1673–1682.

76 Hohlneicher, U., Schafer, M., Fuchs, R. and Budzikiewicz, H. (2001) Ferribactins as the biosynthetic precursors of the *Pseudomonas* siderophores pyoverdins. *Z Naturforsch C*, **56**, 308–310.

77 Dorrestein, P.C., Poole, K. and Begley, T.P. (2003) Formation of the chromophore of the pyoverdine siderophores by an oxidative cascade. *Org Lett*, **5**, 2215–2217.

78 Baysse, C., Budzikiewicz, H., Uria Fernandez, D. and Cornelis, P. (2002) Impaired maturation of the siderophore pyoverdine chromophore in *Pseudomonas fluorescens* ATCC 17400 deficient for the cytochrome *c* biogenesis protein CcmC. *FEBS Lett*, **523**, 23–28.

79 Baysse, C., Matthijs, S., Pattery, T. and Cornelis, P. (2001) Impact of mutations in *hemA* and *hemH* genes on pyoverdine production by *Pseudomonas fluorescens* ATCC17400. *FEMS Microbiol Lett*, **205**, 57–63.

80 Lewenza, S., Gardy, J.L., Brinkman, F.S. and Hancock, R.E. (2005) Genome-wide identification of *Pseudomonas aeruginosa* exported proteins using a consensus computational strategy combined with a laboratory-based PhoA fusion screen. *Genome Res*, **15**, 321–329.

81 Voulhoux, R., Filloux, A. and Schalk, I.J. (2006) Pyoverdine-mediated iron uptake in *Pseudomonas aeruginosa*: the Tat system is required for PvdN but not for FpvA transport. *J Bacteriol*, **188**, 3317–3323.

82 Stintzi, A. *et al.* (1999) The pvc gene cluster of *Pseudomonas aeruginosa*: role in synthesis of the pyoverdine chromophore and regulation by PtxR and PvdS. *J Bacteriol*, **181**, 4118–4124.

83 Spencer, D.H. *et al.* (2003) Whole-genome sequence variation among multiple isolates of *Pseudomonas aeruginosa*. *J Bacteriol*, **185**, 1316–1325.

84 Lamont, I.L. *et al.* (2006) Characterization of a gene encoding an acetylase required for pyoverdine synthesis in *Pseudomonas aeruginosa*. *J Bacteriol*, **188**, 3149–3152.

85 Gaille, C., Kast, P. and Haas, D. (2002) Salicylate biosynthesis in *Pseudomonas aeruginosa*. Purification and characterization of PchB, a novel bifunctional enzyme displaying isochorismate pyruvate-lyase and chorismate mutase activities. *J Biol Chem*, **277**, 21768–21775.

86 Gaille, C., Reimmann, C. and Haas, D. (2003) Isochorismate synthase (PchA), the first and rate-limiting enzyme in salicylate biosynthesis of *Pseudomonas aeruginosa*. *J Biol Chem*, **278**, 16893–16898.

87 Quadri, L.E., Keating, T.A., Patel, H.M. and Walsh, C.T. (1999) Assembly of the *Pseudomonas aeruginosa* nonribosomal peptide siderophore pyochelin: *in vitro* reconstitution of aryl-4,2-bisthiazoline synthetase activity from PchD, PchE, and PchF. *Biochemistry*, **38**, 14941–14954.

88 Patel, H.M. and Walsh, C.T. (2001) *In vitro* reconstitution of the *Pseudomonas aeruginosa* nonribosomal peptide synthesis of pyochelin: characterization of backbone tailoring thiazoline reductase and *N*-methyltransferase activities. *Biochemistry*, **40**, 9023–9031.

89 Reimmann, C., Patel, H.M., Walsh, C.T. and Haas, D. (2004) PchC thioesterase optimizes nonribosomal biosynthesis of the peptide siderophore pyochelin in *Pseudomonas aeruginosa*. *J Bacteriol*, **186**, 6367–6373.

90 Mercado-Blanco, J. *et al.* (2001) Analysis of the *pmsCEAB* gene cluster involved in biosynthesis of salicylic acid and the siderophore pseudomonine in the biocontrol strain *Pseudomonas fluorescens* WCS374. *J Bacteriol*, **183**, 1909–1920.

91 Sepulveda-Torre, L., Huang, A., Kim, H. and Criddle, C.S. (2002) Analysis of regulatory elements and genes required for carbon tetrachloride degradation in *Pseudomonas stutzeri* strain KC. *J Mol Microbiol Biotechnol*, **4**, 151–161.

92 Matthijs, S. *et al.* (2004) The *Pseudomonas* siderophore quinolobactin is synthesized from xanthurenic acid, an intermediate of the kynurenine pathway. *Mol Microbiol*, **52**, 371–384.

93 Poole, K., Heinrichs, D. and Neshat, S. (1993) Cloning and sequence analysis of an EnvCD homologue in *Pseudomonas* aeruginosa: regulation by iron and possible involvement in the secretion of the siderophore pyoverdine. *Mol Microbiol*, **10**, 529–544.

94 Cornelis, P. and Matthijs, S. (2002) Diversity of siderophore-mediated iron uptake systems in fluorescent pseudomonads: not only pyoverdines. *Environ Microbiol*, **4**, 787–798.

95 Ghysels, B. *et al.* (2004) FpvB, an alternative type I ferripyoverdine receptor of *Pseudomonas aeruginosa*. *Microbiology*, **150**, 1671–1680.

96 Ghysels, B. *et al.* (2005) The *Pseudomonas aeruginosa pirA* gene encodes a second receptor for ferrienterobactin and synthetic catecholate analogues. *FEMS Microbiol Lett*, **246**, 167–174.

97 Cobessi, D. *et al.* (2005) The crystal structure of the pyoverdine outer membrane receptor FpvA from *Pseudomonas aeruginosa* at 3.6 angstroms resolution. *J Mol Biol*, **347**, 121–134.

98 Wirth, C., Meyer-Klaucke, W., Pattus, F. and Cobessi, D. (2007) From the periplasmic signaling domain to the extracellular face of an outer membrane signal transducer of *Pseudomonas aeruginosa*: crystal structure of the ferric pyoverdine outer membrane receptor. *J Mol Biol*, **368**, 398–406.

99 Cobessi, D., Celia, H. and Pattus, F. (2005) Crystal structure at high resolution of ferric-pyochelin and its membrane receptor FptA from *Pseudomonas aeruginosa*. *J Mol Biol*, **352**, 893–904.

100 Pawelek, P.D. *et al.* (2006) Structure of TonB in complex with FhuA, *E. coli* outer membrane receptor. *Science*, **312**, 1399–1402.

101 Shultis, D.D., Purdy, M.D., Banchs, C.N. and Wiener, M.C. (2006) Outer membrane active transport: structure of the BtuB:TonB complex. *Science*, **312**, 1396–1399.

102 Moeck, G.S. and Coulton, J.W. (1998) TonB-dependent iron acquisition: mechanisms of siderophore-mediated active transport. *Mol Microbiol*, **28**, 675–681.

103 Ferguson, A.D. and Deisenhofer, J. (2004) Metal import through microbial membranes. *Cell*, **116**, 15–24.

104 Bitter, W., Tommassen, J. and Weisbeek, P.J. (1993) Identification and characterization of the *exbB*, *exbD* and *tonB* genes of *Pseudomonas putida* WCS358, their involvement in ferric-pseudobactin transport. *Mol Microbiol*, **7**, 117–130.

105 Poole, K. *et al.* (1996) The *Pseudomonas aeruginosa tonB* gene encodes a novel TonB protein. *Microbiology*, **142**, 1449–1458.

106 Takase, H., Nitanai, H., Hoshino, K. and Otani, T. (2000) Requirement of the *Pseudomonas aeruginosa tonB* gene for high-affinity iron acquisition and infection. *Infect Immun*, **68**, 4498–4504.

107 Zhao, Q. and Poole, K. (2002) Mutational analysis of the TonB1 energy coupler of *Pseudomonas aeruginosa*. *J Bacteriol*, **184**, 1503–1513.

108 Zhao, Q. and Poole, K. (2000) A second *tonB* gene in *Pseudomonas aeruginosa* is linked to the *exbB* and *exbD* genes. *FEMS Microbiol Lett*, **184**, 127–132.

109 Huang, B. *et al.* (2004) *tonB3* is required for normal twitching motility and extracellular assembly of type IV pili. *J Bacteriol*, **186**, 4387–4389.

110 Postle, K. and Kadner, R.J. (2003) Touch and go: tying TonB to transport. *Mol Microbiol*, **49**, 869–882.

111 Schalk, I.J. *et al.* (2001) Iron-free pyoverdin binds to its outer membrane receptor FpvA in *Pseudomonas aeruginosa*: a new mechanism for membrane iron transport. *Mol Microbiol*, **39**, 351–360.

112 Schalk, I.J., Abdallah, M.A. and Pattus, F. (2002) Recycling of pyoverdin on the FpvA receptor after ferric pyoverdin uptake and dissociation in *Pseudomonas aeruginosa*. *Biochemistry*, **41**, 1663–1671.

113 Koster, W. (2001) ABC transporter-mediated uptake of iron, siderophores, heme and vitamin B12. *Res Microbiol*, **152**, 291–301.

114 Carter, D.M. *et al.* (2006) Interactions between TonB from *Escherichia coli* and the periplasmic protein FhuD. *J Biol Chem*, **281**, 35413–35424.

115 Cuiv, P.O., Clarke, P., Lynch, D. and O'Connell, M. (2004) Identification of *rhtX* and *fptX*, novel genes encoding proteins that show homology and function in the utilization of the siderophores rhizobactin 1021 by *Sinorhizobium meliloti* and pyochelin by *Pseudomonas aeruginosa*, respectively. *J Bacteriol*, **186**, 2996–3005.

116 Michel, L., Bachelard, A. and Reimmann, C. (2007) Ferripyochelin uptake genes are involved in pyochelin-mediated signaling in *Pseudomonas aeruginosa*. *Microbiology*, **153**, 1508–1518.

117 Leach, L.H. and Lewis, T.A. (2006) Identification and characterization of *Pseudomonas* membrane transporters necessary for utilization of the siderophore pyridine-2,6-bis (thiocarboxylic acid) (PDTC). *Microbiology*, **152**, 3157–3166.

118 Andrews, S.C., Robinson, A.K. and Rodriguez-Quinones, F. (2003) Bacterial iron homeostasis. *FEMS Microbiol Rev*, **27**, 215–237.

119 Escolar, L., Perez-Martin, J. and de Lorenzo, V. (1999) Opening the iron box: transcriptional metalloregulation by the Fur protein. *J Bacteriol*, **181**, 6223–6229.

120 Pohl, E. *et al.* (2003) Architecture of a protein central to iron homeostasis: crystal structure and spectroscopic analysis of the ferric uptake regulator. *Mol Microbiol*, **47**, 903–915.

121 Vasil, M.L. (2007) How we learnt about iron acquisition in *Pseudomonas aeruginosa*: a series of very fortunate events. *Biometals*, **20**, 587–601.

122 Ochsner, U.A., Vasil, A.I., Johnson, Z. and Vasil, M.L. (1999) *Pseudomonas aeruginosa fur* overlaps with a gene encoding a novel outer membrane lipoprotein. OmlA. J Bacteriol, **181**, 1099–1109.

123 Wilderman, P.J. *et al.* (2004) Identification of tandem duplicate regulatory small RNAs in *Pseudomonas aeruginosa* involved in iron homeostasis. *Proc Natl Acad Sci USA*, **101**, 9792–9797.

124 Helmann, J.D. (2002) The extracytoplasmic function (ECF) sigma factors. *Adv Microb Physiol*, **46**, 47–110.

125 Missiakas, D. and Raina, S. (1998) The extracytoplasmic function sigma factors: role and regulation. *Mol Microbiol*, **28**, 1059–1066.

126 Hughes, K.T. (1998) The anti-sigma factors. *Annu Rev Microbiol*, **52**, 231–286.

127 Martinez-Bueno, M.A., Tobes, R., Rey, M. and Ramos, J.L. (2002) Detection of multiple extracytoplasmic function (ECF) sigma factors in the genome of *Pseudomonas putida* KT2440 and their counterparts in *Pseudomonas aeruginosa* PA01. *Environ Microbiol*, **4**, 842–855.

128 Oguiza, J.A., Kiil, K. and Ussery, D.W. (2005) Extracytoplasmic function sigma factors in *Pseudomonas syringae*. *Trends Microbiol*, **13**, 565–568.

129 Braun, V. and Mahren, S. (2005) Transmembrane transcriptional control (surface signaling) of the *Escherichia coli* Fec type. *FEMS Microbiol Rev*, **29**, 673–684.

130 Leoni, L., Orsi, N., Lorenzo, V.d. and Visca, P. (2000) Functional analysis of PvdS, an iron starvation sigma factor of *Pseudomonas aeruginosa*. *J Bacteriol*, **182**, 1481–1491.

131 Wilson, M.J. and Lamont, I.L. (2000) Characterization of an ECF sigma factor protein from *Pseudomonas aeruginosa*. *Biochem Biophys Res Commun*, **273**, 578–583.

132 Wilson, M.J. and Lamont, I.L. (2006) Mutational analysis of an extracytoplasmic-function sigma factor to investigate its interactions with RNA polymerase and DNA. *J Bacteriol*, **188**, 1935–1942.

133 Rombel, I.T., McMorran, B.J. and Lamont, I.L. (1995) Identification of a

DNA sequence motif required for expression of iron-regulated genes in pseudomonads. *Mol Gen Genet*, **246**, 519–528.

134 Ochsner, U.A. and Vasil, M.L. (1996) Gene repression by the ferric uptake regulator in *Pseudomonas aeruginosa*: cycle selection of iron-regulated genes. *Proc Natl Acad Sci USA*, **93**, 4409–4414.

135 Barton, H.A. *et al.* (1996) Ferric uptake regulator mutants of *Pseudomonas aeruginosa* with distinct alterations in the iron-dependent repression of exotoxin A and siderophores in aerobic and microaerobic environments. *Mol Microbiol*, **21**, 1001–1017.

136 Cao, H. *et al.* (2001) A quorum sensing-associated virulence gene of *Pseudomonas aeruginosa* encodes a LysR-like transcription regulator with a unique self-regulatory mechanism. *Proc Natl Acad Sci USA*, **98**, 14613–14618.

137 Venturi, V., Ottevanger, C., Leong, J. and Weisbeek, P.J. (1993) Identification and characterization of a siderophore regulatory gene (*pfr*A) of *Pseudomonas putida* WCS358, homology to the alginate regulatory gene *alg*Q of *Pseudomonas aeruginosa*. *Mol Microbiol*, **10**, 63–73.

138 Ambrosi, C. *et al.* (2005) Involvement of AlgQ in transcriptional regulation of pyoverdine genes in *Pseudomonas aeruginosa* PAO1. *J Bacteriol*, **187**, 5097–5107.

139 Beare, P.A., For, R.J., Martin, L.W. and Lamont, I.L. (2003) Siderophore-mediated cell signaling in *Pseudomonas aeruginosa*: divergent pathways regulate virulence factor production and siderophore receptor synthesis. *Mol Microbiol*, **47**, 195–207.

140 Redly, G.A. and Poole, K. (2003) Pyoverdine-mediated regulation of FpvA synthesis in *Pseudomonas aeruginosa*: involvement of a probable extracytoplasmic-function sigma factor. FpvI. J Bacteriol, **185**, 1261–1265.

141 Koster, M. *et al.* (1994) Role for the outer membrane ferric siderophore receptor PupB in signal transduction across the bacterial cell envelope. *EMBO J*, **13**, 2805–2813.

142 Wilderman, P.J. *et al.* (2001) Characterization of an endoprotease (PrpL) encoded by a PvdS-regulated gene *Pseudomonas aeruginosa*. *Infect Immun*, **69**, 5385–5394.

143 Vasil, M.L. and Ochsner, U.A. (1999) The response of *Pseudomonas aeruginosa* to iron: genetics, biochemistry and virulence. *Mol Microbiol*, **34**, 399–413.

144 Ochsner, U.A. *et al.* (1996) Exotoxin A production in *Pseudomonas aeruginosa* requires the iron-regulated *pvdS* gene encoding an alternative sigma factor. *Mol Microbiol*, **21**, 1019–1028.

145 Hunt, T.A., Peng, W.T., Loubens, I. and Storey, D.G. (2002) The *Pseudomonas aeruginosa* alternative sigma factor PvdS controls exotoxin A expression and is expressed in lung infections associated with cystic fibrosis. *Microbiology*, **148**, 3183–3193.

146 Xiong, Y.-Q. *et al.* (2000) The oxygen- and iron-dependent sigma factor *pvdS* of *Pseudomonas aeruginosa* is an important virulence factor in experimental infective endocarditis. *J Infect Dis*, **181**, 1020–1026.

147 Stintzi, A., Evans, K., Meyer, J.-M. and Poole, K. (1998) Quorum-sensing and siderophore biosynthesis in *Pseudomonas aeruginosa: lasR/lasI* mutants exhibit reduced pyoverdine biosynthesis. *FEMS Microbiol Lett*, **166**, 341–345.

148 Whiteley, M., Lee, K.M. and Greenberg, E.P. (1999) Identification of genes controlled by quorum sensing in *Pseudomonas aeruginosa. Proc Natl Acad of Sci USA*, **96**, 13904–13909.

149 Hentzer, M. *et al.* (2003) Attenuation of *Pseudomonas aeruginosa* virulence by quorum sensing inhibitors. *EMBO J*, **22**, 3803–3815.

150 Schuster, M., Lostroh, C.P., Ogi, T. and Greenberg, E.P. (2003) Identification, timing, and signal specificity of *Pseudomonas aeruginosa* quorum-controlled genes: a transcriptome analysis. *J Bacteriol*, **185**, 2066–2079.

151 Wagner, V.E. *et al.* (2003) Microarray analysis of *Pseudomonas aeruginosa* quorum-sensing regulons: effects of growth phase and environment. *J Bacteriol*, **185**, 2080–2095.

152 Juhas, M. *et al.* (2004) Global regulation of quorum sensing and virulence by VqsR in *Pseudomonas aeruginosa*. *Microbiology*, **150**, 831–841.

153 Arevalo-Ferro, C. *et al.* (2003) Identification of quorum-sensing regulated proteins in the opportunistic pathogen *Pseudomonas aeruginosa* by proteomics. *Environ Microbiol*, **5**, 1350–1369.

154 Hentzer, M., Givskov, M. and Eberl, L. (2004) Quorum sensing in biofilms: gossip in slime city, in *Microbial Biofilms* (eds M. Ghannoum and G.A. O'Toole), American Society for Microbiology Press, Washington, DC, pp. 118–140.

155 Bredenbruch, F. *et al.* (2006) The *Pseudomonas aeruginosa* quinolone signal (PQS) has an iron-chelating activity. *Environ Microbiol*, **8**, 1318–1329.

156 Diggle, S.P. *et al.* (2007) The *Pseudomonas aeruginosa* 4-quinolone signal molecules HHQ and PQS play multifunctional roles in quorum sensing and iron entrapment. *Chem Biol*, **14**, 87–96.

157 Shen, J., Meldrum, A. and Poole, K. (2002) FpvA receptor involvement in pyoverdine biosynthesis in *Pseudomonas aeruginosa*. *J Bacteriol*, **184**, 3268–3275.

158 Lamont, I.L. *et al.* (2002) Siderophore-mediated signaling regulates virulence factor production in *Pseudomonas aeruginosa*. *Proc Natl Acad Sci USA*, **99**, 7072–7077.

159 Gensberg, K., Hughes, K. and Smith, A.W. (1992) Siderophore-specific induction of iron uptake in *Pseudomonas aeruginosa*. *J Gen Microbiol*, **138**, 2381–2387.

160 James, H.E., Beare, P.A., Martin, L.W. and Lamont, I.L. (2005) Mutational analysis of a bifunctional ferrisiderophore receptor and signal-transducing protein from *Pseudomonas aeruginosa*. *J Bacteriol*, **187**, 4514–4520.

161 Redly, G.A. and Poole, K. (2005) FpvIR control of *fpvA* ferric pyoverdine receptor gene expression in *Pseudomonas aeruginosa*: demonstration of an interaction between FpvI and FpvR and identification of mutations in each compromising this interaction. *J Bacteriol*, **187**, 5648–5657.

162 Enz, S. *et al.* (2003) Sites of interaction between the FecA and FecR signal transduction proteins of ferric citrate transport in *Escherichia coli* K-12. *J Bacteriol*, **185**, 3745–3752.

163 Heinrichs, D.E. and Poole, K. (1993) Cloning and sequence analysis of a gene (*pch*R) encoding an AraC family activator of pyochelin and ferripyochelin receptor synthesis in *Pseudomonas aeruginosa*. *J Bacteriol*, **175**, 5882–5889.

164 Heinrichs, D.E. and Poole, K. (1996) PchR, a regulator of ferripyochelin receptor gene (*fpt*A) expression in *Pseudomonas aeruginosa*, functions both as an activator and as a repressor. *J Bacteriol*, **178**, 2586–2592.

165 Michel, L. *et al.* (2005) PchR-box recognition by the AraC-type regulator PchR of *Pseudomonas aeruginosa* requires the siderophore pyochelin as an effector. *Mol Microbiol*, **58**, 495–509.

166 Reimmann, C., Serino, L., Beyeler, M. and Haas, D. (1998) Dihydroaeruginoic acid synthetase and pyochelin synthetase, products of the *pchEF* genes, are induced by extracellular pyochelin in *Pseudomonas aeruginosa*. *Microbiology*, **144**, 3135–3148.

167 Dean, C.R. and Poole, K. (1993) Expression of the ferric enterobactin receptor (PfeA) of *Pseudomonas aeruginosa*: involvement of a two-component regulatory system. *Mol Microbiol*, **8**, 1095–1103.

168 Rodrigue, A. *et al.* (2000) Two-component systems in *Pseudomonas aeruginosa*: why so many? *Trends Microbiol*, **8**, 498–504.

169 Dean, C.R., Neshat, S. and Poole, K. (1996) PfeR, an enterobactin-responsive activator of ferric enterobactin receptor gene expression in *Pseudomonas aeruginosa*. *J Bacteriol*, **178**, 5361–5369.

170 Ochsner, A., Johnson, Z. and Vasil, M.L. (2000) Genetics and regulation of two distinct haem-uptake systems, *phu* and has, in *Pseudomonas aeruginosa*. *Microbiology*, **146**, 185–198.

171 Palma, M., Worgall, S. and Quadri, L.E. (2003) Transcriptome analysis of the *Pseudomonas aeruginosa* response to iron. *Arch Microbiol*, **180**, 374–379.

172 Upritchard, H.G. *et al.* (2007) Adsorption to metal oxides of the *Pseudomonas aeruginosa* siderophore pyoverdine and implications for bacterial biofilm formation on metals. *Langmuir*, **23**, 7189–7195.

173 McWhirter, M.J., Bremer, P.J., Lamont, I.L. and McQuillan, A.J. (2003) Siderophore-mediated covalent bonding to metal (oxide) surfaces during biofilm initiation by *Pseudomonas aeruginosa* bacteria. *Langmuir*, **19**, 3573–3577.

174 Tummler, B. and Cornelis, P. (2005) Pyoverdine receptor: a case of positive Darwinian selection in *Pseudomonas aeruginosa*. *J Bacteriol*, **187**, 3289–3292.

175 de Chial, M. *et al.* (2003) Identification of type II and type III pyoverdine receptors from *Pseudomonas aeruginosa*. *Microbiology*, **149**, 821–831.

176 De Vos, D. *et al.* (2001) Study of pyoverdine type and production by *Pseudomonas aeruginosa* isolated from cystic fibrosis patients: prevalence of type II pyoverdine isolates and accumulation of pyoverdine-negative mutations. *Arch Microbiol*, **175**, 384–388.

177 Palmer, K.L., Mashburn, L.M., Singh, P.K. and Whiteley, M. (2005) Cystic fibrosis sputum supports growth and cues key aspects of *Pseudomonas aeruginosa* physiology. *J Bacteriol*, **187**, 5267–5277.

178 Sun, G.X. and Zhong, J.J. (2006) Mechanism of augmentation of organotin decomposition by ferripyochelin: formation of hydroxyl radical and organotin–pyochelin–iron ternary complex. *Appl Environ Microbiol*, **72**, 7264–7269.

179 Cox, C.D. and Adams, P. (1985) Siderophore activity of pyoverdin for *Pseudomonas aeruginosa*. *Infect Immun*, **48**, 130–138.

180 Cox, C.D. (1980) Iron uptake with ferripyochelin and ferric citrate by *Pseudomonas aeruginosa*. *J Bacteriol*, **142**, 581–587.

181 Poole, K., Young, L. and Neshat, S. (1990) Enterobactin-mediated iron transport in *Pseudomonas aeruginosa*. *J Bacteriol*, **172**, 6991–6996.

182 Meyer, J.M. (1992) Exogenous siderophore-mediated iron uptake in *Pseudomonas aeruginosa*: possible involvement of porin OprF in iron translocation. *J Gen Microbiol*, **138**, 951–958.

183 Smith, A.W., Poyner, D.R., Hughes, H.K. and Lambert, P.A. (1994) Siderophore activity of myo-inositol hexakisphosphate in *Pseudomonas aeruginosa*. *J Bacteriol*, **176**, 3455–3459.

184 Matilla, M.A., Ramos, J.L., Duque, E., de Dios Alché, J., Espinosa Urgel, M. and Ramos-González, M.I. (2007) Temperature and pyoverdine-mediated iron acquisition control surface motility of *Pseudomonas putida*. *Environ Microbiol.*, **9**, 1842–1850.

7
Quorum Sensing in Pseudomonads

S.P. Diggle, S. Heeb, J.F. Dubern, M.P. Fletcher, S.A. Crusz, P. Williams, and M. Cámara

7.1
Introduction to Quorum Sensing

Like all living organisms, bacteria depend on the availability of resources from their environment to multiply. Single, planktonic cells are subject to casual encounters with nutrient-rich microenvironments for their development and propagation. However, when such microenvironments are encountered, bacteria will often proliferate to form established multicellular communities that have the potential to adapt to and modify their environment, allowing the exploitation of the nutrient resources that would otherwise be restricted for individual cells. Detecting the moment at which a bacterial cell changes from being part of a dispersed population of individual cells to a community strong enough to invest part of their accumulated resources to modify their environment is crucial. At this point, the behavior of the individual cells within the population must be changed in a coordinated manner through the modulation of the expression of particular genes. The concept of multicellular behavior was introduced in the 1960s with studies on fruiting body formation in *Myxococcus xanthus*, on streptomycin biosynthesis and aerial mycelium formation in *Streptomyces griseus*, on the induction of genetic competence in *Streptococcus pneumoniae*, and also on the control of bioluminescence in marine vibrios [1]. It was later found that there are mechanisms specifically devoted to coordinate this multicellular behavior which rely on the release into the medium of small signaling molecules. These accumulate within the local environment and disperse among the population, ultimately triggering the coordinated expression of a large set of genes which induces the change in group behavior. This concept has been termed quorum sensing (QS) and it is defined as a mechanism by which bacteria regulate specific target genes in response to a critical concentration of signal molecule dedicated to the measurement of the cell population density [1,2].

N-Acyl-homoserine (AHL)-dependent QS was originally described in *Vibrio fischeri*, which symbiotically colonizes the light organ of the bobtail squid (*Euprymna scolopes*), as a means to trigger bioluminescence. It was noted that although *V. fischeri* only

Pseudomonas. Model Organism, Pathogen, Cell Factory. Edited by Bernd H.A. Rehm

ISBN: 978-3-527-31914-5

produced light at high cell population densities, the organism produced an extracellular substance that could induce bioluminescence at low cell densities [3]. The "autoinducer" concerned was subsequently purified and the structure determined as *N*-(3-oxohexanoyl)homoserine lactone (3O-C6-HSL) [4]. Until 1992, this AHL was exclusively associated with *V. fischeri*. However, while working on the biosynthesis of the β-lactam antibiotic, carbapen-2-em-3-carboxylic acid, in the plant pathogen *Erwinia carotovora*, Bainton *et al.* discovered that carbapenem biosynthesis in this terrestrial microbe was also regulated by 3O-C6-HSL and that other Gram-negative bacteria, including *Pseudomonas aeruginosa* and *Serratia marcescens*, also produced the *V. fischeri* autoinducer [5,6]. The genetics of bioluminescence in *V. fischeri* was unraveled and it was shown that 3O-C6-HSL is synthesized by a protein called LuxI. Upon reaching a threshold concentration, 3O-C6-HSL binds to the transcriptional regulator, LuxR, resulting in the activation of the *luxICDABE* promoter. This leads to an autoinduction loop resulting in more AHL synthesis from the *luxI* gene product and the generation of bioluminescence from the *luxCDABE* genes [7]. For the activation of the *luxICDABE* operon, the LuxR/3O-C6-HSL complex interacts with an inverted repeat DNA recognition sequence called a "*lux* box". These *lux* box sequences are frequently found upstream of genes directly controlled by AHL-mediated QS [8]. In addition, homologs of LuxI and LuxR have now been found in many different Gram-negative bacteria [1].

AHLs are not the only family of QS signal molecules, since in Gram-negative bacteria, 2-alkyl-4-quinolones, furanones, fatty acids and fatty acid methyl esters have also been identified as such. In addition Gram-positive bacteria produce QS molecules ranging from γ-butyrolactones to posttranslationally modified peptides [1,9]. The discovery of QS signaling systems in an ever-growing number of bacterial species has led to a more generic definition of QS as a system that requires: (i) the production of specifically dedicated molecules that accumulate in a confined environment with population density, (ii) a dedicated set of receptors and response regulators for the signal molecules which relay the information on signal accumulation from the environment to target genes, in many instances, via transcriptional or posttranscriptional regulators, and (iii) that the products from the target genes are not essential for primary metabolism or the metabolic utilization of the signal molecule [10] (Figure 7.1).

A key point to remember is that QS, as the determinant of cell population density, forms part of wider regulatory networks within the cell that integrate responses to many different environmental signals (e.g. temperature, pH, osmolarity, oxidative stress, nutrient deprivation) in order to develop the appropriate survival strategy [11]. This chapter provides an overview of the QS systems known to be controlling the lifestyles of various pseudomonads (specifically *Pseudomonas aeruginosa*), enabling them to successfully colonize a wide variety of ecological niches.

7.2
AHL QS in *Pseudomonas aeruginosa*

The discovery of a *luxR* homolog (*lasR*) and the production of the AHL molecule 3O-C6-HSL in *P. aeruginosa* demonstrated that this organism possesses an

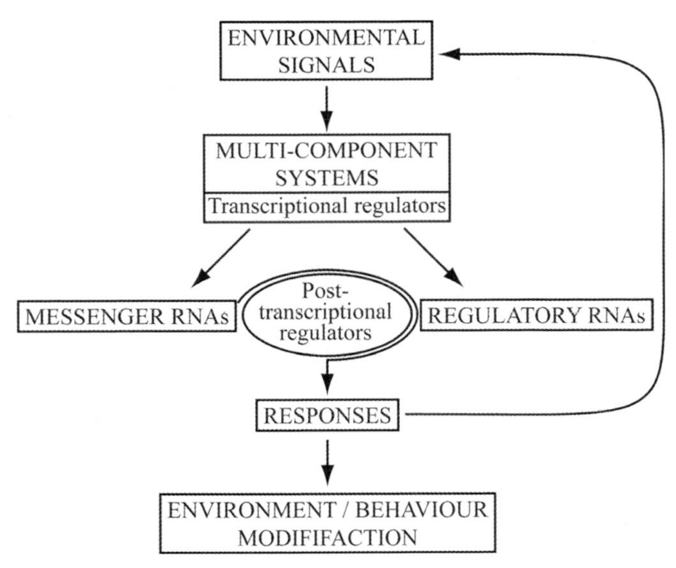

Figure 7.1 Bacterial integration of QS signaling systems. The accumulation of signals in a confined environment results in the activation of bacterial multicomponent systems, such as phosphorelay cascades and/or transcriptional regulators. This can lead to the synthesis of mRNA, which can either be subjected to the action of posttranscriptional regulators or directly drive a bacterial response. In addition, the transcriptional regulators can activate the synthesis of regulatory RNAs that can influence the action of posttranscriptional regulators and hence the final bacterial response. The final outcome of these activation cascades can lead to (i) the generation of more signal molecule as part of the QS autoinduction process and (ii) a coordinated modification in bacterial behavior resulting in the alteration of their environment.

AHL-dependent QS system [6,12]. LasR is a typical member of the LuxR protein family as it contains an N-terminal AHL-binding region showing 36% identity to the AHL-binding region and a DNA-binding region in the C-terminal with 53% identity to the LuxR DNA-binding region [12]. A *luxI* homolog termed *lasI* was discovered downstream from *lasR* and found to direct the synthesis of primarily *N*-(3-oxododecanoyl)-L-homoserine lactone (3O-C12-HSL) (Figure 7.2) together with small amounts of 3O-C6-HSL [13,14]. Unlike the *luxRI* genes, which are divergently transcribed, the *lasRI* genes are transcribed in the same direction [15]. The autoinducer molecule 3O-C12-HSL together with LasR forms the *P. aeruginosa* QS system known as the *las* system. LasR is most sensitively activated by 3O-C12-HSL and has recently been crystallized bound to this signal molecule [16]. The *las* QS system is important in the regulation of the production of elastase, LasA protease, exotoxin A, the general secretory pathway (*xcp*) and biofilm development [12,13,17–20]. The finding that 3O-C12-HSL is necessary for the maturation of *P. aeruginosa* biofilms [17], since *lasI* mutants form abnormal biofilms which are more susceptible to sodium dodecylsulfate (SDS), has important implications, specially in clinical settings (see Section 7.6).

Further complexity was added when a second AHL-dependent QS system was discovered in *P. aeruginosa*. Ochsner *et al.* identified a second LuxR homolog termed

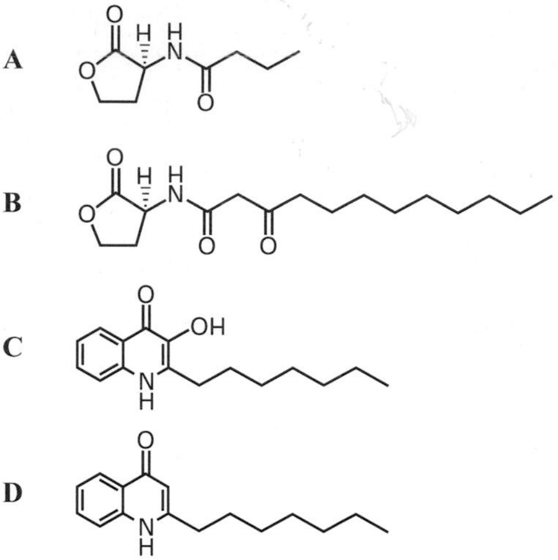

Figure 7.2 Main QS signal molecules in *P. aeruginosa*. (A) *N*-butanoyl-L-homoserine lactone, (B) 3O-C12-HSL, (C) PQS and (D) HHQ.

rhlR as it was found to control production of the biosurfactant, rhamnolipid [21]. The same gene was discovered in parallel by Latifi *et al.*, who named it *vsmR*, and by Brint and Ohman, who retained the original name *rhlR* [22,23]. Latifi *et al.* also described a *luxI* homolog which was termed *vsmI* (but subsequently named *rhlI*) and showed that it directed the synthesis of *N*-butanoyl-L-homoserine lactone (C4-HSL) (Figure 7.2) and *N*-hexanoyl-L-homoserine lactone (C6-HSL) in a molar ratio of approximately 15:1 [24]. RhlR/C4-HSL not only induces the expression of the *rhlAB* operon encoding for a rhamnosyltransferase required for rhamnolipid production [21,25], but also the production of elastase, LasA protease, pyocyanin, alkaline protease, RpoS, and LecA and LecB lectins [22–24,26–29].

The *las* and *rhl* systems do not function independently of each other. Latifi *et al.* showed that LasR/3O-C12-HSL controlled the transcription of *rhlR* placing the *las* system above the *rhl* system in a QS hierarchy [29]. In simplistic terms, this would suggest that the *las* system partially regulates the *rhl* system which in turns controls the expression of virulence factors (Figure 7.3). However, more detailed studies have shown that there are distinct differences between specific promoters for *las* and *rhl* [30], suggesting the existence of distinct *las* and *rhl* regulons.

P. aeruginosa cells are not freely permeable to 3O-C12-HSL. Instead, the *mexA–mexB–oprM* multidrug efflux system actively pumps 3O-C12-HSL out of the bacterial cell [31] and it was found that the cellular content of 3O-C12-HSL was 3 times higher than the external level. In contrast, it was shown that C4-HSL was freely diffusible, and that cellular and external concentrations of this AHL reached a steady state in around 30 s [31]. This demonstrated that the length and/or the degree of

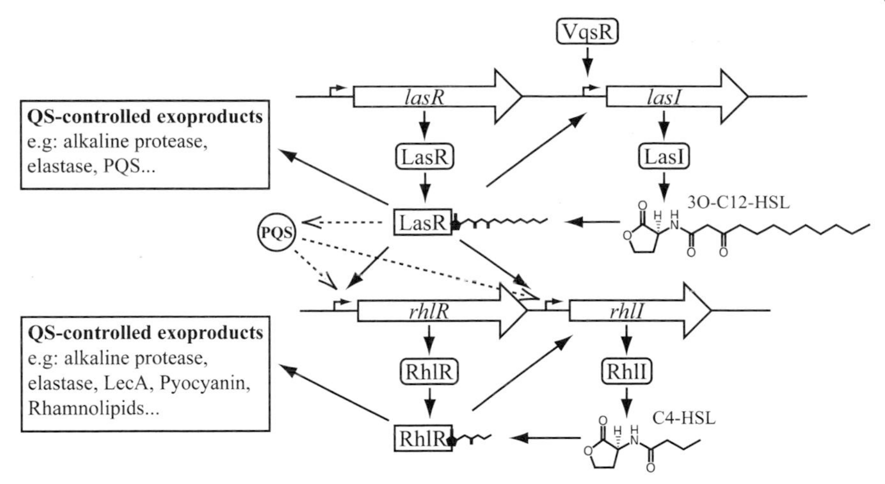

Figure 7.3 AHL-dependent QS in *P. aeruginosa*. The QS cascade is induced at high population cell densities when within the cell, the LasR response protein binds to a critical concentration of 3O C12-HSL signal that has been produced by neighboring cells and taken up from the surrounding environment. This results in activation of the *las* QS system and the production of a number of QS-regulated phenotypes. Activation of the *las* system is also important in the induction of the *rhl* QS system that is also required for the production of *rhl*-controlled phenotypes.

substitution of the AHL side-chain determines whether an AHL is freely diffusible or subject to active transport [31].

The completed *Pseudomonas* genome sequence [32] revealed two further *luxR* homologs. Both of these have no adjacent *luxI* genes and are not responsible directly for controlling/regulating signal generation. The first was termed QscR (QS control repressor) [33]. In a *qscR* mutant, both *lasI* and *rhlI* are prematurely transcribed as is the production of both 3O-C12-HSL and C4-HSL. In addition, QscR serves to repress the synthesis of the virulence determinants pyocyanin and hydrogen cyanide, with pyocyanin production being significantly enhanced in a *qscR* mutant. A possible mechanism for the mode of action of QscR was elucidated using fluorescence anisotropy and *in vivo* cross-linking. It was found that at low AHL concentrations, which would correspond with low population densities, QscR could form inactive heterodimers with LasR and RhlR. With an increase in AHL levels, these heterodimers dissociate, resulting in the formation of LasR and RhlR homodimers and their activation upon binding their cognate AHL [34]. Studies of the QscR transcriptome revealed that it has its own distinct regulon, controlling over 400 genes, and can also act as both an activator and repressor, binding to promoters with similar elements to those found in LasR- and RhlR-dependent promoters such as *lasB* and *hcnAB* [35,36].

The fourth *luxR* homolog identified in *P. aeruginosa* was *vqsR* (gene PA2591) which modulates a number of QS-controlled genes [37]. Disruption of *vqsR* resulted in the loss of AHL and exoproduct production, and diminished bacterial virulence for *Caenorhabditis elegans* [37]. In the absence of *vqsR* there is a significant decrease in *lasI* mRNA levels, but not those from *lasR*, suggesting that the positive effect of VqsR on QS is via *lasI* [37,38]. The *vqsR* gene possesses a *las* box and is controlled by the *las* QS system as LasR directly binds the *vqsR* promoter in the presence of 3O-C12-HSL [39].

Over recent years, there have been a number of studies examining the global influence of AHL-dependent QS in *P. aeruginosa*. The first genome approach described was by Whiteley *et al.*, who used random transposon mutagenesis to identify genes controlled by QS [40]. In total, 39 genes were identified and divided into four classes of QS-regulated genes according to the pattern and timing of their regulation. A further random transposon mutagenesis study searched for "super" regulators of the QS system to integrate the AHL-dependent QS regulons within other regulatory networks in the cell. This study identified the transcriptional regulator MvaT, which was shown to be involved in the growth phase dependency of the QS cascade [41]. Three separate microarray studies have defined the AHL-dependent QS transcriptome in *P. aeruginosa* [42–44]. Amongst the main conclusions from these studies were: (i) around 6% of the entire *P. aeruginosa* genome is controlled by AHL-mediated QS, (ii) the timing of gene expression is on a continuum although the majority of QS control takes place during the transition to stationary phase, (iii) QS has both a positive and negative regulatory effect on the transcriptome with a quarter of the genes identified being repressed by QS, and (iv) environmental factors such as oxygen availability affect the transcripts of many QS-regulated genes.

What is clear from the studies mentioned above is that AHL-dependent QS in *P. aeruginosa* is integrated within a complex regulatory network in the cell, and is under the transcriptional and posttranscriptional control of a large number of other regulators (see Table 7.1). For a more detailed description of the regulators controlling QS in *P. aeruginosa* see Venturi *et al.* [2].

In addition to the AHLs, *P. aeruginosa* produces two other classes of QS molecules: the 2-alkyl-4-quinolones (AHQs) and cyclic dipeptides [62,64,65]. Although the latter have not been extensively studied, a significant amount of work has been done on the role of AHQs in the pseudomonads (Section 7.4).

7.3
AHL-Dependent QS in Other Pseudomonads

In addition to *P. aeruginosa*, AHL QS systems have also been described in several other *Pseudomonas* species (e.g. *P. fluorescens*, *P. putida*, *P. aureofaciens*, *P. chlororaphis* and *P. syringae*). In these organisms AHLs play an important role in controlling key biological functions such as phytopathogenicity and plant growth promotion where they act as plant-beneficial bacteria by antagonizing plant-deleterious microorganisms or through the production of traits that directly influence plant disease resistance and growth. A list of the AHL QS systems identified in *Pseudomonas* spp. and their main roles are summarized in Table 7.2. Further information can be obtained from the references specified in Table 7.2.

Table 7.1 Examples of regulators of QS in *P. aeruginosa*.

Regulator	Description	Link with QS	Phenotypes affected	References
VfR	cyclic AMP receptor protein	*lasR* expression reduced 90% in *vfR* mutant	exotoxin A, proteases	[45]
RsaL	repressor of QS	directly represses the *lasI* gene	elastase	[46–48]
RpoS	stationary-phase σ factor	negative regulator of *rhlI* expression	LecA, elastase, pyocyanin, swarming	[29,41,49–51]
RpoN	alternative σ factor	negative regulator of *lasRI*, *rhlI*; can also be a positive activator of *rhlI*	elastase, rhamnolipid, hydrogen cyanide	[52,53]
MvaT	HNS-like transcriptional regulator	involved in the growth phase dependency of QS	LecA, elastase, pyocyanin, swarming, biofilms, MexEF–OprN efflux pump	[41,54,55]
PprB	response regulator	*lasI*, *rhlI* and *rhlR* all decreased in a *pprB* mutant	protease, elastase, pyocyanin, hemolytic activity, swimming and swarming motility	[56]
VqsM	transcriptional regulator	*lasI* and *rhlI* repressed in a *vqsM* mutant	protease, elastase, swimming and swarming motility	[57]
RsmA	posttranscriptional regulator	negatively controls *lasI* and *rhlI*	protease, pyocyanin, LecA, hydrogen cyanide, swarming	[58]
DksA	posttranscriptional regulator	negatively controls *rhlI*	elastase, rhamnolipid	[59]
GacA	global activator	positive regulator of *lasR* and *rhlR*	pyocyanin, lipase, hydrogen cyanide	[60]
PhoB	response regulator	stimulator of *rhlR* under phosphate-limited conditions	pyocyanin	[61]
PQS	*Pseudomonas* quinolone signal	positive regulator of *rhl* QS	elastase, LecA, pyocyanin, biofilm formation	[62–64]

Table 7.2 AHL-dependent QS systems in the pseudomonads.

***Pseudomonas* strain**	**Genes**	**AHLs produced**	**Phenotype affected**	**References**
P. aeruginosa	*lasI, lasR, rhlI, rhlR*	3O-C12-AHL, C4-AHL	pyocyanin, HCN, exotoxin A, lectins, proteases, elastases, lipases, rhamnolipids, swarming, biofilm formation	[12,21,23,66]
P. putida	*ppuI, ppuR*	3O-C12-AHL	biofilm structure and formation, biosurfactant	[67–69]
P. aureofaciens/ chlororaphis	*phzI, phzR, csaI, csaR*	C4-AHL, C6-AHL, C8-AHL	phenazines, biofilm formation, rhizosphere colonization	[70–72]
P. fluorescens	*mupI, mupR*	3-OH-C14-AHL, C10-AHL, C6-AHL	mupirocin	[73]
	pcoI, pcoR	C6-AHL, C8-AHL, 3O-C6-AHL, 3O-C8-AHL	2,4-diacetylphloroglucinol, HCN, biofilm formation, rhizosphere colonization	[74]
	phzI, phzR	3-OH-C6-AHL	phenazines	[75]
	luxI, luxR	3-OH-C8-AHL	biosurfactant	[76]
	hdtS	3-OH-C14-AHL	unknown	[77]
P. syringae	*ahlI, ahlR*	3O-C6-AHL	cell aggregation	[78]

7.4 AHQ-Dependent QS in Pseudomonads

Pesci *et al.* demonstrated that *P. aeruginosa* utilized a non-AHL-mediated QS signaling pathway via 2-heptyl-3-hydroxy-4-quinolone [named the *Pseudomonas* quinolone signal (PQS)] (Figure 7.2). It was shown that addition of a spent culture medium extract from a *P. aeruginosa* wild-type strain (PAO1) induced *lasB* expression in a PAO1 *lasR* mutant [64]. This was interesting as this induction was unlikely to have been due to the AHLs because the mutant used was AHL deficient. This finding implied the presence of a non-AHL signal which was capable of activating *lasB* expression, and which probably required LasR and 3O-C12-HSL for its production.

Additionally, spent culture supernatant extracts from PAO1 failed to induce *lasB* expression in a *rhlR/rhlI* double mutant, suggesting that the *rhl* system was required for the bioactivity of the new signal [64]. Mass spectrometry and chemical synthesis confirmed that the molecule was an AHQ containing additional oxygenation of the hetero-aromatic core. The AHQ was identified as PQS heralding a new class of *P. aeruginosa* QS signal molecules, whose mechanism of regulation and action was interwoven with that of the two previously discovered AHL-dependent QS systems [62,64]. Although this was the first time that AHQs had been shown to be involved in QS in *Pseudomonas*, they have been known as *P. aeruginosa* metabolites since the 1940s. In addition to PQS, other major AHQs produced by this organism include 2-heptyl-4-quinolone (HHQ) (Figure 7.2), 2-nonyl-4-quinolone (NHQ) and 2-heptyl-4-quinolone *N*-oxide (HHQNO).

Based on *lasB* gene fusion data, AHQs (including PQS) were initially thought to be produced primarily in late log phase with maximal production of PQS occurring in late stationary phase after 30–42 h [79]. However, two independent studies on the timing of PQS production, using more direct detection methods, both demonstrated that substantial levels of PQS are also produced in late logarithmic/early stationary phase cultures [63,80].

7.4.1
Regulation and Biosynthesis of AHQs

The genes responsible for AHQ biosynthesis in *P. aeruginosa* were identified soon after the discovery of PQS [81,82]. Several of the genes were in the region of the *phnAB* operon (PA1001–PA1002), previously identified as coding for proteins responsible for anthranilate biosynthesis [83]. Further investigation revealed a five-gene operon under the control of a putative transcriptional regulator. This operon was termed *pqsABCDE* (PA0996–PA1000) and the transcriptional regulator was named *pqsR* (PA1003). The *pqsA–D* genes are involved in the biosynthesis of the PQS precursor HHQ from anthranilate [81,84,85]. Interestingly, a liquid chromatography (LC)/mass spectroscopy (MS) study revealed around 56 different AHQs in *P. aeruginosa* cultures. These mainly consisted not only of PQS and HHQ, but also HHQNO and NHQ, with other alkyl side-chain length congeners and the mono-unsaturated alkyl side-chain derivatives of these compounds also present [86].

Another gene encoding a predicted FAD-dependent monooxygenase was found to be required for PQS production and termed *pqsH* (PA2587) [81]. PqsH converts HHQ into PQS and, interestingly, *pqsH* expression is, at least in part, controlled by LasR, linking the AHL and AHQ signaling [84]. However, PQS can be produced in significant amounts in a *lasR* mutant at high cell population densities [63], suggesting that PqsH can be controlled in a LasR-independent manner since no other enzymes capable of converting HHQ to PQS have been identified.

The role of PqsE is currently not known, although it does have structural similarities to proteins from the metallo-β-lactamase superfamily. A mutation in *pqsE* does not affect PQS production, but abolishes the production of PQS-dependent exoproducts such as pyocyanin [81]. PqsE was therefore proposed to be required for the cellular response to PQS by generating an as yet unidentified signaling molecule from PQS.

PqsR, a LysR-type regulator initially termed MvfR, is a transcriptional regulator of the *pqsABCDE* and *phnAB* operons [81], and in a *pqsR* mutant expression of *phnAB*, *pqsABCDE* and hence AHQ production is abolished [84]. Interestingly, PQS is responsible for its own autoinduction. A large reduction in *pqsA* expression occurs in a *pqsR* mutant showing that *pqsR* is required for *pqsA* transcription [87]. PqsR binds to the *pqsA* promoter in DNA mobility shift assays in the absence of PQS, but in the presence of this signal molecule this binding is greatly enhanced, indicating that PQS is a ligand for PqsR-driven transcriptional control [87]. Interestingly, PQS is not the only AHQ capable of upregulating AHQ expression via PqsR. HHQ can function as a ligand for PqsR enhancing binding to the *pqsA* promoter, and hence is responsible for its own synthesis and so can be defined as an autoinducer [88]. Interestingly, the EC_{50} for *pqsA* activation by HHQ is lower than that for PQS, indicating that the *pqsA* promoter is more sensitive to the other HHQ and PQS congeners with longer or shorter alkyl chains can also activate *pqsA* expression although the C7 chain length is optimal [173].

7.4.2
Function of AHQs in *P. aeruginosa*

One function of PQS is to upregulate the *rhl* system and cooperatively control expression of various target genes. Several studies have demonstrated this activity. In *pqsE* and *pqsR* mutants, pyocyanin production [63,81,82], *phzA1* expression [89], lectin production [63], elastase production and rhamnolipid levels were considerably reduced compared with wild-type [63,82]. PQS [63,89], HHQ or HHQNO [89] addition could not restore secondary metabolite production in these mutants, presumably because of a lack of a functional PqsE [81]. Whole-genome expression studies have examined the subset of genes regulated by the AHQ system [89]. A *pqsR* mutant displayed 22 genes whose expression had been repressed by *pqsR*, with another 121 genes whose expression was enhanced by *pqsR* compared to the wild-type. Expression of *pqs* and *phn* operons was abolished, and that of *pqsR* reduced. In addition, the transcription of *phz* (phenazine), *hcnAB* (hydrogen cyanide), *chiC* (chitinase), *mexI–opmD* (broad substrate efflux pump), and *lecA* and *lecB* (lectins) was also reduced. Interestingly, a subset of genes which were not coregulated with *rhl* was also found. Other studies analyzed the whole-gene expression profiles of a *pqsH* mutant in the presence and absence of PQS in comparison with that of a *pqsR* mutant [88]. It was shown that there was a less than 2-fold change of all genes regulated by *pqsR* in the *pqsH* mutant and that this was not increased when PQS was added. This therefore implies that HHQ, which, in common with PQS binds to PqsR, is also capable of activating the expression of many of the genes that are down regulated in a *pqsR* mutant [88]. However, HHQ is not as effective as PQS in driving the expression of genes such as *lecA* [89]. These studies reveal the importance of AHQs for full expression of *P. aeruginosa* virulence determinants and secondary metabolites.

AHQs (especially PQS) have not just been shown to act as QS signal molecules since PQS also chelates ferric iron, an essential bacterial nutrient. Brendenbruch *et al.* [90] and Diggle *et al.* [89] demonstrated the induction of siderophore-mediated iron acquisition systems such as the pyochelin and pyoverdin biosynthetic gene clusters and *pvdS* (which codes for an extracytoplasmic function σ factor required for the expression of the pyoverdin biosynthesis genes) when PAO1 is exposed to PQS. This is presumably because PQS induces an iron starvation response by chelating the growth medium iron [90]. Using methyl-PQS because of its increased solubility in methanol, iron-binding affinity constants show that between a pH range of 6–8, PQS binds Fe(III) most favorably in a ratio of 3 or 2:1, even at low iron concentrations, with approximately equal concentrations of each species present [91]. Interestingly, HHQ is unable to bind iron and does not form a complex as it lacks the 3′-OH group necessary for iron chelation. Interestingly, a PQS–Fe(III) complex cannot be used as a siderophore by a pyoverdin/pyochelin double mutant to aid bacterial growth. Hence, it has been speculated that PQS could transfer its bound iron to receptor-associated apo-siderophores within the cell envelope, as they have higher affinity for this metal than PQS. This would enable the cell to utilize the iron bound to PQS [91]. A summary of the contribution of PQS to QS in *P. aeruginosa* is shown in Figure 7.4.

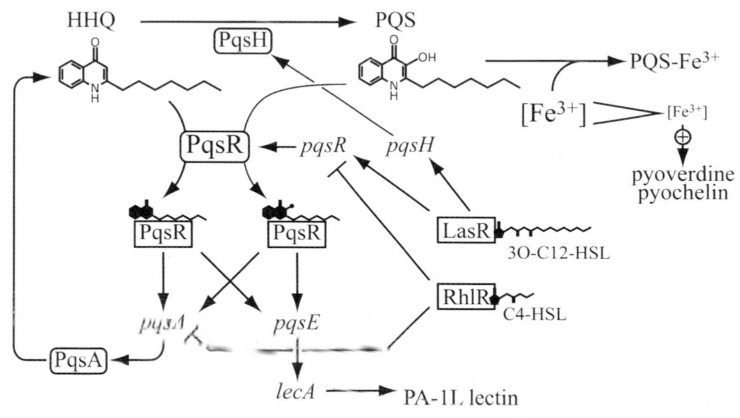

Figure 7.4 A model for AHQ- and AHL-dependent QS in *P. aeruginosa*. HHQ, the immediate precursor of PQS, drives the expression of *pqsA* via PqsR and hence HHQ functions as an autoinducer driving its own biosynthesis. HHQ is converted to PQS by the action of PqsH. PQS in common with HHQ induces the expression of *pqsA* in a PqsR-dependent manner. However, genes such as *lecA* and those required for pyocyanin biosynthesis also require PqsE. Furthermore, PQS released from the cell is capable of binding iron to form a PQS–Fe^{3+} complex. The removal of iron from the extracellular environment by PQS induces iron acquisition genes in both a PqsR- and PqsE-independent manner, and is not related to cell–cell signaling in *P. aeruginosa*. AHQ- and AHL-dependent QS are intimately linked since LasR/3-oxo-C12-HSL is required for full expression of *pqsH*, while *pqsR* is positively regulated by LasR/3-oxo-C12-HSL but repressed by RhlR/C4-HSL. (→) Denotes positive regulation; (⊣) denotes repression.

7.4.3
AHQs in Other Pseudomonads

As production of PQS in *P. aeruginosa* utilizes anthranilate derived from the common tryptophan biosynthetic pathway, other bacterial species may be capable of making similar molecules. This is indeed the case and AHQs have been found in other microorganisms. HHQ and 2-pentyl-4-quinolone (PHQ) are produced by the marine organism *Pseudomonas bromoulitis* [92], whereas NHQ and NHQNO are produced by new species of *Pseudomonas* harvested from the surface of the marine sponge *Homophymia* from New Caledonia [93], and 2-pentyl-4-quinolone PHQ plus HHQ were found in a marine *Alteromonas* SWAT5 strain [94]. Diggle *et al.* demonstrated that *Burkholderia pseudomallei* and *B. thailandensis* have functional *pqsABCDE* operons (termed *hhqABCDE*). An AHQ bioreporter was used to unequivocally identify HHQ in *P. putida* and *B. cenocepacia* and was coupled with LCMS/MS to identify HHQ, NHQ, and 2-heptyl-4-hydroxyquinoline-N-oxide in *B. pseudomallei* [95]. It is currently not known how biologically significant these molecules are to these organisms, but disruption of AHQ signaling in *B. pseudomallei* results in altered colony morphology and increased elastase synthesis [95]. Interestingly, to date, PQS has not been identified in any other organism than *P. aeruginosa* and whether PQS gives *P. aeruginosa* a competitive edge in certain environmental niches is currently unknown. It has recently been suggested that AHQs produced by different microorganisms that share the same niche may be used for cross-talk between them although this remains to be demonstrated [96].

7.5
Gac/Rsm System in Pseudomonads

In addition to controlling genes and operons at the transcriptional level, bacteria also frequently regulate gene expression at the posttranscriptional level, adjusting the rates at which specific transcribed mRNAs are translated or recycled into nucleotides. Global gene regulation following QS also occurs posttranscriptionally, and this is mediated by sets of small, noncoding regulatory RNAs and small RNA-binding proteins [97]. In the pseudomonads, the best known global posttranscriptional regulation consists of the Gac two-component system, the Rsm posttranscriptional regulator elements and a still uncharacterized signal molecule(s) that ultimately controls the positive or negative balance of the entire pathway. Initially discovered in *P. syringae* pv. *syringae* B728a as an essential factor for ecological fitness and plant pathogenicity [98,99], the GacS sensor kinase is anchored in the cytoplasmic membrane and senses a diffusible extracellular signal molecule(s) produced by the cells themselves [100]. Activation of GacS leads to its autophosphorylation, followed by a phosphotransfer cascade that ends with the phosphorylation of the response regulator GacA [101,102]. First described in the plant-beneficial *P. fluorescens* strain CHA0 for controlling global antibiotic and cyanide production, GacA function is essential for the production of several secondary metabolites required for the protection of plant roots against fungal infections [103]. GacS and GacA homologs are not restricted to the pseudomonads, but have been described in a great variety of bacteria belonging essentially to the γ-proteobacteria group [104].

The GacA/S system with its uncharacterized signal molecule(s) exerts its effect on gene regulation indirectly via the RsmA/CsrA family of posttranscriptional regulators, originally discovered in *Escherichia coli* as the carbon storage regulator (CsrA) [105] and the later described in *Erwinia* spp. as RsmA (regulator of secondary metabolism) [106]. Homologs these posttranscriptional regulators are found in a wide variety of eubacteria from both Gram-negative and Gram-positive groups [107]. RsmA, the best known homolog in *Pseudomonas*, is a small, around 7-kDa protein that forms dimers which avidly bind target mRNAs in the cells and either block their translation and promote their degradation or enhance their stability and extend their translatable half-life [107]. The mechanistic differences between positive and negative effects on mRNA stability and translation rates are not yet fully understood, but may involve the binding of the dimers to different places in the transcript, the negative effect usually being observed when RsmA binds close to the ribosome binding sites [108]. The global effects of RsmA are abrogated once phosphorylated GacA activates the transcription of small regulatory RNAs, for which RsmA homologs have a high affinity. Abundant production of these small regulatory RNAs is thought to quickly titrate out the RsmA dimers from their target mRNAs, thus antagonizing their action [109]. In the pseudomonads, depending on the strains, the Gac/Rsm posttranscriptional regulatory systems are found either as the main QS system controlling secondary metabolism or as a parallel mechanism for posttranscriptionally fine-tuning other QS circuits.

7.5.1 Gac/Rsm as the Principal QS System

Some *Pseudomonas* strains which do not possess the AHL- or PQS-dependent QS circuitry rely instead on the Gac/Rsm system to adjust their secondary metabolism according to their population density. This is the case for *P. fluorescens* strain CHA0, in which RsmX, RsmY and RsmZ are three small regulatory RNAs that are produced when the Gac system is activated [110]. These three noncoding RNAs of about 120 nucleotides each share little sequence conservation with each other except for the presence of short, repetitive ribosome-binding site-like sequences frequently occurring on single stranded regions of predicted short hairpin structures, a characteristic of this family of riboregulators [109]. Conserved features among *rsmX*, *rsmY* and *rsmZ*, the genes for the small RNAs antagonizing RsmA in *Pseudomonas* spp., are an upstream activating sequence in their promoter necessary for efficient transcription and thought to be a GacA-binding site, and a rho-independent transcription terminator at their 3′-end. These features have been useful in the prediction of similar regulatory RNAs in other eubacteria [111]. The three small RNA appear to be expressed in a population density-dependent manner and have redundant functions, as in strain CHA0 the simultaneous deletion of any one or two of these genes only results in the partial loss of production of hydrogen cyanide, exoprotease or the antifungal antibiotic 2,4-diacetylphloroglucinol. Complete loss of production of these secondary metabolites and of the biocontrol activity is only achieved in the triple *rsmXYZ* mutant [110]. As transcription of these three genes is highly dependent on a functional Gac system, the lack of production of the three regulatory RNAs explains why these same phenotypes are observed in both *gacA* and *gacS* mutants.

With the exception of *P. aeruginosa*, most other *Pseudomonas* species whose genomes have been sequenced appear to encode more than one RsmA homolog, up to four in the case of *P. syringae* pv. *tomato* DC3000 [107]. All these homologs have highly conserved primary sequences and the few differences may possibly confer some form of target specificity. However, apart from RsmA, only RsmE, the additional homolog in *P. fluorescens* CHA0, has been studied. Overexpression of *rsmA* or *rsmE* in strain CHA0 causes the same marked reductions in secondary metabolite production as do *gacA* or *gacS* mutations [112]. Although the expression levels of *rsmA* and *rsmE* in strain CHA0 are slightly different during growth in that *rsmE* appears to be more strongly induced by increasing population density [110], differences in the specificities of the proteins they encode towards regulatory or target RNAs remain to be determined. Nevertheless, the prevalence and inherent complexity of multiple RsmA homologs in most strains highlights the importance of global posttranscriptional regulation in the biology of the pseudomonads.

Strains belonging to various species including *P. aeruginosa* PAO1, *P. fluorescens* CHA0, *P. fluorescens* SBW25 and *P. corrugata* LMG2172 produce signal molecules that are organic solvent-extractable from liquid culture supernatants. When crude extracts from these organisms are added to *P. fluorescens* CHA0, they stimulate the expression of *rsmX*, *rsmY* and *rsmZ*, and as a consequence enhance the production of secondary metabolites such as antibiotics [113]. These signal molecules are not related to AHL, AI-2 or peptides [114] and unlikely to be AHQ as strain CHA0 lacks the *pqs* biosynthetic genes. Very little Gac signal is produced by *P. aeruginosa* PAO1 or *P. fluorescens* CHA0 when *gacA* is mutated, indicating that the signal biosynthesis genes follow an autoinduction loop in the regulatory cascade [113]. In addition, *gac* mutants are also signal-blind [100]. Although thiamine metabolism is essential for this signal to be produced in strain CHA0, the chemistry and biochemistry of the compounds remain to be elucidated [113].

7.5.2
Gac/Rsm as a QS Fine-tuning System

In the Pseudomonas genomes so far sequenced, up to three small regulatory RNAs that are thought to antagonize up to four RsmA homologs have been found; however, some species such as *P. aeruginosa* PAO1 or PA14 only produce two such RNAs, RsmY and RsmZ, and one RsmA homolog [107]. In this organism, proteomic analysis has shown that the Rsm system mediates most of the functions of the Gac system, which is required for full virulence in burnt mice, nematodes, insects and plants [115–117]. The reduced virulence of *gac* mutants results from a reduced production of virulence factors that are positively regulated by the parallel AHL- and PQS-dependent QS systems [60,115,118–120]. As in *P. fluorescens* CHA0, the Gac system of *P. aeruginosa* regulates the production of certain virulence factors directly at the level of their biosynthetic gene mRNA translation. However, an additional level of control is applied as the Gac system also positively controls the AHL-dependent QS system [58,60,118] and as a consequence it indirectly controls the transcription of a wide set of extracellular virulence factors [38,121] (Figure 7.5). As an example, the *hcnABC* operon coding for the HCN synthase complex is simultaneously controlled at the transcriptional and at the posttranscriptional levels by the Gac/Rsm system [118,122]. In addition to being phosphorylated by GacS upon stimulation by an autoinduced signal, activation of the response regulator GacA activity is also considered to be modulated by two other inner membrane-bound sensor kinases. On one hand, RetS, involved in the regulation of exopolysaccharide production and type III secretion, has opposite effects to GacS and is thought to influence the phosphorylation state of GacA [123]. On the other hand, LadS, which causes loss of adhesion when mutated, has a positive effect on GacA activation [124]. Both RetS and LadS could potentially be sensing additional QS signals which may or may not be related to the molecule(s) sensed by GacS. These complex regulatory networks and their interactions are likely to allow the rapid fine-tuning of specific gene expression

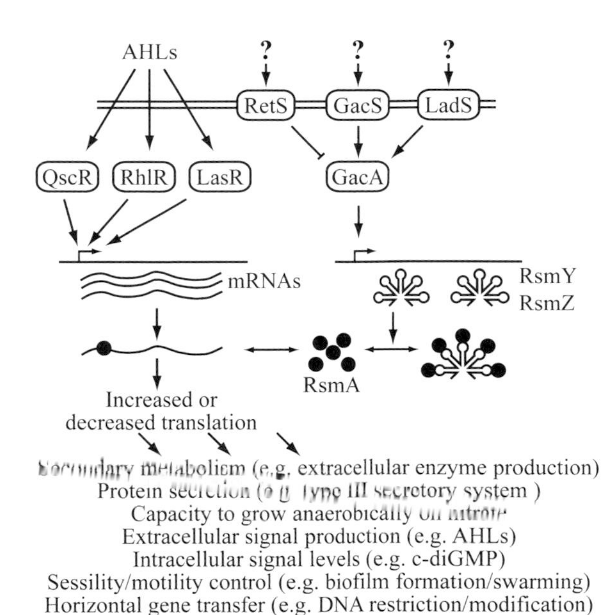

Figure 7.5 The Gac/Rsm system in *P. aeruginosa*. The activation of the GacS sensor kinase takes place upon interaction with an unknown diffusible signal molecule(s) leading to its autophosphorylation and the phosphotransfer to GacA. The activated GacA exerts its effect on gene regulation via the regulatory RNAs RsmY and RsmZ, which in turn interact with the posttranscriptional regulatory protein RsmA antagonizing its effects on target mRNAs. The sensor kinases RetS and LadS are thought to be activated by additional signal molecules and channel their action through GacA with opposite effects. The Gac/Rsm systems are interlinked with the AHL QS system as some of the mRNAs transcribed upon activation of QscR, RhlR and LasR by AHLs are posttranscriptionally regulated by RsmA.

at the transcriptional and posttranscriptional levels by integrating simultaneously various QS signals from the environment, and converging to drive the appropriate cellular responses.

7.6
QS and Infection

P. aeruginosa is widely regarded as a versatile and highly adaptable organism, capable of survival in a variety of habitats. This ability is attributed to the possession of a large and complex genome which is regulated by a number of integrated systems, including cell population density-dependent QS. Within the human host, *P. aeruginosa* is an aggressive opportunistic pathogen and QS plays a key role in the coordination of an arsenal of virulence determinants, which lead to a range of acute and chronic disease states. These include acute pneumonia, septicemia, burn wound infections and chronic lung infections in patients with cystic fibrosis (CF).

7.6.1
Role of QS in the Pathogenesis of *P. aeruginosa* Infections

By regulating gene expression and virulence determinant production in response to cell population density, QS enables the invading *P. aeruginosa* population to maximize its chances of establishing infection in the host. Using QS this organism can evade host immune detection and clearance whilst low in number, and launch a unified attack to overwhelm the host when a sufficient population size has been attained.

7.6.2
Evidence for QS *In vivo*

The first line of evidence for the relevance of QS in infection comes from the detection of active QS systems in clinical samples. Sputum from CF patients has been found to contain mRNA transcripts for *lasR* and *lasI*, the levels of which correlate with the levels of transcripts for various QS-controlled genes, suggesting that AHLs regulate the expression of various virulence factors during human infections [125]. Furthermore, AHL molecules have been directly detected in experimental animal models and human infections [126,127].

In addition to AHLs, a role for AHQs in establishing infection, regulating virulence factors and enhancing the severity of infection has been suggested in several studies. Clinical isolates from CF patients produce several AHQs including HHQNO, NHQNO and 2-undecyl-3-hydroxy-4-quinoline-N-oxide plus their unsaturated derivatives, HHQ [128] and also PQS [129]. PQS was also found in isolates from infants with CF and early-stage *Pseudomonas* infections [130].

Further evidence of a role for QS in the regulation of virulence *in vivo* comes from evaluating strains of *P. aeruginosa* that contain deletions in one or more QS genes using various experimental infection models. These include a neonatal mouse model of acute pneumonia and a burned-mouse model of infection [131–133]. In these studies, the mutant strains were found to induce less tissue destruction and reduced cases of pneumonia than wild-type *P. aeruginosa*. Interestingly, the AHL QS mutants were not completely avirulent. This emphasizes the multifactorial nature of virulence and the fact that other factors play a role in regulating pathogenesis. AHQs have also been shown to be important in the regulation of virulence and severity of infections in moth larvae, nematode assays and a burned-mouse model [81,82,89,116].

7.6.3
QS and Biofilms in *P. aeruginosa*

Acute *P. aeruginosa* infections are caused by free-swimming planktonic cells that multiply and spread rapidly within the host tissues. However, it is also apparent that *P. aeruginosa* is capable of growing as surface-associated biofilms, and this mode of growth is a strategy for causing chronic, persistent infections since biofilms are highly tolerant of antibiotics, disinfectants and the actions of the immune system. Evidence for the occurrence of *P. aeruginosa* biofilms in the CF lung has been obtained using both light and electron microscopy of sputum and postmortem samples [134,135]. The observation that the ratio of AHLs produced by CF sputum infected with *P. aeruginosa* was the same as biofilm grown organisms has been taken as further physiological evidence for the occurrence of the biofilm lifestyle in the CF lung [136]. Free-swimming organisms grown as broth cultures produced more 3O-C12-HSL in contrast to the biofilm grown organisms and sputum samples, in which more C4-HSL was detected.

Given that biofilms represent communities of cells, it was proposed that bacterial communication in the form of QS may play a role in their development [137]. The first evidence linking QS to biofilm development was based on the hypothesis that as QS requires a sufficient population density, signals would not be expected to participate in the initial stages of biofilm formation, attachment and proliferation. However, they may be involved in differentiation. In support of this, it was demonstrated that *lasI* mutants incapable of synthesizing 3O-C12-HSL formed biofilms with abnormal structure (thin and undifferentiated) that were more sensitive to the detergent biocide SDS [17].

Alongside this, O'Toole and Kolter [138] conducted a simple genetic screen in which random transposon mutants of *P. aeruginosa* were grown in microtiter plates and those that did not form biofilms were investigated further. They found that flagellar motility was required for primary adhesion and that type IV pili were essential for cellular aggregation. In a following paper, Glessner *et al.* reported that both the *las* and *rhl* QS systems were required for type IV pilus-dependent twitching motility. C4-HSL in particular appeared to influence both the export and surface assembly of surface type IV pili, while 3O-C12-HSL played a role in maintaining cell–cell spacing and associations required for effective twitching motility [139]. However, subsequent work revealed that AHL-dependent QS does not regulate twitching motility [140]. Stable twitching defective variants accumulate during the culture of *lasI* and *rhlI* mutants as a consequence of secondary mutations in *vfr* and *algR*. These two genes encode key regulators which control a variety of phenotypes, including twitching motility. Consequently, the authors suggested that mutations in one regulatory system may create distortions that select during subsequent culturing for compensatory mutations in other regulatory genes within the cellular network [140].

Subsequent research has demonstrated that many different QS-regulated functions may affect biofilm development. For example, *P. aeruginosa* produces two sugar-binding lectins, LecA and LecB, which are expressed in biofilms. The production of these lectins is QS regulated, and both *lecA* and *lecB* mutant strains formed abnormal biofilms [141,142]. There is evidence that the QS-regulated surfactant rhamnolipid is necessary for maintaining the open spaces between cell aggregates in structured biofilms [143]. Rhamnolipid production may aid in the formation of mature mushroom structures [144]. Recently, Sakuragi and Kolter have shown that AHL-dependent QS directly regulates expression of the *pel* operon. This operon is responsible for the synthesis of a glucose-rich exopolysaccharide component of the biofilm matrix. Transcription of the *pel* operon is greatly reduced in *lasI* and *rhlI* mutants, and could be restored by provision of exogenous 3O-C12-HSL [145].

However, QS is not essential for biofilm development. The initial studies demonstrating a role for *las*-regulated functions were challenged by the report that wild-type and QS mutant strains formed flat biofilms that were structurally indistinguishable under certain environmental growth conditions [146]. It is increasingly evident that biofilm formation and maturation and the role of QS are dynamic processes which adapt in response to environmental cues such as nutrient availability [147]. For some conditions, QS may play a crucial role in the development of biofilm communities; yet in others, biofilm formation itself may represent a means to achieve a quorum.

Within an established biofilm, the close proximity of cells and diffusion limited environment would seem the ideal setting for the functioning of QS, and as such there is evidence that it plays an active role. The intrinsic antibiotic resistance of *P. aeruginosa* is well recognized, yet it is apparent that bacteria embedded in biofilms are much more tolerant of antimicrobials than their planktonic counterparts [148]. Several mechanisms have been proposed for this observation, including the failure of antibiotics to penetrate the full depth of the biofilm, an altered chemical microenvironment within the biofilm affecting antimicrobial action and speculation that some biofilm bacteria may adopt a protective, "spore-like" state [149]. In addition, there are several reports of QS-deficient *P. aeruginosa* mutants which show reduced tolerance to antimicrobial agents when cultured as a biofilm. For example, using viability staining, a QS-deficient biofilm was almost entirely eradicated by tobramycin treatment in contrast with the wild-type, in which only cells in the top layer of the biofilm were killed [150].

AHQs have been shown to be important in maintaining and establishing biofilms. Addition of exogenous PQS to a growing culture of PAO1 enhanced biofilm development, possibly due to an increased effect on lectin A production [63]. An interesting contribution of AHQ signaling to biofilm development has also been proposed [151]. *P. aeruginosa* produces extracellular DNA which can function as a cell–cell interconnecting matrix component in biofilms [152]. This DNA is similar to whole-genome DNA and therefore could be generated via lysis of a bacterial subpopulation. It is released maximally in late log phase, which correlates with maximal PQS release [63,80]. Furthermore, strains with mutations in *pqsA* release only low levels of DNA and their biofilms contain less extracellular DNA than biofilms formed by the wild-type strain [149].

A further role for *P. aeruginosa* QS has been demonstrated in mixed species biofilms, which are thought to occur widely in nature. In the CF lung, *P. aeruginosa* and the emerging pathogen *B. cepacia* may coexist in biofilms, and this environment may provide the means for interaction between these two organisms. Both species are know to employ QS and there is evidence for unidirectional cross-talk with *B. cepacia* capable of perceiving the signals of *P. aeruginosa*, but the latter not responding to the signals of *B. cepacia* [153]. Interestingly, although *Staphylococcus aureus* uses a different peptide-based QS system, the exogenous addition of 3O-C12-HSL to cultures of *S. aureus* affected virulence-determinant production by this organism, suggesting that QS may promote the displacement of *S. aureus* by *P. aeruginosa* in the CF lung [154].

7.6.4
QS Molecules and the Host

QS enables pathogenic bacteria to coordinate their disease-causing strategies in response to interactions with each other and their environment. Increasingly, it is recognized that the QS signal molecules themselves can interact with eukaryotic cells and hence directly affect the host. For example, there are numerous reports that 3O-C12-HSL is a potent immune modulator which impacts on cytokine production and immune cell activity *in vitro* and *in vivo* [155]. In addition, it this AHL has been shown to exert a pharmacological effect on the cardiovascular system, inhibiting vasoconstrictor tone of both pulmonary and coronary blood vessels from the pig [156], and triggering bradycardial effects in rats [157]. Hence, this QS signal molecule may function as a virulence determinant in its own right, and enhance *P. aeruginosa* survival in the host by increasing nutrient supply via the bloodstream and down-regulating host defenses. Furthermore, PQS is a more potent immune modulator than 3O-C12-HSL with which it is capable of acting synergistically to inhibit T cell proliferation [158].

Whilst QS signal molecules may exert a direct effect on host tissues, there is evidence that QS-controlled functions also play a role. When human polymorphonuclear leukocytes were incubated with supernatants from QS-competent *P. aeruginosa* there was rapid necrosis of these immune cells both *in vivo* and *in vitro*. The cytotoxin responsible was identified as rhamnolipid B, which belongs to a class of well-known biosurfactants from *P. aeruginosa* and is QS regulated [159]. In addition, QS may allow *P. aeruginosa* to respond to host immune activation. It was observed that the cytokine interferon-γ binds to the outer membrane protein OprF resulting in the induction of a QS-dependent virulence determinant LecA lectin [160]. Further work has demonstrated that the endogenous opioid, dynorphin, released during severe host stress, directly activates AHQ signaling in *P. aeruginosa* and enhances virulence [161]. Hence, this opportunistic pathogen can exploit a weakened host by actively sensing and responding to alterations in immune function and the presence of stress-induced physiological disturbance via QS mechanisms.

7.6.5 Disruption of QS Represents a Novel Approach for Managing *P. aeruginosa* Infections

The discovery of antibiotics in the early 20th century marked the beginning of a revolution in the treatment of infectious diseases. However, today, many species of bacteria once easily treated with antibiotics exhibit resistance to many classes of these potent drugs. It is therefore desirable to develop new treatment strategies for bacterial infections. The fact that bacterial cells use signal molecules to transmit information between each other presents a unique opportunity to interfere with QS by removing the signal by a process known as quorum quenching (QQ) [162]. Given the significant contribution of QS to the virulence of *P. aeruginosa*, disruption of cell–cell communication represents a novel approach to managing infections caused by this pathogen. This is a particularly attractive option for a number of reasons. As virulence rather than viability is attenuated, this approach would not exert such a great selection pressure for the occurrence of resistant strains, which is the major drawback to conventional antibiotic-mediated killing or growth arrest. In addition, systems controlled by extracellular small molecules are ideal targets for which novel drugs can be designed.

Inspiration can be sought from the action of eukaryotic host tissues themselves which have evolved mechanisms to disarm the *P. aeruginosa* QS systems as part of their innate defense against this invading pathogen. Differentiated human airway epithelia inactivate 3O-C12-HSL by a cell-associated enzymatic activity, [163]. Further work has shown that this degradation is due to the lactonase activity of the human paraoxonases (encoded by the *PON* genes) which inactivate 3O-C12-HSL by opening the homoserine lactone ring. It has also been shown that in mouse serum PON1, PON2 and PON3 hydrolyze 3O-C12-HSL,while PON1 can prevent QS and biofilm development [164]. It therefore appears that the mammalian host has developed systems for interrupting bacterial cell–cell communication and is an area worthy of future study. However, it is not just mammalian systems that provide strategies for inactivating QS and degrading QS signal molecules (see below).

So far, QQ has focused on AHLs and three main mechanisms of degradation have been described. First, it was shown that AHL signals are rendered biologically inactive in alkaline environments due to chemical pH-dependent lactonolysis and, therefore, in certain environmental niches signaling may be ineffective [165]. Lactonolysis involves hydrolysis of the homoserine lactone ring. In addition, lactonolysis can be enzymatically driven and AHL lactonases have now been isolated from several bacterial species and their potential to block AHL QS-mediated responses established. The first lactonase described was AiiA from *Bacillus* sp. 240B1. Expression of the *aiiA* gene in *P. aeruginosa* significantly reduced the levels of AHLs and reduced the production of virulence determinants, although it failed to interfere with surface colonization [120]. Another way of inactivating an AHL signal is via acylase activity where there is cleavage of the amide bond, thus releasing the homoserine lactone and the fatty acid side-chain. A number of bacterial species have been shown to degrade AHL signals by hydrolyzing the amide bond and specific acylases have been identified in *Ralstonia*, *P. aeruginosa* and *Streptomyces*. The *aiiD* gene was cloned from a *Ralstonia* sp. and introduced into *P. aeruginosa* where it quenched QS signals, and decreased its ability to swarm and to produce elastase and pyocyanin [166]. *P. aeruginosa* produces at least two endogenous acylases, PvdQ and QuiP. PvdQ degrades AHLs with side-chains ranging in length from 11 to 14 carbons at physiologically relevant, low, concentrations. Of the two main AHL signal molecules of *P. aeruginosa* PAO1 only 3-oxo-C12-HSL is susceptible to the enzyme. Addition of purified PvdQ to *P. aeruginosa* PAO1 cultures completely inhibited accumulation of 3-oxo-C12-HSL,and consequently reduced production of the virulence factors elastase and pyocyanin. Similar results were obtained when *pvdQ* was overexpressed in *P. aeruginosa* [167].

In addition to this "QQ" activity, there has been much interest in the discovery of QS antagonists. The first naturally occurring agents found to possess QS antagonistic activity were brominated furanones produced by the Australian macroalga *Delisea pulchra* [168]. Biofilms grown in the presence of synthetic furanone derivatives were rendered susceptible to antimicrobial killing with the antibiotic tobramycin and detergent SDS in contrast to the untreated biofilms [43]. Filamentous fungi produce a large variety of secondary metabolites including β-lactam antibiotics. During a screening of extracts from different *Penicillium* species, the phytotoxins patulin and penicillic acid were identified as being biologically active QS-inhibitory compounds [169]. Garlic, renowned for its antifungal, anticancer and antimicrobial activities, has also been shown to specifically inhibit QS-regulated gene expression in *P. aeruginosa*, as demonstrated by DNA microarray-based transcriptomic analysis [170]. The mechanism by which garlic blocks QS is unknown; however, clear *in vitro* improvement of *P. aeruginosa* biofilm susceptibility to antibiotics and immune cells and *in vivo* clearance of pulmonary infection in a mouse model have been demonstrated [171].

Modulation of the host immune system is an additional potential mechanism for tackling *P. aeruginosa* QS regulated infections. Synthetic analogs of the 3O-C12-HSL signal molecule have been shown to possess immune modularity activity. Experimental vaccine development has demonstrated that active immunization with a 3O-C12-HSL–carrier protein conjugate protected mice from lethal *P. aeruginosa* lung infection. A specific antibody to the 3O-C12-HSL molecule played a protective role in the acute infection, probably by blocking of host inflammatory responses without altering the lung bacterial burden [172].

7.7 Conclusions

Studies of QS-mediated signaling processes in bacteria have revealed the existence of intricate interlinked regulatory networks that enable bacterial populations to fine tune their responses to environmental changes and increase their chances of survival. Within the pseudomonads three major QS signaling systems have been described; these are the AHL, AHQ and Gac systems. These systems are interlinked, and play an important role in controlling the production of virulence determinants, secondary metabolism and biofilm development. Understanding how these QS-controlled mechanisms enable some pseudomonads to behave as cunning and sophisticated pathogens provides novel opportunities for managing infection by attenuating their virulence.

References

1 Williams, P., Winzer, K., Chan, W.C. and Camara, M. (2007) *Philos Trans Roy Soc B*, **362**, 1119–1134.

2 Venturi, V. (2006) *FEMS Microbiol Rev*, **30**, 274–291.

3 Nealson, K.H. and Markovit, A. (1970) *J Bacteriol*, **104**, 300–312.

4 Eberhard, A., Burlingame, A.L., Eberhard, C., Kenyon, G.L., Nealson, K.H. and Oppenheimer, N.J. (1981) *Biochemistry*, **20**, 2444–2449.

5 Bainton, N.J., Stead, P., Chhabra, S.R., Bycroft, B.W., Salmond, G.P.C., Stewart, G.S.A.B. and Williams, P. (1992) *Biochem J*, **288**, 997–1004.

6 Bainton, N.J., Bycroft, B.W., Chhabra, S.R., Stead, P., Gledhill, L., Hill, P.J., Rees, C.E.D., Winson, M.K., Salmond, G.P.C., Stewart, G.S.A.B. and Williams, P. (1992) *Gene*, **116**, 87–91.

7 Engebrecht, J., Nealson, K. and Silverman, M. (1983) *Cell*, **32**, 773–781.

8 Egland, K.A. and Greenberg, E.P. (1999) *Mol Microbiol*, **31**, 1197–1204.

9 Camara, M., Williams, P. and Hardman, A. (2002) *Lancet Infect Dis*, **2**, 667–676.

10 Winzer, K. and Williams, P. (2001) *Int J Med Microbiol*, **291**, 131–143.

11 Withers, H., Swift, S. and Williams, P. (2001) *Curr Opin Microbiol*, **4**, 186–193.

12 Gambello, M.J. and Iglewski, B.H. (1991) *J Bacteriol*, **173**, 3000–3009.

13 Pearson, J.P., Gray, K.M., Passador, L., Tucker, K.D., Eberhard, A., Iglewski, B.H. and Greenberg, E.P. (1994) *Proc Natl Acad Sci USA*, **91**, 197–201.

14 Passador, L., Cook, J.M., Gambello, M.J., Rust, L. and Iglewski, B.H. (1993) *Science*, **260**, 1127–1130.

15 Seed, P.C., Passador, L. and Iglewski, B.H. (1995) *J Bacteriol*, **177**, 654–659.

16 Bottomley, M.J., Muraglia, E., Bazzo, R. and Carfi, A. (2007) *J Biol Chem*, **282**, 13592–13600.

17 Davies, D.G., Parsek, M.R., Pearson, J.P., Iglewski, B.H., Costerton, J.W. and Greenberg, E.P. (1998) *Science*, **280**, 295–298.

18 ChaponHerve, V., Akrim, M., Latifi, A., Williams, P., Lazdunski, A. and Bally, M. (1997) *Mol Microbiol*, **24**, 1169–1178.

19 Toder, D.S., Gambello, M.J. and Iglewski, B.H. (1991) *Mol Microbiol*, **5**, 2003–2010.

20 Gambello, M.J., Kaye, S. and Iglewski, B.H. (1993) *Infect Immun*, **61**, 1180–1184.

21 Ochsner, U.A., Koch, A., Fiechter, A. and Reiser, J. (1994) *J Bacteriol*, **176**, 2044–2054.

22 Brint, J.M. and Ohman, D.E. (1995) *J Bacteriol*, **177**, 7155–7163.

23 Latifi, A., Winson, M.K., Foglino, M., Bycroft, B.W., Stewart, G.S.A.B., Lazdunski, A. and Williams, P. (1995) *Mol Microbiol*, **17**, 333–343.

24 Winson, M.K., Camara, M., Latifi, A., Foglino, M., Chhabra, S.R., Daykin, M., Bally, M., Chapon, V., Salmond, G.P.C., Bycroft, B.W., Lazdunski, A.,

Stewart, G.S.A.B. and Williams, P. (1995) *Proc Natl Acad Sci USA*, **92**, 9427–9431.

25 Pearson, J.P., Pesci, E.C. and Iglewski, B.H. (1997) *J Bacteriol*, **179**, 5756–5767.

26 Winzer, K., Falconer, C., Garber, N.C., Diggle, S.P., Camara, M. and Williams, P. (2000) *J Bacteriol*, **182**, 6401–6411.

27 Ochsner, U.A. and Reiser, J. (1995) *Proc Natl Acad Sci USA*, **92**, 6424–6428.

28 Pearson, J.P., Passador, L., Iglewski, B.H. and Greenberg, E.P. (1995) *Proc Natl Acad Sci USA*, **92**, 1490–1494.

29 Latifi, A., Foglino, M., Tanaka, K., Williams, P. and Lazdunski, A. (1996) *Mol Microbiol*, **21**, 1137–1146.

30 Schuster, M., Urbanowski, M.L. and Greenberg, E.P. (2004) *Proc Natl Acad Sci USA*, **101**, 15833–15839.

31 Pearson, J.P., Van Delden, C. and Iglewski, B.H. (1999) *J Bacteriol*, **181**, 1203–1210.

32 Stover, C.K., Pham, X.Q., Erwin, A.L., Mizoguchi, S.D., Warrener, P., Hickey, M.J., Brinkman, F.S.L., Hufnagle, W.O., Kowalik, D.J., Lagrou, M., Garber, R.L., Goltry, L., Tolentino, E., Westbrock-Wadman, S., Yuan, Y., Brody, L.L., Coulter, S.N., Folger, K.R., Kas, A., Larbig, K., Lim, R., Smith, K., Spencer, D., Wong, G.K.S., Wu, Z., Paulsen, I.T., Reizer, J., Saier, M.H., Hancock, R.E.W., Lory, S. and Olson, M.V. (2000) *Nature*, **406**, 959–964.

33 Chugani, S.A., Whiteley, M., Lee, K.M., D'Argenio, D., Manoil, C. and Greenberg, E.P. (2001) *Proc Natl Acad Sci USA*, **98**, 2752–2757.

34 Ledgham, F., Ventre, I., Soscia, C., Foglino, M., Sturgis, J.N. and Lazdunski, A. (2003) *Mol Microbiol*, **48**, 199–210.

35 Lequette, Y., Lee, J.H., Ledgham, F., Lazdunski, A. and Greenberg, E.P. (2006) *J Bacteriol*, **188**, 3365–3370.

36 Lee, J.H., Lequette, Y. and Greenberg, E.P. (2006) *Mol Microbiol*, **59**, 602–609.

37 Juhas, M., Wiehlmann, L., Huber, B., Jordan, D., Lauber, J., Salunkhe, P., Limpert, A.S., von Gotz, F., Steinmetz, I., Eberl, L. and Tummler, B. (2004) *Microbiology*, **150**, 831–841.

38 Juhas, M., Wiehlmann, L., Salunkhe, P., Lauber, J., Buer, J. and Tummler, B. (2005) *FEMS Microbiol Lett*, **242**, 287–295.

39 Li, L.L., Malone, J.E. and Iglewski, B.H. (2007) *J Bacteriol*, **189**, 4367–4374.

40 Whiteley, M., Lee, K.M. and Greenberg, E.P. (1999) *Proc Natl Acad Sci USA*, **96**, 13904–13909.

41 Diggle, S.P., Winzer, K., Lazdunski, A., Williams, P. and Camara, M. (2002) *J Bacteriol*, **184**, 2576–2586.

42 Wagner, V.E., Bushnell, D., Passador, L., Brooks, A.I. and Iglewski, B.H. (2003) *J Bacteriol*, **185**, 2080–2095.

43 Hentzer, M., Wu, H., Andersen, J.B., Riedel, K., Rasmussen, T.B., Bagge, N., Kumar, N., Schembri, M.A., Song, Z.J., Kristoffersen, P., Manefield, M., Costerton, J.W., Molin, S., Eberl, L., Steinberg, P., Kjelleberg, S., Hoiby, N. and Givskov, M. (2003) *EMBO J*, **22**, 3803–3815.

44 Schuster, M., Lostroh, C.P., Ogi, T. and Greenberg, E.P. (2003) *J Bacteriol*, **185**, 2066–2079.

45 Albus, A.M., Pesci, E.C., RunyenJanecky, L.J., West, S.E.H. and Iglewski, B.H. (1997) *J Bacteriol*, **179**, 3928–3935.

46 De Kievit, T., Seed, P.C., Nezezon, J., Passador, L. and Iglewski, B.H. (1999) *J Bacteriol*, **181**, 2175–2184.

47 Rampioni, G., Bertani, I., Zennaro, E., Polticelli, F., Venturi, V. and Leoni, L. (2006) *J Bacteriol*, **188**, 815–819.

48 Rampioni, G., Polticelli, F., Bertani, I., Righetti, K., Venturi, V., Zennaro, E. and Leoni, L. (2007) *J Bacteriol*, **189**, 1922–1930.

49 Suh, S.J., Silo-Suh, L., Woods, D.E., Hassett, D.J., West, S.E.H. and Ohman, D.E. (1999) *J Bacteriol*, **181**, 3890–3897.

50 Whiteley, M., Parsek, M.R. and Greenberg, E.P. (2000) *J Bacteriol*, **182**, 4356–4360.

51 Schuster, M., Hawkins, A.C., Harwood, C.S. and Greenberg, E.P. (2004) *Mol Microbiol*, **51**, 973–985.

52 Heurlier, K., Denervaud, V., Pessi, G., Reimmann, C. and Haas, D. (2003) *J Bacteriol*, **185**, 2227–2235.

53 Thompson, L.S., Webb, J.S., Rice, S.A. and Kjelleberg, S. (2003) *FEMS Microbiol Lett*, **220**, 187–195.

54 Vallet, I., Diggle, S.P., Stacey, R.E., Camara, M., Ventre, I., Lory, S., Lazdunski, A., Williams, P. and Filloux, A. (2004) *J Bacteriol*, **186**, 2880–2890.

55 Westfall, L.W., Carty, N.L., Layland, N., Kuan, P., Colmer-Hamood, J.A. and Hamood, A.N. (2006) *FEMS Microbiol Lett*, **255**, 247–254.

56 Dong, Y.H., Zhang, X.F., Soo, H.M.L., Greenberg, E.P. and Zhang, L.H. (2005) *Mol Microbiol*, **56**, 1287–1301.

57 Dong, Y.H., Zhang, X.F., Xu, J.L., Tan, A.T. and Zhang, L.H. (2005) *Mol Microbiol*, **58**, 552–564.

58 Pessi, G., Williams, F., Hindle, Z., Heurlier, K., Holden, M.T.G., Camara, M., Haas, D. and Williams, P. (2001) *J Bacteriol*, **183**, 6676–6683.

59 Branny, P., Pearson, J.P., Pesci, E.C., Kohler, T., Iglewski, B.H. and Van Delden, C. (2001) *J Bacteriol*, **183**, 1531–1539.

60 Reimmann, C., Beyeler, M., Latifi, A., Winteler, H., Foglino, M., Lazdunski, A. and Haas, D. (1997) *Mol Microbiol*, **24**, 309–319.

61 Jensen, V., Lons, D., Zaoui, C., Bredenbruch, F., Meissner, A., Dieterich, G., Munch, R. and Haussler, S. (2006) *J Bacteriol*, **188**, 8601–8606.

62 Diggle, S.P., Cornelis, P., Williams, P. and Camara, M. (2006) *Int J Med Microbiol*, **296**, 83–91.

63 Diggle, S.P., Winzer, K., Chhabra, S.R., Chhabra, S.R., Worrall, K.E., Camara, M. and Williams, P. (2003) *Mol Microbiol*, **50**, 29–43.

64 Pesci, E.C., Milbank, J.B.J., Pearson, J.P., McKnight, S., Kende, A.S., Greenberg, E.P. and Iglewski, B.H. (1999) *Proc Natl Acad Sci USA*, **96**, 11229–11234.

65 Holden, M.T.G., Chhabra, S.R., de Nys, R., Stead, P., Bainton, N.J., Hill, P.J., Manefield, M., Kumar, N., Labatte, M., England, D., Rice, S., Givskov, M., Salmond, G.P.C., Stewart, G.S.A.B., Bycroft, B.W., Kjelleberg, S.A. and Williams, P. (1999) *Mol Microbiol*, **33**, 1254–1266.

66 Jones, S., Yu, B., Bainton, N.J., Birdsall, M., Bycroft, B.W., Chhabra, S.R., Cox, A.J.R., Golby, P., Reeves, P.J., Stephens, S., Winson, M.K., Salmond, G.P.C., Stewart, G.S.A.B. and Williams, P. (1993) *EMBO J*, **12**, 2477–2482.

67 Steidle, A., Allesen-Holm, M., Riedel, K., Berg, G., Givskov, M., Molin, S. and Eberl, L. (2002) *Appl Environ Microbiol*, **68**, 6371–6382.

68 Dubern, J.F., Lugtenberg, B.J.J. and Bloemberg, G.V. (2006) *J Bacteriol*, **188**, 2898–2906.

69 Bertani, I. and Venturi, V. (2004) *Appl Environ Microbiol*, **70**, 5493–5502.

70 Chin-A-Woeng, T.F.C., van den Broek, D., de Voer, G., van der Drift, K.M.G.M., Tuinman, S., Thomas-Oates, J.E., Lugtenberg, B.J.J. and Bloemberg, G.V. (2001) *Mol Plant Microbe Interact*, **14**, 969–979.

71 Mavrodi, D.V., Ksenzenko, V.N., Bonsall, R.F., Cook, R.J., Boronin, A.M. and Thomashow, L.S. (1998) *J Bacteriol*, **180**, 2541–2548.

72 Pierson, L.S., Keppenne, V.D. and Wood, D.W. (1994) *J Bacteriol*, **176**, 3966–3974.

73 El-Sayed, A.K., Hothersall, J. and Thomas, C.M. (2001) *Microbiology*, **147**, 2127–2139.

74 Wei, H.L. and Zhang, L.Q. (2006) *Antonie Van Leeuwenhoek*, **89**, 267–280.

75 Khan, S.R., Mavrodi, D.V., Jog, G.J., Suga, H., Thomashow, L.S. and Farrand, S.K. (2005) *J Bacteriol*, **187**, 6517–6527.

76 Cui, X., Harling, R., Mutch, P. and Darling, D. (2005) *Eur J Plant Pathol*, **111**, 297–308.

77 Laue, R.E., Jiang, Y., Chhabra, S.R., Jacob, S., Stewart, G.S.A.B., Hardman, A., Downie, J.A., O'Gara, F. and Williams, P. (2000) *Microbiology*, **146**, 2469–2480.

78 Quinones, B., Dulla, G. and Lindow, S.E. (2005) *Mol Plant Microbe Interact*, **18**, 682–693.

79 McKnight, S.L., Iglewski, B.H. and Pesci, E.C. (2000) *J Bacteriol*, **182**, 2702–2708.

80 Lepine, F., Deziel, E., Milot, S. and Rahme, L.G. (2003) *Biochem Biophys Acta*, **1622**, 36–41.

81 Gallagher, L.A., McKnight, S.L., Kuznetsova, M.S., Pesci, E.C. and Manoil, C. (2002) *J Bacteriol*, **184**, 6472–6480.

82 Cao, H., Krishnan, G., Goumnerov, B., Tsongalis, J., Tompkins, R. and Rahme, L.G. (2001) *Proc Natl Acad Sci USA*, **98**, 14613–14618.

83 Essar, D.W., Eberly, L., Hadero, A. and Crawford, I.P. (1990) *J Bacteriol*, **172**, 884–900.

84 Deziel, E., Lepine, F., Milot, S., He, J.X., Mindrinos, M.N., Tompkins, R.G. and Rahme, L.G. (2004) *Proc Natl Acad Sci USA*, **101**, 1339–1344.

85 Bredenbruch, F., Nimtz, M., Wray, V., Morr, M., Muller, R. and Haussler, S. (2005) *J Bacteriol*, **187**, 3630–3635.

86 Lepine, F., Milot, S., Deziel, E., He, J.X. and Rahme, L.G. (2004) *J Am Soc Mass Spectrom*, **15**, 862–869.

87 McGrath, S., Wade, D.S. and Pesci, E.C. (2004) *FEMS Microbiol Lett*, **230**, 27–34.

88 Xiao, G.P., Deziel, E., He, J.X., Lepine, F., Lesic, B., Castonguay, M.H., Milot, S., Tampakaki, A.P., Stachel, S.E. and Rahme, L.G. (2006) *Mol Microbiol*, **62**, 1689–1699.

89 Deziel, E., Gopalan, S., Tampakaki, A.P., Lepine, F., Padfield, K.E., Saucier, M., Xiao, G.P. and Rahme, L.G. (2005) *Mol Microbiol*, **55**, 998–1014.

90 Bredenbruch, F., Geffers, R., Nimtz, M., Buer, J. and Haussler, S. (2006) *Environ Microbiol*, **8**, 1318–1329.

91 Diggle, S.P., Matthijs, S., Wright, V.J., Fletcher, M.P., Chhabra, S.R., Lamont, I.L., Kong, X.L., Hider, R.C., Cornelis, P., Camara, M. and Williams, P. (2007) *Chem Biol*, **14**, 87–94.

92 Wratten, S.J., Wolfe, M.S., Andersen, R.J. and Faulkner, D.J. (1977) *Antimicrob Agents Chemother*, **11**, 411–414.

93 Bultel-Ponce, V., Brouard, J.P., Vacelet, J. and Guyot, M. (1999) *Tetrahedron Lett*, **40**, 2955–2956.

94 Long, R.A., Qureshi, A., Faulkner, D.J. and Azam, F. (2003) *Appl Environ Microbiol*, **69**, 568–576.

95 Diggle, S.P., Lumjiaktase, P., Dipilato, F., Winzer, K., Kunakorn, M., Barrett, D.A., Chhabra, S.R., Camara, M. and Williams, P. (2006) *Chem Biol*, **13**, 701–710.

96 Eberl, L. (2006) *Int J Med Microbiol*, **296**, 103–110.

97 Bejerano-Sagie, M. and Xavier, K.B. (2007) *Curr Opin Microbiol*, **10**, 189–198.

98 Hrabak, E.M. and Willis, D.K. (1992) *J Bacteriol*, **174**, 3011–3020.

99 Kitten, T., Kinscherf, T.G., McEvoy, J.L. and Willis, D.K. (1998) *Mol Microbiol*, **28**, 917–929.

100 Zuber, S., Carruthers, F., Keel, C., Mattart, A., Blumer, C., Pessi, G., Gigot-Bonnefoy, C., Schnider-Keel, U., Heeb, S., Reimmann, C. and Haas, D. (2003) *Mol Plant Microbe Interact*, **16**, 634–644.

101 Pernestig, A.K., Melefors, O. and Georgellis, D. (2001) *J Biol Chem*, **276**, 225–231.

102 Rich, J.J., Kinscherf, T.G., Kitten, T. and Willis, D.K. (1994) *J Bacteriol*, **176**, 7468–7475.

103 Laville, J., Voisard, C., Keel, C., Maurhofer, M., Defago, G. and Haas, D. (1992) *Proc Natl Acad Sci USA*, **89**, 1562–1566.

104 Heeb, S. and Haas, D. (2001) *Mol Plant Microbe Interact*, **14**, 1351–1363.

105 Romeo, T., Gong, M., Liu, M.Y. and Brun-Zinkernagel, A.M. (1993) *J Bacteriol*, **175**, 4744–4755.

106 Cui, Y., Chatterjee, A., Liu, Y., Dumenyo, C.K. and Chatterjee, A.K. (1995) *J Bacteriol*, **177**, 5108–5115.

107 Heeb, S., Kuehne, S.A., Bycroft, M., Crivii, S., Allen, M.D., Haas, D., Camara, M. and Williams, P. (2006) *J Mol Biol*, **355**, 1026–1036.

108 Blumer, C., Heeb, S., Pessi, G. and Haas, D. (1999) *Proc Natl Acad Sci USA*, **96**, 14073–14078.

109 Babitzke, P. and Romeo, T. (2007) *Curr Opin Microbiol*, **10**, 156–163.

110 Kay, E., Dubuis, C. and Haas, D. (2005) *Proc Natl Acad Sci USA*, **102**, 17136–17141.

111 Kulkarni, P.R., Cui, X., Williams, J.W., Stevens, A.M. and Kulkarni, R.V. (2006) *Nucleic Acids Res*, **34**, 3361–3369.

112 Reimmann, C., Valverde, C., Kay, E. and Haas, D. (2005) *J Bacteriol*, **187**, 276–285.

113 Dubuis, C. and Haas, D. (2007) *Appl Environ Microbiol*, **73**, 650–654.

114 Heeb, S., Blumer, C. and Haas, D. (2002) *J Bacteriol*, **184**, 1046–1056.

115 Gallagher, L.A. and Manoil, C. (2001) *J Bacteriol*, **183**, 6207–6214.

116 Jander, G., Rahme, L.G. and Ausubel, F.M. (2000) *J Bacteriol*, **182**, 3843–3845.

117 Mahajan-Miklos, S., Rahme, L.G. and Ausubel, F.M. (2000) *Mol Microbiol*, **37**, 981–988.

118 Kay, E., Humair, B., Denervaud, V., Riedel, K., Spahr, S., Eberl, L., Valverde, C. and Haas, D. (2006) *J Bacteriol*, **188**, 6026–6033.

119 Mahajan-Miklos, S., Tan, M.W., Rahme, L.G. and Ausubel, F.M. (1999) *Cell*, **96**, 47–56.

120 Reimmann, C., Ginet, N., Michel, L., Keel, C., Michaux, P., Krishnapillai, V., Zala, M., Heurlier, K., Triandafillu, K., Harms, H., Defago, G. and Haas, D. (2002) *Microbiology*, **148**, 923–932.

121 Lazdunski, A.M., Ventre, I. and Sturgis, J.N. (2004) *Nat Rev Microbiol*, **2**, 581–592.

122 Pessi, G. and Haas, D. (2001) *FEMS Microbiol Lett*, **200**, 73–78.

123 Goodman, A.L., Kulasekara, B., Rietsch, A., Boyd, D., Smith, R.S. and Lory, S. (2004) *Dev Cell*, **7**, 745–754.

124 Ventre, I., Goodman, A.L., Vallet-Gely, I., Vasseur, P., Soscia, C., Molin, S., Bleves, S., Lazdunski, A., Lory, S. and Filloux, A. (2006) *Proc Natl Acad Sci USA*, **103**, 171–176.

125 Storey, D.G., Ujack, E.E., Rabin, H.R. and Mitchell, I. (1998) *Infect Immun*, **66**, 2521–2528.

126 Middleton, B., Rodgers, H.C., Camara, M., Knox, A.J., Williams, P. and Hardman, A. (2002) *FEMS Microbiol Lett*, **207**, 1–7.

127 Wu, H., Song, Z.J., Hentzer, M., Andersen, J.B., Heydorn, A., Mathee, K., Moser, C., Eberl, L., Molin, S., Hoiby, N. and Givskov, M. (2000) *Microbiology*, **146**, 2481–2493.

128 Machan, Z.A., Taylor, G.W., Pitt, T.L., Cole, P.J. and Wilson, R. (1992) *J Antimicrob Chemother*, **30**, 615–623.

129 Collier, D.N., Anderson, L., McKnight, S.L., Noah, T.L., Knowles, M., Boucher, R., Schwab, U., Gilligan, P. and Pesci, E.C. (2002) *FEMS Microbiol Lett*, **215**, 41–46.

130 Guina, T., Purvine, S.O., Yi, E.C., Eng, J., Goodlett, D.R., Aebersold, R. and Miller, S.I. (2003) *Proc Natl Acad Sci USA*, **100**, 2771–2776.

131 Rumbaugh, K.P., Griswold, J.A., Iglewski, B.H. and Hamood, A.N. (1999) *Infect Immun*, **67**, 5854–5862.

132 Pearson, J.P., Feldman, M., Iglewski, B.H. and Prince, A. (2000) *Infect Immun*, **68**, 4331–4334.

133 Tang, H.B., DiMango, E., Bryan, R., Gambello, M., Iglewski, B.H., Goldberg, J.B. and Prince, A. (1996) *Infect Immun*, **64**, 37–43.

134 Worlitzsch, D., Tarran, R., Ulrich, M., Schwab, U., Cekici, A., Meyer, K.C., Birrer, P., Bellon, G., Berger, J., Weiss, T., Botzenhart, K., Yankaskas, J.R., Randell, S., Boucher, R.C. and Doring, G. (2002) *J Clin Invest*, **109**, 317–325.

135 Lam, J., Chan, R., Lam, K. and Costerton, J.W. (1980) *Infect Immun*, **28**, 546–556.

136 Singh, P.K., Schaefer, A.L., Parsek, M.R., Moninger, T.O., Welsh, M.J. and Greenberg, E.P. (2000) *Nature*, **407**, 762–764.

137 Williams, P. and Stewart, G.S.A.B. (1994) Cell density-dependent control of gene expression in bacteria - implications for biofilm development and control, in *Bacterial Biofilms and their Control in Medicine and Industry* (eds J. Wimpenny,

W. Nichols, D. Sticker, H.M. Lappin-Scott), Bioline, Cardiff, pp. 9–12.

138 O'Toole, G.A. and Kolter, R. (1998) *Mol Microbiol*, **30**, 295–304.

139 Glessner, A., Smith, R.S., Iglewski, B.H. and Robinson, J.B. (1999) *J Bacteriol*, **181**, 1623–1629.

140 Beatson, S.A., Whitchurch, C.B., Semmler, A.B.T. and Mattick, J.S. (2002) *J Bacteriol*, **184**, 3598–3604.

141 Tielker, D., Hacker, S., Loris, R., Strathmann, M., Wingender, J., Wilhelm, S., Rosenau, F. and Jaeger, K.E. (2005) *Microbiology*, **151**, 1313–1323.

142 Diggle, S.P., Stacey, R.E., Dodd, C., Camara, M., Williams, P. and Winzer, K. (2006) *Environ Microbiol*, **8**, 1095–1104.

143 Davey, M.E., Caiazza, N.C. and O'Toole, G.A. (2003) *J Bacteriol*, **185**, 1027–1036.

144 Lequette, Y. and Greenberg, E.P. (2005) *J Bacteriol*, **187**, 37–44.

145 Sakuragi, Y. and Kolter, R. (2007) *J Bacteriol.*, **189**, 5383–5386.

146 Heydorn, A., Ersboll, B., Kato, J., Hentzer, M., Parsek, M.R., Tolker-Nielsen, T., Givskov, M. and Molin, S. (2002) *Appl Environ Microbiol*, **68**, 2008–2017.

147 Shrout, J.D., Chopp, D.L., Just, C.L., Hentzer, M., Givskov, M. and Parsek, M.R. (2006) *Mol Microbiol*, **62**, 1264–1277.

148 Donlan, R.M. (2002) *Emerg Infect Dis*, **8**, 881–890.

149 Stewart, P.S. and Costerton, J.W. (2001) *Lancet*, **358**, 135–138.

150 Bjarnsholt, T., Jensen, P.O., Burmolle, M., Hentzer, M., Haagensen, J.A.J., Hougen, H.P., Calum, H., Madsen, K.G., Moser, C., Molin, S., Hoiby, N. and Givskov, M. (2005) *Microbiology*, **151**, 373–383.

151 Allesen-Holm, M., Barken, K.B., Yang, L., Klausen, M., Webb, J.S., Kjelleberg, S., Molin, S., Givskov, M. and Tolker-Nielsen, T. (2006) *Mol Microbiol*, **59**, 1114–1128.

152 Whitchurch, C.B., Tolker-Nielsen, T., Ragas, P.C. and Mattick, J.S. (2002) *Science*, **295**, 1487–11487.

153 Riedel, K., Hentzer, M., Geisenberger, O., Huber, B., Steidle, A., Wu, H., Hoiby, N., Givskov, M., Molin, S. and Eberl, L. (2001) *Microbiology*, **147**, 3249–3262.

154 Qazi, S., Middleton, B., Muharram, S.H., Cockayne, A., Hill, P., O'Shea, P., Chhabra, S.R., Camara, M. and Williams, P. (2006) *Infect Immun*, **74**, 910–919.

155 Pritchard, D.I. (2006) *Int J Med Microbiol*, **296**, 111–116.

156 Lawrence, R.N., Dunn, W.R., Bycroft, B., Camara, M., Chhabra, S.R., Williams, P. and Wilson, V.G. (1999) *Br J Pharmacol*, **128**, 845–848.

157 Gardiner, S.M., Chhabra, S.R., Harty, C., Williams, P., Pritchard, D.I., Bycroft, B.W. and Bennett, T. (2001) *Br J Pharmacol*, **133**, 1047–1054.

158 Hooi, D.S.W., Bycroft, B.W., Chhabra, S.R., Williams, P. and Pritchard, D.I. (2004) *Infect Immun*, **72**, 6463–6470.

159 Jensen, P.O., Bjarnsholt, T., Phipps, R., Rasmussen, T.B., Calum, H., Christoffersen, L., Moser, C., Williams, P., Pressler, T., Givskov, M. and Hoiby, N. (2007) *Microbiology*, **153**, 1329–1338.

160 Wu, L.C., Estrada, O., Zaborina, O., Bains, M., Shen, L., Kohler, J.E., Patel, N., Musch, M.W., Chang, E.B., Fu, Y.X., Jacobs, M.A., Nishimura, M.I., Hancock, R.E.W., Turner, J.R. and Alverdy, J.C. (2005) *Science*, **309**, 774–777.

161 Zaborina, O., Lepine, F., Xiao, G., Valuckaite, V., Chen, Y., Li, T., Ciancio, M., Zaborin, A., Petroff, E., Turner, J.R., Rahme, L.G., Chang, E. and Alverdy, J.C. (2007) *Plos Pathog*, **3**, e35.

162 Dong, Y.H., Wang, L.H. and Zhang, L.H. (2007) *Philos Trans R Soc Lond B Biol Sci*, **362**, 1201–1211.

163 Chun, C.K., Ozer, E.A., Welsh, M.J., Zabner, J. and Greenberg, E.P. (2004) *Proc Natl Acad Sci USA*, **101**, 3587–3590.

164 Ozer, E.A., Pezzulo, A., Shih, D.M., Chun, C., Furlong, C., Lusis, A.J., Greenberg, E.P. and Zabner, J. (2005) *FEMS Microbiol Lett*, **253**, 29–37.

165 Yates, E.A., Philipp, B., Buckley, C., Atkinson, S., Chhabra, S.R., Sockett, R.E., Goldner, M., Dessaux, Y., Camara, M.,

Smith, H. and Williams, P. (2002) *Infect Immun*, **70**, 5635–5646.

166 Lin, Y.H., Xu, J.L., Hu, J.Y., Wang, L.H., Ong, S.L., Leadbetter, J.R. and Zhang, L.H. (2003) *Mol Microbiol*, **47**, 849–860.

167 Sio, C.F., Otten, L.G., Cool, R.H., Diggle, S.P., Braun, P.G., Bos, R., Daykin, M., Camara, M., Williams, P. and Quax, W.J. (2006) *Infect Immun*, **74**, 1673–1682.

168 Givskov, M., DeNys, R., Manefield, M., Gram, L., Maximilien, R., Eberl, L., Molin, S., Steinberg, P.D. and Kjelleberg, S. (1996) *J Bacteriol*, **178**, 6618–6622.

169 Rasmussen, T.B., Skindersoe, M.E., Bjarnsholt, T., Phipps, R.K., Christensen, K.B., Jensen, P.O., Andersen, J.B., Koch, B., Larsen, T.O., Hentzer, M., Eberl, L., Hoiby, N. and Givskov, M. (2005) *Microbiology*, **151**, 1325–1340.

170 Rasmussen, T.B., Bjarnsholt, T., Skindersoe, M.E., Hentzer, M., Kristoffersen, P., Kote, M., Nielsen, J., Eberl, L. and Givskov, M. (2005) *J Bacteriol*, **187**, 1799–1814.

171 Bjarnsholt, T., Jensen, P.O., Rasmussen, T.B., Christophersen, L., Calum, H., Hentzer, M., Hougen, H.P., Rygaard, J., Moser, C., Eberl, L., Hoiby, N. and Givskov, K. (2005) *Microbiology*, **151**, 3873–3880.

172 Miyairi, S., Tateda, K., Fuse, E.T., Ueda, C., Saito, H., Takabatake, T., Ishii, Y., Horikawa, M., Ishiguro, M., Standiford, T.J. and Yamaguchi, K. (2006) *J Med Microbiol*, **55**, 1381–1387.

173 Fletcher, M.P., Diggle, S.P., Crusz, S.A., Chhabra, S.R., Canara, M. and Williams, P. (2007) *Environ Microbiol*, **9**, 2683–2693.

8
Regulatory Networks in *Pseudomonas aeruginosa*: Role of Cyclic-di(3′,5′)-Guanylic Acid

Ute Römling and Susanne Häussler

8.1
Introduction – The History of Cyclic-di(3′,5′)-Guanylic Acid Detection

Very often, major discoveries are made by chance and their impact is not immediately obvious, but suddenly becomes apparent after a period of time. The history of the identification of the molecule cyclic-di(3′,5′)-guanylic acid (c-di-GMP; Figure 8.1) is no exception. Comparison of the *in vivo* and *in vitro* rates of cellulose biosynthesis using the model organism *Gluconacetobacter xylinus* (previously named *Acetobacter xylinum*) suggested that an unknown factor is required to activate cellulose biosynthesis *in vitro* [1,2]. After having identified this factor to be GTP based [3], it did not take Benziman's group too long to identify the molecule c-di-GMP as the activator of cellulose biosynthesis with the help of a series of elegant biochemical experiments in the mid-1980s [4]. Benziman's group later also demonstrated the binding of c-di-GMP to components of the cellulose synthase, suggesting for c-di-GMP a role as an allosteric activator [5].

Since the ability of bacteria to produce cellulose was considered to be an exception, the discovery of the allosteric activator c-di-GMP awoke interest, almost without exception, among plant scientists who, unsuccessfully, tried to connect the molecule to the biosynthesis of cellulose in plants [6,7].

The breakthrough for c-di-GMP, again from Benziman's group, came with the identification of c-di-GMP metabolizing proteins, di-guanylate cyclases and phosphodiesterase, which each possessed one GGDEF and one EAL domain as common domains [8]. However, since not only the overall protein structures, but also the amino acid sequences in the GGDEF and EAL domains were highly similar, although the proteins catalyzed opposite biochemical reactions, the assignment of the synthesizing and degrading function to a particular domain turned out to be an impossible task.

c-di-GMP finally gained recognition as a common bacterial signaling molecule at the beginning of this millennium [9–11]. Although the genome of the first bacterium which was sequenced, *Haemophilus influenzae*, did not encode GGDEF and EAL domain proteins, the abundance of GGDEF and EAL domain proteins in bacterial genomes makes the GGDEF and EAL domains now two of the largest protein

Pseudomonas. Model Organism, Pathogen, Cell Factory. Edited by Bernd H.A. Rehm

ISBN: 978-3-527-31914-5

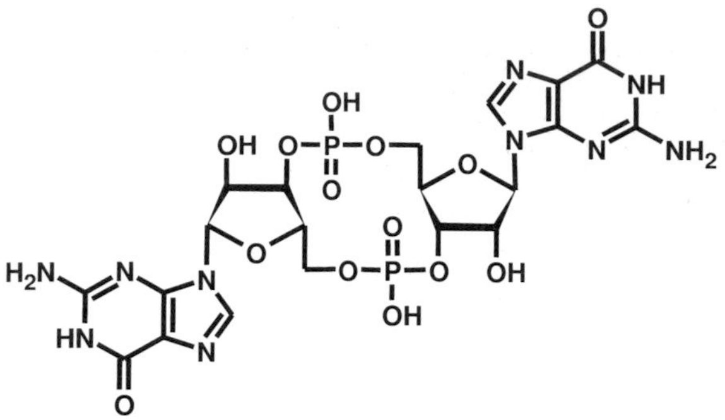

Figure 8.1 Chemical structure of c-di-GMP.

superfamilies with members only in bacteria. In parallel, GGDEF domain proteins were recognized as playing a role in developmental behavior in several bacteria, e.g. biofilm formation, demonstrating a role for c-di-GMP as a global secondary signaling molecule rather than an allosteric activator of one specific enzymatic reaction [12–14].

8.2 Principles of c-di-GMP Signaling

8.2.1 Transition between Motility and Sessility

According to the experimental information currently available, c-di-GMP is involved in a broad variety of cellular processes [15–18]. Despite this diversity, however, a good fraction of the phenotypes detected are associated with two fundamentally different lifestyles of bacteria, i.e. the sessile and the motile lifestyle, respectively, whereby c-di-GMP levels influence the transition between the two lifestyles [17] (Figure 8.2). The role of c-di-GMP in the transition between the sessile and motile lifestyle has been demonstrated in fundamental experiments [19]. Thereby, saturation of c-di-GMP-directed processes was achieved by expressing a di-guanylate cyclase leading to a high c-di-GMP production, while c-di-GMP depletion was achieved by the overexpression of a cytoplasmic phosphodiesterase. High c-di-GMP concentrations triggered the sessile lifestyle, favoring phenotypes such as extended biofilm formation that is associated with the expression of adhesive matrix components, exopolysaccharides and fimbriae. On the other hand, low c-di-GMP concentrations lead to the promotion of all forms of motility such as swimming, swarming and twitching motility. This correlation has so far not only been demonstrated for *Pseudomonas aeruginosa, Salmonella typhimurium* and *Escherichia coli* [19], but also for many other bacteria [20–22].

c-di-GMP concentrations also regulate the expression of virulence factors. This fact was demonstrated for the first time in *Vibrio cholerae*, where the expression of cholera toxin was shown to be dependent on low c-di-GMP concentrations [18]. Primarily, based on experiments conducted in *P. aeruginosa*, it is considered that c-di-GMP mediates the transition between acute and chronic infections (Figure 8.2).

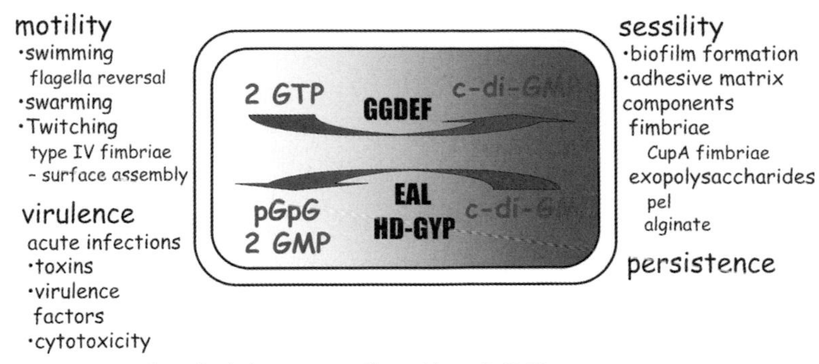

Figure 8.2 Well-studied phenotypes affected by c-di-GMP signaling in many bacteria with particular attention to *P. aeruginosa*-specific phenotypes.

Although the basic principles of c-di-GMP signaling described above hold true, there is an enormous complexity of c-di-GMP signaling in individual bacteria [16]. This complexity has its foundation in the presence of multiple paralogous GGDEF and EAL domain proteins in the individual bacterial genome, which function as diguanylate cyclases and phosphodiesterases, respectively (see below). As far as experimental evidence indicates, entire overlap of function does not seem to exist for individual GGDEF and EAL domain proteins, presumably due to the modular nature of the proteins, the unique expression and activation pattern, the distinct activity and localization of certain proteins, and the presence of downstream targets such as c-di-GMP-binding proteins (Figure 8.3) [16,17,23–25].

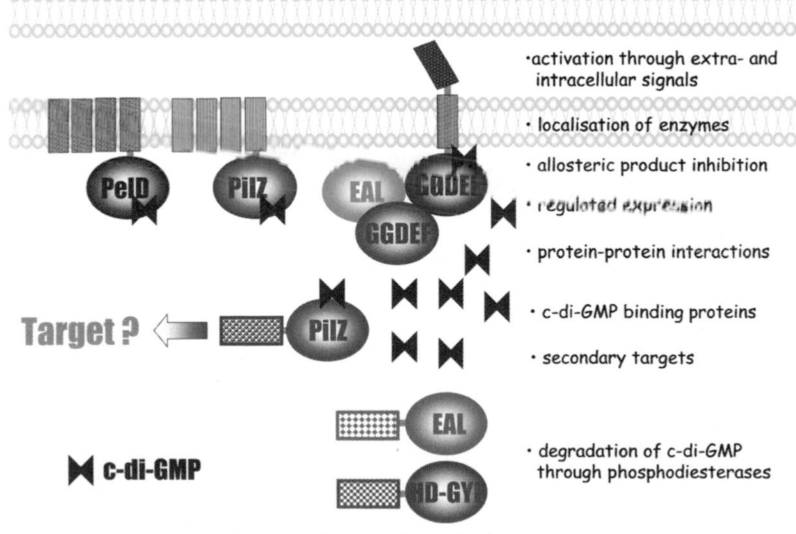

Figure 8.3 Factors contributing to the specificity of c-di-GMP signaling.

8.2.2
Di-guanylate Cyclases

Biochemical studies have shown that the GG(D/E)EF domain catalyzes the condensation of two molecules of GTP to form c-di-GMP and consequently functions as a di-guanylate cyclase [26,27]. This enzymatic specificity did not come entirely as a surprise, since genetic studies had already indicated that not only GGDEF-EAL domain-containing di-guanylate cyclases from *G. xylinus*, but also unrelated proteins with only a GGDEF domain were able to stimulate cellulose biosynthesis in *G. xylinus* [28]. In addition, bioinformatic analysis had demonstrated the structural similarity between GGDEF domains and adenylate cyclases [29]. The GGDEF domain superfamily can be classified in three subgroups according to sequence homology. Group I domains mostly contain a conserved GG(D/E)EF motif together with additional conserved amino acids [29]. Experimental evidence has shown that there exist enzymatically active and inactive domains; however, the amino acids which determine enzymatic functionality are not entirely defined. Nevertheless, it is clear that the GG(D/E)EF amino acid motif is not the only determinant of enzymatic specificity, since three highly homologous phosphodiesterases in *G. xylinus* possess the GGDEF motif [8].

8.2.3
Phosphodiesterases

It has been demonstrated in several independent *in vitro* studies that the EAL domain possesses c-di-GMP-specific phosphodiesterase activity [30–32]. In contrast to the di-guanylate cyclase activity of the GGDEF domain, the enzymatic function of the EAL domain does not require N- or C-terminal domains for activation [31]. This is also demonstrated by the fact that there exist c-di-GMP-dependent phosphodiesterases with only an EAL domain, but (almost) no di-guanylate cyclases with only a GGDEF domain (unpublished results).

There is also a significant fraction of proteins that harbor a GGDEF/EAL domain constellation. In this constellation both domains can be active [33]; alternatively, only one domain acts as an activating domain [30]. Binding of GTP to a GG(D/E)EF domain with the core motif GEDEF has previously been shown to activate the phosphodiesterase activity of the C-terminal EAL domain in *Caulobacter crescentus* CC3396 [30]. In *P. aeruginosa*, a positive allosteric effector site for GTP was suggested to be present in the GGDEF domains FimX and BifA, which harbor both a GGDEF and an EAL domain, but exclusively display phosphodiesterase activity [34,35].

Apart from the EAL domain, the nonhomologous HD-GYP domain also possesses phosphodiesterase activity [36]. Actually, the function of HD-GYP as a c-di-GMP-specific phosphodiesterase was predicted by bioinformatic analysis, since the number of EAL domains did not match the number of GGDEF domains in certain bacterial chromosomes [10]. The evolutionary significance of two nonhomologous c-di-GMP-specific phosphodiesterases is not clear. *P. aeruginosa* possesses two HD-GYP domain proteins and one YD-GYP domain protein, which have not been characterized so far.

8.2.4
Principle Concepts of c-di-GMP Signaling

One central question with respect to c-di-GMP signaling is how bacteria coordinate the specificity of c-di-GMP signaling. Although obviously useful for the identification of the whole range of c-di-GMP-regulated pathways, simple global changes of the cytoplasmic c-di-GMP levels are unlikely to account for the downstream effect of c-di-GMP in many circumstances. However, specificity of c-di-GMP signaling seems to be a multifactorial process [16,17]. Regulated expression, localization of proteins, activation of c-di-GMP-metabolizing enzymes by extra- and intracellular stimuli, protein–protein interactions, and c-di-GMP-binding proteins were proposed and partially shown to contribute to the specificity of c-di-GMP signaling [24,26,27,37–39] (see Figure 8.3). To what extend c-di-GMP is a freely diffusible molecule remains to be investigated. It is a major task to elucidate the contributions of the above mentioned mechanisms to the specificity of c-di-GMP signaling.

8.3
GGDEF and EAL Domain Proteins in *P. aeruginosa*

Clinical and environmental isolates of *P. aeruginosa* code for 16 GGDEF, four EAL and 16 GGDEF/EAL domain proteins on the core genome [40] (Figure 8.4). A variable number of strains, among them the genetic reference strain *P. aeruginosa* PAO1, also harbor the GGDEF domain protein PA2771 and/or the EAL domain protein PA2818 on the core genome, yet close to variable genetic loci. In addition, GGDEF and EAL domain proteins can be encoded by pathogenicity islands, such as *pvrR* in *P. aeruginosa* PA14. Potentially all of these proteins play a role in c-di-GMP metabolism, thereby creating complex regulatory networks [24,25,35].

Most of these proteins show complex, multimodular structures, where the GGDEF and/or EAL domains are combined with, sometimes multiple, signal sensing and/or signal transduction domains. In *P. aeruginosa*, signals which influence the enzymatic activity of di-guanylate cyclases and phosphodiesterases have not been identified. However, sequence homologies allow some conclusions. For example, is the PAS domain, which is frequently found in conjunction with GGDEF and EAL domains in *P. aeruginosa*, a sensor for oxygen and/or the redox potential [37]. The GAF domain, which is also present in cGMP-specific phosphodiesterases of eukaryotes, can bind small molecules like cGMP and cAMP [41]. Some GGDEF and EAL domains are coupled C-terminal to a classical receiver domain of two-component signaling systems, thus being embedded in the signaling cascade by phosphor relays [26,42].

8.4
c-di-GMP Signaling in Biofilm Formation

Intracellular signaling via c-di-GMP is a key factor for the development and maintenance of biofilm structures in *P. aeruginosa* [19,43,44]. As outlined above,

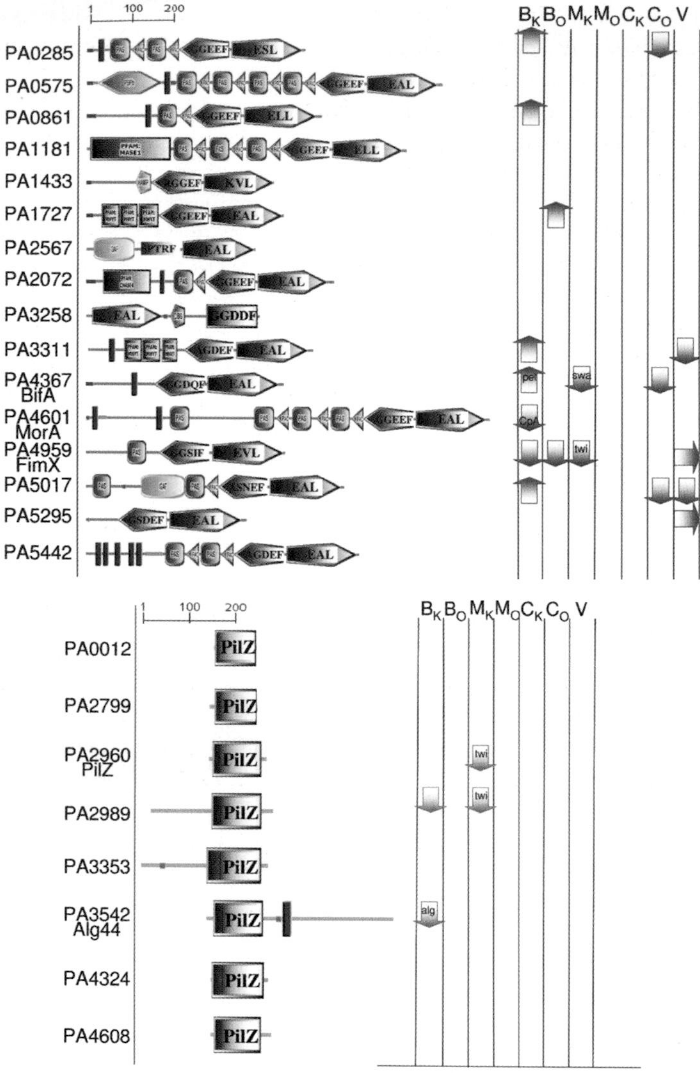

Figure 8.4 Domain structures and elucidated functions of GGDEF, EAL and PilZ domain proteins. B_k; knock-out mutant shows a phenotype in biofilm formation; B_o; overexpression of protein shows a phenotype in biofilm formation; M_k; knock-out mutant shows a phenotype in motility; M_o; overexpression of protein shows a phenotype in motility; C_k; knock-out mutant shows a phenotype in the cytotoxicity assay; C_o; overexpression of protein shows a phenotype in the cytotoxicity assay; V; knock-out mutant shows a phenotype in virulence *, phenotype dependent on strain background; Pel = biosynthesis of PEL exopolysaccharide affected; swa = swarming motility affected; CpA = biosynthesis of CupA fimbriae affected; twi = twitching motility affected; alg = alginate biosynthesis affected. Data were collected primarily from Kulasakara *et al.*, PNAS USA, 103, 2839-2844, 2006 and Merighi *et al.*, Mol Microbiol, 65, 876-895; but other sources were also considered.

low concentrations of c-di-GMP are associated with motile cells, while increasing concentrations of c-di-GMP promote the expression of adhesive matrix components, which results in increased biofilm formation.

Biofilm formation in *P. aeruginosa* is affected by multiple GGDEF-EAL domain proteins. A comprehensive study investigated the effect of knock-out mutants and overexpression of GGDEF/EAL domain proteins on biofilm formation [40]. Thereby knock-out mutants and overexpression strains showed complex phenotypes with respect to biofilm formation. In fact, only four GGDEF domain proteins (PA 11107, PA1120, PA3702 and PA5487) showed the classical phenotypic behavior of a knock-out and overexpression strain, i.e. reduced biofilm formation in the knock-out mutant, with biofilm formation enhanced in the wild-type strain overexpressing the protein. While the knock-out of other GGDEF, EAL and GGDEF-EAL domain proteins (e.g. PA0338 and PA2133) did not show a phenotype, overexpression analysis displayed altered biofilm formation, suggesting that the proteins are involved in biofilm formation, although not under the conditions used to examine the strains. There were also situations where the knock-out mutant showed a phenotype, yet the wild type strain overexpressing the protein did not show a phenotype (like with PA0861, PA3311 and PA3343). This situation can most likely be explained by a spatially restricted activity of the GGDEF-EAL domain protein, which, although overexpressed, did not display extended activity composed to the chromosomally encoded copy of the protein. However, many proteins did not show a phenotype, neither when overexpressed or knocked-out, suggesting that they affect biofilm formation only in the early stages, affect biofilm formation under conditions not investigated in this work or play a role in physiological processes other than biofilm formation in *P. aeruginosa*. For example, PA4601 (MorA), which was not identified in the above-mentioned study, was recently reported to affect biofilm formation up to the 10 h time point [45]. RetS is a hybrid sensor kinase/response regulator with an unconventional arrangement of functional domains that was demonstrated to suppress biofilm formation [46]. The GGDEF domain protein PA4332 abolished enhanced biofilm formation in a *retS* mutant of *P. aeruginosa*, suggesting that *retS* suppresses PA4332 function.

Eight GGDEF or GGDEF/EAL domain proteins were observed to increase the total intracellular c-di-GMP concentrations when overexpressed [40]. In most cases observation of di-guanylate cyclase activity correlated with the increase in biofilm formation when the proteins were overexpressed. However, the increase in c-di-GMP levels did not correlate with the amount of biofilm observed. In addition, two of eight GGDEF domain proteins showed a significant increase in c-di-GMP levels without an increase in biofilm formation. This data indicate that, although overexpressed, the GGDEF domain proteins produce c-di-GMP pools which perform a specific function in the cell. Five EAL domain proteins lead to breakdown of synthetic c-di-GMP when cell lysates with overexpressed proteins are used. None of these proteins lead to a decrease in biofilm formation when overexpressed.

The three component system RocS1–RocA1–RocR affects biofilm formation through the regulation of expression of CupB and CupC, but not CupA fimbriae and the type III secretion system (TSS) [35,42]. The RocS1–RocA1–RocR three-component system consists of the unorthodox composite sensor kinase RocS1 and two response regulators. While RocA1 is a conventional response regulator with a DNA-binding domain, RocR, which is an antagonist of RocA1, harbors an EAL domain N-terminal to the receiver domain. This three-component system is also known as the *sadARS* system [35]. RocR (PA3947), where the presence of the EAL domain suggests a role in c-di-GMP degradation, specifically suppresses expression of CupC fimbriae. Since Cup fimbriae are only poorly expressed under laboratory conditions, the RocS1–RocA1–RocR system was not identified in the above-mentioned high-throughput screen investigating the influence on GGDEF/EAL domain proteins on biofilm formation.

Biofilm formation in *P. aeruginosa* is influenced by antibiotics [47]. Thus, subinhibitory concentrations of tobramycin, an aminoglycoside produced by *Streptomyces tenebrarius*, lead to enhanced biofilm formation. Signaling is conducted thought the EAL domain protein *arr* (PA2818). Deletion of *arr* leads to reduced phosphodiesterase activity of membrane preparations.

It has been demonstrated that dispersal of biofilms in a continuous flow cell system is caused by a sudden drop in c-di-GMP levels [20,22]. In *P. aeruginosa* PAO1 BdlA (biofilm dispersion locus), a putative chemotaxis transducer protein, was shown to mediate biofilm dispersal in response to different environmental cues such as heavy metals and hydrogen peroxide [48]. The *bdlA* mutant showed an approximately 6-fold higher intracellular c-di-GMP level in a 5-day-old biofilm.

8.5
c-di-GMP Signaling in Fimbrial Biogenesis

Mutants in *wspF* lead to a highly autoaggregative phenotype in *P. aeruginosa* PAO1 [43]. *wspF* is located in a gene cluster predicted to code for a chemotaxis-like signal transduction cascade homologous to the *wsp* gene cluster in *P. fluorescens* previously demonstrated to be involved in autoaggregation and the evolution of a wrinkled colony morphology [49]. Mutations in *wspF* result in the constitutive activation of the Wsp system in *P. fluorescens*, which consequently leads to the phosphorylation of the response regulator WspR [50]. Once activated, WspR exhibits di-guanylate cyclase activity required for the expression of a wrinkled colony morphology [51]. In *P. aeruginosa* the *wspF* mutant shows increased c-di-GMP levels [44]. Furthermore, suppressor mutations of the *wspF* autoaggregative phenotype were located in *wspR* [43]. Another suppressor mutation that partially restored the wild-type phenotype lies within the *cupA* gene cluster. This gene cluster encodes for fimbrial adhesins previously demonstrated to be essential for bacterial biofilm formation [52]. Interestingly, mutations in *wspF* were recently shown to be one of the most frequent genetic adaptations of *P. aeruginosa* during 8 years of chronic cystic fibrosis (CF) lung infection [53].

8.6 c-di-GMP Signaling in Exopolysaccharide Production

P. aeruginosa has the potential capability to produce a variety of exopolysaccharides, whereby, PEL and PSL alginate, exopolysaccharides have been implicated in biofilm formation [54–60]. The production of the above mentioned exopolysaccharides was demonstrated to be dependent on c-di-GMP signaling [44,61–63].

The *pel* gene products perform the biosynthesis of a glucose-rich exopolysaccharide [55]. A regulatory network of c-di-GMP signaling was described for the regulation of PEL exopolysaccharide in *P. aeruginosa* PA14. The di-guanylate cyclase SadC (surface attachment defective; PA4332) activates the biosynthesis of the PEL exopolysaccharide on a posttranscriptional level [64], while the phosphodiesterase BifA (biofilm formation; PA4367) counteracts this activity [35]. The PEL exopolysaccharide is almost exclusively responsible for the wrinkled colony morphology on Congo Red agar plates in the *sadC* mutant. A *sadC bifA* double mutant still expressed a colony morphology more wrinkled than wild-type PA14, indicating that SadC is not the only di-guanylate cyclase for which the c-di-GMP affects PEL biosynthesis and is degraded by BifA. Candidates di-guanylate cyclases are WspR (PA3702), PA1120 and PA1107, which activate biosynthesis of the PEL polysaccharide when overexpressed [44,62]. Thereby WspR was shown to affect the *pel* genes on a transcriptional level [44].

The *psl* gene locus, which is present in *P. aeruginosa* PAO, but not PA14, encodes a exopolysaccharide rich in mannose and glucose [55,65]. Using transcriptional profiling, WspR was shown to control the expression of genes required for the production of PSL polysaccharides [44]. Moreover, there is evidence that sodium dodecyl sulfate-induced aggregation involves a c-di-GMP pathway that eventually activates synthesis of PSL exopolysaccharide [66].

Alginate biosynthesis is controlled by c-di-GMP signaling. Overexpression of the di-guanylate cyclases PA1120 and PA1107 was shown to increase alginate production in the wild-type strain *P. aeruginosa* PA14, whereas overexpression of the phosphodiesterases PA2133 and PA2200 reduced alginate production in a mucoid variant of PA14 [63]. However, the chromosomally encoded di-guanylate cyclase and phosphodiesterase, which affect alginate biosynthesis, have not yet been identified.

8.7 c-di-GMP Signaling and Motility

P. aeruginosa displays several modes of motility such as swimming, swarming and twitching motility. While flagella are required for swimming and swarming motility, type IV pili display twitching motility. Motility often plays a dual role in biofilm formation, being required for biofilm formation and the development of the three-dimensional architecture, but also for biofilm dissolution. This implies that motility is highly regulated in biofilms. Fundamental experiments have shown that all three types of motility are regulated by c-di-GMP signaling [19,43].

Twitching motility is achieved by type IV pili, which extend and retract to pull the cell forward. The assembly of functional type IV pili on the bacterial pole requires approximately 40 gene products [67]. *fimX* mutants exhibit reduced twitching motility [34,68]. More detailed analysis showed that FimX does not influence the amount of pilin monomers, but the levels of surface pili [68]. FimX is a multimodular protein; it possesses a response regulator and a PAS/PAC domain N-terminal to an imperfect GGDEF and an EAL domain. FimX was demonstrated to be a phosphodiesterase, but the imperfect GGDEF domain of FimX likely serves to activate phosphodiesterase activity when GTP is bound [34], as it has recently been described for the *C. crescentus* composite GGDEF/EAL protein, CC3396 [30]. The localization of FimX to a single pole is dependent on an intact GGDEF and EAL domain [68]; while the polar localization is dependent on the response regulator domain, since surface pili originated from nonpolar sites when a FimX protein with a deleted response regulator domain is expressed [34]. Twitching motility in the *fimX* mutant responds like the wild-type to a wide range of environmental conditions such as bovine serum albumin (stimulation) and high osmolarity (inhibition) [68]. Exceptions are tryptone and mucin, whereby the *fimX* mutant does not respond, while the wild-type strain is stimulated by those substances.

Swarming motility is inhibited by the di-guanylate cyclase SadC and stimulated by the phosphodiesterase BifA in *P. aeruginosa* PA14 [35,64]. Indeed, this c-di-GMP signaling network seems to regulate the transition between sessility and motility, since it affects PEL exopolysaccharide biosynthesis and swarming motility. The PEL exopolysaccharide synthesis is not required for the inhibition of swarming motility in the *bifA* mutant [35]. SadC and BifA also do not affect flagella biosynthesis as swimming motility is not affected by the two proteins. However, the flagella reversal rate is affected, demonstrating its importance for swarming motility.

Constitutive activation of the *wsp* chemotaxis-like system by *wspF* mutations does affect swimming and swarming motility [43]. Suppressor mutations in the di-guanylate cyclase WspR and CupA fimbriae entirely or partially restored the wild-type phenotypes, demonstrating that WspR- and partially WspR-regulated expression of CupA fimbriae affects swimming and swarming motility.

8.8
c-di-GMP Signaling in the Formation of Small Colony Variants

As opposed to planktonic *P. aeruginosa* communities, a very typical finding of *P. aeruginosa* biofilm cultures is the rapid establishment of a morphologically diverse population [69,70]. It is assumed that a diverse population is more likely to survive environmental perturbations than a homogenous population. Intriguingly, not only *P. aeruginosa* cultures grown *in vitro* within biofilms give rise to a variety of bacterial phenotypes, but this morphological diversification is also a typical microbiological finding in CF specimens [71,72].

A variety of different colony morphotypes can be found from CF specimens; hyper-biofilm-forming autoaggregative *P. aeruginosa* small colony variants (SCVs) are of special interest since they display enhanced antibiotic resistance in comparison to fast-growing revertants [73]. SCVs have been isolated not only *in vivo* from the chronically infected CF lung, but also *in vitro* after exposure to antibiotics [74]. Drenkard and Ausubel [74] described a phase variation mechanism between the antibiotic-susceptible smooth wild-type colony morphology and an antibiotic-resistant, autoaggregative rough SCV in *P. aeruginosa* strain PA14. The phenotype variant regulator PvrR, that was later shown to exhibit phosphodiesterase activity [40], greatly enhanced the rate of reversion from a rough SCV to the wild-type colony morphology. Indeed, a clinical autoaggregative *P. aeruginosa* SCV strain isolated from the respiratory tract of a chronically infected CF patient was recently demonstrated to display higher intracellular c-di-GMP concentrations in contrast to its fast growing parent strain [75]. A screen for mutants that have lost the auto-aggregative SCV biofilm phenotype identified seven independent insertions, whereby all mutants, independent of the target gene, showed reduced expression of CupA fimbriae. While two insertions were found within the *cupA* gene cluster, two other insertions were in genes encoding membrane proteins with a GGDEF domain (PA1120 and *morA*). A decrease in the level of c-di-GMP was observed in the SCV transposon mutant with an insertion in the GGDEF domain protein PA1120. Finally, overexpression of the PvrR phosphodiesterase in the clinical SCV led to a reduction in c-di-GMP level, a reduced expression of CupA fimbriae and a switch to the wild-type colony morphology.

8.9
c-di-GMP Signaling in Virulence

Biofilm formation and virulence are coupled in *P. aeruginosa*, since a variety of components such as type IV pili and the TSS play a role in *P. aeruginosa* biofilm formation and virulence [76–78]. This finding also implies that c-di-GMP signaling plays a role in *P. aeruginosa* virulence.

A comprehensive screen of GGDEF and EAL mutants for TSS-mediated cytotoxicity against CHO cells showed that mutants with insertions in *PA3947* (also called *rocR*), *fimX* (*PA4959*) and *pvrR* were defective for cytotoxicity, while an insertion in *wspR* (*PA3702*) caused a partial cytotoxic defect [40]. The *fimX* mutant had previously been reported to be defective in cytotoxicity [34]. On the other hand, additional phenotypes were observed by the overexpression of GGDEF and EAL domain proteins. PA14 overexpressing PA2133, PA2200, PA3947, PA4367, PA4396, PA5017 and PA5442 were significantly impaired in their ability to kill CHO cells. Although PA4396 codes for a protein with an unconventional GGDEF domain, the other genes encode EAL or GGDEF/EAL domain proteins. Interestingly, three of five EAL domain proteins that showed obvious phosphodiesterase activity in cell extracts lead to a defect in cytotoxicity. Although it seems evident that the reduced cytotoxicity by the above-mentioned genes is most likely a consequence of reduced c-di-GMP levels, the effect of reduced c-di-GMP levels on bacterial–host interaction is not obvious from the experiments performed.

The effects of GGDEF and EAL domain proteins on cytotoxicity and biofilm formation are not coupled [40]. Although it was demonstrated that overexpression of a number of genes harboring an EAL domain leads to reduced cytotoxicity, the ability of the bacteria to form biofilms was not altered. On the other hand, overexpression of a number of genes encoding a GGDEF domain resulted in enhanced biofilm formation without affecting cytotoxicity, demonstrating that expression of extracellular matrix components does not interfere with the TTSS-mediated cytotoxic activity.

Analysis of a selected number of mutants in the acute infection model of murine thermal injury revealed no entire correlation between cytotoxicity and virulence [40]. For example, consistent with the outcome on cytotoxicity, knock-out of *rocR* and *pvrR* abolished the ability to cause lethal infection. A contribution of *pvrR* to animal and plant virulence had also been reported earlier [79]. However, the virulence phenotypes of mutants in *fimX* and *wspR* were identical to wild-type PA14, despite a strong or partial defect in CHO cell cytotoxicity, respectively. Consistent with the above findings, the *fimX* mutant is also as virulent as the wild-type strain in a murine model of acute pneumonia [34]. The lack of correlation between cytotoxicity and virulence in mice indicates that the cytotoxic phenotype is not the sole determinant of virulence. Genes encoding di-guanylate cyclases and phosphodiesterases could play a broad role in acute infections as they also regulate virulence factors such as type IV pili [34] and possibly other factors.

8.10
c-di-GMP-Binding Proteins in *P. aeruginosa*

The identification of the downstream targets of c-di-GMP is a first step to elucidate the molecular mechanisms of c-di-GMP signalling. In fact, the cellulose synthase in *G. xylinus* was the first target of c-di-GMP that was identified [5,23]. However, only recently, with the help of bioinformatics, was the very C-terminal part of the cellulose synthase recognized as a distinct domain also present in proteins that could be connected to c-di-GMP-regulated pathways [23]. Consequently, this domain, the PilZ domain, was predicted to bind c-di-GMP. It took less than 1 year until binding of c-di-GMP to PilZ domains was experimentally confirmed [39,80,81].

In *P. aeruginosa*, eight PilZ domain-containing proteins were identified including PilZ itself (Figure 8.4) [63]. Mutations in the genes revealed a phenotype for *pilZ* (PA2960) and *PA2989* in twitching motility, and PA2989 also showed poor biofilm formation Curiously, although seven PilZ domain proteins showed binding of c-di-GMP, the name-giving protein itself lacked c-di-GMP binding. Since some of the consensus residues of the PilZ signature are not present in PilZ [23], it seems likely that PilZ indeed does not bind c-di-GMP or displays a low affinity to c-di-GMP [63]. In addition, the structure of the PilZ domain of PA4608 was recently solved and c-di-GMP binding was confirmed [82].

The PilZ domain protein Alg44, which is required for alginate biosynthesis, was demonstrated to bind c-di-GMP *in vitro* and *in vivo* [63]. When expressed in *Escherichia coli* Alg44 was able to sequester c-di-GMP from the cellulose synthase $BcsA_{PilZ}$, which abolished cellulose production. Lack of c-di-GMP binding by Alg44 containing single amino acid substitutions in consensus residues of the PilZ domain leads to impairment of alginate production. Alg44 is a bipartite modular protein with a N-terminal PilZ domain and a C-terminal NolF domain. The NolF domain has been proposed to couple cytoplasmic and outer membrane components of a transport apparatus [83], suggesting a role for Alg44 in alginate transport rather than biosynthesis. Concurrent with this prediction, Alg44 and mutant proteins not able to bind c-di-GMP have been found membrane-associated. However, another work reported Alg44 to be located soluble in the periplasm [84]. Although the precise localization of Alg44 needs to be defined, the latter finding would imply an extra-cytoplasmic function of c-di-GMP which requires a mechanism for c-di-GMP secretion.

PEL exopolysaccharide biosynthesis is regulated by c-di-GMP on a posttranscriptional level [35,62]. Among the gene products of the *pel* operon, PelD, an inner membrane protein with a large cytosolic domain, was shown to specifically bind c-di-GMP [62]. Searching for homology to known c-di-GMP binding sites identified sequence similarity to the inhibitory (I)-site of certain di-guanylate cyclases with the RxxD-binding motif which mediates feedback inhibition [85]. Amino acid substitutions in this motif abolished c-di-GMP binding and production of the PEL exopolysaccharide.

8.11 Perspectives

Although there seem to be some straightforward principles, c-di-GMP signaling in *P. aeruginosa* is complex, since there are a variety of c-di-GMP-synthesizing and -degrading proteins as well as c-di-GMP-binding proteins. Therefore, the mechanisms whereby the specificity of c-di-GMP signaling is achieved are diverse, ranging from regulation of gene expression, and the enzymatic activity of the protein over signal transduction cascades and protein–protein interactions [85a] to degradation of c-di-GMP and c-di-GMP-binding proteins. Much is left to be discovered here, in particular to what extent different regulatory mechanisms contribute to the specificity of the signaling molecule produced by a particular di-guanylate cyclase.

c-di-GMP-binding proteins such as Alg44 are part of a macromolecular complex [63]. Binding of c-di-GMP to Alg44 and other c-di-GMP-binding proteins or domains in macromolecular biosynthesis complexes represents a new posttranslational regulatory mechanism carried out by c-di-GMP signaling. The challenge remains to determine the precise molecular mechanisms whereby c-di-GMP-binding proteins exert their function in macromolecular complexes and to determine the structure–function relationships in relation to other biosynthetic components.

In addition, considering the number of known c-di-GMP-binding proteins in relation to the number of di-guanylate cyclases and phosphodiesterases (Figure 8.4), some components seem to be lacking. It will be of immense interest to identify novel c-di-GMP-binding proteins, possibly with novel binding motifs. In addition, targets downstream of c-di-GMP-binding proteins remain to be discovered.

As described in this chapter, there are already a range of phenotypes reported to be influenced by c-di-GMP signaling. Such phenotypes include the biosynthesis of alginate, PEL and PSL exopolysaccharide, the biosynthesis of type IV pili, CupA, CupB and CupC fimbriae, and the flagella reversal rate (Figure 8.4) [35,42,44,63,64,86]. Other phenotypes are related to virulence such as the global virulence of bacteria in animals and plants, cytotoxicity against CHO cells, and, possibly, the TSS (Figure 8.4) [34,40,79]. However, as one can conclude from the number of proteins involved in c-di-GMP metabolism and preliminary experimental data [40], there must exist yet undiscovered phenotypes influenced by the c-di-GMP signaling system.

P. aeruginosa has a variety of sophisticated intra- and extracellular signaling pathways such as two quorum sensing (QS) signaling systems and phosphor transduction by two-component systems. There is evidence that the c-di-GMP signaling system is tightly connected to two-component signaling; in fact, some of the putative phosphodiesterases and di-guanylate cyclases are an integral part of two-component signaling systems [40,42]. A connection between QS signaling and c-di-GMP metabolism has been detected in bacterial species other than *P. aeruginosa* [21,87,88]. How c-di-GMP signaling and regulation by small RNAs is connected needs to be elucidated [46].

In summary, although we have already obtained a significant glimpse on how c-di-GMP signaling works, it is expected that the future will lead to the discovery of novel phenotypes affected by c-di-GMP signaling and to the elucidation of the molecular mechanisms within the c-di-GMP signaling system.

Acknowledgments

The work of U.R. on c-di-GMP signaling was supported by the Karolinska Institutet (eliktforskartjänst to U.R.), Vetenskapsrådet, the European Commission and Mukoviszidose eV. Financial support of S.H. by the Helmholtz-Gemeinschaft is gratefully acknowledged.

References

1 Aloni, Y. and Benziman, M. (1982) Intermediates of cellulose synthesis in *Acetobacter xylinum*, in *Cellulose and Other Natural Polymer Systems* (ed. R.M. Brown, Jr.), Plenum, New York, pp. 341–359.

2 Glaser, L. (1958) The synthesis of cellulose in cell-free extracts of *Acetobacter xylinum*. *J Biol Chem*, **232**, 627–636.

3 Aloni, Y., Delmer, D.P. and Benziman, M. (1982) Achievement of high rates of *in vitro* synthesis of 1, 4-beta-D-glucan: activation by cooperative interaction of the *Acetobacter xylinum* enzyme system with GTP, polyethylene glycol, and a protein factor. *Proc Natl Acad Sci USA*, **79**, 6448–6452.

4 Ross, P., Weinhouse, H., Aloni, Y., Michaeli, D., Weinberger-Ohana, P., Mayer, R., Braun, S., de Vroom, E., van der Marel, G.A., van Boom, J.H. and Benziman, M. (1987) Regulation of cellulose synthesis in *Acetobacter xylinum* by cyclic diguanylic acid. *Nature*, **325**, 279–281.

5 Weinhouse, H., Sapir, S., Amikam, D., Shilo, Y., Volman, G., Ohana, P. and Benziman, M. (1997) c-di-GMP-binding protein, a new factor regulating cellulose synthesis in *Acetobacter xylinum*. *FEBS Lett*, **416**, 207–211.

6 Amor, Y., Mayer, R., Benziman, M. and Delmer, D. (1991) Evidence for a cyclic diguanylic acid-dependent cellulose synthase in plants. *Plant Cell*, **3**, 989–995.

7 Delmer, D.P. and Amor, Y. (1995) Cellulose biosynthesis. *Plant Cell*, **7**, 987–1000.

8 Tal, R., Wong, H.C., Calhoon, R., Gelfand, D., Fear, A.L., Volman, G., Mayer, R., Ross, P., Amikam, D., Weinhouse, H., Cohen, A., Sapir, S., Ohana, P. and Benziman, M. (1998) Three *cdg* operons control cellular turnover of cyclic di-GMP in *Acetobacter xylinum*: genetic organization and occurrence of conserved domains in isoenzymes. *J Bacteriol*, **180**, 4416–4425.

9 D'Argenio, D.A. and Miller, S.I. (2004) Cyclic di-GMP as a bacterial second messenger. *Microbiology*, **150**, 2497–2502.

10 Galperin, M.Y., Nikolskaya, A.N. and Koonin, E.V. (2001) Novel domains of the prokaryotic two-component signal transduction systems. *FEMS Microbiol Lett*, **203**, 11–21.

11 Jenal, U. (2004) Cyclic di-guanosine-monophosphate comes of age: a novel secondary messenger involved in modulating cell surface structures in bacteria? *Curr Opin Microbiol*, **7**, 185–191.

12 Hecht, G.B. and Newton, A. (1995) Identification of a novel response regulator required for the swarmer-to-stalked-cell transition in *Caulobacter crescentus*. *J Bacteriol*, **177**, 6223–6229.

13 Jones, H.A., Lillard, J.W. Jr. and Perry, R.D. (1999) HmsT, a protein essential for expression of the haemin storage (Hms^{+}) phenotype of *Yersinia pestis*. *Microbiology*, **145**, 2117–2128.

14 Römling, U., Rohde, M., Olsen, A., Normark, S. and Reinköster, J. (2000) AgfD, the checkpoint of multicellular and aggregative behaviour in *Salmonella typhimurium* regulates at least two independent pathways. *Mol Microbiol*, **36**, 10–23.

15 Dow, J.M., Fouhy, Y., Lucey, J.F. and Ryan, R.P. (2006) The HD-GYP domain, cyclic di-GMP signaling, and bacterial virulence to plants. *Mol Plant Microbe Interact*, **19**, 1378–1384.

16 Römling, U., Gomelsky, M. and Galperin, M.Y. (2005) c-di-GMP: the dawning of a novel bacterial signalling system. *Mol Microbiol*, **57**, 629–639.

17 Römling, U. and Amikam, D. (2006) Cyclic di-GMP as a second messenger. *Curr Opin Microbiol*, **9**, 218–228.

18 Tamayo, R., Pratt, J.T. and Camilli, A. (2007) Roles of cyclic diguanylate in the regulation of bacterial pathogenesis. *Annu Rev Microbiol*, **61**, 131–148.

19 Simm, R., Morr, M., Kader, A., Nimtz, M. and Römling, U. (2004) GGDEF and EAL domains inversely regulate cyclic di-GMP levels and transition from sessility to motility. *Mol Microbiol*, **53**, 1123–1134.

20 Gjermansen, M., Ragas, P., Sternberg, C., Molin, S. and Tolker-Nielsen, T. (2005) Characterization of starvation-induced dispersion in *Pseudomonas putida* biofilms. *Environ Microbiol*, **7**, 894–906.

21 Rahman, M., Simm, R., Kader, A., Basseres, E., Römling, U. and Möllby, R. (2007) The role of c-di-GMP signaling in an

Aeromonas veronii biovar *sobria* strain. *FEMS Microbiol Lett*, **273**, 172–179.

22 Thormann, K.M., Duttler, S., Saville, R.M., Hyodo, M., Shukla, S., Hayakawa, Y. and Spormann, A.M. (2006) Control of formation and cellular detachment from *Shewanella oneidensis* MR-1 biofilms by cyclic di-GMP. *J Bacteriol*, **188**, 2681–2691.

23 Amikam, D. and Galperin, M.Y. (2006) PilZ domain is part of the bacterial c-di-GMP binding protein. *Bioinformatics*, **22**, 3–6.

24 Kader, A., Simm, R., Gerstel, U., Morr, M. and Römling, U. (2006) Hierarchical involvement of various GGDEF domain proteins in rdar morphotype development of *Salmonella enterica* serovar Typhimurium. *Mol Microbiol*, **60**, 602–616.

25 Simm, R., Lusch, A., Kader, A., Andersson, M. and Römling, U. (2007) Role of EAL-containing proteins in multicellular behavior of *Salmonella enterica* serovar Typhimurium. *J Bacteriol*, **189**, 3613–3623.

26 Paul, R., Weiser, S., Amiot, N.C., Chan, C., Schirmer, T., Giese, B. and Jenal, U. (2004) Cell cycle-dependent dynamic localization of a bacterial response regulator with a novel di-guanylate cyclase output domain. *Genes Dev*, **18**, 715–727.

27 Ryjenkov, D.A., Tarutina, M., Moskvin, O.V. and Gomelsky, M. (2005) Cyclic diguanylate is a ubiquitous signaling molecule in bacteria: insights into biochemistry of the GGDEF protein domain. *J Bacteriol*, **187**, 1792–1798.

28 Ausmees, N., Mayer, R., Weinhouse, H., Volman, G., Amikam, D., Benziman, M. and Lindberg, M. (2001) Genetic data indicate that proteins containing the GGDEF domain possess diguanylate cyclase activity. *FEMS Microbiol Lett*, **204**, 163–167.

29 Pei, J. and Grishin, N.V. (2001) GGDEF domain is homologous to adenylyl cyclase. *Proteins*, **42**, 210–216.

30 Christen, M., Christen, B., Folcher, M., Schauerte, A. and Jenal, U. (2005) Identification and characterization of a cyclic di-GMP-specific phosphodiesterase and its allosteric control by GTP. *J Biol Chem*, **280**, 30829–30837.

31 Schmidt, A.J., Ryjenkov, D.A. and Gomelsky, M. (2005) The ubiquitous protein domain EAL is a cyclic diguanylate-specific phosphodiesterase: enzymatically active and inactive EAL domains. *J Bacteriol*, **187**, 4774–4781.

32 Tamayo, R., Tischler, A.D. and Camilli, A. (2005) The EAL domain protein VieA is a cyclic diguanylate phosphodiesterase. *J Biol Chem*, **280**, 33324–33330.

33 Tarutina, M., Ryjenkov, D.A. and Gomelsky, M. (2006) An unorthodox bacteriophytochrome from *Rhodobacter sphaeroides* involved in turnover of the second messenger c-di-GMP. *J Biol Chem*, **281**, 34751–34758.

34 Kazmierczak, B.I., Lebron, M.B. and Murray, T.S. (2006) Analysis of FimX, a phosphodiesterase that governs twitching motility in *Pseudomonas aeruginosa*. *Mol Microbiol*, **60**, 1026–1043.

35 Kuchma, S.L., Brothers, K.M., Merritt, J.H., Liberati, N.T., Ausubel, F.M. and O'Toole, G.A. (2007) BifA, a c-di-GMP phosphodiesterase, inversely regulates biofilm formation and swarming motility by *Pseudomonas aeruginosa* PA14. *J Bacteriol.*, **189**, 8165–8178.

36 Ryan, R.P., Fouhy, Y., Lucey, J.F., Crossman, L.C., Spiro, S., He, Y.W., Zhang, L.H., Heeb, S., Camara, M., Williams, P. and Dow, J.M. (2006) Cell–cell signaling in *Xanthomonas campestris* involves an HD-GYP domain protein that functions in cyclic di-GMP turnover. *Proc Natl Acad Sci USA*, **103**, 6712–6717.

37 Chang, A.L., Tuckerman, J.R., Gonzalez, G., Mayer, R., Weinhouse, H., Volman, G., Amikam, D., Benziman, M. and Gilles-Gonzalez, M.A. (2001) Phosphodiesterase A1, a regulator of cellulose synthesis in *Acetobacter xylinum*, is a heme-based sensor. *Biochemistry*, **40**, 3420–3426.

38 Delgado-Nixon, V.M., Gonzalez, G. and Gilles-Gonzalez, M.A. (2000) Dos, a heme-binding PAS protein from *Escherichia coli*, is a direct oxygen sensor. *Biochemistry*, **39**, 2685–2691.

39 Ryjenkov, D.A., Simm, R., Römling, U. and Gomelsky, M. (2006) The PilZ domain is a receptor for the second messenger c-di-GMP. The PilZ domain protein YcgR controls motility in enterobacteria. *J Biol Chem*, **41**, 30310–30314.

40 Kulesekara, H., Lee, V., Brencic, A., Liberati, N., Urbach, J., Miyata, S., Lee, D.G., Neely, A.N., Hyodo, M., Hayakawa, Y., Ausubel, F.M. and Lory, S. (2006) Analysis of *Pseudomonas aeruginosa* diguanylate cyclases and phosphodiesterases reveals a role for bis-(3′–5′)-cyclic-GMP in virulence. *Proc Natl Acad Sci USA*, **103**, 2839–2844.

41 Zoraghi, R., Corbin, J.D. and Francis, S.H. (2004) Properties and functions of GAF domains in cyclic nucleotide phosphodiesterases and other proteins. *Mol Pharmacol*, **65**, 267–278.

42 Kulasekara, H.D., Ventre, I., Kulasekara, B.R., Lazdunski, A., Filloux, A. and Lory, S. (2005) A novel two-component system controls the expression of *Pseudomonas aeruginosa* fimbrial *cup* genes. *Mol Microbiol*, **55**, 368–380.

43 D'Argenio, D.A., Calfee, M.W., Rainey, P.B. and Pesci, E.C. (2002) Autolysis and autoaggregation in *Pseudomonas aeruginosa* colony morphology mutants. *J Bacteriol*, **184**, 6481–6489.

44 Hickman, J.W., Tifrea, D.F. and Harwood, C.S. (2005) A chemosensory system that regulates biofilm formation through modulation of cyclic diguanylate levels. *Proc Natl Acad Sci USA*, **102**, 14422–14427.

45 Choy, W.K., Zhou, L., Syn, C.K., Zhang, L.H. and Swarup, S. (2004) MorA defines a new class of regulators affecting flagellar development and biofilm formation in diverse *Pseudomonas* species. *J Bacteriol*, **186**, 7221–7228.

46 Goodman, A.L., Kulasekara, B., Rietsch, A., Boyd, D., Smith, R.S. and Lory, S. (2004) A signaling network reciprocally regulates genes associated with acute infection and chronic persistence in *Pseudomonas aeruginosa*. *Dev Cell*, **7**, 745–754.

47 Hoffman, L.R., D'Argenio, D.A., MacCoss, M.J., Zhang, Z., Jones, R.A. and Miller, S.I. (2005) Aminoglycoside antibiotics induce bacterial biofilm formation. *Nature*, **436**, 1171–1175.

48 Morgan, R., Kohn, S., Hwang, S.H., Hassett, D.J. and Sauer, K. (2006) BdlA, a chemotaxis regulator essential for biofilm dispersion in *Pseudomonas aeruginosa*. *J Bacteriol*, **188**, 7335–7343.

49 Spiers, A.J., Kahn, S.G., Bohannon, J., Travisano, M. and Rainey, P.B. (2002) Adaptive divergence in experimental populations of *Pseudomonas fluorescens*. I. Genetic and phenotypic bases of wrinkly spreader fitness. *Genetics*, **161**, 33–46.

50 Bantinaki, E., Kassen, R., Knight, C.G., Robinson, Z., Spiers, A.J. and Rainey, P.B. (2007) Adaptive divergence in experimental populations of *Pseudomonas fluorescens*. III. Mutational origins of wrinkly spreader, diversity. *Genetics*, **176**, 441–453.

51 Goymer, P., Kahn, S.G., Malone, J.G., Gehrig, S.M., Spiers, A.J. and Rainey, P.B. (2006) Adaptive divergence in experimental populations of *Pseudomonas fluorescens*. II. Role of the GGDEF regulator WspR in evolution and development of the wrinkly spreader phenotype. *Genetics*, **173**, 515–526.

52 Vallet, I., Olson, J.W., Lory, S., Lazdunski, A. and Filloux, A. (2001) The chaperone/usher pathways of *Pseudomonas aeruginosa*: identification of fimbrial gene clusters (*cup*) and their involvement in biofilm formation. *Proc Natl Acad Sci USA*, **98**, 6911–6916.

53 Smith, E.E., Buckley, D.G., Wu, Z., Saenphimmachak, C., Hoffman, L.R., D'Argenio, D.A., Miller, S.I., Ramsey, B.W., Speert, D.P., Moskowitz, S.M., Burns, J.L., Kaul, R. and Olson, M.V. (2006) Genetic adaptation by *Pseudomonas aeruginosa* to the airways of cystic fibrosis patients. *Proc Natl Acad Sci USA*, **103**, 8487–8492.

54 Boyd, A. and Chakrabarty, A.M. (1995) *Pseudomonas aeruginosa* biofilms: role of the alginate exopolysaccharide. *J Ind Microbiol*, **15**, 162–168.

55 Friedman, L. and Kolter, R. (2004a) Two genetic loci produce distinct carbohydrate-rich structural components of the *Pseudomonas aeruginosa* biofilm matrix. *J Bacteriol*, **186**, 4457–4465.

56 Friedman, L. and Kolter, R. (2004b) Genes involved in matrix formation in *Pseudomonas aeruginosa* PA14 biofilms. *Mol Microbiol*, **51**, 675–690.

57 Jackson, K.D., Starkey, M., Kremer, S., Parsek, M.R. and Wozniak, D.J. (2004) Identification of *psl*, a locus encoding a potential exopolysaccharide that is essential for *Pseudomonas aeruginosa* PAO1 biofilm formation. *J Bacteriol*, **186**, 4466–4475.

57a Matsukawa, M. and Greenberg, E.P. Putative exopolysaccharide synthesis genes influence *pseudomonas aeruginosa*, biofilm development. *J Bacteriol*, **186**, 4449–4456.

58 Meluleni, G.J., Grout, M., Evans, D.J. and Pier, G.B. (1995) Mucoid *Pseudomonas aeruginosa* growing in a biofilm *in vitro* are killed by opsonic antibodies to the mucoid exopolysaccharide capsule but not by antibodies produced during chronic lung infection in cystic fibrosis patients. *J Immunol*, **155**, 2029–2038.

59 Overhage, J., Schemionek, M., Webb, J.S. and Rehm, B.H. (2005) Expression of the *psl* operon in *Pseudomonas aeruginosa* PAO1 biofilms: PslA performs an essential function in biofilm formation. *Appl Environ Microbiol*, **71**, 4407–4413.

60 Vasseur, P., Vallet-Gely, I., Soscia, C., Genin, S. and Filloux, A. (2005) The *pel* genes of the *Pseudomonas aeruginosa* PAK strain are involved at early and late stages of biofilm formation. *Microbiology*, **151**, 985–997.

61 Caiazza, N.C., Merritt, J.H., Brothers, K.M. and O'Toole, G.A. (2007) Inverse regulation of biofilm formation and swarming motility by *Pseudomonas aeruginosa* PA14. *J Bacteriol*, **189**, 3603–3612.

62 Lee, V.T., Matewish, J.M., Kessler, J.L., Hyodo, M., Hayakawa, Y. and Lory, S. (2007) A cyclic-di-GMP receptor required for bacterial exopolysaccharide production. *Mol Microbiol*, **65**, 1474–1484.

63 Merighi, M., Lee, V.T., Hyodo, M., Hayakawa, Y. and Lory, S. (2007) The second messenger bis-(3′–5′)-cyclic-GMP and its PilZ domain-containing receptor Alg44 are required for alginate biosynthesis in *Pseudomonas aeruginosa*. *Mol Microbiol*, **65**, 876–895.

64 Merritt, J.H., Brothers, K.M., Kuchma, S.L. and O'Toole, G.A. (2007) SadC reciprocally influences biofilm formation and swarming motility via modulation of exopolysaccharide production and flagellar function *J Bacteriol*, **189**, 8154–8164.

65 Ma, L., Lu, H., Sprinkle, A., Parsek, M.R. and Wozniak, D. (2007) *Pseudomonas aeruginosa* Psl is a galactose- and mannose-rich exopolysaccharide. *J Bacteriol*. **189**, 8353–8356.

66 Klebensberger, J., Lautenschlager, K., Bressler, D., Wingender, J. and Philipp, B. (2007) Detergent-induced cell aggregation in subpopulations of *Pseudomonas aeruginosa* as a preadaptive survival strategy. *Environ Microbiol*, **9**, 2247–2259.

67 Mattick, J.S., Whitchurch, C.B. and Alm, R.A. (1996) The molecular genetics of type-4 fimbriae in *Pseudomonas aeruginosa* – a review. *Gene*, **179**, 147–155.

68 Huang, B., Whitchurch, C.B. and Mattick, J.S. (2003) FimX, a multidomain protein connecting environmental signals to twitching motility in *Pseudomonas aeruginosa*. *J Bacteriol*, **185**, 7068–7076.

69 Boles, B.R., Thoendel, M. and Singh, P.K. (2004) Self-generated diversity produces "insurance effects" in biofilm communities. *Proc Natl Acad Sci USA*, **101**, 16630–16635.

70 Deziel, E., Comeau, Y. and Villemur, R. (2001) Initiation of biofilm formation by

Pseudomonas aeruginosa 57RP correlates with emergence of hyperpiliated and highly adherent phenotypic variants deficient in swimming, swarming, and twitching motilities. *J Bacteriol*, **183**, 1195–1204.

71 Komiyama, K., Tynan, J.J., Habbick, B.F., Duncan, D.E. and Liepert, D.J. (1985) *Pseudomonas aeruginosa* in the oral cavity and sputum of patients with cystic fibrosis. *Oral Surg Oral Med Oral Pathol*, **59**, 590–594.

72 Seale, T.W., Thirkill, H., Tarpay, M., Flux, M. and Rennert, O.M. (1979) Serotypes and antibiotic susceptibilities of *Pseudomonas aeruginosa* isolates from single sputa of cystic fibrosis patients. *J Clin Microbiol*, **9**, 72–78.

73 Häussler, S., Tümmler, B., Weissbrodt, H., Rohde, M. and Steinmetz, I. (1999) Small-colony variants of *Pseudomonas aeruginosa* in cystic fibrosis. *Clin Infect Dis*, **29**, 621–625.

74 Drenkard, E. and Ausubel, F.M. (2002) *Pseudomonas* biofilm formation and antibiotic resistance are linked to phenotypic variation. *Nature*, **416**, 740–743.

75 Meissner, A., Wild, V., Simm, R., Rohde, M., Erck, C., Bredenbruch, F., Morr, M., Römling, U. and Häussler, S. (2007) *Pseudomonas aeruginosa cupA*-encoded fimbriae expression is regulated by a GGDEF and EAL domain-dependent modulation of the intracellular level of cyclic diguanylate. *Environ Microbiol*, **9**, 2475–2485.

76 Farinha, M.A., Conway, B.D., Glasier, L.M., Ellert, N.W., Irvin, R.T., Sherburne, R. and Paranchych, W. (1994) Alteration of the pilin adhesin of *Pseudomonas aeruginosa* PAO results in normal pilus biogenesis but a loss of adherence to human pneumocyte cells and decreased virulence in mice. *Infect Immun*, **62**, 4118–4123.

77 Kang, P.J., Hauser, A.R., Apodaca, G., Fleiszig, S.M., Wiener-Kronish, J., Mostov, K. and Engel, J.N. (1997) Identification of *Pseudomonas aeruginosa* genes required for epithelial cell injury. *Mol Microbiol*, **24**, 1249–1262.

78 O'Toole, G.A. and Kolter, R. (1998) Flagellar and twitching motility are necessary for *Pseudomonas aeruginosa* biofilm development. *Mol Microbiol*, **30**, 295–304.

79 He, J., Baldini, R.L., Deziel, E., Saucier, M., Zhang, Q., Liberati, N.T., Lee, D., Urbach, J., Goodman, H.M. and Rahme, L.G. (2004) The broad host range pathogen *Pseudomonas aeruginosa* strain PA14 carries two pathogenicity islands harboring plant and animal virulence genes. *Proc Natl Acad Sci USA*, **101**, 2530–2535.

80 Christen, M., Christen, B., Allan, M.G., Folcher, M., Jeno, P., Grzesiek, S. and Jenal, U. (2007) DgrA is a member of a new family of cyclic diguanosine monophosphate receptors and controls flagellar motor function in *Caulobacter crescentus*. *Proc Natl Acad Sci USA*, **104**, 4112–4117.

81 Pratt, J.T., Tamayo, R., Tischler, A.D. and Camilli, A. (2007) PilZ domain proteins bind cyclic diguanylate and regulate diverse processes in *Vibrio cholerae*. *J Biol Chem*, **282**, 12860–12870.

82 Ramelot, T.A., Yee, A., Cort, J.R., Semesi, A., Arrowsmith, C.H. and Kennedy, M.A. (2007) NMR structure and binding studies confirm that PA4608 from *Pseudomonas aeruginosa* is a PilZ domain and a c-di-GMP binding protein. *Proteins*, **66**, 266–271.

83 Binet, R., Letoffe, S., Ghigo, J.M., Delepelaire, P. and Wandersman, C. (1997) Protein secretion by Gram-negative bacterial ABC exporters – a review. *Gene*, **192**, 7–11.

84 Remminghorst, U. and Rehm, B.H. (2006) Alg44, a unique protein required for alginate biosynthesis in *Pseudomonas aeruginosa*. *FEBS Lett*, **580**, 3883–3888.

85a Andrade, MO., Alegria, MC., Guzzo, CR., Docena, C., Rosa, MC., Ramos CH. and Farah, CS. (2006) The HD-GYP domain of

RpfG mediates a direct linkage between the Rpf quorumsensing pathway and a subset of diguanylate cyclase proteins in the phytopathogen *Xanthomonas axonopodis* pv citri. *Mol Microbiol.*, **62**, 537–551.

85 Chan, C., Paul, R., Samoray, D., Amiot, N.C., Giese, B., Jenal, U. and Schirmer, T. (2004) Structural basis of activity and allosteric control of diguanylate cyclase. *Proc Natl Acad Sci USA*, **101**, 17084–17089.

86 Häussler, S. (2004) Biofilm formation by the small colony variant phenotype of *Pseudomonas aeruginosa*. *Environ Microbiol*, **6**, 546–551.

87 Camilli, A. and Bassler, B.L. (2006) Bacterial small-molecule signaling pathways. *Science*, **311**, 1113–1116.

88 Lim, B., Beyhan, S. and Yildiz, F.H. (2007) Regulation of *Vibrio* polysaccharide synthesis and virulence factor production by CdgC, a GGDEF-EAL domain protein, in *Vibrio cholerae*. *J Bacteriol*, **189**, 717–729.

9
Pseudomonas aeruginosa: A Model for Biofilm Formation

Diane McDougald, Janosch Klebensberger, Tim Tolker-Nielsen, Jeremy S. Webb, Tim Conibear, Scott A. Rice, Sylvia M. Kirov, Carsten Matz, and Staffan Kjelleberg

9.1
Introduction

Much of the fundamental understanding of microbial physiology is based on laboratory studies of freely suspended cells. While these studies have been essential for our foundational understanding of the genetics, physiology and behavior of microbes, it is now recognized that a majority of bacterial cells in nature exist in biofilms [1] associated with surfaces or as floating cell aggregates. In fact, it has recently been proposed that microbial communities originally developed on surfaces, including the first bacterial and archael cells, and that the planktonic cell phenotype evolved as a dispersal mechanism [2]. Hallmarks of cells residing in biofilm communities are increased metabolic efficiency [3] as well as increased resistance to environmental stresses such as desiccation, ultraviolet radiation and oxidative stress [4–6]. This correlation has dramatic consequences as residing in aggregates has been shown to confer increased resistance of bacterial cells also to biocides such as antibiotics, disinfectants and detergents [7–9]. In addition, once established, these biofilms are able to resist invasion by other organisms and predation by protozoans in nature or host immune cells in the human body [5,6,10]. This is especially problematic as it is also recognized that the majority of bacterial infections involve biofilms [11]. The recent explosion of research in the field of biofilm biology has led to an enhanced appreciation for the multicellular aspects of microbiology and has resulted in the general acceptance of a model of the biofilm mode of life.

Pseudomonas aeruginosa has become a model organism for the study of biofilms due to its metabolic versatility and variability in its response to environmental signals, which promotes successful colonization of different habitats and growth under varying environmental conditions [12,13]. This ability is likely a reflection of its large genome, allowing for metabolic plasticity and quick responses to varying stimuli. *P. aeruginosa* is also a human pathogen that causes infection in burn patients, and is the predominant cause of lung infections and mortality in patients with cystic fibrosis

Pseudomonas. Model Organism, Pathogen, Cell Factory. Edited by Bernd H.A. Rehm

ISBN: 978-3-527-31914-5

(CF) [14,15]. This chapter will address various aspects of biofilm development, dispersal and resistance, and its role in the infection process.

9.2 Biofilm Development

Our understanding of *P. aeruginosa* biofilm development has in large been derived from the study of monospecies biofilms established in flow cell systems. Development of a biofilm occurs via a series of well-defined stages which include; (i) translocation to the surface and initial and reversible attachment of cells, (ii) irreversible attachment, (iii) microcolony formation, (iv) maturation and differentiation of the biofilm, and (v) dispersal of single cells from the biofilm (Figure 9.1). Several factors are required for biofilm development, including attachment via adhesive protein, cell aggregation via proteins, extracellular DNA (eDNA) and polysaccharides, and cell motility [16,17].

Moreover, during biofilm maturation, cell aggregates form mushroom-shaped microcolonies with interconnecting void channels that are thought to be important for the delivery of nutrient to and removal of waste from the cells within the microcolonies. These processes are likely to be actively maintained and in the case of *P. aeruginosa*, this process has been linked to quorum sensing (QS) [18]. The development of mature biofilms of *P. aeruginosa* also involves multiple cell types. For example, formation of the mushroom-shaped microcolonies involves both a nonmotile subpopulation that forms the stalk of the microcolony and a motile subpopulation expressing type IV pili that form the cap [19]. Similarly, the production of the surfactant rhamnolipid

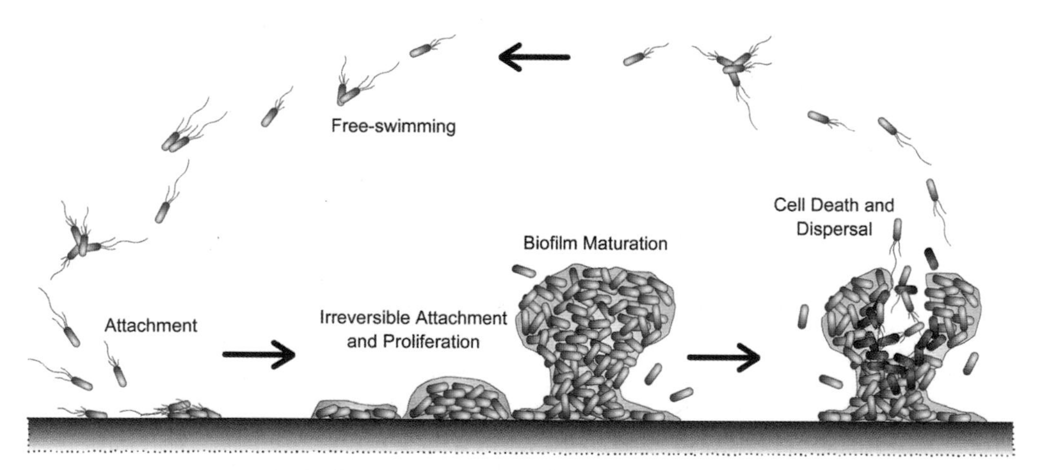

Figure 9.1 The biofilm life cycle. Biofilm formation is a multistage process that involves (i) reversible attachment of cells to a surface, (ii) secretion of adhesins and EPS resulting in irreversible attachment of the biofilm and cell proliferation, (iii) microcolony formation and maturation of the biofilm, and (iv) cell death in microcolonies and dispersal of single cells that return to a planktonic phase, completing the biofilm lifecycle.

and the eDNA that aids in matrix stability is limited to subpopulations in the stalk [19,20], while subpopulations in the cap show increased resistance to membrane-targeting antimicrobials [21]. Investigation of dispersal cell populations also indicates that different subpopulations of cells exist, with some exhibiting phenotypic variation [7]. Phenotypic variants can be stable and thus are likely due to mutation of the cells within the microcolony prior to dispersal, resulting in a dispersal population that exhibits a variety of phenotypic responses which may serve to insure that a subpopulation of cells will be adapted to the newly encountered environment [22].

A number of studies have focused on the identification of a developmental program which regulates the phenotypes of biofilm formation and resistance. Transcriptomic analysis showed that early biofilm cells express genes more similar to stationary-phase cells than to actively growing cells [23], which explains their increased resistance to a number of biocides. In addition, QS has been shown to be important for biofilm development [18] and resistance [5,10]. However, the lack of a core set of genes that are consistently expressed during biofilm formation and the observation that genes involved in general adaptive responses are induced [23] indicates that a specific genetic program for biofilm development may not exist [17].

9.2.1
Early Attachment and Colonization Events

9.2.1.1 Translocation and Attachment to Surfaces

The first stages in the model of the biofilm life cycle (Figure 9.1) describe the approach and attachment events of cells to a surface. These stages of attachment can be defined as (i) the transport of a bacterial cell to a surface by diffusion, convection and/or active movement, (ii) the initial and reversible attachment of the cell, and (iii) irreversible attachment. There are three different mechanisms by which cells reach a surface, i.e. Brownian motion, flagella-driven motility and physiochemical attraction. The diffusion of cells by Brownian motion is very slow compared to the convective transport via liquid flow. Nevertheless, in most cases the final transport to a surface is diffusion controlled, because the flow velocity immediately adjacent to the surface–liquid interface is negligible (hydrodynamic boundary layer) [24]. In this context, active movement by flagella might be an advantage for overcoming the boundary layer more quickly than nonmotile cells. In addition, once motile bacteria are in close contact with a surface, they exhibit the possibility of site-directed motility towards the surface in a chemotactic response to any concentration gradient which exists within the boundary layer. The same parameters that attract chemicals and nonliving particles to surfaces are also responsible for the initial and reversible attachment of the cell to a surface, because a bacterial suspension can be defined as living colloidal system [25]. Thus, the initial attachment of a cell can be described by colloidal chemical theories such as the Derjaguin, Landau, Verwey and Overbeek (DLVO) theory [26,27]. In general, these theories describe the total interaction force/energy between charged surfaces interacting through a liquid medium as a summation of van der Waals, electrostatic and hydrophobic interactions, finally leading to attraction or repulsion.

9.2.1.2 Role of Flagella and Fimbriae in Reversible and Irreversible Attachment

In a broad screen for attachment-deficient mutants in *P. aeruginosa* PA14, O'Toole and Kolter [28] identified several transposon insertion mutants with reduced attachment to surfaces in a microtiter plate assay. Several of these mutants were shown to harbor the transposon in genes responsible for flagellum biosynthesis. The authors speculated that flagellum-derived motility might be responsible for overcoming the repulsive forces at the surface–water interface in order to reach the substratum and may also contribute to initial attachment by specific interactions. For example, it has been proposed that the flagellar cap protein, FliD, interacts with mucin, one of the most abundant host polymers found in CF patients (see also Section 9.4.1), resulting in the immobilization of cells [29,30]. Another set of mutants they identified was defective in twitching motility, which enables the bacterial cells to migrate along a surface via extension and retraction of type IV pili [31,32]. It was postulated that after the flagella-mediated, reversible attachment, twitching motility is needed for microcolony formation and further biofilm development. While other studies arrived at similar results [33–35], a study by Klausen *et al.* [36] did not establish a significant influence on initial surface attachment of *P. aeruginosa* by either flagella or type IV pili in flow chamber setup with citrate as a sole carbon source. However, the authors demonstrated that flagella and type IV pili are involved in biofilm architecture, and that twitching motility is essential for the development of highly structured biofilms during growth on glucose [36,37]. These somehow contradictory results were explained by strain-specific and nutritional differences, and it was suggested that biofilm initiation and development can occur via different routes which are strongly dependent on the experimental setup. This idea is further supported by the observation by O'Toole and Kolter [38] that several attachment-deficient *P. fluorescens* mutants could be rescued in their ability to attach surfaces by varying nutrient and experimental conditions. In addition, a nutrient-dependent correlation between the extents of swarming motility, another surface associated flagella-dependent motility, and biofilm architecture in *P. aeruginosa* has been observed [39].

Interestingly, the cytoplasmic protein SadB has been recently found to be an important component for the transition from reversible to irreversible attachment in *P. aeruginosa* [40]. SadB has been shown to inversely regulate swarming motility and biofilm formation by altering flagellar reversal rates and the expression of extracellular polymeric substances (EPSs), respectively [40]. As similar phenotypes could be found in mutants deficient in the flagellar stators MotAB and MotCD, the authors suggested that SadB, as well as the flagellar stators, could be part of a regulatory system involved in the inverse control of swarming motility and biofilm formation in response to an unknown signal [41,42]. Another important component in the transition from reversible to irreversible attachment is the LapA protein in *P. fluorescens*. LapA is a large adhesion that is loosely associated with the outer membrane and is secreted by the ABC transporter, LapEBC [43]. Recent studies revealed that secretion of LapA is modulated by the inner membrane protein, LapD, and that the LapAEBC operon is under the control of the Pho regulon [44,45].

Adhesive fimbrial structures, synthesized and assembled by the chaperone–usher pathway (Cup), represent another class of important surface associated structures which have been found to be involved in early stages of biofilm development.

P. aeruginosa harbors three different cluster of genes (*cupA–C*) encoding components of a Cup responsible for fimbriae expression [46]. Vallet *et al.* [47] demonstrated the essential function of the *cup* gene cluster in adhesion to abiotic surfaces and the formation of biofilms. They also demonstrated that the expression levels of the *cup* genes are low under standard culturing conditions as a result of a complex regulation involving the HNS -like protein MvaT [48], which acts as a transcriptional repressor in a phase-variable manner [49]. The authors speculated that this phase-variable expression could contribute to the overall fitness of a population under varying conditions and postulated specific, but unknown environmental stimuli that trigger this expression.

As illustrated above, the mechanisms that are involved in the early colonization events of reversible and irreversible attachment of bacterial cells to a surface include physicochemical parameters and specific interactions with biological molecules which are either surface expressed or excreted. The variety of adhesive structures expressed by microorganisms and their potential to adapt the adhesive structures to specific environmental conditions highlights the complex regulation of this process and the flexibility of bacteria to attach to a variety of surfaces for niche exploitation.

9.2.2 Extracellular Matrix

The formation and maintenance of structured microbial communities critically depends on the presence of cell–cell interconnecting extracellular substances. The cell–cell interconnecting network in biofilms is usually referred to as the extracellular biofilm matrix and is composed of a variety of biopolymers, including polysaccharide, protein and DNA. In addition to the exopolymers, outer membrane proteins and a variety of cell appendages, such as fimbriae, pili and flagella, may also have cell–cell interconnecting functions, and can therefore be considered part of the biofilm matrix. A single bacterial species can produce several different biofilm matrix components and it appears that not all of these components are expressed during biofilm formation in a particular environment. It is anticipated that the capacity of bacteria to produce different biofilm matrix components allows colonization of different niches through different biofilm development pathways, and contributes to the resistance of the biofilm to environmental stresses and host defenses [5,11]. In the present section, we address polysaccharide, protein and DNA as components of the *P. aeruginosa* biofilm matrix.

9.2.2.1 Polysaccharide

P. aeruginosa can, depending on the strain and growth conditions, produce at least three different exopolysaccharides: alginate, PEL and PSL [50–53]. In addition, two other gene clusters that may be involved in exopolysaccharide biosynthesis (PA1381–PA1398 and PA3552–PA3558) are present in the genome of the reference strain *P. aeruginosa* PAO1.

Mucoid forms of *P. aeruginosa* that overproduce alginate, an exopolysaccharide containing mannuronic acid and guluronic acid, are often found in the lungs of CF patients [51]. The mucoid conversion is in many cases associated with inactivating mutations in the *mucA* gene, which encodes an anti-σ factor of AlgT. AlgT is required for expression of the *alg* alginate biosynthetic genes [54–56]. Alginate is thought to have a protective function in the relatively harsh host environment where the bacteria are continually subjected to attack by the immune system [57–59] (see Section 9.4.1 for a detailed discussion on mucoidy in CF).

In nonmucoid *P. aeruginosa* strains, the exopolysaccharides PEL and PSL appear to play critical roles as matrix components [50,52,53,60]. PEL and PSL are evidently both branched heteropolysaccharides, the main component of PEL being glucose, whereas the main component of PSL is mannose [50,60]. High-level expression of PEL and PSL in *P. aeruginosa* was shown to lead to the formation of wrinkly colonies on agar plates, a phenotype often associated with overexpression of biofilm matrix components [50] (see also Section 9.3.1). Synthesis of PEL was shown to be required for *P. aeruginosa* PA14 to form pellicle biofilms at the air/liquid interface of broth cultures [60]. Studies using a static attachment assay provided evidence that PSL is important in the early stages of *P. aeruginosa* biofilm development, whereas the synthesis of PEL seems to be important in later stages of biofilm development [50,61]. In a continuous culture flow chamber setup, a *P. aeruginosa* strain deficient in PSL production was found to be impaired in biofilm formation [52,53], supporting a role of PSL in the early stages of biofilm development.

The *pelA–G* and *pslA–O* gene clusters encode machinery for PEL and PSL biosynthesis [50,52,53,60,61]. The exact functions of the gene products are not described yet, but sequence analysis has shown that they contain domains which are present in sugar-processing proteins in other organisms. Many of the proteins contain transmembrane domains, indicating they may be membrane localized.

Recent data indicate that synthesis of the PEL and PSL polysaccharides in *P. aeruginosa* are controlled via regulatory proteins with GGDEF and EAL domains [62]. As discussed in detail below (Section 9.3.1), such proteins modulate the intracellular levels of cyclic di-guanosine-monophosphate (c-di-GMP), which serves as a second messenger and affects the synthesis or transport of polysaccharide and protein biofilm matrix components [63]. Transcription of the *pel* and *psl* loci were found to be regulated through the *wsp* (PA3708–PA3702) chemosensory system, of which the gene product WspR contains the catalytic GGDEF domains. Whereas high levels of c-di-GMP were found to stimulate transcription of the *pel* and *psl* loci and induce biofilm formation, low intracellular c-di-GMP levels were found to decrease biofilm formation in a microtiter plate assay and in a flow chamber system [62].

In addition to the regulation of PEL and PSL synthesis via GGDEF/EAL domain proteins, it has been shown that PEL synthesis is regulated through the *las* QS system in *P. aeruginosa* PA14 at room temperature [64]. As addressed below (Section 9.3.2) it appears that the production of other biofilm matrix components such as eDNA [19] and lectins [65] are under QS control as well.

9.2.2.2 **Protein**

As discussed earlier, flagellum-driven motility can promote initial biofilm formation by facilitating transport of the bacteria to the surface. In addition, flagella can act as both cell–surface adhesins and cell–cell adhesins [28,38,66]. However, for both flagella and type IV pili, this dependency is due to specific growth conditions tested [36]. *In vitro* studies have revealed that type IV pili of *P. aeruginosa* display specific affinity towards asialo-GM_1 and asialo-GM_2 residues of mammalian cell surfaces [67–69]. Recently, the type IV pili of *P. aeruginosa* and *Neisseria gonorrhoeae* were shown to bind with high affinity to DNA [70,71]. eDNA has been shown to be part of the matrix in *P. aeruginosa* biofilms [19,53,72,73], and type IV pili might act as cross-linkers between the cells and eDNA. In addition to attachment, type IV pili are used by *P. aeruginosa*, and a number of other bacteria, to perform surface associated motility [74]. Sauer and Camper [75] have shown that expression of the major structural component of type IV pili, PilA, is downregulated in the initial stages of attachment, but upregulated during biofilm formation.

As discussed earlier, mutations in genes encoding Cup fimbriae were shown to affect the ability of *P. aeruginosa* to attach to surfaces in microtiter plates [47]. In addition, evidence is accruing that Cup fimbriae appear to also play a role as cell–cell linkers in the later stages of biofilm development. Mutations in any of these genes were shown to result in the formation of *P. aeruginosa* biofilms with altered mature structure [76].

P. aeruginosa is able to produce lectins which have affinity towards carbohydrate moieties on eukaryotic cell surfaces [77]. However, it has recently been shown that the *P. aeruginosa* lectins may also bind to carbohydrates of the biofilm matrix and thereby promote biofilm formation, and thus it is possible that these lectins have dual roles [78,79]. Cell fractionation experiments suggested that LecB was exported and bound to the outer membrane through interaction with fucose-containing ligands [79]. Staining of *P. aeruginosa* biofilms with fluorescently labeled LecB protein confirmed the presence of lectin binding sites in the biofilms [79]. Recently, the galactophilic lectin LecA has also been shown to have a role in *P. aeruginosa* biofilm development [80].

9.2.2.3 DNA

Murakawa [81,82] characterized the extracellular "slime" produced by *P. aeruginosa*. The chemical composition of slimes from 20 clinical *P. aeruginosa* isolates was investigated and it was found that slimes from 18 strains consisted primarily of DNA, while two strains with a mucoid phenotype produced slimes composed primarily of polyuronic acid (which most likely was alginate). More recently, eDNA has been shown to be a key matrix component for both the *P. aeruginosa* PAO1 reference strain and for clinical *P. aeruginosa* isolates [72,73]. In addition, Matsukawa and Greenberg [53] investigated the composition of the extracellular matrix of mature *P. aeruginosa* PAO1 biofilms and found that eDNA was by far was the most abundant polymer. Experiments involving comparative PCR and Southern analysis have suggested that the eDNA released from *P. aeruginosa* PAO1 is similar to whole-genome DNA [19].

P. aeruginosa PAO1 biofilm formation in microtiter plates and flow chambers was found to be attenuated by the presence of DNase I [73]. In addition, early *P. aeruginosa* PAO1 biofilms, which had been grown in flow chambers without exogenous DNase, were dispersed rapidly after addition of DNase I to the medium, whereas mature *P. aeruginosa* PAO1 biofilms were not [73]. Nemoto *et al.* [72] found that mature biofilms formed by four different clinical *P. aeruginosa* isolates were also dispersed by DNase treatment. While it has been proposed that exopolysaccharide encoded by the *psl* genes may be the main matrix component in mature *P. aeruginosa* PAO1 biofilms [53], the eDNA in mature *P. aeruginosa* PAO1 biofilms appears to have a stabilizing effect. Mature *P. aeruginosa* PAO1 biofilms that were pretreated with DNase I for a short time were more susceptible to sodium dodecylsulfate treatment than biofilms that had not been pretreated with DNase I [19].

A basal level of eDNA present in early *P. aeruginosa* PAO1 biofilms appears to be generated via a pathway which is not linked to QS, whereas the generation of large amounts of eDNA in mature *P. aeruginosa* biofilms evidently depends on the *las*, *rhl* and *pqs* QS systems [19]. The increased level of eDNA in *P. aeruginosa* wild-type biofilms in comparison to in *P. aeruginosa lasIrhlI* biofilms appears to be linked to QS via a mechanism that results in lysis of a small subpopulation of the cells [19]. In support of a role of QS in cell lysis, D'Argenio *et al.* (see Section 9.3.2) reported that mutants overproducing the *Pseudomonas* quinolone signal (PQS) displayed high levels of autolysis, whereas mutants that could not produce PQS did not show autolysis. In addition, Heurlier *et al.* [83] demonstrated that *P. aeruginosa* QS mutants, unlike the wild-type, did not undergo cell lysis in stationary phase cultures. The QS-regulated lysis of a subpopulation of cells has implications for biofilm dispersal (see Section 9.2.4.3).

The eDNA in *P. aeruginosa* biofilms appears to be organized in distinct patterns [19]. In 4-day-old flow-chamber-grown *P. aeruginosa* biofilms, which contained mushroom-shaped structures, the eDNA was found to be located primarily in the stalk portion of the mushroom-shaped structures with the highest concentration in the outer part of the stalks [19]. It has been shown that the formation of the mushroom-shaped structures in *P. aeruginosa* biofilms occurs in a sequential process involving a nonmotile bacterial subpopulation that forms the stalks (initial microcolonies) by growth in certain foci of the biofilm and a migrating bacterial subpopulation that subsequently forms the caps on top of the stalks via a process that requires type IV pili [36] (Figure 9.2). It is currently not understood how the accumulation of

Figure 9.2 Mushroom-shaped multicellular biofilm structures with yellow caps and cyan stalks. Confocal laser scanning microscopic images were acquired in a 4-day-old biofilm which was initiated with a 1:1 mixture of yellow fluorescent *P. aeruginosa* PAO1 wild-type and cyan fluorescent *P. aeruginosa pilA* derivative and grown on citrate minimal medium. Bars = 20 mm.

the type IV pili-expressing cells is coordinated so that they form mushroom caps. However, because type IV pili bind to DNA [70,71], it is tempting to speculate that the high concentration of eDNA on the outer part of the stalks might cause accumulation of the bacteria, which in combination with bacterial growth may result in the formation of the mushroom caps. In agreement with this suggestion, type IV pili-mediated migration of Myxobacteria during fruiting body formation has been shown to depend on the presence of exopolymer material [84].

9.2.3 Phenotypic and Genotypic Diversification in *Pseudomonas* Biofilms

A recent observation in biofilm research is that many stable phenotypic and genotypic variant strains are generated during the process of biofilm formation. This phenomenon is thought to enhance the ability of biofilm cells to adapt, persist and spread under diverse environmental conditions and stresses [22,85]. Commonly documented phenotypes arising in *Pseudomonas* biofilms include mucoidy [86,87] (see also Section 9.4.1), small colony variants (SCVs) [34,87–90], hyper-piliation [34,89], lipopolysaccharide deficiency [91] and antibiotic resistance [7]. Such changes are often observed over successive generations, implying that these are heritable genetic changes [34,85,89,92]. This genetic variation has implications for a wide range of phenomena, from the adaptive evolution of bacterial communities, to the metabolic capability of consortia utilized for biodegradation, to the fouling of industrial equipment. In this section we will review some of the key causes and mechanisms for the production of phenotypic variants within biofilms.

9.2.3.1 Ecological and Genetic Causes of Diversity in *Pseudomonas* Biofilms

From an ecological perspective, biofilms are highly heterogeneous ecosystems, exhibiting complex architectures and steep chemical gradients. As such, biofilms are likely to contain many different microenvironments and niches within which strong selection can act on resident bacteria. Competition between cells for nutrients is likely to be intense and selection will favor mutants that can gain access to limiting resources. The complexity of such interactions in biofilms may further be enhanced by, for example, competitive interactions with other species or by the impact of predatory grazing organisms on biofilm cells. Consequently, biofilms are thought to favor the emergence of diverse "ecological specialists", i.e. different variants that are competitively dominant within different niches and selective forces within biofilms [93–96].

In addition to the ecological causes of phenotypic variation, the diversification that is observed within *Pseudomonas* biofilms must ultimately be driven by underlying genetic changes that generate variation among biofilm bacteria. It is now clear that rapid evolution occurs in *Pseudomonas* biofilms leading to the divergence of clones with altered genotypes [97,98], occurring principally by processes of recombination and mutation. In CF lung biofilms, progressive genetic adaptations can occur over time and have been shown to be caused, for example, by point mutations in structural or regulatory genes such as *mutS*, *mucA* or *rpoN* [99–101]. Larger chromosomal rearrangements and inversions also occur during *P. aeruginosa* adaptation to the CF lung [102]. Many other recombination events may contribute to genetic change during biofilm development. For example, processes of transposition, transformation, and bacteriophage induction and transduction all occur prominently within biofilms [103–105]. Mutation rates in bacteria residing within biofilms can increase under stress because of a reduced ability to deal with DNA-damaging free radicals generated by metabolism [106]. Indeed phenotypic variation linked to the production of damaging reactive oxygen or nitrogen species has recently been documented for *P. aeruginosa* and other biofilm-forming bacteria [85,92]. Other processes, such as defects in mismatch repair systems, may also enhance mutation rates in biofilms. For example, mutator strains of *P. aeruginosa* have been found in long-term chronic infections within the CF lung [101], and may allow for rapid diversification and adaptation within biofilms. Ultimately, *P. aeruginosa* strains that have resided within CF biofilms over extended periods of time become increasingly genetically distinct from the parental strain (see also Section 9.4.1).

9.2.3.2 Colony Morphology Variation in *P. aeruginosa* Biofilms

In the laboratory environment, phenotypic changes in bacterial cells are most easily visualized on solid agar media by changes in colony morphology. For *P. aeruginosa* biofilms, subpopulations of slow-growing bacteria with distinctive small, rough colony morphologies are commonly reported in the literature. Although terminology varies, this phenotype is most commonly referred to as a SCV. Bacterial cells that produce SCVs exhibit an autoaggregative phenotype which is usually a consequence of the overexpression of either adhesive surface structures or EPSs resulting in increased adherence of the cell [34,95]. These cells are often hyper-piliated, and defective in swimming, twitching, swarming, chemotaxis and pigment production [107,108]. SCVs may also exhibit reduced activity of the electron transport chain, causing slowed metabolism of sugars and decreased production of ATP [108,109]. Some SCVs autoaggregate in liquid culture and adhere strongly to solid surfaces. Once attached to a new surface, many SCVs have been shown to form biofilms with increased biomass and three-dimensional structure [89,110], and are associated with a generalized increase in tolerance to antimicrobial agents [7,88]. Often, colonies exhibiting the SCV phenotype display differences in other traits (including motility, expression of cell surface components, auxotrophy), suggesting that SCV phenotypes may reflect a broad range of genetic changes that occur during *P. aeruginosa* biofilm development. The best characterized genetic mechanisms linked to phenotypic diversification in *P. aeruginosa* biofilms are described below.

9.2.3.3 GacS Sensor Kinase

A recent study by Davies *et al.* [110] showed that the GacS/GacA regulatory system of *P. aeruginosa* was in part responsible for controlling stable phenotypic change. Biofilms of the *gacS* mutant gave rise to SCVs with increasing frequency when exposed to antimicrobial compounds. When cultured, the SCV produced thicker biofilms with greater cell density and greater antimicrobial resistance than did the wild-type or parental *gacS* mutant strains. As has been described previously for other colony morphology variants, the *gacS*-derived SCVs were less motile than the wild-type strain and autoaggregated in broth culture. Complementation with *gacS* in *trans* restored the ability of the SCV to revert to a normal colony morphotype. These findings indicate that mutation of *gacS* is associated with the occurrence of stress-resistant SCV cells in *P. aeruginosa* biofilms and suggest that in some instances GacS may be necessary for reversion of these variants to a wild-type state [110].

9.2.3.4 **Bacteriophage and SCVs**

A number of studies have suggested a role for bacteriophage in host diversification and phenotypic variation, including the production of SCV phenotypes, in *P. aeruginosa*. Microarray studies have shown that bacteriophage genes are among the most highly upregulated groups of genes during *P. aeruginosa* biofilm development [23,105] and genes of a Pf1-like filamentous bacteriophage (designated Pf4), which exists as a prophage in the genome of *P. aeruginosa*, showed up to 83.5-fold activation during biofilm development compared with expression in planktonic cells [105]. Other studies have shown that activation of Pf4 genes in biofilms is regulated by QS in *P. aeruginosa* [23], and that activity of Pf4 phage is linked to killing and lysis of a subpopulation of *P. aeruginosa* cells within biofilms [111]. Therefore, induction of the Pf1-like phage in *P. aeruginosa* appears to be biofilm specific, and to play an important role in the physiology and development of *P. aeruginosa* biofilms. Recently, it was shown that induction of Pf4 correlated with the emergence of SCV phenotypes from biofilms and that infection of *P. aeruginosa* with biofilm-derived Pf4 caused the emergence of SCVs within the culture [92]. These SCVs exhibited enhanced attachment and accelerated biofilm development similar to other SCVs derived from biofilms, thus it appears that Pf4 can mediate phenotypic variation and the production of an SCV phenotype in *P. aeruginosa*.

Other bacteriophage have been implicated in regulating phenotypic variation in *P. aeruginosa*. Bacteriophage PP7, a lytic RNA-containing bacteriophage, was shown to cause diversification in and the evolution of a small-rough colony phenotype in *P. aeruginosa* [112]. Lysogenic bacteriophage, among other factors (see above and Section 9.4.1), have also been suggested to be involved in mucoid conversion of *P. aeruginosa* in the CF lung [113]. Thus, bacteriophage appear to play an important role in phenotypic diversification in *P. aeruginosa*.

While genetic mechanisms are linked to the emergence of stable genetic and phenotypic variation that arises during biofilm development, important adaptive genetic changes can also occur without visible differences in gross colony phenotype [114]. Thus, the full extent to which genomic diversification occurs within *P. aeruginosa* biofilm populations, and its role within the complex, multicellular lifestyle of biofilms, remains poorly understood. In this respect, approaches (such as genomics) that generate an understanding of mutations occurring in an entire biofilm population would greatly improve our knowledge on the key mutational processes that drive the generation of genetic variants, thereby enhancing our understanding of evolutionary process occurring within biofilms, and lead to novel approaches to modulate their development.

9.2.4 Biofilm Dispersal

The need to disperse to and colonize new environments is a fundamental constraint on the lifestyle of all sessile organisms, and it is now clear that *P. aeruginosa* biofilms, among other biofilm-forming bacteria, display dispersal stages as a feature of their development. *P. aeruginosa* biofilms can undergo active dispersal events where sessile, matrix-encased biofilm cells convert *en masse* to free-swimming, planktonic bacteria. Until recently, the mechanisms by which active bacterial dispersal from biofilms occurs remained almost completely unexplored and little was known about the functions or regulatory pathways involved in release of bacteria from biofilms. However, recently there has been a surge of interest in biofilm dispersal phenomena, perhaps in part because strategies to manipulate biofilm dispersal would find broad application in the control of microbial biofilms in medical, environmental and industrial environments.

Biofilm dispersal in *P. aeruginosa* appears to be a complex and dynamic process involving multiple environmental and genetic determinants. In many cases, biofilm dispersion can be induced by specific changes in the physicochemical environment, e.g. by changes in nutrient availability, including carbon sources [115–118], iron availability [119] and oxygen [120]. In addition, several physiological processes and regulatory mechanisms are involved in *P. aeruginosa* biofilm dispersal, including intracellular c-di-GMP levels [121] (see Section 9.3.1 below), alginate lyase-mediated breakdown of the biofilm matrix [122,123], production of surfactants which loosen cells from the biofilm [124] and production of free radical species [111,125]. In this section, we will examine recent key findings in the area of the mechanisms and regulation of biofilm dispersal in *P. aeruginosa*.

9.2.4.1 Role of c-di-GMP in Dispersal

As discussed below in Section 9.3.1, there is considerable evidence that cellular adhesiveness, aggregation and dispersal within biofilms of diverse bacteria, including *Pseudomonas*, are modulated by c-di-GMP. For example, in *P. aeruginosa*, mutation of a chemotaxis and biofilm dispersal regulator, designated BdlA, led to increased c-di-GMP levels, reduced dispersal and enhanced cell adherence [126]. In *P. putida*, dispersal was found to be induced by starvation conditions and was characterized by a rapid dissolution of the entire biofilm within minutes after the flow of nutrients to the biofilm was stopped [116]. In the latter study, genetic analysis of mutants defective in nutrient-induced dispersal revealed a key role for a *P. putida* gene PP0615, which contains both GGDEF and EAL domains, in the biofilm dissolution process. Thus, intracellular signaling via c-di-GMP appears to be an important regulatory mechanism responsible for modulating biofilm formation and dispersal in *Pseudomonas*.

9.2.4.2 Denitrification, Nitric Oxide and Dispersal in *P. aeruginosa*

In *P. aeruginosa* biofilms, steep oxygen gradients can occur, leading to anaerobic regions within the biofilm, and biofilms predominantly exhibit gene expression profiles consistent with anaerobic growth [23,127]. Anaerobic respiratory metabolism in *P. aeruginosa* uses nitrate (NO_3^-), nitrite (NO_2^-) or nitrous oxide (N_2O) as terminal electron acceptors and nitric oxide (NO) is generated in this process through the activity of the enzyme nitrite reductase. Thus, anaerobic physiology is an important feature of biofilm development.

A number of recent studies have linked anaerobic metabolism with dispersal processes in *P. aeruginosa*. Recently it was found that NO is able to induce biofilm dispersal at concentrations that are nontoxic to *P. aeruginosa* [125]. Using the NO donor sodium nitroprusside (SNP), at low (nanomolar), nongrowth-inhibitory concentrations were found to cause dispersal of biofilm cells from glass surfaces. Addition of SNP to established *P. aeruginosa* biofilms on glass slides caused up to 80% reduction in the amount of biomass, and enhanced swimming and swarming motility. It was demonstrated that anaerobic denitrification occurs inside *P. aeruginosa* biofilms and that levels of peroxynitrite were enhanced in regions of the biofilm that had undergone dispersal. A *nirS* mutant strain of *P. aeruginosa*, lacking the only enzyme capable of generating metabolic NO, did not undergo dispersal in the study. Microarray studies have also revealed that genes involved in adherence are down-regulated in *P. aeruginosa* upon exposure to NO ([128] and N. Barraud *et al.*, unpublished). This suggests a mechanism by which NO-exposed bacteria detach from the biofilm leading to reduced biofilm biomass and increased number of planktonic organisms. Moreover, a recent study [129] has demonstrated a role for nitrate- as well as NO-sensing response regulators in regulating motility and dispersal of this organism.

In addition, there is data suggesting that NO- and c-di-GMP-mediated dispersal are interrelated. Several signal transduction pathways are known to regulate the activity of GGDEF and EAL domains, including those sensing oxygen, pH, temperature and other environmental stimuli [121,130,131]. Intriguingly, Aravind *et al.* [132] also found that NO sensing proteins, called heme NO-binding (HNOB) domains, are frequently associated with GGDEF and EAL domains in diverse bacteria, suggesting a link between NO sensing and c-di-GMP turnover. Low nanomolar levels of NO were found to induce differential expression of genes encoding proteins harboring GGDEF and/or EAL in transcriptomic experiments in *P. aeruginosa* (N. Barraud *et al.*, unpublished) Furthermore, addition of the NO donor SNP caused reduced intracellular levels of c-di-GMP and NO-mediated dispersal was greatly reduced in the presence of exogenous GTP – a phosphodiesterase inhibitor.

Bacterial responses to NO have been studied extensively in planktonic bacteria in the context of physiological responses to nitrosative stress [128,133–135]. However, there is relatively little information on the cellular targets or signaling mechanisms of NO with relevance to biofilm development and differentiation processes. Analyses of microbial genomes have suggested that homologous NO-sensing receptor domains are common to both prokaryotic and eukaryotic regulatory proteins [132,136]. In eukaryotes, NO signaling is known to play an important role in the regulation of diverse processes, including apoptosis, cell proliferation and differentiation. Intriguing similarities exist between the signaling role of NO in eukaryotes, and its control of biofilm cell differentiation, death and dispersal, as demonstrated by our research.

Biofilms are suggested to exhibit developmental analogies with multicellular eukaryotes [137,138] and therefore it may be relevant to examine these bacterial biofilm populations for the origins of key regulatory processes observed in more complex organisms. This work on NO-mediated control of biofilm development in *P. aeruginosa* may point to a conserved role for NO signaling in the regulation of differentiation and developmental events across bacterial and eukaryotic physiology.

9.2.4.3 **Microcolony-based Biofilm Dispersal**

A characteristic of mature *Pseudomonas* biofilms is the formation of multicellular three-dimensional microcolonies that are often distinct from the bulk biofilm biomass. Several recent studies have reported pronounced activity and motility localized to the center of microcolonies, leading to the dispersal of cells from inside the structure, and leaving behind large transparent cavities or hollow "shells" made up of nonmotile cells [138–140] (Figure 9.3). This process of evacuation of cells from microcolonies has been termed "seeding dispersal" in order to differentiate it from the process of erosion, which is the passive removal of cells from the biofilm by fluid shear and has clinical implications for CF patients [141] (see Section 9.4.1).

Such processes are likely to involve enzymes that degrade the extracellular polysaccharide matrix (e.g. polysaccharide lyases) which are known to be of importance in *Pseudomonas* dispersal [122,123]. As these enzymes are not thought to be transported across the cell membrane, their release to the extracellular polysaccharide matrix may rely primarily upon lysis of cells within the biofilm [142]. Indeed, one process that is often observed in association with microcolony seeding dispersal is the death and lysis of subpopulations of cells within the biofilm. In *P. aeruginosa*, cell lysis during biofilm development was linked to the production of oxidative and nitrosative

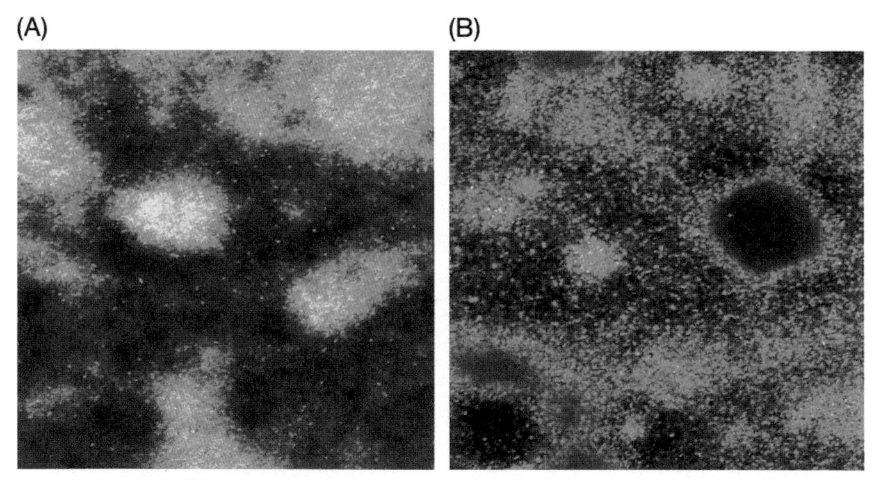

Figure 9.3 Confocal laser scanning microscopic image of *P. aeruginosa* PAO1 wild-type showing cell death inside the microcolonies (A) followed by dispersal, leaving hollow colonies of live cells (B). Cells were grown in M9 in flow cells and stained with BacLight Live/Dead stain at days 5 and 6.

radicals in microcolonies, and the activation of a lytic form of the filamentous prophage Pf4 of *P. aeruginosa* [138] as described previously. Genes that affected expression of receptors for the phage (RpoN-mediated regulation of T4P and flagellae) controlled cell death during development [111]. However, the molecular mechanism of lysis in *P. aeruginosa* biofilms remains to be fully elucidated as these symbiotic filamentous phage are generally thought to be harmless to host cells. Intriguingly, the Pf1-like prophage of *P. aeruginosa* encodes homologues of proteins from two different *Escherichia coli* toxin–antitoxin elements [138,143] which may play a role in cell killing within biofilms. It is proposed that autolysis impacts on dispersal processes in biofilms by disrupting the biofilm architecture and that surviving cells in the biofilm benefit from the nutrients released by their dead siblings [138], which facilitates conversion of surviving cells to the motile dispersal phenotype. It has been observed that autolysis was correlated with over production of the QS signal PQS and it was suggested that lysis was linked to activation of the phage of PAO1 [144].

Active seeding dispersal, the final stage of the biofilm life cycle, has received increased attention in recent years. A number of environmental and genetic cues and their downstream alteration in gene expression have been identified for this process. Intracellular c-di-GMP has emerged as an important regulator of planktonic versus biofilm lifestyles in *Pseudomonas* and other bacteria. In addition, there is now also evidence of a role for anaerobic metabolism and NO signaling in c-di-GMP-mediated control of biofilm dispersal and spatially organized patterns of seeding dispersal and lysis in microcolonies.

9.3 Regulatory Mechanisms of *P. aeruginosa* Biofilms

9.3.1 c-di-GMP

As described above, the development of cell aggregates or biofilms is a succession of several complex events, which obviously requires the involvement of regulatory circuits. Recently, the novel second messenger c-di-GMP has been found to be involved in the regulation of biofilm formation in *P. aeruginosa* and other bacteria [145,146]. This intracellular signaling molecule was originally found in *Gluconacetobacter xylinus* where it acts as an allosteric regulator of cellulose synthase [147,148]. The global impact of c-di-GMP as a second messenger was recognized by the identification of conserved domains responsible for c-di-GMP turnover and their wide distribution among Eubacteria [130,149] (Figure 9.4). The biosynthesis of c-di-GMP from two GTPs is catalyzed by specific diguanylate cyclases (DGCs) containing a characteristic GG[D/E]EF domain as the active center [63,150]. Hydrolysis of c-di-GMP is catalyzed by specific phosphodiesterases (PDEs) containing either a conserved E[A/V]L or a HD-GYP domain [151,152]. Although many proteins possess both GGDEF and EAL domains, all of the composite proteins tested to date, showed either DGC or PDE activity, but not both. Partly, this observation can be explained by the fact that many of these dual proteins

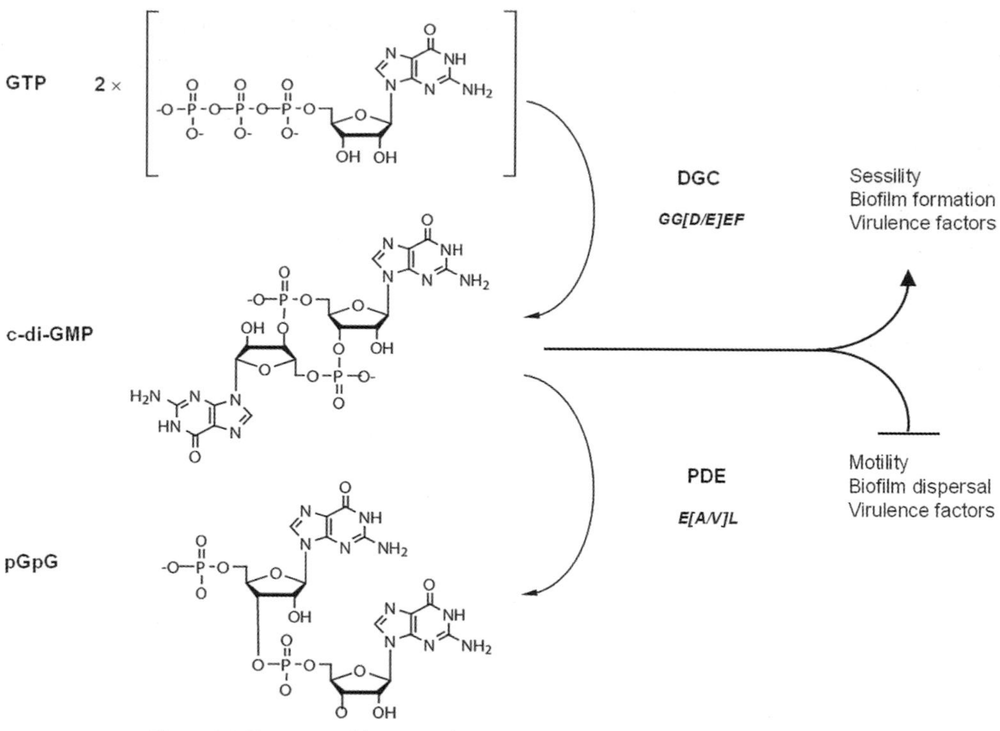

Figure 9.4 Turnover of the second messenger c-di-GMP. Biosynthesis of c-di-GMP is catalyzed by specific DGCs and degradation catalyzed by specific PDEs. The corresponding conserved motifs of the catalytic active sites are indicated in italic. Several cellular functions regulated by c-di-GMP signaling pathways are shown.

contain an imperfect (degenerate) GGD[D/E]F motif, which has been shown to disrupt the corresponding DGC activity [153]. These degenerate domains have, however, been shown to have important regulatory functions, such as the allosteric control of neighboring PDE activity [151].

Since its discovery, increasing evidence of a key role for c-di-GMP in the regulation of the transition from sessility to motility and *vice versa* has been obtained. The huge variety of sensory and regulatory domains associated with GGDEF and EAL domains and their multiple occurrence in organisms suggests that a multitude of environmental signals can be perceived and transmitted by c-di-GMP signaling pathways to regulate cellular functions [130]. In general, it has been postulated that c-di-GMP-dependent regulation could influence transcription, translation and posttranslational mechanisms [145].

9.3.1.1 **c-di-GMP Regulation of SCVs**

The most prominent example for the involvement of c-di-GMP regulatory system represents the occurrence of the highly adherent SCV phenotypic variants among pseudomonads. It has recently been shown that the two-component response regulator, PvrR, is involved in setting the frequency of switching from wild-type morphology to the SCV's phenotype [7]. Interestingly, the PvrR protein harbors a conserved EAL domain with PDE activity [154] and it is therefore likely that PvrR influences this phenotypic switch by the turnover of the signaling molecule c-di-GMP. This hypothesis is further supported by Meissner *et al.* [155], who demonstrated decreased levels of c-di-GMP and the loss of the autoaggregative phenotype in a SCV by deletion of the DGC PA1120 or the GGDEF/EAL containing motility regulator MorA (see below).

9.3.1.2 **Autoaggregation and c-di-GMP**

Spiers *et al.* [96] identified the Wsp chemosensory pathway by transposon mutagenesis in an effort to identify genes involved in the development of the WS phenotype of *P. fluorescens* [94]. The WS phenotype is due to the overexpression of cellulose encoded by the *wss* operon which is regulated by the *wsp* operon [95,156]. In the corresponding signal transduction pathway, the activity of the sensor kinase WspE positively regulates the response regulator WspR, which harbors a GGDEF domain, resulting in the biosynthesis of c-di-GMP [62,153,157]. Kinase activity of WspE depends on activation of the membrane-bound receptor-signaling complex composed of WspABDE in response to an unknown environmental signal. Further regulation of WspE activity is mediated by the methyltranferases WspC and the methylesterase WspF, both of which are involved in a feedback mechanism leading to adaptation to the signal [156].

Separately, this pathway has also been found to be responsible for an autoaggregative phenotype in *P. aeruginosa* where it has been suggested to regulate the expression of two other gene clusters (*psl* and *pel*) involved in the biosynthesis of a mannose and glucose-rich polysaccharide of unknown structure [50,52,60,144]. Interestingly, this pathway was also linked to the expression of adhesive fimbriae, encoded by the *cupA* gene cluster in *P. aeruginosa* [47,144]. Recently, two other studies identified a two-component system which controls the expression of the *cupB* and *cupC* gene cluster [76,158]. These genes encode fimbriae responsible for adhesion to surfaces as described in Section 9.2.1.2. In this regulatory system, the sensor kinase RocS1 (SadS) interacts with two response regulators, i.e. the DNA-binding protein RocA1 (SadA), which acts as a transcriptional activator of the *cupB* and *cupC* genes, and RocR (SadR), which is an antagonist of RocA1 activity. The response regulator RocR (SadR) harbors a conserved EAL domain, suggesting that this protein might exhibit enzymatic activity by hydrolyzing the second messenger c-di-GMP [130,151,159].

9.3.1.3 **Motility**

In addition to the regulation of adhesive surface structures and EPSs, c-di-GMP-dependent regulation also has been linked to the motility of cells. As already mentioned (see Section 9.2.1), swimming motility, driven by flagella, and twitching motility, driven by type IV pili, have both been shown to be involved in attachment of cells and the development of biofilm structure. Recent studies implicate the involvement of two GGDEF/EAL-containing proteins in the regulation of these cell surface structures, i.e. MorA and FimX [160,161]. A MorA mutant exhibited constitutive flagella expression during all growth phases in *P. putida*, and decreased biofilm formation in *P. putida* and *P. aeruginosa*, respectively. While the involvement in c-di-GMP turnover by MorA is predicted based on structural analysis, PDE activity of the FimX protein has been shown experimentally in a recent study [162]. In this as well as a previous study [161], the authors demonstrated that loss of function of the polarly localized FimX protein affected the density of expressed pili. In addition, the N-terminal receiver domain of FimX was shown to be essential for the localization of the FimX protein and its loss resulted in disruption of the polarity of pili expression. Finally, the PDE BifA together with the DGC SadC are involved in the inverse regulation of biofilm formation and swarming motility [76,163]. It is suggested that the BifA and SadC activity modulate the intracellular levels of c-di-GMP to an unknown environmental signal. The signal is transduced by the SadB protein (see Section 9.1) to finally regulate EPS production and swarming motility.

9.3.1.4 **Environmental Stimuli and Targets of c-di-GMP Signaling**

Apart from the various regulatory effects of c-di-GMP-dependent signaling on biofilm development and dispersal described above, information about specific environmental triggers, target genes and the corresponding regulatory circuits of c-di-GMP-dependent signaling is still limited. Only a few of the environmental stimuli triggering c-di-GMP-dependent signaling in pseudomonads, such as increased biofilm formation in response to sublethal concentrations of aminoglycoside antibiotics [164], the induction of biofilm dispersal in response to carbon starvation, sudden changes in nutrient availability [116,126] and NO (Barraud *et al.*, unpublished data), and inhibition of biofilm formation in response to phosphate limitation [45], are known. Knowledge about the mechanistic effects of c-di-GMP for expression of various phenotypes in response to stimuli discussed above is even more limited. The only downstream targets for c-di-GMP signaling identified so far are the cellulose synthase complex of *G. xylinus*, as mentioned above, and the recently identified PilZ domain encoded by at least three different genes in *P. aeruginosa* [147,148,165]. Recently, two studies confirmed that one of those genes (PA4608) from *P. aeruginosa* indeed encodes a PilZ domain protein that exhibits high-affinity binding of c-di-GMP [166,167]. Based on these important findings, the authors postulated that most, if not all, proteins containing a PilZ domain function as a receptor for the signaling molecule c-di-GMP. This result may have a major impact on the understanding of the regulatory mechanism of downstream c-di-GMP targets.

For a broader understanding of the global regulatory function of c-di-GMP signaling systems, the identification of more environmental triggers and their corresponding regulated targets is needed. In addition, questions regarding the interaction of these second messenger pathways with other global regulatory systems such as QS, as well as the mechanisms by which microorganisms harboring multiple c-di-GMP signaling pathways manage to regulate the relevant target, will be key aspects of future research in this field. The potential to extend our current knowledge of this signaling pathway will enhance our understanding of biofilm development and virulence of *P. aeruginosa* as well as other bacteria.

9.3.2
QS and *P. aeruginosa* Biofilms

Biofilms are more resistant to antimicrobials and antibiotics than their free-living counterparts, and are therefore difficult to remove. Traditional biocidal agents are failing to control biofilm-based infections and there is a clear need for a new approach to this problem. One such strategy relies on subtle manipulation of bacterial behaviors rather than killing by biocidal agents which strongly select for resistant mutants. QS is one of the key regulatory systems controlling biofilm formation. This section will present the current understanding of how QS coordinates biofilm formation in *P. aeruginosa*.

There are a number of excellent reviews on QS covering specific details of the gene families involved in biofilm formation and development, and their mechanisms of action [168–171]; therefore, the QS system of *P. aeruginosa* will only be briefly described here. QS in *P. aeruginosa* is complex, as it relies on a *N*-acylated homoserine lactone (AHL) QS system, which is composed two sets of I and R genes, *lasRI* and *rhlRI*, where the *las* system is hierarchically dominant over the *rhl* system. The LasI and RhlI are AHL synthases that produce the AHL signals, 3-oxo-dodecanoyl-homoserine lactone (OdDHL) and *N*-butanoyl-L-homoserine lactone (BHL). *P. aeruginosa* utilizes PQS as a second QS signal [172], which plays a role in the regulation of the *rhl* system [173], where its regulatory role is between the *las* and *rhl* systems, and is partially responsible for regulatory control of *rhl*. QS controls a range of responses in *P. aeruginosa*, many of which are virulence factors such as elastase, exotoxin A, hydrogen cyanide, pyocyanin, rhamnolipid, Xcp secretion, lectins and biofilm formation [18,65,174,175].

With respect to biofilm formation, it is perhaps logical that QS is involved as these high-density populations would represent an ideal environment for signals to accumulate to a concentration sufficient for the QS system to operate. Indeed, Charlton *et al.* [176] demonstrated that biofilms of *P. aeruginosa* accumulate high concentrations of AHL signals, up to 630 μM OdDHL in the biofilm compared to 14 μM in the effluent. This signal concentration is well above the half maximal concentration (3–5 nM OdDHL) required for the induction of the QS circuit in *P. aeruginosa* [177], suggesting that the AHL concentration within the biofilm is well above the threshold required to induce a QS response. Moreover, it has been shown that many QS genes are expressed within the biofilm [35,64,178]. For example, using reporter gene constructions, De Kievit *et al.* [178] showed that the *lasI* and *rhl* were expressed within the biofilm, although their expression was localized both spatially and temporally, with *lasI* gene expression decreasing over time while *rhlI* was expressed in the region of the biofilm near the substratum, throughout the biofilm lifecycle.

9.3.2.1 Effect of Nutrient on QS Regulation of Biofilm Formation

While it seems accepted that biofilm formation and maintenance is QS regulated, the literature would suggest that this is not an absolute requirement, as some studies clearly show effects of QS on biofilm formation and others demonstrate that wild-type and QS mutants form identical biofilms [10,179]. In the study by De Kievit *et al.* [178] it was shown that the wild-type and QS mutant formed different biofilms when glucose was used as a carbon source, while the comparison of biofilms formed with citrate as the sole carbon source showed that there was no effect of the QS system on biofilm formation. This effect seen when glucose is the sole carbon source may be linked to the QS control of the *pel* genes which are involved in the production of a glucose-rich exopolysaccharide [64] (see Section 9.2.2). Interestingly, expression of the *pel* genes is also regulated by the Wsp system (see Section 9.3.1). Other studies have shown that the wild-type and QS mutant strains formed identical biofilms when glucose or glutamate were used as a carbon sources, but when grown on succinate, the QS mutant generated microcolonies, whilst the wild-type remained as a flat, unstructured biofilm [39]. In this study, QS mutants were reduced in swarming motility, which contributed to the formation of microcolonies when grown on succinate. As we begin to unravel the relationship of QS with other global regulators and cellular biochemical processes, a picture is emerging that suggests that QS is intimately connected to central metabolic processes which naturally complicates our ability to define the role of QS on biofilm formation.

Carbon is not the only nutrient that may affect the role of QS in biofilm formation. Jensen *et al.* [180] showed that under phosphate-limiting conditions the expression of *rhlR* and PQS is dependent on *phoB*. It was also observed that *rhlI* and *lasR* contain putative *phoB* promoters. Similarly, Rashid *et al.* [181] have demonstrated a link between biofilm formation, QS and phosphate. A polyphosphate kinase (*ppk*) mutant was shown to have significantly reduced biofilm formation and reduced production of both OdDHL and BHL. Nitrogen metabolism has also been shown to regulate QS. An *rpoN* mutant exhibited reduced expression of RhlI, elastase and pyocyanin, and the *rpoN* mutant formed an aberrant biofilm, similar to that of the *rhlI* mutant [143]. In contrast, an *rpoN* mutant described by Heulier *et al.* [182] had increased expression of *lasRI* and *rhlRI* as well as increased amounts of elastase, pyocyanin and hydrogen cyanide. While these two studies appear to be inconsistent with each other, it is clear that the RpoN regulon influences QS which may link QS to nitrogen levels and it is possible that the differences in effects observed are linked to the different media and growth conditions used in the respective studies.

Early biofilm studies investigating the physical nature of biofilms using microelectrodes [183], showed that oxygen concentrations decreased rapidly within microcolonies and recent observations have underscored the importance of these observations. For example, in one study comparing gene expression in an artificial sputum medium at different oxygen concentrations [184], *lasI, rhlI, rhlR, rsaL* (a repressor of *lasI*) and hydrogen cyanide were induced at 0.4% oxygen, but not at 2% oxygen or under anaerobic conditions when compared to control cultures cultivated at 20% oxygen [184]. This suggests that the QS system may be particularly active under a very narrow range of oxygen concentrations, which has implications for QS expression in the lung.

Toyfuku *et al.* [185] have shown that *rhlI* mutants grew as well as the wild-type under anaerobic conditions. However, the exogenous addition of BHL to the *rhlI* mutant inhibited growth under these conditions, indicating that production of

BHL can inhibit anaerobic growth, and that this activity appears to be through the repression of *nirS*, *norC* and *nosR*. Thus, as suggested by the study by Alvarez-Ortega above [184], *rhl* expression may be decreased under anaerobic conditions, which could lead to increased activity of the denitrification pathway. As discussed above in Section 9.2.4, NO has been shown to accumulate in microcolonies of *P. aeruginosa* at the time of hollow colony formation, thus linking dispersal and QS in the biofilm, suggesting that this specific effect may be limited to the anoxic or anaerobic centers of the microcolonies. Thus, QS may be active under a range of different oxygen concentrations, but may affect specific genes depending on the oxygen tension.

9.3.2.2 **QS, Central Metabolism and Biofilm Gradients**

The QS signal PQS has been shown to be derived from anthranilate which can be synthesized from chorismate and or by tryptophan degradation [186]. Interestingly, it has been shown that anthranilate synthase genes are positively controlled by AHL-mediated QS, linking AHL control of metabolic genes, which are in turn important for the production of PQS [187,188]. AHL production is directly linked to methionine metabolism via *S*-adenosyl methionine (SAM), which plays a role in methyl transfer reactions and polyamine synthesis. Both AHL and polyamine synthesis generate methylthioadenosine (MTA) which is detoxified by conversion to adenosine via adenine. Adenosine is subsequently converted into hypoxanthine via inosine. The conversion of inosine to hypoxanthine is mediated by the nucleoside hydrolase, Nuh, which is positively controlled by LasR [189], linking AHL-mediated QS to central metabolism. Another example of QS control of key metabolic processes in *P. aeruginosa* is the observation that genes in the Entner–Doudoroff pathway are upregulated by QS as well as *coxAB*, encoding cytochromes, as shown by microarray studies [187,188,190].

The above information on the interrelationship of QS and metabolism highlights the overall complexity of the regulation of QS, and provides an important context for unraveling how QS works in a high-density environment such as a biofilm. In particular, it is helpful in understanding how cells can control QS in a biofilm where, as noted above, signal concentrations can be well in excess of that needed to induce gene expression or where it has been shown that *lasI* and *rhlI* are expressed in spatially and temporally defined patterns within the biofilm [178]. The implication is that gene expression is coordinately controlled by multiple systems, of which the QS system is one, providing the cells with both fine control and flexibility. Another important point when studying QS is that the biofilm should not be considered a homogenous population. Localized differences in oxygen concentrations, for example, occur where cells near the biofilm–liquid interface will experience different oxygen concentrations than those in the centers of microcolonies, which could result in the two populations of cells having different QS responses. This is supported by work by Xu *et al.* [191] who demonstrated that oxygen gradients occurred within *P. aeruginosa* biofilms and these gradients corresponded to zones of metabolic activity, where active protein synthesis was localized to the aerobic zone and was absent in the anaerobic zone. In particular, the cells within the center of the microcolonies may experience anaerobic conditions, which could reduce *rhl* expression, leading to derepression of *nirS*, to generate NO, the accumulation of which initiates the process of cell death within the microcolonies (see Section 9.2.4), but does not appear to lead to killing of the cells on the outer edges of the biofilm, presumably due to oxygen mediated repression of *nirS*.

There are several examples of QS genes where their expression within the biofilm is spatially dependent, highlighting this concept of multiple regulators fine-tuning their expression. For example, Boles *et al.* [192] observed that the QS-controlled rhamnolipid production was involved in at least one form of dispersal of *P. aeruginosa* from the center of the microcolony only and not at the edges. Similarly, the PQS controlled secretion of DNA into the biofilm matrix is limited to the cells that make up the stalk of the microcolony and is absent in the cap. This is supported by experiments using a *pqsA–gfp* fusion that showed expression only on the outer part of the stalks and not in the cap [193]. Thus, QS-controlled genes within the biofilm are differentially expressed across the biofilm, which may be due to coregulation by regulators that respond to very specific environmental and physiological cues.

9.3.2.3 **QS and Chronic Infection**

It has been shown that QS plays a role in biofilm maintenance and resistance to antibiotics and surfactants, and in the host immune response, and is therefore a strong target for the development of a therapeutic strategy to control *P. aeruginosa* infections [10,187,194,195]. AHLs can be detected in CF sputum [196–200] and bacteria isolated from CF sputa produce them in ratios indicative of biofilm rather than planktonic mode of growth [201]. The observation that AHLs and PQS can be recovered from the sputum of *P. aeruginosa*-infected CF patients strongly implies that biofilm formation is a key feature of the disease process. The QS system has been shown to be integral to both acute and chronic *in vivo* infections in various animal models [187,202–204]. Moreover, the recently discovered clinical efficacy in CF patients of macrolide antibiotics, such as azithromycin, normally not included in antipseudomonal therapy because of their lack of bactericidal or bacteriostatic activity for *P. aeruginosa*, has been attributed to their demonstrated QS antagonistic activities [205–208]. Microenvironmental influences may, however, influence the extent to which QS exerts its effects in all bacterial biofilms in the lung and QS-deficient strains of *P. aeruginosa* appear capable of causing infections [39,209,210]. The development of a biofilm control strategy necessitates an understanding of the microenvironment experienced by the cells within the biofilm. Thus, the question of where the biofilms occur in these patients and whether they are experiencing anaerobic or aerobic conditions is relevant to consider in the QS context.

The study of QS and its role in biofilm formation, maintenance and physiology of the community is an exciting area with significant challenges in the ability to study localized gene expression by different populations within the community. Meeting these challenges will depend on new techniques that allow for the probing of global gene expression by single cells or small populations and advances in bioinformatics and systems biology. Such advances which will help to present a unified picture of the interrelationship of QS and other regulators.

9.4
Biofilm/Eukaryote Interactions

9.4.1
Clinical *P. aeruginosa* Biofilm Infections

P. aeruginosa causes a wide range of acute and chronic infections in immunocompromised and hospitalized individuals [211,212]. Chronic infections are now widely accepted to involve biofilm growth [11,213,214]. For *P. aeruginosa*, such biofilm-related infections can result from the use of medical devices (mechanical ventilators, central venous catheters, urinary catheters, contact lenses) and also frequently occur in the respiratory tract of patients suffering from bronchiectasis, chronic obstructive bronchopulmonary disease or CF. Biofilm-related infections also include persistent infections of wounds, ear (chronic suppurative otitis media) and eye (keratitis/corneal ulcer), and chronic prostatitis and diarrhea (rarely reported) [211,215–219]. Once established, these infections are refractory to host defenses and resist eradication with even the most aggressive conventionally used therapies, and hence pose a significant clinical challenge.

9.4.1.1 Chronic Lung Infections in CF

P. aeruginosa lung infections in CF are considered the archetype of human biofilm infections and have provided a major impetus for the study of *P. aeruginosa* as a model organism for elucidating bacterial biofilm development and the factors determining acute versus chronic disease outcome [220]. In recent years, considerable progress has been made in understanding the pathogenic events that lead to chronic infection in the CF lung.

CF is an inherited genetic disorder where patients have mutations in the CF transmembrane regulator gene that affects chloride channels, resulting in altered electrolyte secretion across epithelial surfaces and multisystem disease [221–223]. In the lung, paraciliary fluid in the lower respiratory tract is decreased leading to impaired mucociliary clearance of inhaled microbes. Although a number of bacterial pathogens may be acquired, often in age-dependent sequence [222], and the CF lung microbial flora is recognized as being very diverse [224], it is *P. aeruginosa* that eventually dominates and establishes intractable infections that result in lung failure and premature death [14,225]. The airways (bronchi) of most CF individuals become colonized in childhood, commonly within 3 years of birth and invariably by adolescence, with strains of this organism, usually acquired from environmental sources [226]. Such colonizing strains are initially planktonic, motile and nonmucoid. They express type IV fimbriae that attach to mucin/mucin-coated epithelial cells, secrete AHLs, express virulence determinants under their control (e.g. toxins, proteases, pyocyanin) that likely contribute to tissue damage, and facilitate colonization through aggregation and the formation of microcolonies. These early infections are sensitive to antibiotics and can be eradicated with aggressive antimicrobial therapy [221,222,227]. Early prophylactic antibiotic therapy has proved effective at delaying *P. aeruginosa* initial colonization and the establishment of chronic infection [228,229]. Ultimately, however, conditions in the lung induce bacterial conversion to a mucoid form in which the organism produces vast amounts of exopolysaccharide, with concomitant loss of flagella and type IV fimbriae, and the formation of large bacterial communities which adopt a biofilm mode of growth and cannot be cleared by even the highest antibiotic doses.

9.4.1.2 **Conversion to Mucoidy and Anaerobic Growth**

P. aeruginosa grows in different niches in the CF lung. Autopsy analysis of the respiratory zone (respiratory bronchioles, alveolar ducts and sacs) shows a diversity of biofilm aggregates [230]. This zone (95% of the lung volume) is initially aerobic, but if mucus plugs the airflow it can become microaerophilic, and if abscesses form, anaerobic. Mucoid biofilms in the respiratory bronchioles are generally surrounded with numerous polymorphonuclear leukocytes (PMNs). The smaller (5% of the lung volume) conductive zone of the lung (trachea, bronchi and terminal bronchioles) becomes anaerobic in advanced disease due to mucus plugging. It is thought that mucoid biofilms from the respiratory zone are transported to the conductive zone where they grow in large mucopurulent (dead PMNs) "macrocolonies" (around 100-μm conglomerates) from which nonmucoid revertants can emerge [230–233].

Conversion to mucoidy (mutation in the *mucA* gene) is one of the best prognostic indicators of chronic lung infection [234,235]. Alginate confers increased resistance to host immune responses [236–238], may impede antibiotic penetration into biofilm colonies [239], protects against reactive oxygen intermediates [59], and plays an important role in the development and architecture of biofilms [240,241]. Important factors driving mucoid conversion have been identified as hypoxic conditions in the mucus and the release of reactive oxygen species (e.g. H_2O_2, O_2^-) from the infiltrating PMNs [100,233]. PMN activity and death is integral to the development of chronic infection and not only for the above role. DNA and actin released from damaged airway epithelial cells and necrotic PMNs provide important scaffolding for biofilm microcolony development [73,222,242,243]. Not all the DNA in the biofilm matrix is derived from PMN destruction, however. As discussed above, DNA released from bacterial cells themselves also makes a significant contribution to the extracellular matrix. *P. aeruginosa* rhamnolipid production seems mainly responsible for the necrotic effects on host cells [244]. Rhamnolipids also affect/abolish PMN migration in *in vitro* models and induce similar effects in the lungs of mice infected with *P. aeruginosa* [244]. Hence, not only are rhamnolipids thought to make important contributions to biofilm initiation, architecture and dispersal [124,192,245], but also *P. aeruginosa* uses them to defend itself aggressively against PMNs. Nonmucoid isolates produce more rhamnolipids than mucoid strains [246], consistent with the increased death of PMN observed in the conductive zone.

P. aeruginosa adapts well to an anaerobic environment. Although widely considered an aerobic organism, it is able to respond to oxygen concentration along a continuum [184] and it can grow under anaerobic conditions if sufficient terminal electron receptors (NO_3^-, NO_2^- or arginine) are available, as they are in the CF lung setting [127,247,248]. Robust *P. aeruginosa* biofilm formation has been shown in model systems using strain PAO1 under these conditions [249,250]. The viability of anaerobic biofilms is dependent on *rhl* QS and NO reductase to prevent the accumulation of toxic NO, a byproduct of anaerobic respiration [127,185]. Universal stress proteins and an outer membrane protein, OprF, have also been identified as essential for *P. aeruginosa* surviving anaerobic energy stress in CF mucus [127,251]. To reduce the effects of other hazardous byproducts of anaerobic respiration (e.g. O_2^- and H_2O_2), the organism possesses two superoxide dismutases (SODs), cofactored by iron or manganese, and two heme-containing catalases. QS reportedly controls the expression of SOD and catalase genes, and hence influences *P. aeruginosa* susceptibility to H_2O_2 [252].

9.4.1.3 Phenotypic and Genotypic Analyses of *P. aeruginosa* in CF Infection

Longitudinal phenotypic and genotypic analyses of *P. aeruginosa* isolated from CF patients have shown that most patients harbor one or only a few unique strain genotypes [253]. Notwithstanding this, considerable phenotypic diversity is seen, not only among different patient strains, but also within isolates from an individual patient [35,249,254] (Figure 9.5). As mentioned earlier, heterogenous environments exist at the sites of chronic infection and these, as well as individual patient factors, differing antibiotic therapies, bacterial species competition and the duration of colonization, are likely contributing factors to the range of phenotypes seen [255]. Nevertheless, some traits are consistently acquired during chronic infection in different CF patients, suggesting there is a conserved pattern of evolution by which *P. aeruginosa* adapts to the CF airway. *P. aeruginosa* phenotypes in chronic infection are generally characterized by lack of swimming motility [99], overproduction of alginate and decreased antibiotic susceptibility [51], lack of expression of the type III secretion system and an associated decrease in cytotoxicity [256], lack of production of virulence factors and factors such as the siderophore, pyoverdine and pyocins [257], deficiency in O-antigen polysaccharide [258], increased phage resistance and structural alterations in lipid A component of the outer membrane [259], amino acid auxotrophy [260,261], and increased incidence of the mutator phenotype [97,101] (as mentioned earlier in Section 9.2.3). The mucoid phenotype and nonmucoid

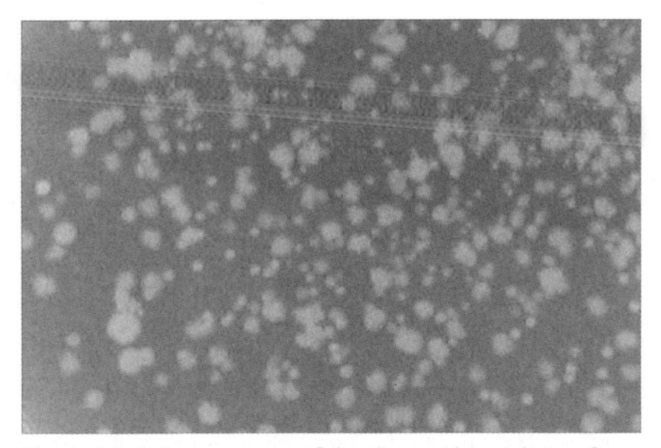

Figure 9.5 Colony variants of the dispersal population from a mature biofilm of a clinical *P. aeruginosa* isolate from a chronically infected CF individual.

revertants (containing mutations in *mucA*) of the same strain are frequently present in sputum simultaneously [262,263].

Consistent with the above phenotypic observations, recent studies have highlighted that *P. aeruginosa* undergoes substantial genetic change during chronic infection in CF patient airways [98,255]. In addition to the well-studied mutation in the *mucA* gene, there is an accumulation of "loss-of-function" mutations in specific genes. Thus, comparison of the genome sequence of a *P. aeruginosa* strain early in a CF infection with that of the descendent of that strain 7.5 years later revealed 68 mutations in the late isolate [98]. Most were single base pair changes and many were predicted to result in change or loss of protein function. A timeline for this strain's evolution was constructed by examining isolates at intermediate times. CF isolates from other patients were also examined and it was found that common genes were mutated in most CF patients. Overall, virulence genes required for the initiation of acute infections were selected against [98,255,264]. One of the common target genes was *lasR* (mutated in 19 of 30 CF patients) whose elimination would result in the loss of many acute virulence determinants in one hit. Possibly, such virulence factor loss helps the bacteria evade the host immune response. Loss of LasR also confers a selective growth advantage with carbon and nitrogen sources, such as amino acids, and an increased tolerance to a β-lactam antibiotic widely used in CF treatment [255]. In addition, LasR mutants are significantly more resistant to cell lysis and death than the wild-type strain [83]. This resistance may be linked to dispersal processes and the generation of additional variation via PQS involvement [83,144] as indicated in Section 9.2.3. Mutations in *lasR* included point as well as those caused by an insertion sequence (IS) element identical to one implicated in loss of O-antigen biosynthesis [255,258]. Such IS elements that actively transpose in the genome may contribute to faster evolution in *P. aeruginosa* lineages in individual CF patients [102].

Multidrug-efflux pump genes are another common target for adaptive mutations [98]. These pumps provide *P. aeruginosa* with natural resistance to antibiotics by exporting them from the cytoplasm. The most frequently mutated gene in this category is *mexZ*, a negative regulator of *mexX* and *mexY*, components of one of the now four described *P. aeruginosa* efflux pumps, MexXY–OprM [265–267]. This mutation is likely selected as a result of the commonly used treatment with the aminoglycoside tobramycin, as increased *mexX* and *mexY* are associated with resistance to this drug [268]. Efflux pumps can accommodate a variety of molecules in addition to antibiotics, including QS effector molecules, and thus may have additional roles in ensuring *Pseudomonas* survival in chronic infections. Different efflux pumps have been implicated in the transport of PQS and BHL [265] and OdDHL [269,270], possibly providing another mechanism for the observed decreased in acute virulence factor expression that seems to be advantageous for persistence in the CF lung.

The above insights into how *P. aeruginosa* adapts from an acute to chronic lifestyle have greatly increased the understanding of this organism's pathogenesis in the CF lung and the mechanisms by which it becomes refractory to antibiotic therapy. However, the adaptations that occur in chronic infection, and their links to biofilm growth and dispersal are only just starting to be understood. Many additional factors (e.g. the role of predisposing viral and other infections in initial colonization and exacerbations, mixed species biofilms and QS "cross-talk" by AHLs from *Burkholderia* spp. and AI-2 signaling from other oropharyngeal flora) still remain largely unexplored [51,271–273]. Nevertheless, the research discussed in the preceding sections is already leading to additional strategies for the improved management and treatment of *P. aeruginosa* pulmonary infections in CF.

9.4.2 Similarities Across *P. aeruginosa* Biofilm–Host Interactions

9.4.2.1 Interactions with Host Immune Cells

Hallmarks of *P. aeruginosa* virulence in the CF lung include bacterial growth in microcolonies encased by an exopolymeric matrix and the conversion from non-mucoid to mucoid alginate-overproducing strains as described above. The inflammatory response to chronic *P. aeruginosa* lung infections is mainly characterized by the constant influx of neutrophils or PMNs. Although PMNs have been described to efficiently phagocytose adherent *P. aeruginosa* [274], they appear ineffective in combating *P. aeruginosa* infections of the lung. In fact, large-scale oxidative bursts released by "frustrated" PMNs contribute substantially to the damage of lung tissue, rather than the bacterial infection itself. Apparently, the EPS matrix plays a key role in the reduced ability of phagocytic immune cells to instantly penetrate and phagocytose biofilms. There is evidence that *P. aeruginosa* mutant biofilms lacking the exopolymer alginate are more susceptible to leukocyte killing [275]. Apart from forming a physical barrier against attacking phagocytes, EPS may also function as a chemically reactive barrier by scavenging reactive oxygen species, which are charged by PMNs during the oxygen-dependent attack. For example, EPS isolated from a mucoid *Burkholderia cenocepacia* strain was found to inhibit neutrophil chemotaxis and scavenge neutrophil-derived reactive oxygen species [276]. Similar findings have been presented for alginate from *P. aeruginosa* [59,277]. The EPS matrix may also interfere with the antimicrobial peptides (AMPs) charged during oxygen-independent phagocytic attack. Alginate appears to promote the aggregation and sequestration of cationic AMPs (CAMPs) by inducing conformational changes in the peptide structure and removing the AMP from the target site, i.e. the cytoplasmic membrane [278]. Hence, the interaction with specific extracellular biofilm polymers seems to be instrumental for the tolerance of *P. aeruginosa* biofilms to high amounts of CAMPs.

Resistance of *P. aeruginosa* biofilm against phagocytic attack may not only be due to the physical or diffusional barrier that EPS forms against host immune cells. Observations of inactive or "frustrated" leukocytes sitting on or in biofilms without clearing the biofilm [279,280] suggest that *P. aeruginosa* biofilms interfere chemically with phagocytic activity and cellular processes via biofilm-secreted effectors. Bacterial life in the EPS matrix promotes the localization of cells in close proximity and thus communication and cooperation via QS. Bjarnsholt *et al.* [10] recently demonstrated that the limited penetration and elimination of *P. aeruginosa* wild-type biofilms by PMNs is dependent on a functional *las/rhl* QS system. The two QS-regulated proteases alkaline protease and elastase have been described to inhibit chemotaxis, oxidative burst, phagocytosis and other microbicidal activities of phagocytes [281,282]. QS controlled inhibitors secreted by *P. aeruginosa* further include cyanide [283] with its general activity against most eukaryotic cells as it inhibits mitochondrial cytochrome oxidase *c*. The pigment pyocyanin was found to induce neutrophil apoptosis and to impair neutrophil-mediated host defenses *in vivo* [284], while rhamnolipid has been described to inhibit macrophage membrane function and thus phagocytosis *in vitro* and *in vivo* [285]. A recent study has reported that exposure to 100 μg rhamnolipid B ml^{-1} purified from *P. aeruginosa* cultures causes rapid necrosis of PMNs [244].

Besides regulating the expression of exoenzymes and toxins, QS molecules themselves may also act as modulators of phagocytic activity and inflammatory response. For the OdDHL signal of *P. aeruginosa*, a variety of effects on immunocompetent cells have been described, including inhibition of tumor necrosis factor-α synthesis by activated macrophages, inhibition of proliferation of T lymphocytes and modulation of the B cell response [286–288]. Many reports point to OdDHL as a regulator of interleukin-8 production, a proinflammatory cytokine involved in attraction and activation of PMNs. First evidence for a direct effect of AHL signal molecules on leucocytes was reported by Tateda *et al.* [289] showing that OdDHL, in concentrations of 12–50 μM, accelerate apoptosis in macrophages and PMNs. Lower concentrations of *P. aeruginosa* QS signals were found to block the activation of PMNs and their oxidative burst, supporting the observation that QS mutants cause a faster activation of the host defense system *in vivo* [10].

9.4.2.2 **Interactions with Free-living Protozoa**

Interestingly, phagocytosis constitutes not only the primary line of host immune defense against microbial pathogens, but is also employed by protozoa – the primary consumers of bacteria in most soil, freshwater and marine ecosystems. Hence, growth and survival of bacterial pathogens such as *P. aeruginosa* in the environment is thought to heavily depend on the antagonistic interaction with free-living protozoa. Upon exposure to bacterivorous flagellates during early surface colonization, *P. aeruginosa* forms distinct grazing-protected microcolonies [5]. While vast surface areas become cleared by flagellate grazers such as *Rhynchomonas nasuta*, only *P. aeruginosa* cells organized in microcolonies survive. Flagella and type IV pili play a key role in the formation of inedible microcolonies. Pili-deficient cells (*pilA* mutant) formed considerably fewer microcolonies in the presence of grazers, while cells lacking functional flagella (*fliM* mutant) were completely unable to form microcolonies. The alginate-overproducing strain PDO300 formed microcolonies without grazers in early biofilm development, which remained largely unaffected or even grew in size when exposed to flagellate grazing [5]. Being embedded in alginate, bacterial cells may be less edible by flagellate grazers, hence offering an explanation for significantly lower flagellate numbers in the presence of PDO300 biofilms. Expression of genes encoding flagella, type IV pili, and alginate is regulated by the alternative σ factor RpoN. Accordingly, the *rpoN* mutant formed fewer microcolonies in response to flagellate grazing than the wild-type strain [5]. Similar to the interactions between *P. aeruginosa* biofilms and host immune cells, growing in microcolonies and being encased in an exopolymeric matrix is essential to evade environmental predators like the protozoa. However, the formation of inedible microcolonies is only beneficial against a certain size class of feeding predators. Size-selective predators, such as the flagellate *R. nasuta* and *Bodo designis*, can handle only one or a few prey bacteria at a time, so that cells glued together lead to longer handling times and reduced feeding efficiencies, which favors bacteria growing in microcolonies [290]. However, when challenged with non-size-selective protozoan grazers, microcolonies and alginate production by *P. aeruginosa* did not suffice to acquire grazing protection against the amoeba *Acanthamoeba polyphaga* and the biofilm-browsing ciliate *Tetrahymena* sp. [290].

Later in biofilm development, however, mature biofilms of *P. aeruginosa* were found to exhibit acute toxicity to flagellate and amoebal predators [5,290]. Interestingly, biofilms of the *rhlR/lasR* mutant showed a complete loss of antiprotozoal activity. Agar plate-based studies on the social amoeba *Dictyostelium discoideum* confirmed the involvement of the *rhl/las* QS system in the killing of protozoa by *P. aeruginosa* [291,292]. The study by Cosson *et al.* reported RhlR-regulated rhamnolipids to be responsible for the killing of *D. discoideum*. Similar to the recent observations on the rhamnolipid-mediated lysis of PMNs [244], purified rhamnolipids lysed amoebae in minutes, as did *P. aeruginosa* culture supernatant [291]. Pukatzki *et al.* found that LasR-regulated virulence factors contribute to the pathogenicity of *P. aeruginosa* toward *D. discoideum*. While the effect of pyocyanin was ruled out in this study, the QS-regulated antiprotozoal factor remained unidentified. Notably, the same study found indications for the involvement of a QS-independent virulence pathway in the killing of *D. discoideum*, the type III secretion system. *P. aeruginosa* strains defective in the lipase cytotoxin ExoU and structural components of the secretion apparatus were less virulent and allowed for amoebal growth [292].

9.4.2.3 **Interactions with Fungi**

Similar to the interaction with protozoa, bacteria are found to compete and coexist with fungi in a myriad of environments, ranging from soils to the human skin and CF lung. While the predominant colonizer of the CF lung is *P. aeruginosa*, the opportunistic fungal pathogen *Candida albicans* is also commonly observed. *C. albicans* is a dimorphic fungus that exists as yeast-form and filamentous cells. *In vitro* experiments with *C. albicans* revealed that *P. aeruginosa* readily colonizes the surface of *C. albicans* hyphae and forms dense biofilms [293]. *P. aeruginosa* cells adhered to the hyphal surface in a polarized fashion with the pole opposite that of the flagella. This interaction was stimulated by nutrient limitation which favors the hyphal growth form of *C. albicans* and enhances *P. aeruginosa* attachment. Notably, *P. aeruginosa* did not form biofilms on yeast-form cells. Whether the differential attachment to hyphae and yeast is a result of specific surface ligands or factors that inhibit bacterial attachment on yeast cells is not yet known. In the process of biofilm formation, a number of *P. aeruginosa* mutants that are known to be defective in attachment to abiotic surfaces were also impaired in biofilm formation on [293] fungi, similar to the interaction observed between *P. aeruginosa* and other host surfaces, as discussed in this chapter. For example, mutants defective in type IV pili (*pilB* and *pilC*) were delayed in biofilm formation on *C. albicans* hyphae, and biofilms produced by a flagellar mutant (*flgK*) were less robust than the wild type.

During biofilm development, the consortium of *P. aeruginosa* biofilms on *C. albicans* hyphae became encased in an extracellular matrix. Within 24–48 h of biofilm formation, *P. aeruginosa* killed *C. albicans* hyphae [293]. *P. aeruginosa* mutants lacking type IV pili (*pilB* and *pilC*) were delayed in both biofilm formation and fungal killing. Mutants lacking the alternative σ factor RpoN or QS-related transcription factors, such as LasR and RhlR, were both defective in biofilm formation and decreased in their ability to kill *C. albicans* hyphae. Although the virulence factors responsible were not identified, the correlation observed between the ability of *P. aeruginosa* to form biofilms and to kill *C. albicans* suggests that factors may be differentially produced within biofilms on eukaryotic surfaces or in a density-dependent fashion, or that these secreted factors are more efficacious when produced in close proximity to the target cells. *P. aeruginosa* biofilms on *C. albicans* hyphae occurred much more readily under nutrient-limiting conditions, so that it is likely that *P. aeruginosa* colonizes and kills *C. albicans* to obtain nutrients from the fungus. Interestingly, the secreted phospholipase C, which degrades eukaryotic membrane lipids, was found to participate in fungal lysis [293].

Taken together, the three examples of interactions between *P. aeruginosa* and eukaryotic host cells discussed in this chapter demonstrate the adaptive advantage of bacterial life in biofilms, particularly as it allows growth and survival of high cell densities and the creation of chemical microenvironments that may interfere with eukaryotic cells. Future studies will particularly increase our understanding of the chemical antagonisms and synergisms between biofilms and eukaryotic host cells.

9.5 Future Perspectives

Contemporary research in the field of biofilms has offered a glimpse into many aspects of hitherto poorly studied aspects of bacterial biology. In particular, studies using model biofilm systems, including the use of strains of *P. aeruginosa*, have been highly informative for unraveling novel phenotypic and genotypic traits that appear to be hallmarks of biofilm biology and more broadly reflective of bacterial behaviors in multicellular consortia and in a sessile lifestyle. These responses, as discussed in the present chapter, suggest that there are striking similarities between multicellular consortia of bacteria and eukaryotes – a finding that will strongly facilitate our understanding of fundamental aspects of microbiology.

Clearly, while many new insights of how high-density multicellular bacterial populations operate have been gained in the last few years, our research efforts to better understand biofilm biology have also revealed many unknowns of these systems. For example, the majority of genes that are expressed in biofilms are of unknown function, probably due to the fact that they have no role in planktonic biology. Also, regulatory circuits such as those that involve extracellular communication, biofilm specific differentiation, depend on structural constraints accommodated by biofilm microcolonies, and those that engage host specific responses, appear to operate in a biofilm-specific fashion. The identity and function of such genes, regulatory processes and biofilm unique phenotypic responses should be scrutinized using relevant biofilm and, when appropriate, biofilm–host interaction-based experimental approaches. These studies will improve our understanding of the inherent resistance of biofilms, give us new tools for biofilm control, allow for a better understanding of disease processes by pathogens and, more broadly, improve our understanding of how bacteria handle various selection pressures in their environments.

Yet, methodological shortcomings and the lack of appropriate systems for the study of biofilms are the biggest challenge faced by biofilm researchers at present. While the application of new techniques has allowed for exciting progress to date, as reported in this chapter, the issues of scale and resolution are often of concern. For example, transcriptomic and proteomic assays on biofilm-derived cells have been instrumental in the identification of novel genes and differential gene expression. This outcome was highlighted by the finding that 25% of all proteins were upregulated in 1-day-old biofilm cells when compared to planktonic cells [35,294]. However, such transcriptomic and proteomic approaches yield information on activities of an average population, and cannot give us specific information on the different subpopulations of cells residing within the biofilm community. The ideal scenario would allow quick and efficient separation of the various subpopulations within a biofilm for analysis by these techniques. While cell sorting is one possibility for the separation of subpopulations expressing different traits, the sample size collected is not large enough for these analyses at the present time. It is envisaged that the field of genomics will provide exciting new possibilities for the study of biofilms through sequencing of individual cells and allow us to identify genetic changes occurring within the various subpopulations of biofilms at different stages of development.

It is likely that the challenge for biofilm research goes beyond the need to introduce novel high-resolution technologies. Future biofilm research, also for advancing our understanding of the biology of specific organisms such as *P. aeruginosa*, should also embrace the extension of currently used laboratory-based studies to the study of mixed-species microbial consortia. It is clear that most bacterial biofilm populations, in environmental as well as medical settings, are those of mixed species communities. Culture-independent molecular community analysis of biofilm samples from natural habitats in which our target bacterial species or strain reside should precede the design of mixed-species, defined-laboratory-reared consortia for studies of the behavior of our model biofilm organisms.

Acknowledgment:

The authors would like to express their sincere thanks to Nic Barraud for Figure 9.1. All authors contributed equally to this chapter. Research in the authors' laboratories is supported by the Australian Research Council, National Health and Medical Research Council (Australia), Center for Marine Bio-Innovation, Environmental Biotechnology CRC, Danish Biotechnological Research Program, Biotechnology and Biological Sciences Research Council (UK), and the University of Tasmania Institutional Research Grants Scheme.

References

1 Costerton, J.W., Geesey, G.G. and Cheng, K.J. (1978) *Sci Am*, **238**, 86–95.
2 Stoodley, P., Sauer, K., Davies, D.G. and Costerton, J.W. (2002) *Annu Rev Microbiol*, **56**, 187–209.
3 Christensen, B.B., Haagensen, J.A.J., Heydorn, A. and Molin, S. (2002) *Appl Environ Microbiol*, **68**, 2495–2505.
4 Jefferson, K.K. (2004) *FEMS Microbiol Lett*, **236**, 163–173.
5 Matz, C., Bergfeld, T., Rice, S.A. and Kjelleberg, S. (2004) *Environ Microbiol*, **6**, 218–226.
6 Matz, C. *et al.* (2005) *Proc Natl Acad Sci USA*, **102**, 16819–16824.
7 Drenkard, E. and Ausubel, F.M. (2002) *Nature*, **416**, 740–743.
8 Fux, C.A., Costerton, J.W., Stewart, P.S. and Stoodley, P. (2005) *Trends Microbiol*, **13**, 34–40.
9 Lewis, K. (2001) *Antimicrob Agents Chemother*, **45**, 999–1007.
10 Bjarnsholt, T. *et al.* (2005) *Microbiology*, **151**, 373–383.
11 Costerton, J.W., Stewart, P.S. and Greenberg, E.P. (1999) *Science*, **284**, 1318–1322.
12 Clarke, P.H. (1982) *Antonie Van Leeuwenhoek*, **48**, 105–130.
13 Rodrigue, A., Quentin, Y., Lazdunski, A., Méjean, V. and Foglino, M. (2000) *Trends Microbiol*, **8**, 498–504.
14 Lyczak, J.B., Cannon, C.L. and Pier, G.B. (2002) *Clin Microbiol Rev*, **15**, 194–222.
15 Tredget, E.E., Shankowsky, H.A., Rennie, R., Burrell, R.E. and Logsetty, S. (2004) *Burns*, **30**, 3–26.
16 Ghannoum, M.A. and O'Toole, G.A. (2004) *Microbial Biofilms*, ASM Press, Washington, DC.
17 Kjelleberg, S. and Givskov, M. (2007) *The Biofilm Mode of Life: Mechanisms and Adaptations* (eds. S. Kjelleberg and M. Givskov), Horizon Bioscience, Wymondham, 5–21.
18 Davies, D.G. *et al.* (1998) *Science*, **280**, 295–298.
19 Allesen-Holm, M. *et al.* (2006) *Mol Microbiol*, **59**, 1114–1128.
20 Lequette, Y. and Greenberg, E.P. (2005) *J Bacteriol*, **187**, 37–44.
21 Haagensen, J.A.J. *et al.* (2007) *J Bacteriol*, **189**, 28–37.
22 Boles, B.R., Thoendel, M. and Singh, P.K. (2004) *Proc Natl Acad Sci USA*, **101**, 16630–16635.
23 Hentzer, M., Eberl, L. and Givskov, M. (2005) *Biofilms*, **2**, 37–61.
24 van Loosdrecht, M.C., Norde, W. and Zehnder, A.J. (1990) *J Biomater Appl*, **5**, 90–106.
25 Marshall, K.C. (1976) *Interfaces in Microbial Ecology* (ed. K.C. Marshall), Harvard University Press, Cambridge, MA, pp. 21–38.
26 Derjaguin, B.V. and Landau, L.D. (1941) *Acta Phys Chim URSS*, **14**, 633–662.
27 Verwey, E.J.W. and Overbeek, J.T.G. (1948) *Theory of the Stability of Lyophobic Colloids. The Interaction of Particles Having an Electric Double Layer*, Elsevier, Amsterdam.
28 O'Toole, G.A. and Kolter, R. (1998) *Mol Microbiol*, **30**, 295–304.

29 Arora, S.K., Ritchings, B.W., Almira, E.C., Lory, S. and Ramphal, R. (1998) *Infect Immun*, **66**, 1000–1007.

30 Landry, R.M., An, D., Hupp, J.T., Singh, P.K. and Parsek, M.R. (2006) *Mol Microbiol*, **59**, 142–151.

31 Semmler, A.B., Whitchurch, C.B. and Mattick, J.S. (1999) *Microbiology*, **145**, 2863–2873.

32 Skerker, J.M. and Berg, H.C. (2001) *Proc Natl Acad Sci USA*, **98**, 6901–6904.

33 Chiang, P. and Burrows, L.L. (2003) *J Bacteriol*, **185**, 2374–2378.

34 Deziel, E., Comeau, Y. and Villemur, R. (2001) *J Bacteriol*, **183**, 1195–1204.

35 Sauer, K., Camper, A.K., Ehrlich, G.D., Costerton, J.W. and Davies, D.G. (2002) *J Bacteriol*, **184**, 1140–1154.

36 Klausen, M. *et al.* (2003) *Mol Microbiol*, **48**, 1511–1524.

37 Klausen, M., Aaes-Jorgensen, A., Molin, S. and Tolker-Nielsen, T. (2003) *Mol Microbiol*, **50**, 61–68.

38 O'Toole, G.A. and Kolter, R. (1998) *Mol Microbiol*, **28**, 449–461.

39 Shrout, J.D. *et al.* (2006) *Mol Microbiol*, **62**, 1264–1277.

40 Caiazza, N.C., Merritt, J.H., Brothers, K.M. and O'Toole, G. (2007) *J Bacteriol*, **189**, 3603–3612.

41 Toutain, C.M., Caiazza, N.C., Zegans, M.E. and O'Toole, G. (2007) *Res Microbiol*, **158**, 471–477.

42 Toutain, C.M., Zegans, M.E. and O'Toole, G. (2005) *J Bacteriol*, **187**, 771–777.

43 Hinsa, S.M., Espinosa-Urgel, M., Ramos, J.L. and O'Toole, G.A. (2003) *Mol Microbiol*, **49**, 905–918.

44 Hinsa, S.M. and O'Toole, G. (2006) *Microbiology*, **152**, 1375–1383.

45 Monds, R.D., Newell, P.D., Gross, R.H. and O'Toole, G. (2007) *Mol Microbiol*, **63**, 656–679.

46 Ruer, S., Stender, S., Filloux, A. and de Bentzmann, S. (2007) *J Bacteriol*, **189**, 3547–3555.

47 Vallet, I., Olson, J.W., Lory, S., Lazdunski, A. and Filloux, A. (2001) *Proc Natl Acad Sci USA*, **98**, 6911–6916.

48 Vallet, I. *et al.* (2004) *J Bacteriol*, **186**, 2880–2890.

49 Vallet-Gely, I., Donovan, K.E., Fang, R., Joung, J.K. and Dove, S.L. (2005) *Proc Natl Acad Sci USA*, **102**, 11082–11087.

50 Friedman, L. and Kolter, R. (2004) *J Bacteriol*, **186**, 4457–4465.

51 Govan, J.R. and Deretic, V. (1996) *Microbiol Rev*, **60**, 539–574.

52 Jackson, K.D., Starkey, M., Kremer, S., Parsek, M.R. and Wozniak, D.J. (2004) *J Bacteriol*, **186**, 4466–4475.

53 Matsukawa, M. and Greenberg, E.P. (2004) *J Bacteriol*, **186**, 4449–4456,

54 DeVries, C.A. and Ohman, D.E. (1994) *J Bacteriol*, **176**, 6677–6687.

55 Hershberger, C.D., Ye, R.W., Parsek, M.R., Xie, Z.-D. and Chakrabarty, A.M. (1995) *Proc Natl Acad Sci USA*, **92**, 7941–7945.

56 Martin, D.W. *et al.* (1993) *Proc Natl Acad Sci USA*, **90**, 8377–8381.

57 Krieg, D.P., Helmke, R.J., German, V.F. and Mangos, J.A. (1988) *Infect Immun*, **56**, 3173–3179.

58 Simpson, J.A., Smith, S.E. and Dean, R.T. (1988) *J Gen Microbiol*, **134**, 29–36.

59 Simpson, J.A., Smith, S.E. and Dean, R.T. (1989) *Free Radic Biol Med*, **6**, 347–353.

60 Friedman, L. and Kolter, R. (2004) *Mol Microbiol*, **51**, 675–690.

61 Vasseur, P., Vallet-Gely, I., Soscia, C., Genin, S. and Filloux, A. (2005) *Microbiology*, **151**, 985–997.

62 Hickman, J.W., Tifrea, D.F. and Harwood, C.S. (2005) *Proc Natl Acad Sci USA*, **102**, 14422–14427.

63 Ryjenkov, D.A., Tarutina, M., Moskvin, O.V. and Gomelsky, M. (2005) *J Bacteriol*, **187**, 1792–1798.

64 Sakuragi, Y. and Kolter, R. (2007) *J Bacteriol*, **189**, 5383–5386.

65 Winzer, K. *et al.* (2000) *J Bacteriol*, **182**, 6401–6411.

66 Yamada, M., Ikegami, A. and Kuramitsu, H.K. (2005) *FEMS Microbiol Lett*, **250**, 271–277.

67 Gupta, S.K., Berk, R.S., Masinick, S. and Hazlett, L.D. (1994) *Infect Immun*, **62**, 4572–4579.

68 Hahn, H.P. (1997) *Gene*, **192**, 99–108.

69 Ramphal, R. *et al.* (1991) *Infect Immun*, **59**, 700–704.

70 Aas, F.E. *et al.* (2002) *Mol Microbiol*, **46**, 749–760.

71 van Schaik, E.J. *et al.* (2005) *J Bacteriol*, **187**, 1455–1464.

72 Nemoto, K. *et al.* (2003) *Chemotherapy*, **49**, 121–125.

73 Whitchurch, C.B., Tolker-Nielsen, T., Ragas, P.C. and Mattick, J.S. (2002) *Science*, **295**, 1487–11487.

74 Mattick, J.S. (2002), *56*, 289–314.

75 Sauer, K. and Camper, A.K. (2001) *J Bacteriol*, **183**, 6579–6589.

76 Kuchma, S.L., Connolly, J.P. and O'Toole, G.A. (2005) *J Bacteriol*, **187**, 1441–1454.

77 Wentworth, J.S. *et al.* (1991) *Biofouling*, **4**, 94–104.

78 Loris, R., Tielker, D., Jaeger, K.E. and Wyns, L. (2003) *J Mol Biol*, **331**, 861–870.

79 Tielker, D. *et al.* (2005) *Microbiology*, **151**, 1313–1323.

80 Diggle, S.P. *et al.* (2006) *Environ Microbiol*, **8**, 1095–1104.

81 Murakawa, T. (1973) *Jpn J Microbiol*, **17**, 273–281.

82 Murakawa, T. (1973) *Jpn J Microbiol*, **17**, 513–520.

83 Heurlier, K. *et al.* (2005) *J Bacteriol*, **187**, 4875–4883.

84 Lu, A. *et al.* (2005) *Mol Microbiol*, **55**, 206–220.

85 Mai-Prochnow, A., Webb, J.S., Ferrari, B.C. and Kjelleberg, S. (2006) *Appl Environ Microbiol*, **72**, 5414–5420.

86 Deretic, V., Schurr, M.J., Boucher, J.C. and Martin, D.W. (1994) *J Bacteriol*, **176**, 2773–2780.

87 Martin, C., Ichou, M.A., Massicot, P., Goudeau, A. and Quentin, R. (1995) *J Clin Microbiol*, **33**, 1461–1466.

88 Häussler, S., Tümmler, B., Weissbrodt, H., Rohde, M. and Steinmetz, I. (1999) *Clin Infect Dis*, **29**, 621–625.

89 Haussler, S. *et al.* (2003), 52, 295–301.

90 von Götz, F. *et al.* (2004) *J Bacteriol*, **186**, 3837–3847.

91 Dasgupta, T. *et al.* (1994) *Infect Immun*, **62**, 809–817.

92 Webb, J.S., Lau, M. and Kjelleberg, S. (2004) *J Bacteriol*, **186**, 8066–8073.

93 Rainey, P.B. (1999) *Curr Biol*, **9**, R371–R373.

94 Rainey, P.B. and Travisano, M. (1998) *Nature*, **394**, 69–72.

95 Spiers, A.J., Bohannon, J., Gehrig, S.M. and Rainey, P.B. (2003) *Mol Microbiol*, **50**, 15–27.

96 Spiers, A.J., Kahn, S.G., Bohannon, J., Travisano, M. and Rainey, P.B. (2002) *Genetics*, **161**, 33–46.

97 Kresse, A.U., Dinesh, S.D., Larbig, K. and Romling, U. (2003) *Mol Microbiol*, **47**, 145–158.

98 Smith, E.E. *et al.* (2006) *Proc Natl Acad Sci USA*, **103**, 8487–8492.

99 Mahenthiralingam, E., Campbell, M.E. and Speert, D.P. (1994) *Infect Immun*, **62**, 596–605.

100 Mathee, K. *et al.* (1999) *Microbiology*, **145**, 1349–1357.

101 Oliver, A., Cantón, R., Campo, P., Baquero, F. and Blázquez, J. (2000) *Science*, **288**, 1251.

102 Kresse, A.U., Blöcker, H. and Römling, U. (2006) *Arch Microbiol*, **185**, 245–254.

103 Hausner, M. and Wuertz, S. (1999) *Appl Environ Microbiol*, **65**, 3710–3713.

104 Molin, S. and Tolker-Nielsen, T. (2003) *Curr Opin Biotechnol*, **14**, 255–261.

105 Whiteley, M. *et al.* (2001) *Nature*, **413**, 860–864.

106 Nystrom, T. (2003) *Mol Microbiol*, **48**, 17–23.

107 Bayer, A.S., Norman, D.C. and Kim, K.S. (1987) *Antimicrob Agents Chemother*, **31**, 70–75.

108 Bryan, L.E., Kwan, S. and Godfrey, A.J. (1984) *Antimicrob Agents Chemother*, **25**, 382–384.

109 Proctor, R.A. *et al.* (2006) *Nature Rev Microbiol*, **4**, 295–305.

110 Davies, J.A. *et al.* (2007) *FEMS Microbiol Ecol*, **59**, 32–46.

111 Webb, J.S. *et al.* (2003) *J Bacteriol*, **185**, 4585–4592.

112 Brockhurst, M.A., Buckling, A. and Rainey, P.B. (2005) *Proc Biol Sci*, **272**, 1385–1391.

113 Tejedor, C., Foulds, J. and Zasloff, M. (1982) *Am Soc Microbiol*, **36**, 440–441.

114 Kirisits, M.J., Prost, L., Starkey, M. and Parsek, M.R. (2005) *Appl Environ Microbiol*, **71**, 4809–4821.

115 Delaquis, P.J., Caldwell, D.E., Lawrence, J.R. and McCurdy, A.R. (1989) *Microb Ecol*, **18**, 199–210.

116 Gjermansen, M., Ragas, P., Sternberg, C., Molin, S. and Tolker-Nielsen, T. (2005) *Environ Microbiol*, **7**, 894–906.

117 Hunt, S.M., Werner, E.M., Huang, B., Hamilton, M.A. and Stewart, P.S. (2004) *Appl Environ Microbiol*, **70**, 7418–7425.

118 Sauer, K. *et al.* (2004) *J Bacteriol*, **186**, 7312–7326.

119 Banin, E., Brady, K.M. and Greenberg, E.P. (2006) *Appl Environ Microbiol*, **72**, 2064–2069.

120 Applegate, D.H. and Bryers, J.D. (1991) *Biotechnol Bioeng*, **37**, 17–25.

121 Simm, R., Morr, M., Kader, A., Nimtz, M. and Romling, U. (2004) *Mol Microbiol*, **53**, 1123–1134.

122 Allison, D.G., Ruiz, B., SanJose, C., Jaspe, A. and Gilbert, P. (1998) *FEMS Microbiol Lett*, **167**, 179–184.

123 Boyd, A. and Chakrabarty, A.M. (1994) *Appl Environ Microbiol*, **60**, 2355–2359.

124 Davey, M.E., Caiazza, N.C. and O'Toole, G.A. (2003) *J Bacteriol*, **185**, 1027–1036.

125 Barraud, N. *et al.* (2006) *J Bacteriol*, **188**, 7344–7353.

126 Morgan, R., Kohn, S., Hwang, S.H., Hassett, D.J. and Sauer, K. (2006) *J Bacteriol*, **188**, 7335–7343.

127 Hassett, D.J. *et al.* (2002) *Adv Drug Deliv Rev*, **54**, 1425–1443.

128 Firoved, A.M., Wood, S.R., Ornatowski, W., Deretic, V. and Timmins, G.S. (2004) *J Bacteriol*, **186**, 4046–4050.

129 Van Alst, N.E., Picardo, K.F., Iglewski, B.H. and Haidaris, C.G. (2007) *Infect Immun*, **75**, 3780–3790.

130 Galperin, M.Y., Nikolskaya, A.N. and Koonin, E.V. (2001) *FEMS Microbiol Lett*, **203**, 11–21.

131 Romling, U., Gomelsky, M. and Galperin, M.Y. (2005) *Mol Microbiol*, **57**, 629–639.

132 Aravind, L., Anantharaman, V. and Iyer, L.M. (2003) *Curr Opin Microbiol*, **6**, 490–497.

133 Mills, P.C., Richardson, D.J., Hinton, J.C. and Spiro, S. (2005) *Biochem Soc Trans*, **33**, 198–199.

134 Poole, R.K. (2005) *Biochem Soc Trans*, **33**, 176–180.

135 Tucker, N.P., D'Autreaux, B., Spiro, S. and Dixon, R. (2006) *Biochem Soc Trans*, **34**, 191–194.

136 Iyer, L.M., Anantharaman, V. and Aravind, L. (2003) *BMC Genomics*, **4**, 5.

137 Branda, S.S. and Kolter, R. (2004) *Microbial Biofilms* (eds M. Ghannoum and G. O'Toole), ASM Press, Washington, DC, pp. 20–29.

138 Webb, J.S., Givskov, M. and Kjelleberg, S. (2003) *Curr Opin Microbiol*, **6**, 578–585.

139 Purevdorj-Gage, B., Costerton, W.J. and Stoodley, P. (2005) *Microbiology*, **151**, 1569–1576.

140 Tolker-Nielsen, T. *et al.* (2000) *J Bacteriol*, **182**, 6482–6489.

141 Kirov, S.M., Webb, J.S. and Kjelleberg, S. (2005) *Microbiology*, **151**, 3452–3453.

142 Sutherland, I.W. (1999) *Microbial Extracellular Polymeric Substances: Characterization, Structure and Function* (eds J. Wingender, T.R. Neu and F. Hans-Curt), Springer, Berlin, pp. 73–92.

143 Thompson, L.S., Webb, J.S., Rice, S.A. and Kjelleberg, S. (2003) *FEMS Microbiol Lett*, **220**, 187–195.

144 D'Argenio, D.A., Calfee, M.W., Rainey, P.B. and Pesci, E.C. (2002) *J Bacteriol*, **184**, 6481–6489.

145 Jenal, U. and Malone, J. (2006) *Annu Rev Genet*, **40**, 385–407.

146 Römling, U. and Amikam, D. (2006) *Curr Opin Microbiol*, **9**, 218–228.

147 Ross, P. *et al.* (1990) *J Biol Chem*, **265**, 18933–18943.

148 Weinhouse, H. *et al.* (1997) *FEBS Lett*, **416**, 207–211.

149 Tal, R. *et al.* (1998) *J Bacteriol*, **180**, 4416–4425.

150 Chan, C. *et al.* (2004) *Proc Natl Acad Sci USA*, **101**, 17084–17089.

151 Christen, M., Christen, B., Folcher, M., Schauerte, A. and Jenal, U. (2005) *J Biol Chem*, **280**, 30829–30837.

152 Ryan, R.P. *et al.* (2006) *Proc Natl Acad Sci USA*, **103**, 6712–6717.

153 Malone, J.G. *et al.* (2007) *Microbiology*, **153**, 980–994.

154 Kulasakara, H. *et al.* (2006) *Proc Natl Acad Sci USA*, **103**, 2839–2844.

155 Meissner, A. *et al.* (2007) *Environ Microbiol*, **9**, 2475–2485.

156 Bantinaki, E. *et al.* (2007) *Genetics*, **176**, 441–453.

157 Goymer, P. *et al.* (2006) *Genetics*, **173**, 515–526.

158 Kulasekara, H.D. *et al.* (2005) *Mol Microbiol*, **55**, 368–380.

159 Christen, M., Christen, B., Folcher, M., Schauerte, A. and Jenal, U. (2005), 280, 30829–30837.

160 Choy, W.K., Zhou, L., Syn, C.K., Zhang, L. H. and Swarup, S. (2004), 186, 7221–7228.

161 Huang, B., Whitchurch, C.B. and Mattick, J.S. (2003) *J Bacteriol*, **185**, 7068–7076.

162 Kazmierczak, B.I., Lebron, M.B. and Murray, T.S. (2006) *Mol Microbiol*, **60**, 1026–1043.

163 Merritt, J.H., Brothers, K.M., Kuchma, S.L. and O'Toole, G. (2007) *J Bacteriol*, **189**, 8154–8161.

164 Hoffman, L.R. *et al.* (2005) *Nature*, **436**, 1171–1175.

165 Amikam, D. and Galperin, M.Y. (2006) *Bioinformatics*, **22**, 3–6.

166 Christen, M. *et al.* (2007) *Proc Natl Acad Sci USA*, **104**, 4112–4117.

167 Ramelot, T.A. *et al.* (2007) *Proteins*, **66**, 266–271.

168 Bjarnsholt, T. and Givskov, M. (2007) *Anal Bioanal Chem*, **387**, 409–414.

169 Kirisits, M.J. and Parsek, M.R. (2006) *Cell Microbiol*, **8**, 1841–1849.

170 Nasser, W. and Reverchon, S. (2007) *Anal Bioanal Chem*, **387**, 381–390.

171 Venturi, V. (2006) *FEMS Microbiol Rev*, **30**, 274–291.

172 Pesci, E.C. *et al.* (1999) *Proc Natl Acad Sci USA*, **96**, 11229–11234.

173 McKnight, S.L., Iglewski, B.H. and Pesci, E.C. (2000) *J Bacteriol*, **182**, 2702–2708.

174 Brint, J.M. and Ohman, D.E. (1995) *J Bacteriol*, **177**, 7155–7163.

175 Chapon-Herve, V. *et al.* (1997) *Mol Micriobiol*, **24**, 1169–1178.

176 Charlton, T.S. *et al.* (2000) *Environ Microbiol*, **2**, 530–541.

177 Pearson, J.P. *et al.* (1994) *Proc Natl Acad Sci USA*, **91**, 197–201.

178 De Kievit, T.R., Gillis, R., Marx, S., Brown, C. and Iglewski, B.H. (2001) *Appl Environ Microbiol*, **67**, 1865–1873.

179 Heydorn, A. *et al.* (2002) *Appl Environ Microbiol*, **68**, 2008–2017.

180 Jensen, V. *et al.* (2006) *J Bacteriol*, **188**, 8601–8606.

181 Rashid, M.H. *et al.* (2000) *Proc Natl Acad Sci USA*, **97**, 9636–9641.

182 Heurlier, K., Dénervaud, V., Pessi, G., Reimmann, C. and Haas, D. (2003) *J Bacteriol*, **185**, 2227–2235.

183 Debeer, D., Stoodley, P., Roe, F. and Lewandowski, Z. (1994) *Biotechnol Bioeng*, **43**, 1131–1138.

184 Alvarez-Ortega, C. and Harwood, C.S. (2007) *Mol Microbiol*, **65**, 153–165.

185 Toyofuku, M. *et al.* (2007) *J Bacteriol*, **189**, 4969–4972.

186 Farrow, J.M., III and Pesci, E.C. (2007) *J Bacteriol*, **189**, 3425–3433.

187 Hentzer, M. *et al.* (2003) *EMBO J*, **22**, 3803–3815.

188 Schuster, M., Lostroh, C.P., Ogi, T. and Greenberg, E.P. (2003) *J Bacteriol*, **185**, 2066–2079.

189 Heurlier, K., Denervaud, V. and Haas, D. (2006) *Int J Med Microbiol*, **296**, 93–102.

190 Wagner, V.E., Bushnell, D., Passador, L., Brooks, A.I. and Iglewski, B.H. (2003) *J Bacteriol*, **185**, 2080–2095.

191 Xu, K., Stewart, P.S., Xia, F., Huang, C.-T. and McFetters, G.A. (1998) *Appl Environ Microbiol*, **64**, 4035–4039.

192 Boles, B.R. and Thoendel, M. and Singh, P.K. (2005) *Mol Microbiol*, **57**, 1210–1223.

193 Yang, L. *et al.* (2007) *Microbiology*, **153**, 1318–1328.

194 Bjarnsholt, T. *et al.* (2005) *Microbiology*, **151**, 3873–3880.

195 Shih, P.-C. and Huang, C.-T. (2002) *J Antimicrob Chemother*, **49**, 309–314.

196 Chambers, C.E., Visser, M.B., Schwab, U. and Sokol, P.A. (2005) *FEMS Microbiol Lett*, **244**, 297–304.

197 Duan, K. and Surette, M.G. (2007) *J Bacteriol*, **189**, 4827–4836.

198 Erickson, D.L. *et al.* (2002) *Infect Immun*, **70**, 1783–1790.

199 Middleton, B. *et al.* (2002) *FEMS Microbiol Lett*, **207**, 1–7.

200 Storey, D.G., Ujack, E.E., Rabin, H.R. and Mitchell, I. (1998) *Infect Immun*, **66**, 2521–2528.

201 Singh, P.K. *et al.* (2000) *Nature*, **407**, 762–764.

202 Christensen, L.D. *et al.* (2007) *Microbiology*, **153**, 2312–2320.

203 Rumbaugh, K.P., Griswold, J.A., Iglewski, B.H. and Hamood, A.N. (1999) *Infect Immun*, **67**, 5854–5862.

204 Wu, H. *et al.* (2001) *Microbiology*, **147**, 1105–1113.

205 Favre-Bonté, S., Kohler, T. and Van Delden, C. (2003) *J Antimicrob Chemother*, **52**, 598–604.

206 Gillis, R.J. and Iglewski, B.H. (2004) *J Clin Microbiol*, **42**, 5842–5845.

207 Nalca, Y. *et al.* (2006) *Antimicrob Agents Chemother*, **50**, 1680–1688.

208 Tateda, K. *et al.* (2001) *Antimicrob Agents Chemother*, **45**, 1930–1933.

209 Kjelleberg, S. and Molin, S. (2002) *Curr Opin Microbiol*, **5**, 254–258.

210 Schaber, J.A. *et al.* (2004) *J Med Microbiol*, **53**, 841–853.

211 Mesaros, N. *et al.* (2007) *Clin Microbiol Infect*, **13**, 560–578.

212 Van Delden, C. and Iglewski, B.H. (1998) *Emerg Infect Dis*, **4**, 551–560.

213 Fux, C.A., Stoodley, P., Hall-Stoodley, L. and Costerton, J.W. (2003) *Expert Rev Anti Infect Ther*, **1**, 667–683.

214 Hall-Stoodley, L., Costerton, J.W. and Stoodley, P. (2004) *Nat Rev Microbiol*, **2**, 95–108.

215 Adlard, P.A., Kirov, S.M., Sanderson, K. and Cox, G.E. (1998) *Epidemiol Infect*, **121**, 237–241.

216 Donlan, R.M. and Costerton, J.W. (2002) *Clin Microbiol Rev*, **15**, 167–193.

217 Kobayashi, H. (2005) *Treat Respir Med*, **4**, 241–253.

218 Koch, C., Frederiksen, B. and Høiby, N. (2003) *Semin Respir Crit Care Med*, **24**, 703–716.

219 Sadikot, R.T., Blackwell, T.S., Christman, J.W. and Prince, A.S. (2005) *Am J Respir Crit Care Med*, **171**, 1209–1223.

220 Furukawa, S., Kuchma, S.L. and O'Toole, G.A. (2006) *J Bacteriol*, **188**, 1211–1217.

221 Boucher, R.C. (2004) *Eur Respir J*, **23**, 146–158.

222 Gibson, R.L., Burns, J.L. and Ramsey, B.W. (2003) *Am J Respir Crit Care Med*, **168**, 918–951.

223 Tatterson, L.E., Poschet, J.F., Firoved, A., Skidmore, J. and Deretic, V. (2001) *Front Biosci*, **6**, D890–897.

224 Rogers, G.B. *et al.* (2004) *J Clin Microbiol*, **42**, 5176–5183.

225 Deretic, V., Schurr, M.J. and Yu, H. (1995) *Trends Microbiol*, **3**, 351–356.

226 Burns, J.L. *et al.* (2001) *J Infect Dis*, **183**, 444–452.

227 Döring, G. *et al.* (2000) *Eur Respir J*, **16**, 749–767.

228 Frederiksen, B., Koch, C. and Høiby, N. (1997) *Pediatr Pulmonol*, **23**, 330–335.

229 Valerius, N.H., Koch, C. and Høiby, N. (1991) *Lancet*, **338**, 725–726.

230 Høiby, N. (2006) *Microbe*, **1**, 571–577.

231 Baltimore, R.S., Christie, C.D. and Smith, G.J. (1989) *Am Rev Respir Dis*, **140**, 1650–1661.

232 Lam, J., Chan, R., Lam, K. and Costerton, J.W. (1980) *Infect Immun*, **28**, 546–556.

233 Worlitzsch, D. *et al.* (2002) *J Clin Invest*, **109**, 317–325.

234 Bragonzi, A. *et al.* (2006) *Microbiology*, **152**, 3261–3269.

235 Ramsey, D.M. and Wozniak, D.J. (2005) *Mol Microbiol*, **56**, 309–322.

236 Mai, G.T., Seow, W.K., Pier, G.B., McCormack, J.G. and Thong, Y.H. (1993) *Infect Immun*, **61**, 559–564.

237 Pedersen, S.S., Hoiby, N., Espersen, F. and Koch, C. (1992) *Thorax*, **47**, 6–13.

238 Pier, G.B., Small, G.J. and Warren, H.B. (1990) *Science*, **249**, 537–540.

239 Stewart, P.S. and Costerton, J.W. (2001) *Lancet*, **358**, 135–138.

240 Hentzer, M. *et al.* (2001) *J Bacteriol*, **183**, 5395–5401.

241 Stapper, A.P. *et al.* (2004) *J Med Microbiol*, **53**, 679–690.

242 Walker, T.S. *et al.* (2005) *Infect Immun*, **73**, 3693–3701.

243 Watt, A.P., Courtney, J., Moore, J., Ennis, M. and Elborn, J.S. (2005) *Thorax*, **60**, 659–664.

244 Jensen, P.O. *et al.* (2007) *Microbiology*, **153**, 1329–1338.

245 Deziel, E., Lepine, F., Milot, S. and Villemur, R. (2003) *Microbiology*, **149**, 2005–2013.

246 McClure, C.D. and Schiller, N.L. (1992) *J Leukoc Biol*, **51**, 97–102.

247 Jones, K.L. *et al.* (2000) *Pediatr Pulmonol*, **30**, 79–85.

248 Palmer, K.L., Brown, S.A. and Whiteley, M. (2007) *J Bacteriol*, **189**, 4449–4455.

249 O'May, C.Y., Reid, D.W. and Kirov, S.M. (2006) *FEMS Immunol Med Microbiol*, **48**, 373–380.

250 Yoon, S.S. *et al.* (2002) *Dev Cell*, **3**, 593–603.

251 Boes, N., Schreiber, K., Hartig, E., Jaensch, L. and Schobert, M. (2006) *J Bacteriol*, **188**, 6529–6538.

252 Hassett, D.J. *et al.* (1999) *Mol Microbiol*, **34**, 1082–1093.

253 Burns, J.L. *et al.* (1998) *Clin Infect Dis*, **27**, 158–163.

254 Head, N.E. and Yu, H. (2004) *Infect Immun*, **72**, 133–144.

255 D'Argenio, D.A. *et al.* (2007) *Mol Microbiol*, **64**, 512–533.

256 Jain, M. *et al.* (2004) *J Clin Microbiol*, **42**, 5229–5237.

257 De Vos, D. *et al.* (2001) *Arch Microbiol*, **175**, 384–388.

258 Spencer, D.H. *et al.* (2003) *J Bacteriol*, **185**, 1316–1325.

259 Ernst, R.K. *et al.* (2006) *J Bacteriol*, **188**, 191–201.

260 Barth, A.L. and Pitt, T.L. (1995) *J Clin Microbiol*, **33**, 37–40.

261 Barth, A.L. and Pitt, T.L. (1996) *J Med Microbiol*, **45**, 110–119.

262 Jelsbak, L. *et al.* (2007) *Infect Immun*, **75**, 2214–2224.

263 Moyano, A.J., Lujan, A.M., Argarana, C.E. and Smania, A.M. (2007) *Mol Microbiol*, **64**, 547–559.

264 Nguyen, D. and Singh, P.K. (2006) *Proc Natl Acad Sci USA*, **103**, 8305–8306.

265 Köhler, T., van Delden, C., Curty, L.K., Hamzehpour, M.M. and Pechere, J.-C. (2001) *J Bacteriol*, **183**, 5213–5222.

266 Morita, Y., Sobel, M.L. and Poole, K. (2006) *J Bacteriol*, **188**, 1847–1855.

267 Poole, K., Krebes, K., McNally, C. and Neshat, S. (1993) *J Bacteriol*, **175**, 7363–7372.

268 Sobel, M.L., McKay, G.A. and Poole, K. (2003) *Antimicrob Agents Chemother*, **47**, 3202–3207.

269 Evans, K. *et al.* (1998) *J Bacteriol*, **180**, 5443–5447.

270 Pearson, J.P., van Delden, C. and Iglewski, B.H. (1999) *J Bacteriol*, **181**, 1203–1210.

271 Duan, K., Dammel, C., Stein, J., Rabin, H. and Surette, M.G. (2003) *Mol Microbiol*, **50**, 1477–1491.

272 Eberl, L. and Tummler, B. (2004) *Int J Med Microbiol*, **294**, 123–131.

273 Harrison, F. (2007) *Microbiology*, **153**, 917–923.

274 Lee, D.A., Hoidal, J.R., Clawson, C.C., Quie, P.G. and Peterson, P.K. (1983) *J Immunol Methods*, **63**, 103–114.

275 Leid, J.G. *et al.* (2005) *J Immunol*, **175**, 7512–7518.

276 Bylund, J., Burgess, L.A., Cescutti, P., Ernst, R.K. and Speert, D.P. (2006) *J Biol Chem*, **281**, 2526–2532.

277 Learn, D.B., Brestel, E.P. and Seetharama, S. (1987) *Infect Immun*, **55**, 1813–1818.

278 Chan, C., Burrows, L.L. and Deber, C.M. (2004) *J Biol Chem*, **279**, 38749–38754.

279 Jesaitis, A.J. *et al.* (2003) *J Immunol*, **171**, 4329–4339.

280 Leid, J.G., Shirtliff, M.E., Costerton, J.W. and Stoodley, P. (2002) *Infect Immun*, **70**, 6339–6345.

281 Kharazmi, A., Doring, G., Hoiby, N. and Valerius, N.H. (1984) *Infect Immun*, **43**, 161–165.

282 Kharazmi, A., Hoiby, N., Doring, G. and Valerius, N.H. (1984) *Infect Immun*, **44**, 587–591.

283 Pessi, G. and Haas, D. (2000) *J Bacteriol*, **182**, 6940–6949.

284 Allen, L. *et al.* (2005) *J Immunol*, **174**, 3643–3649.

285 McClure, C.D. and Schiller, N.L. (1996) *Curr Opin Microbiol*, **33**, 109–117.

286 Chhabra, S.R. *et al.* (2003) *J Med Chem*, **46**, 97–104.

287 Ritchie, A.J., Yam, A.O.W., Tanabe, K.M., Rice, S.A. and Cooley, M.A. (2003) *Infect Immun*, **71**, 4421–4431.

288 Telford, G. *et al.* (1998) *Infect Immun*, **66**, 36–42.

289 Tateda, K. *et al.* (2003) *Infect Immun*, **71**, 5785–5793.

290 Weitere, M., Bergfeld, T., Rice, S.A., Matz, C. and Kjelleberg, S. (2005) *Environ Microbiol*, **7**, 1593–1601.

291 Cosson, P. *et al.* (2002) *J Bacteriol*, **184**, 3027–3033.

292 Pukatzki, S., Kessin, R.H. and Mekalanos, J.J. (2002) *Proc Natl Acad Sci USA*, **99**, 3159–3164.

293 Hogan, D.A. and Kolter, R. (2002) *Science*, **296**, 2229–2232.

294 Southey-Pillig, C.J., Davies, D.G. and Sauer, K. (2005) *J Bacteriol*, **187**, 8114–8126.

10
Bacteriophages of *Pseudomonas*

Kirsten Hertveldt and Rob Lavigne

10.1
Introduction

The start of bacteriophage (phage) research dates back to the beginning of the 20th century, when Fredrick Twort and Felix d'Herelle independently discovered these biological entities. The French researcher Felix d'Herelle ultimately described them as bacteriophages ("bacteria eaters") in 1917. Over a century of phage research has revealed these viruses as extremely diversified and ubiquitously present in the biosphere, preying on Eubacteria and Archaea in a wide range of biological niches.

10.1.1
Bacteriophage Propagation Cycle

10.1.1.1 Temperate versus Lytic Phages

All phages are true parasites, implying the need of a bacterial host for progeny production. Based on their propagation cycle, most phages – including those infecting *Pseudomonas* – can be divided in the two major groups: virulent and temperate phages.

After adsorption to their host, virulent phages immediately inject their genomes, and, following transcription, redirect the host metabolism towards phage genome replication and the production of new phage virions, which are released within minutes to hours after the initial infection event

In contrast, temperate phages have two replication modes open to them: they either enter a lytic propagation cycle or initiate a lysogenic cycle. In the lysogenic, "prophage" state, the phage sequence is quiescent. In this state, specific repressor proteins are expressed from the phage genome regulating transcription in a way which inhibits the lytic cycle. In the lysogenic cycle, the phage DNA usually integrates into the host genome or resides as a plasmid in the bacterial cytoplasm, as reported for a few phages including coliphage P1 [1]. Host cells harboring prophages are called lysogenic or lysogenized and upon host cell division these prophage sequences are transmitted to the daughter cells. The level, condition and ability of induction of

Pseudomonas. Model Organism, Pathogen, Cell Factory. Edited by Bernd H.A. Rehm

ISBN: 978-3-527-31914-5

prophage sequences to their lytic cycle are variable, but ultraviolet light, mitomycin and prolonged starvation have been reported as inducers of the lytic cycle.

The completion of many bacterial genome sequences (among which nine completed *Pseudomonas* sequences deposited in the public databases) revealed the presence of several complete and partial prophage sequences within the host chromosome, corroborating the very intimate relation between phage and host genome (reviewed in Refs. [2,3]). Due to phage integration and imprecise DNA excision events from the host chromosome, temperate phages often transduce bacterial sequences from one cell to another [1]. Transduction is defined as phage-mediated transfer of host DNA from a donor to an acceptor host cell. Two major types of transduction can be distinguished: generalized and specialized. In generalized transduction, almost any bacterial sequence can be transmitted to a donor cell. This can occur in phage particles which contain only host DNA as a byproduct during normal phage production in the lytic cycle. Alternatively, in case of temperate phages, this can be mediated by incorporation of heterogeneous DNA into a generalized transducing phage, obtained after random phage genome integration and excision from the host chromosome (e.g. D3112, Section 10.2.3.1). The transducing capacity of virulent phages is limited and occurs at lower frequency. In specialized transduction, a particular DNA sequence can be introduced into other cells by a specialized transducing phage. In this case, a phage with a single (or a limited number of) specific integration position(s) in the bacterial genome is involved. Upon excision, transmission is limited to specific host sequences surrounding the integration site (e.g. D3, Section 10.2.3.2).

Transduction events can drastically influence bacterial evolution by lateral gene transfer. Novel properties, acquired by the phage and further evolved, often provide regulation mechanisms aimed at an optimal phage production. Examples of such beneficial mechanisms include phage-encoded factors that render host cells immune to superinfection, and the acquisition of virulence factors and mechanisms which support the photosynthesis capacity in certain cyanobacteria during phage infection, thus conferring a symbiotic interaction between prophage and host [4].

10.1.1.2 **Lytic Infection Cycle**

A lytic infection cycle occurs according to the following scheme: phage adsorption to (a) host receptor(s), internalization of the phage genome, transcription and translation of phage proteins, phage genome replication, morphogenesis, genome packaging, and phage progeny release from the host cell.

Phage adsorption is achieved by interaction between specific phage proteins or envelope components and bacterial receptor molecules. For Gram-negative species, specific lipopolysaccharide (LPS) components, outer membrane proteins, capsule and pili structures have been reported as receptor molecules. After irreversible adhesion, the phage genome must cross three major barriers in Gram-negative bacteria (the outer membrane, the peptidoglycan layer and the inner membrane) to obtain access to the transcription/translation metabolism of its host. For this, phages employ various strategies, generally varying with phage particle morphology. Usually, metabolic energy is required to achieve this, based on available ATP, specific enzymatic activity or membrane potential [5]. The infection initiation process of both T4 and Ff filamentous phages has been described in more detail in Refs. [6] and [7], respectively.

Bacteriophages employ different replication modes which have been thoroughly reviewed in Ref. [8]. Replication modes studied for coliphages also apply for *Pseudomonas* phages. The *Pseudomonas* phages, discussed in more detail in Section 10.2, most likely replicate according to the replication mode of the closest related model phage mentioned between brackets. Filamentous *Pseudomonas* phages Pf1 and Pf3 (Section 10.2.7.2) and ϕCTX (Section 10.2.2.1) likely replicate in rolling circle mode (model phages: Ff filamentous phages and phage P2); phage D3 (Section 10.2.3.2) in θ (and σ) replication mode (model phage: λ); ϕKMV, LKD16, LKA1 and gh-1 by initiation of DNA replication by transcription (model phages: T7, early replication of T4); and bacteriophages D3112/B3 (Section 10.2.3.1) by recombination-dependent DNA replication (model phages: Mu and T4). The replication modes of the model phages are briefly introduced in the following paragraph. Illustrations of these replication modes are presented in Ref. [8].

The rolling circle replication mode in filamentous phages encompasses the conversion of the (+) strand phage DNA genome to a double-stranded (ds) DNA molecule by help of the host replisome. Negative supercoils are introduced by host gyrase. Subsequently, nicking occurs in the positive strand, followed by displacement synthesis by the host replisome. After full circle replication, covalent linkage in the original (+) strand is restored. Once the amount of single-stranded (ss) DNA-binding proteins is high enough, the (+) strand DNA is covered by ssDNA-binding proteins and prevented from further replication cycles.

In case of phage P2, the phage DNA enters the cell as linear dsDNA with cohesive (*cos*) ends (19-nucleotide complementary 5′-overhangs) circularized at the *cos* sites inside the host cell. Nicking occurs on the (+) strand and subsequently the host replisome synthesizes a new strand by strand displacement. Second-strand synthesis occurs on the displaced ssDNA during strand displacement. Once the replisome reaches the nick site, the displaced DNA is liberated from the dsDNA circle. After gap sealing by the host DNA ligase and introduction of negative supercoils by the host gyrase, the dsDNA may undergo a new replication cycle or be the substrate for packaging into phage heads.

Phage λ enters the cell as linear dsDNA with overhanging *cos* ends (12-nucleotide complementary 5′-overhangs). This DNA is converted to a closed supercoiled dsDNA molecule. A phage initiator protein O binds to the *ori*λ, located in the *O* gene, which triggers local unwinding. Subsequently, leading and lagging strand synthesis are completed with the help of host proteins, among which DnaB, DnaG and DNA polymerase III in a bidirectional replication mode. After primer removal and gap sealing, the two dsDNA molecules are resolved by DNA topoisomerase. New dsDNA molecules may resume the θ mode of replication or initiate a σ mode of replication, the latter being unidirectional as opposed to the bidirectional mode in θ replication.

Both early replication of T4 and replication of *Escherichia coli* phage T7 are initiated by transcription. In this case, an RNA polymerase binds to a promoter on the dsDNA and synthesizes a short (untranslated) transcript that serves as a primer for displacement synthesis of the leading strand. Upon assembly of the primosome (primase and helicase complex), strand-displacement synthesis switches from a unidirectional synthesis to bidirectional replication after creating access of the primosome to the opposite strand.

In the case of bacteriophage Mu, genome replication is recombination dependent. After injection of the phage genome in the host cell, it integrates at a random position in the host genome, which can lead to the establishment of lysogeny. Upon lytic development replicative transposition of the Mu genome occurs – a process that depends on the host replication apparatus [9].

At the end of the genome replication cycle, formation of filamentous phage particles takes place at the membrane [10]. Most other bacteriophages assemble in the cytoplasm by a head-filling mechanism. In the case of phage P2, closed circular dsDNA serves as a substrate for DNA packaging into phage heads. In this case, a terminase consisting of two subunits performs linearization of the circular replication intermediates at the *cos* sites during packaging of the phage DNA [11]. Packaging of λ DNA into phage (pro)heads requires genome concatemers which are processed at the *cos* sites [12]. Packaging of the phage T7 genome starts from concatemeric DNA as template. This concatemeric DNA is cut by site-specific nucleases which generate single-genome length fragments with 5′-overhangs [13]. Packaging starts with the right end and nuclease activity is associated with capsid proteins [14]. A fill-in reaction is required to restore the terminal repeats.

In the final stage of an infection event, new progeny phages are released by extrusion or, in most cases, cell lysis. Essential in cell lysis is the peptidoglycan degradation by a phage-encoded muralytic enzyme or endolysin after permeabilization and destabilization of the inner membrane by the holin. Holins are small, hydrophobic proteins that accumulate and oligomerize in the cell membrane during the late-gene expression phase. A time-based permeabilization of the membrane is achieved due to a balanced expression of holins and antiholins at the transcriptional level [15], thus providing access to the peptidoglycan for the lysin at a predetermined moment.

A second, holin-independent lysin transport system has also been described [16,17]. These phages use a lysin which contains an N-terminal signal sequence, transporting the enzyme to the periplasm by the host Sec system, thus achieving lysis in a holin-independent manner after cleavage of the signal sequence.

Recently, a hybrid between these two systems was identified in phage P1, called the signal anchor and release (SAR) mechanism [18]. Phages using SAR encode both a holin and a lysin with a noncleavable N-terminal signal (type II signal anchor). Hence, the N-terminal of the lysin remains part of the mature lysin and stays embedded in the inner cell membrane in an inactive form (due to the close proximity of the catalytic residues to the membrane or because of rearranged disulfide bonds in the lysin [19]). Triggering of the lysin is accomplished by the holins, accumulating in the inner membrane to a critical concentration which punctures the membrane, destroying membrane potential and releasing the lysin to the periplasm.

In addition to the lysin/holin cassette, a number of phages (λ, T7) contain an Rz/Rz1 system, located directly downstream from the lysis cassette. Orthologs of Rz/Rz1 are found in many Gram-negative phages. It is suggested that these gene products destabilize the outer membrane by forming a complex after periplasmic translocation which cleaves the oligopeptide links between the murein and the outer membrane protein Lpp [20].

10.1.2
Phage Classification

In 1966, an international committee was formed to address classification of viruses of both eukaryotes and prokaryotes. From these initiatives, the International Committee on Taxonomy of viruses (ICTV) was established in 1973 [21], cataloguing and describing the various physiological and morphological characteristics of viruses. In order to condense and categorize virus data, the ICTV is essentially occupied with classification to allow easier management of phage collections and databases, identification of new phages, and comparative studies. Ideally, such a classification system should also reflect evolutionary relationships, but for bacteriophages this remains an issue of great debate. Generally, phages are subdivided based on the nature of their genome (dsDNA, ssDNA, dsRNA and ssRNA) and the overall virion morphology (tail type, polyhedral, filamentous and pleomorphic). A partial and limited hierarchical structure is established for tailed dsDNA phages, as a single order, the *Caudovirales*, and 13 families [22]. The advent of the genome and proteome era is fueling debate on the reassessment of this subdivision, and the integration of genome data, sometimes leading to suggestions of radically different taxonomic classification methods [23–25].

Over 5500 phages have been examined after negative staining in the electron microscope [22], among which 511 described phages infect the *Pseudomonas* genus. Tailed phages constitute the vast majority (more than 97%) of the phages observed on *Pseudomonas* while only a small fraction (less than 3%) have a polyhedral, filamentous or pleomorphic morphology (denominated as PFP phages) [22]. Tailed phages are divided in three families based on their tail morphology. Phages belonging to the *Myoviridae* have a long contractile tail, *Siphoviridae* have a long, noncontractile tail and *Podoviridae* have a short tail. The majority of phage observations on *Pseudomonas* belong to the *Siphoviridae* family (47%) [22]. The PFP phages are classified in 10 families, of which only for the polyhedral and filamentous shapes are phages infecting *Pseudomonas* [26]. Polyhedral phages are icosahedra or related bodies with cubic symmetry, filamentous phages have helical symmetry and pleomorphic phages apparently lack obvious symmetry axes. They can be enveloped or not. In total, 427 phage genome sequences are currently available, 27 of which are annotated genomes of phages infecting *Pseudomonas*. Annotated sequences are listed in Table 10.1 under ICTV family classification and major representatives are discussed further in Section 10.2.

10.1.3
Phages of *Pseudomonas* Spp.

Phages have been found in several *Pseudomonas* species, including *P. acidovorans, P. aeruginosa, P. cepacia, P. facilis, P. fluorescens, P. malthophilia, P. mallei, P. plecoglossicida,*

Table 10.1 Overview of phages infecting *Pseudomonas* (major representatives with annotated genome sequences are listed together with their morphological and genomic characteristics).

Shape	Nucleic acid	Virus family	Morphology[a]	Phage[b]	Natural host	Genome properties[c]	Predicted genes	Genome organization	Particular features
Tailed	dsDNA, linear	*Myoviridae*	icosahedral head (60–120 nm ∅) long, contractile tail (135–200 nm)	ϕKZ NC_004629	*P. aeruginosa*	280334 bp – 36.8% G + C circularly permuted and terminally redundant	306 ORFs 6 tRNAs	ϕKZ-like	virulent; large capsid (120 nm ∅)
				EL NC_007623	*P. aeruginosa*	211215 bp – 49.3% G + C circularly permuted and terminally redundant	201 ORFs 1 tRNA	ϕKZ-like	virulent; large capsid
				ϕCTX NC_003278	*P. aeruginosa*	35580 bp – 62% G + C 21 nt 5′-extruding cohesive ends	74 ORFs	P2-like	temperate; cytotoxin-converting
	dsDNA, linear	*Siphoviridae*	icosahedral head (50–60 nm ∅) long, noncontractile tail (140–190 nm)	D3112 NC_005178	*P. aeruginosa*	37611 bp – 64% G + C flanking host sequences	55 ORFs	Mu-like	temperate; random integration
				DMS3 NC_008717	*P. aeruginosa*	36415 bp – 64% G + C	52 ORFs	Mu-like	temperate; random integration
				B3 NC_006548	*P. aeruginosa*	38439 bp – 63.3% G + C Flanking host sequences	59 ORFs	Mu-like	temperate; random integration

				D3 NC_002484	*P. aeruginosa*	56426 bp – 57.8% G + C 9 nt 3′-extruding cohesive ends	90 ORFs 4 tRNA s	λ-like	temperate; serotype-converting; specific integration
	dsDNA, linear	*Podoviridae*	icosahedral head (50–70 nm ∅), short tail	gh-1 NC_004665	*P. putida*	37359 bp – 57% G + C TR: 216 bp	42 ORFs	T7-like	lytic
				ϕKMV NC_005045	*P. aeruginosa*	42519 bp – 62% G + C TR: 414 bp	49 ORFs	ϕKMV-like	lytic
				LKD16 AM265638	*P. aeruginosa*	43200 bp – 62.6% G + C TR: 428 bp	54 ORFs	ϕKMV-like	lytic
				LKA1 AM265639	*P. aeruginosa*	41593 bp – 60.9% G + C TR: 298 bp	56 ORFs	ϕKMV-like	lytic
				PaP2 NC_005884	*P. aeruginosa*	43783 bp – 45% G + C	58 ORFs		temperate
				PaP3 NC_004466	*P. aeruginosa*	45503 bp – 52% G + C 20 nt 5′-extruding cohesive ends	71 ORFs 4 tRNA s		temperate
				F116 NC_006552	*P. aeruginosa*	65195 bp – 63% G + C	70 ORFs		temperate
Polyhedral	ssRNA, linear	*Leviviridae*	icosahedral capsid (23 nm ∅), unenveloped	PP7 NC_001628	*P. aeruginosa*	3,588 nt – 54% G + C positive-stranded	4 ORFs	MS2-like	lytic
	dsRNA, linear	*Cystoviridae*	icosahedral capsid (80 nm ∅), enveloped	ϕ6 NC_003714 NC_003715 NC_003716	*P. syringae*	13385 bp – 55% G + C 3 genome segments (S, L, M)	13 ORFs		lytic

Table 10.1 *(Continued)*

Shape	Nucleic acid	Virus family	Morphology[a]	Phage[b]	Natural host	Genome properties[c]	Predicted genes	Genome organization	Particular features
				ϕ8 NC_003299 NC_003300 NC_003301	P. syringae	14984 bp – 54% G + C 3 genome segments (S, L, M)	16 ORFs		lytic
				ϕ12 NC_004173 NC_004174 NC_004175	P. syringae	13173 bp – 55% G + C 3 genome segments (S, L, M)	15 ORFs		lytic
				ϕ13 NC_004170 NC_004171 NC_004172	P. syringae	13652 bp – 57% G + C 3 genome segments (S, L, M)	13 ORFs		lytic
Filamentous	ssRNA, circular	*Inoviridae*	long filaments 2000 nm × 7 nm	Pf1 NC_001331	P. aeruginosa	7349 nt – 61% G + C	14 ORFs	Ff-like	carrier state; extrusion
			long filaments 700 nm × 7 nm	Pf3 NC_001418	P. aeruginosa	5833 nt – 45% G + C covalently closed ssDNA	9 ORFs	Ff-like	carrier state; extrusion

[a] The sizes indicated reflect approximate dimensions obtained from Ackermann [25], Ackermann and DuBow [27] and their specific genome papers.
[b] Numbers refer to GenBank accession numbers at http://www.ncbi.nlm.nih.gov.
[c] TR = terminal repeats.

P. pseudomallei, P. putida, P. stutzeri, P. syringae and *P. testosteroni* [27–30]. The most encountered host species in the literature of *Pseudomonas* phages is *P. aeruginosa*. This is partly due to both its ubiquitous presence as soil and water bacterium, and to the search for new antibacterial agents against *P. aeruginosa* as an opportunistic pathogen. This importance is reflected by the deposition of over 25 whole-genome sequences for *P. aeruginosa*-infecting phages – belonging to different morphotypes – in the public databases to date. In addition, the genome sequence of one P. *putida*-infecting phage (gh-1) and four phages infecting the phytopathogen *P. syringae* (ϕ6, ϕ8, ϕ13 and ϕ12) have been described (see Section 10.2.6). Moreover, partial genome sequences of phages infecting *P. fluorescens, P. putida* and *P. chlororaphis* have been published [28,31].

10.1.4
Applications of *Pseudomonas* Bacteriophages

Historically, *Pseudomonas* phages have been used for the epidemiological tracing of specific strains (e.g. the *P. aeruginosa* typing phages of the Lindberg set [32–34]) and as tools in molecular biology. Examples of phage applications in biotechnology include the D3-derived cosmid vector [35], the ϕCTX integration vector [36] and the D3112-based system introducing new genetic material stably into the chromosome of *P. aeruginosa* [37].

In recent years, increasing interest in *Pseudomonas* phages and phage genome sequences in general has been in part driven by a renewed interest in phage therapy and the promise of using phage proteins (lysins) or phage-inspired proteins in the treatment of *Pseudomonas* infections. The threat to patients suffering from diseases like cystic fibrosis, cancer, AIDS and burn injuries – all vulnerable to *P. aeruginosa* infections has increased dramatically due to exceptional *Pseudomonas* resistance against commercial antibiotics. Apart from numerous clinical case studies in Eastern Europe and Russia [38], hopeful results regarding phage treatment of infections caused by *P. aeruginosa*, the fish pathogen *P. plecoglossicida* and the plant pathogen *P. syringae* (on human/animals, fish and plants, respectively), indicate the potential value of phages as a treatment approach [38–42].

10.2
Pseudomonas Phage Representatives

10.2.1
Introduction

Phages infecting *Pseudomonas* for which the genome sequence is annotated and deposited in GenBank are listed in Table 10.1. In addition to morphological and nucleic acid properties, information on the genome size, coding potential, genome organization and life cycle is presented. Detailed information on phage morphology and phage dimensions is given in Refs. [27,34]. Within each family, phages with distinct properties are discussed in more detail. Insight in molecular strategies and evolutionary properties of *Pseudomonas* phages has drastically increased since whole-phage genome sequences have become available. Comparative analysis and DNA and protein similarity studies to well-documented model phages allowed speculation on *Pseudomonas* phage biology and their propagation cycle. In addition, relevant aspects towards phage therapy and antibacterial treatment (such as tracing potentially toxic sequences, endolysins, integrases, transposases) can be explored more readily. Knowledge of phage genome sequences can reveal evolutionary relationships among phages, give an indication for the acquisition of novel sequences and locate the intimate relationship with their bacterial host species. Insight in the phage genome content provides a new source of information, but also an additional challenge for the current phage classification system.

10.2.1.1 **Tailed Phages**

Tailed phages encompass the vast majority of described phages in general and also of those infecting *Pseudomonas*, and are classified in the order of the *Caudovirales*. Virions of phages belonging to the *Caudovirales* consist of a protein shell and dsDNA. The virions have an icosahedral head and helical tails. As stated above, the tailed phages are divided in three families (*Myoviridae*, *Siphoviridae* and *Podoviridae*). Within these three families, 15 genera have been defined, based on criteria related to phage genome organization and replication strategy [43,44]. However, classification into genera has only started recently and only a few of the phages infecting *Pseudomonas* are classified into proposed genera.

10.2.1.2 **Tailless Phages**

Only a few tailless phages are described for *Pseudomonas*. Some representatives with a polyhedral and filamentous shape have been sequenced so far. They belong to the *Leviviridae* (ssRNA), *Cystoviridae* (dsRNA) and *Inoviridae* (ssDNA) families (Table 10.1). *Leviviridae* are unenveloped, have a positive-stranded genome (genomic RNA acts as mRNA) and are morphologically similar to polioviruses. *Cystoviridae* have icosahedral capsids surrounded by a lipid-containing envelope. They contain three molecules of dsRNA. *Inoviridae* group phages have a long, rigid or flexible helical tube, the length of which is determined by the phage genome size.

10.2.2 ***Pseudomonas*-Infecting *Myoviridae***

Pseudomonas-infecting *Myoviridae* phages with published genome sequences are ϕCTX, and the giant phages ϕKZ and EL [45–47].

10.2.2.1 **Phage ϕCTX**

Phage ϕCTX is a temperate, cytotoxin (CTX)-converting phage, propagated from *P. aeruginosa* strain PAS10 [48]. Upon lysogenization, the phage-encoded *ctx* gene enables *P. aeruginosa* to produce CTX, a pore-forming cytotoxin which increases host virulence [49]. ϕCTX belongs to the R pyocin-related phage family. R pyocins are bacteriocins that resemble contractile phage tails [50]. R pyocin-related phages are serologically and genetically related to R pyocins, and were probably derived from a common ancestor [47]. Morphologically and at the level of genome organization and open reading frame (ORF) homology, ϕCTX is very comparable to bacteriophage P2 [47]. Both aspects hint at a similar life cycle to bacteriophage P2. In addition, lysogenic strains of both phages are noninducible and both phages display similar adhesion characteristics using LPS as receptor in a Ca^{2+}-dependent manner [51].

The linear, dsDNA ϕCTX genome sequence comprises 35538 bp and has 5′-extended 21-nucleotide single-stranded *cos* ends. Of the 47 predicted ORFs 25 show 28–65% identity to P2 ORFs [47]. From a host integration perspective, the *attP–cos–ctx* region contains several repeats and integration host factor-binding sequences [52]. The *cos* sequence is similar to that of the P2 family, while the position of the *attP* region is different compared to P2 [47]. Indeed, the *ctx* gene is located very close (361 bp) to the left genome end (*cosL*) and is one of the hotspots for acquisition of foreign DNA [47]. The attachment core sequence of the phage (*attP*) of 30 bp is located close (647 bp) to the right end (*cosR*). As seen in many temperate phages, the integrase (*int*) gene of ϕCTX is located immediately upstream of the *attP* site [53] and mediates site-specific recombination. In contrast to the preferred bacterial integration site (*attB*) of phage P2 (*locI* in *E. coli*) [54], the *attB* site for ϕCTX is the 3′-end of the serine tRNA gene [52].

Similar gene arrangements and homology to P2 proteins are most pronounced in the late region, comprising genes involved in capsid synthesis, DNA packaging and tail morphogenesis, reflected in the common morphology between both phages. Phage ϕCTX has an icosahedral head and a contractile tail with tail fibers (Figure 10.1A). Also, a number of the predicted lysis genes within the late region resemble those of phage P2, which by analogy may be involved in the control of the timing of lysis [55,56].

Homology to P2 in the early region is limited to part of the integrase gene and one ORF of unknown function. Only the C-terminal domain of the ϕCTX integrase shows significant similarity to the P2 integrase. This C-terminal domain is in general well conserved in various site-specific recombinases [47].

10.2.2.2 **Giant Virulent *Pseudomonas* Phages**

Within the *Myoviridae* family, the ϕKZ-like phages constitute a newly proposed genus containing a number of *Pseudomonas*-infecting giant phages that share an identical morphology and significant similarity at the protein level [57]. The hallmark representative phage within this genus is ϕKZ, originally isolated in 1975 in Kazakhstan. It has a large icosahedral head (123 nm between opposite apices) and a long contractile tail (196 × 21 nm), consisting of a neck, a collar, a contractile sheath and 30-nm terminal tail fibers [57] (Figure 10.1B). In interaction with its host, *P. aeruginosa* PAO1, ϕKZ behaves as a virulent phage, giving rise to small plaques. ϕKZ is reported to have a broad strain spectrum and is thus attractive for phage therapy specific for *P. aeruginosa* [58]. Some ϕKZ-like phages are already part of therapeutic phage mixtures to treat *P. aeruginosa* infections.

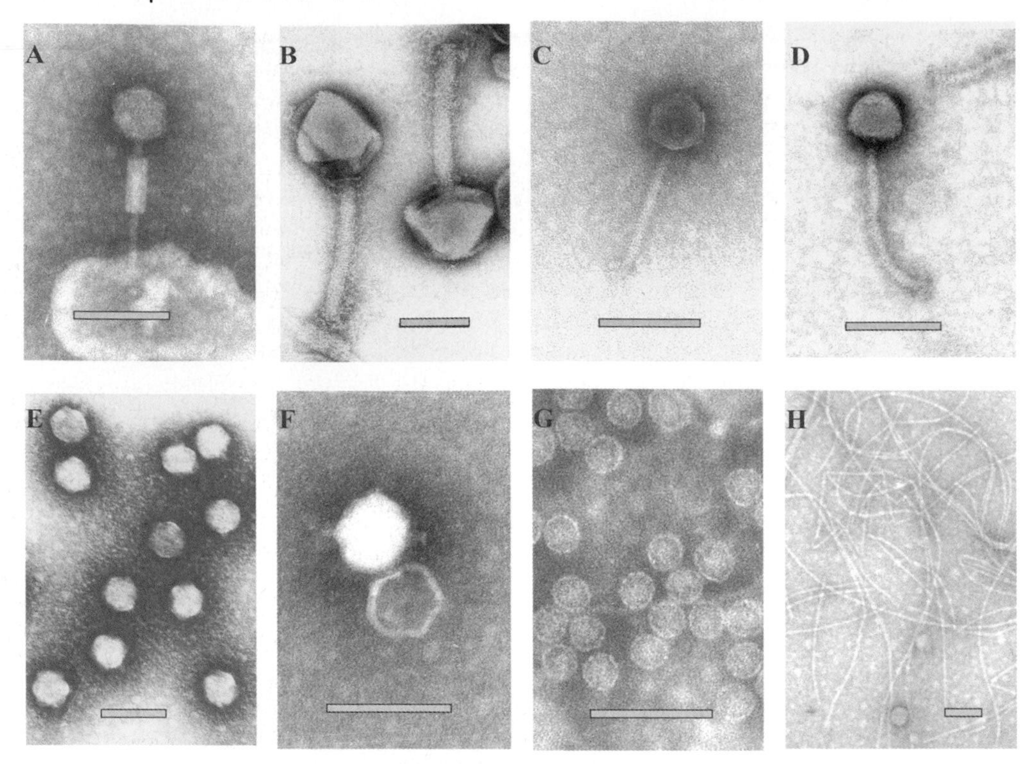

Figure 10.1 Electron microscopy imaging of the *P. aeruginosa* bacteriophages. *Myoviridae* ϕCTX and ϕKZ are shown in (A) and (B), respectively. Notice the contracted tail in this ϕCTX particle. Phages D3 (C) and 73 (D) belong to the *Siphoviridae* and have a long noncontractile tail. Phage 73 has a longer tail and a slightly smaller capsid compared to phage D3. Phage 73 is morphologically identical to phage D3112. *Podoviridae* member ϕKMV (E) is representative for phage LKD16 and LKA1, and corresponds morphologically to the T7-like phage gh-1. (F–H) Representatives for the *Cystoviridae* (ϕ8), the *Leviviridae* (PP7) and the *Inoviridae* (Pf1), respectively. For size comparison, scale markers = 100 nm. With the exception of the electron micrograph (E), all images were kindly provided by courtesy of Dr. Hans-W. Ackermann, Faculty of Medicine, Laval University, Quebec, Canada. Micrograph (E) was kindly provided by Vadim Mesyanzhinov.

The large capsid contains a linear dsDNA genome that is circularly permuted and terminally redundant. The genomic DNA appears to be wound around an inner body inside the head [59]. Cryo-electron microscopy imaging of the capsid revealed that the dsDNA inside the capsid is packaged in a highly condensed series of layers, separated by 24 Å that follows the contour of the inner wall of the capsid [60]. This manner of packaging is similar to that observed in T4 [61,62].

The ϕKZ genome sequence was published in 2002 [46] and comprises 280 334 bp. In contrast, *Myoviridae* phage T4 contains 168 903 bp [63]. With the exception of the *Bacillus* phage G genome which comprises 498 kbp, ϕKZ is the largest phage genome sequence reported. In comparison with the host G + C content (65%), the average G + C content of ϕKZ is low (36.8%). Six tRNAs are encoded on ϕKZ, most likely to optimize codon usage during expression of phage genes. In total, 306 ORFs are predicted, varying in size from 52 to 2237 amino acids. No DNA similarity, but extensive protein similarity exists with phage EL (one-third of the proteins shares 17–55% amino acid identity), another giant phage infecting *P. aeruginosa* for which the genome sequence has been determined [45].

The phage EL genome sequence comprises 211 215 bp and 201 predicted ORFs. In comparison with the ϕKZ G + C content (36.8%), the EL G + C content (49.3%) is more adapted to the G + C content of the host. This difference is reflected in the presence of only one predicted tRNA gene in the EL genome. Transcription of ϕKZ and EL has not been studied experimentally, although conserved promoter motifs and potential terminator sequences have been predicted for both phages. In contrast to the predicted EL terminator sequences, putative terminators in ϕKZ are highly conserved [45,46]. Despite differences in genome size and G + C content between ϕKZ and EL, comparative analysis of both genome sequences revealed a largely consistent gene arrangement, as observed from the number of consecutive blocks of similar genes. A single, large region lacks significant similarity between both phages at the DNA and protein level. The size of these regions is 67 and 17.7 kbp for ϕKZ and EL, respectively, and there is currently no experimental evidence towards the origin and mechanism of acquisition of these large insertions.

Both ϕKZ and EL share very little protein homology with other phages and mainly encode proteins of unknown function. They lack significant DNA sequence similarity to any other known phage species infecting bacteria other than *Pseudomonas*. A limited set of gene products of both ϕKZ and EL show similarity to gene products of a diverse set of phages preying on different host bacteria (including *Lactobacillus planetarium* phage LP65, *Streptococcus thermophilus* phage Sfi19, and *Enterobacteria* bacteriophages T5 and P22).

Consistent with properties of larger virulent phages (among which T4), ϕKZ and EL appear less dependent on host nucleotide metabolism, as they encode some proteins involved in these processes (e.g. thymidylate kinase). In this respect, ϕKZ even seems less dependent on host proteins compared to phage EL, since it encodes additional enzymes involved in nucleotide metabolism in the large unique genome region. Based on protein similarity there is no evidence for the presence of a phage-encoded DNA polymerase of the known DNA polymerase families.

Recent analysis of the EL and ϕKZ virion particles revealed the presence of more than 50 different structural proteins (manuscript in preparation). A previous, limited analysis identified five structural proteins, among which the major capsid protein gp120. Evidence for posttranslational proteolysis of structural proteins was obtained [46].

ϕKZ head structure analysis by cryo-electron microscopy confirmed the icosahedral symmetry and measured a head dimension of 1455 Å along the 5-fold axes and a unique portal vertex to which the contractile tail of around 1800 Å is attached [60]. The capsid consists of a lattice of hexamers composed of the major capsid protein gp120 and pentameric vertices composed of special vertex proteins. Comparison with the hexameric building blocks of the bacteriophages T4, ϕ29, P22 and HK97 reveals a similar shape and size of the hexamers.

Recent visualization of two other giant phages, OBP and Lu11, infecting *P. fluorescens* and *P. putida* var. *manila*, respectively, has provided evidence for the more widespread presence of giant phages among *Pseudomonas* species. Partial genome sequence analysis allowed insight into the coding potential of both phages, allowing grouping of OBP among the ϕKZ-like genus, whereas Lu11 remains a tentative species within the ϕKZ-like genus as it lacks amino acid sequence similarity with the other ϕKZ-like phages [28,57]. A completed genome analysis of Lu11 will show the extent of the molecular correlation to the members of the ϕKZ-like phages.

10.2.3 *Pseudomonas*-Infecting *Siphoviridae*

Pseudomonas phages belonging to the *Siphoviridae* family have an icosahedral head and a flexible, noncontractile tail. All *Pseudomonas* phages within this family described to date have a temperate character, implying the possibility to undergo a lytic or lysogenic interaction with their host. The best known examples of *P. aeruginosa* phages in this family include D3112, B3 and D3, all resembling λ-like virion particles from a morphological perspective (Figure 10.1C and D). In view of their genome organization and propagation cycle, both D3112 and B3 resemble the Myovirus Mu. In contrast, phage D3 relates more to a phage λ type genome organization. Less investigated, although interesting due to an alternative genome organization and gene content, are the closely related *P. aeruginosa* phages M6 (GenBank accession code: NC_007809) and YuA, of which genome sequence analysis is currently under investigation by our group.

10.2.3.1 D3112/B3-like *Pseudomonas* Phages

Phages B3 and D3112 are the best studied temperate, transposable phages of *P. aeruginosa*. Many transposable phages have been isolated from natural populations and clinical isolates of *P. aeruginosa* [64]. B3 was isolated from a pathogenic strain of *P. aeruginosa* (strain 3), containing four different prophages: A3, B3, C3 and D3 [65], whereas D3112 was originally isolated in Kazakhstan. Recently, the genome sequence of DMS3 was determined (GenBank accession code: NC_008717) revealing very close relatedness (DNA similarity) to D3112. Phage DMS3 was isolated from a clinical *P. aeruginosa* strain and enters the lytic cycle by prolonged starvation [66]. D3112, B3 and DMS3 require pili and not LPS for host infection [66,67]. This is in contrast to phage Mu, which is dependent on LPS as receptor. Furthermore, DMS3 infection was shown to be type IV pilus dependent [66]. Hence, the conservation of pilin proteins in *P. aeruginosa* may explain the wide host range of these phages among clinical isolates [67].

D3112, B3 and DMS3 share the transposable nature and propagation cycle of phage Mu, a Myovirus with a broad host range that encompasses *E. coli*, *Salmonella*, *Citrobacter* and *Erwinia* [1]. The broad host range of phage Mu is due to a G segment which can occur in two orientations [G(+)) and G(−)] in the genome by inversion and give rise to alternative tail fibers. Genome similarity of D3112, B3 and DMS3 with phage Mu is reflected by a similar gene organization, gene content and regulatory elements (Figure 10.2). A Mu-like life cycle means that the phage genome integrates into the host genome, regardless of whether a lytic or lysogenic cycle is destined. The integration mechanism used by Mu is not the site-specific recombinase model [68] as in phage λ, but occurs by transposition instead. After induction Mu does not replicate in a free state, but by repeated transposition in the host genome. Upon excision from the host chromosome it takes along host sequences at both genome ends. This process gives rise to mutations and rearrangements in the bacterial sequence, which has led to its nomenclature Mu, from "mutator" [69]. Transposable phages are generalized transducers, possessing the ability to transduce various heterogeneous host genome sequences. Phage Mu DNA is packaged by a headful mechanism, keeping the amount of heterogeneous DNA at the left end constant and with a variable amount of heterogeneous DNA at the right end, based on internal phage genome length changes due to insertions and deletions [70]. The presence of variable host sequences flanking the B3 and D3112 linear genome sequence was confirmed by genome sequence analysis [71,72]. D3112 and B3 are known as generalized transducing phages between identical *P. aeruginosa* strains, whereas DMS3 is the first *P. aeruginosa* phage for which generalized transduction within and between *P. aeruginosa* strains PA14 and PAO1 was reported [66].

Genome sequence analysis of the phages D3112 [72], B3 [71] and DMS3 (GenBank accession code: NC_008717) revealed an average genome size of around 37 kbp, and between 52 and 59 predicted ORFs (Table 10.1). The average G + C content of the phage genomes (63–64%) is very similar to the 67% G + C content of their host, *P. aeruginosa*, suggesting a long history of interaction. Also, codon usage matches well between host and phage, explaining why phages D3112 and B3 do not encode their own tRNA molecules (in contrast to D3). In the case of phage D3112, some ORFs with deviating G + C content (less than 56%) were identified, suggesting a more recent acquisition from another host. DNA hybridization and recombination studies have shown that B3 is somewhat related to D3112 [73]. Comparative analysis of both D3112 and B3 genome sequences reveals several regions (ranging from 100 to 2500 bp) at the right genome end that share DNA identity up to 82% conservation at the DNA level, while both D3112 and B3 lack DNA similarity with phage Mu. A substantial amount of D3112 and B3 proteins share amino acid sequence similarity with enterobacteriophage Mu proteins and proteins encoded by Mu-like prophages (defined as prophages that encode transposase homologs and a substantial number of other Mu protein homologs in early, lysis and head gene regions). Based on comparative analysis of phage Mu, D3112 and B3, the major difference in genome organization between phage D3112/Mu and B3 is a larger early region oriented in the opposite direction in B3, and variation in the presence and location of a DNA modification cluster. In addition, phage Mu contains an invertible G segment, reported in neither phage D3112 nor B3.

Based on the organizational similarity to the Mu genome, the genomes of bacteriophages D3112 and DMS3 are divided into three functional regions: (i) the left end or early region, which is responsible for genome integration, modulation of

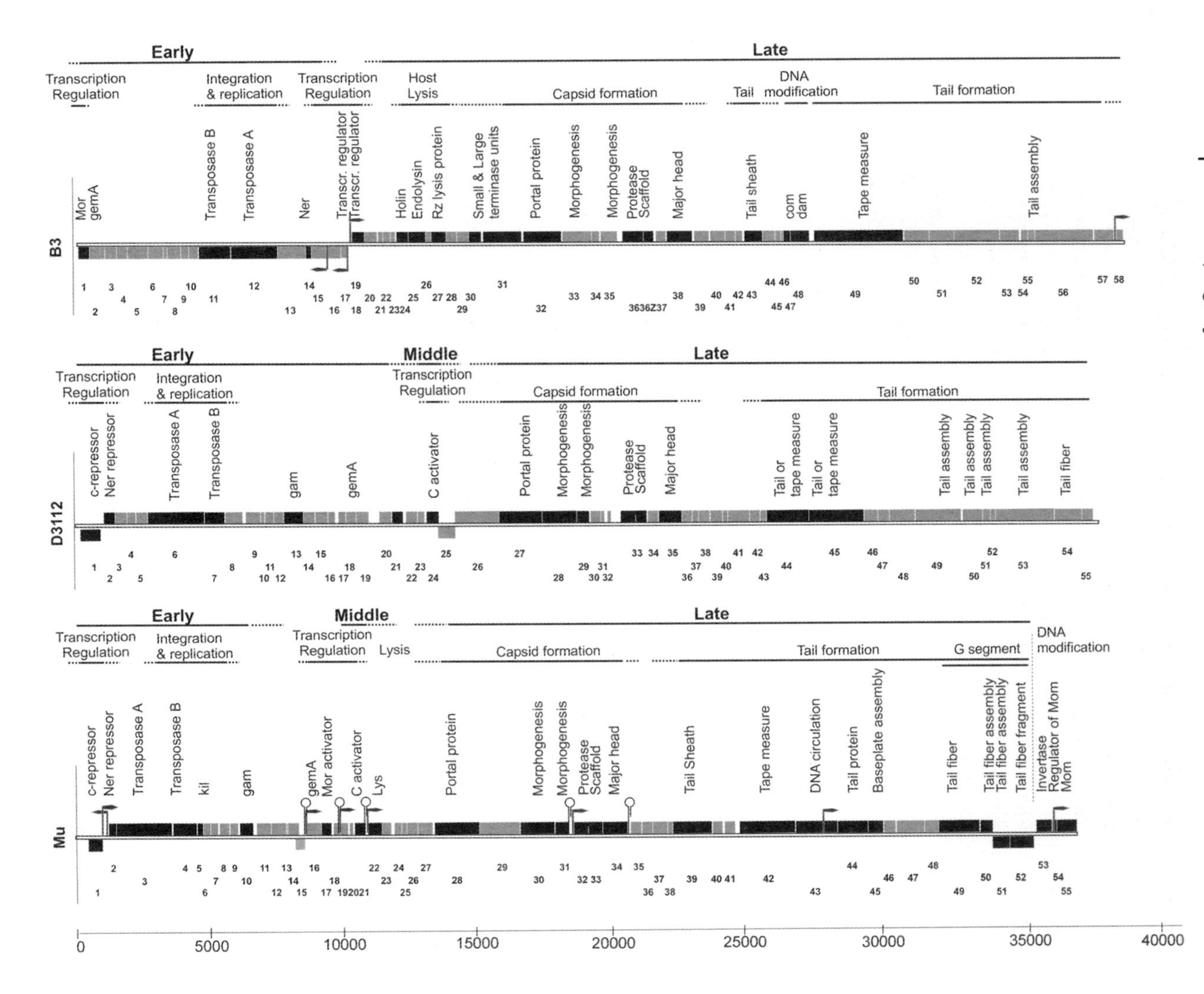

B3
Early
Late
Transcription Regulation
Integration & replication
Transcription Regulation
Host Lysis
Capsid formation
Tail
DNA modification
Tail formation
Mor
gemA
Transposase B
Transposase A
Ner
Transcr. regulator
Transcr. regulator
Holin
Endolysin
Rz lysis protein
Small & Large terminase units
Portal protein
Morphogenesis
Morphogenesis
Protease
Scaffold
Major head
Tail sheath
com
dam
Tape measure
Tail assembly
D3112
Early
Middle
Late
Transcription Regulation
Integration & replication
Transcription Regulation
Capsid formation
Tail formation
c-repressor
Ner repressor
Transposase A
Transposase B
gam
gemA
C activator
Portal protein
Morphogenesis
Morphogenesis
Protease
Scaffold
Major head
Tail or tape measure
Tail or tape measure
Tail assembly
Tail assembly
Tail assembly
Tail assembly
Tail fiber
Mu
Early
Middle
Late
Transcription Regulation
Integration & replication
Transcription Regulation
Lysis
Capsid formation
Tail formation
G segment
DNA modification
c-repressor
Ner repressor
Transposase A
Transposase B
kil
gam
gemA
Mor activator
C activator
Lys
Portal protein
Morphogenesis
Morphogenesis
Protease
Scaffold
Major head
Tail Sheath
Tape measure
DNA circulation
Tail protein
Baseplate assembly
Tail fiber
Tail fiber assembly
Tail fiber assembly
Tail fiber fragment
Invertase
Regulator of Mom
Mom
0
5000
10000
15000
20000
25000
30000
35000
40000

phage gene expression and modulation of host response, (ii) the middle region, which is responsible for control of late gene transcription, and (iii) the right end or late region, which is responsible for virion morphogenesis. These regions correspond to the three transcription phases of phage Mu: early, middle and late [74]. Transcription in D3112 occurs from left to right, with the exception of the c repressor gene and one hypothetical gene (ORF25). Braid *et al.* [71] suggest dividing phage B3 in early and late genome regions only, as there is yet no obvious evidence for the presence of a middle region.

Characteristic for phages D3112, DMS3, B3 and Mu is the presence of both a c repressor and Ner repressor (and predicted homologs), and the transposases A and B in the early genome region. The repressors are involved in the control of the lytic–lysogenic switch, while the transposases are involved in phage integration and transposition [69]. The leftmost ORF in phage D3112, DMS3 and Mu sequences encodes the well-characterized lysogenic c repressor [75]. Expression of this D3112 ORF in *P. aeruginosa* renders the cells immune to superinfection with D3112 [76]. Based on amino acid similarity, the D3112 c repressor is more similar to the cI repressor of the lambdoid phage D3 than to the c repressor of Mu, but the structure and function reflect more an evolutionary relationship to that of phage Mu [75]. The Ner protein in phage Mu is responsible for downregulating early transcription and repressor transcription at late times during the lytic cycle [77]. In phage B3, ORF17 and ORF18 are candidate c repressor genes [71]. Part of one D3112 ORF (ORF11) is similar to the host nuclease inhibitor protein Gam in phage Mu. Conservation of this protein among pathogens suggests that this protein has a role in overcoming host defense and establishing infection. In line with an inversely oriented early region in phage B3, the position of both the putative Ner protein in B3 (ORF14), the candidate c repressor (ORF17 or ORF18), and the Mor and GemA proteins (ORF1 and ORF2) differ from positions in phage D3112 and Mu [71].

In phage Mu, transcription of the middle region is activated by the Mor protein [78] and this region contains the *C* gene. The C protein activates transcription of the late region [79]. The middle region of D3112 (and DMS3) contains the *C* gene and was shown to have a positive effect on the transcription of late viral genes in D3112.

Figure 10.2 Genome organization overview of *Siphoviridae* representatives D3112 and B3. Genome organization of bacteriophages B3, D3112 and phage Mu are represented. For each genome, gene numbers are shown corresponding to the reading frame in which they are expressed. Grey blocks indicate the different genes along the total length of the genome (outlined bar), above the bar for genes expressed on the plus strand, below the bar for minus strand genes. Genes for which functional indications exist are shown in black, annotated and placed in their organizational context. Regulatory elements like promoters (arrows) and terminators (open loops) are shown. Both B3 and D3112 are temperate phages sharing their transposable nature with phage Mu. All three genomes contain the transposase A and B genes involved in host genome integration and replication by transposition. Both D3112 and phage Mu are organized in early, middle and late transcription regions. Compared to phage Mu and D3112, the early region in B3 is inverted. In contrast to phage D3112 and B3, which have a flexible tail, the late region of phage Mu encodes for proteins necessary for formation of a rigid, contractile tail.

In general, the late region encodes morphogenesis and cell lysis genes. In many phage genomes, a cluster of tail genes follows the head genes, the tail sheath and tail tube genes, and a tail-length tape measure gene. D3112 and B3 head (morphogenesis) proteins display amino acid sequence and gene order similarity to the head genes of phage Mu and Mu-like prophages [71,72]. D3112 ORF35, for example, has 61% similarity to the major head subunit gpT in phage Mu. Consistent with the electron microscopy data, the tail morphogenesis proteins of phage D3112 and B3 show similarity to lambdoid phages with a noncontractile tail and not to Mu-like phages, which have a contractile tail. In phage B3, the terminase genes (encoding for the small and large subunits) share similarity with those of phage Mu [71]. Phage B3 has the two-gene DNA modification cluster (*com–dam*) immediately upstream of the predicted B3 tail-length tape measure gene. The Mu DNA modification cluster contains two genes: *com* and *mom*, located at the extremely right end of the genome [69]. The Mom proteins modify about 15% of the adenine residues in the Mu DNA to acetamidoadenine, protecting it from cleavage by a variety of restriction enzymes. The Com protein stabilizes the *mom* mRNA. In phage D3112, this DNA modification cluster seems absent. ORF22 encodes the endolysin of phage Mu directing release of phage particles from the cell. The B3 gene products involved in host cell lysis encompass an endolysin, a holin and Rz homolog.

Homologs of D3112, B3 and Mu genes are found in many phages, prophages and phage-related elements. The level of similarity varies across their genomes. The genes with the highest frequency of homologs occur in the early and head regions [71]. The ecological niche of these phages and their host bacteria are quite diverse and include – among others – strains of *E. coli, Salmonella enterica, Vibrio cholerae, Haemophilus influenzae, Burkholderia cenocepacia* and *Xylella fastidiosa* [71]. This diversity and the highly mosaic structure of both D3112 and B3 suggest that extensive horizontal gene transfer has played an important role in their evolution, thereby acquiring sequences from both λ-like and Mu-like phages.

10.2.3.2 **Bacteriophage D3**

D3 is a temperate, serotype-converting phage, obtained from a clinical isolate [65]. It is called "serotype converting", since lysogenization of host cells causes a change in serological properties. Kuzio and Kropinski [80] showed that the LPS of D3 lysogens lacks receptor activity for phage D3 and that the O-antigenic polysaccharide side-chains are altered, suggesting LPS involvement for the phage receptor. Lysogenization of host cells results in the change of serotype from O5 to O16, according to the International Antigenic Typing Scheme [81].

Morphologically, phage D3 resembles phage λ and also the genome organization is reminiscent of lambdoid phages – with exception of the position of the endolysin gene (Figures 10.1 Figures 10.3). D3 is a member of the B1 (isometric head) subgroup of the family *Siphoviridae*. The head diameter measures 55 nm and the long flexible tail is 7 × 113 nm. The tail has six tail fibers with terminal knobs [82]. Contrary to D3, phage λ uses the LamB protein as cellular receptor [80,81].

In contrast to phage D3112, B3 and Mu-related phages, phage D3 is not a transposable phage. Whereas multiple integration is reported for D3112/B3-like phages, Cavenagh and Miller [83] demonstrated phage integration at two distinct loci for phage D3. The mechanism of integration is most likely based on the Campbell model of insertion [68]. The D3-encoded integrase (ORF35) mediates recombination between a short sequence of the phage DNA (*attP*) and a short sequence of bacterial DNA (*attB*). Induction of the D3 prophage results in low-frequency-transducing lysates in which genes adjacent to the two *attB* sites are transduced [83].

Most likely, the DNA replication mode in phage D3 during lytic development switches from a θ to a σ replication, as is the case for λ [8]. Different host factors including a primase and helicase are required at the *ori*λ to initiate the phage DNA replication. Genome sequencing of phage D3 revealed that it encodes a helicase homolog [84].

The dsDNA genome comprises 56426 bp, encodes 95 putative ORFs and contains four genes specifying tRNAs [84] (Table 10.1). These tRNAs may help in translation of highly transcribed genes with a lower G + C content. Compared to the G + C content in phages D3112 and B, the average G + C content of phage D3 (57.8%) is low. The linear genome has 9-nucleotide 3′-extended cohesive ends [35]. In comparison, coliphage λ possesses 12-nucleotide complementary 5′-overhangs, and coliphages HK97 and HK022 have 10-nucleotide 3′-extended cohesive ends. The *cos* sites allow the DNA to circularize rapidly after insertion. Thus, nicked circles are formed, which are subsequently ligated to form supercoiled molecules. In phage λ, RNA transcription begins immediately and can be divided into three temporal classes designated (i) immediate early, (ii) delayed early and (iii) late. The mechanisms of regulation of both the lytic and temperate life cycle have been intensively studied (reviewed in Ref. [70]). The immunity region of phage D3 is clearly homologous in structure and function to the immunity region of lambdoid phages. It contains the repressor gene (*c1*) and antirepressor gene (*cro*), together with the left operator–promoter (O_LP_L) and right operator–promoter (O_RP_R) complexes. Both in λ and D3, the repressor mRNA lack [originating from the repressor maintenance promoter (P_{RM})] the typical prokaryotic ribosome-binding sites (SD box) and the first three nucleotides code for formylmethionine [84]. The sequence similarity between λ cI and D3 cI proteins is poor, but essential residues are strongly conserved.

The morphogenesis genes show highest similarity to coliphages HK022 and HK97. Four structural proteins of D3 show between 32 and 62% amino acid identity with proteins of HK022. The capsid morphogenesis genes (portal–protease–major head protein) are organized similarly to coliphage HK97. As in the coliphages, D3 capsid proteins undergo proteolytic processing and cross-linking during head morphogenesis.

The D3 ORF31 gene product is a polypeptide of 160 amino acids, with strong homology to λ, P2, 186, KH022 and HK97 endolysins, involved in host cell lysis at the end of the replication cycle. In comparison with phage λ, the genomic position of the phage D3 endolysin is different. It is located near the serotype conversion gene module, also absent in phage λ. Newton *et al.* [81] identified three ORFs within this module responsible for the serotype conversion. These include an α-polymerase inhibitor (*iap*), an O-acetylase (*oac*) and a β-polymerase (wzy_β). The G + C content of this region has a deviant value, suggesting acquisition by horizontal gene transfer.

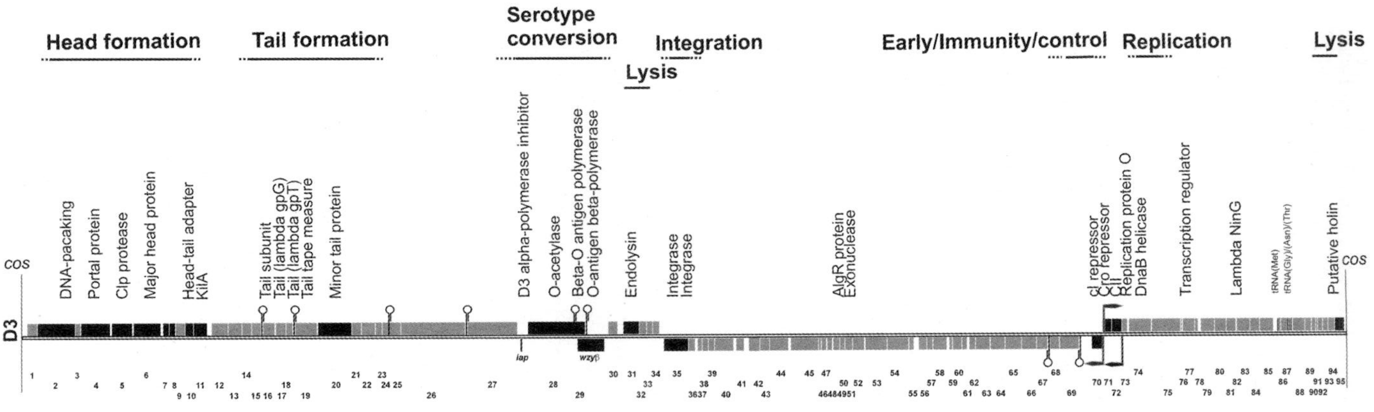
Head formation
Tail formation
Serotype conversion
Integration
Lysis
Early/Immunity/control
Replication
Lysis
cos
D3
DNA-pacaking
Portal protein
Clp protease
Major head protein
Head-tail adapter
KilA
Tail subunit
Tail (lambda gpG)
Tail (lambda gpT)
Tail tape measure
Minor tail protein
D3 alpha-polymerase inhibitor
lap
O-acetylase
Beta-O antigen polymerase
O-antigen beta-polymerase
wzyβ
Endolysin
Integrase
Integrase
AlgR protein
Exonuclease
cI repressor
Cro repressor
CII
Replication protein O
DnaB helicase
Transcription regulator
Lambda NinG
tRNA(Met)
tRNA(Gly)/(Asn)/(Thr)
Putative holin
cos

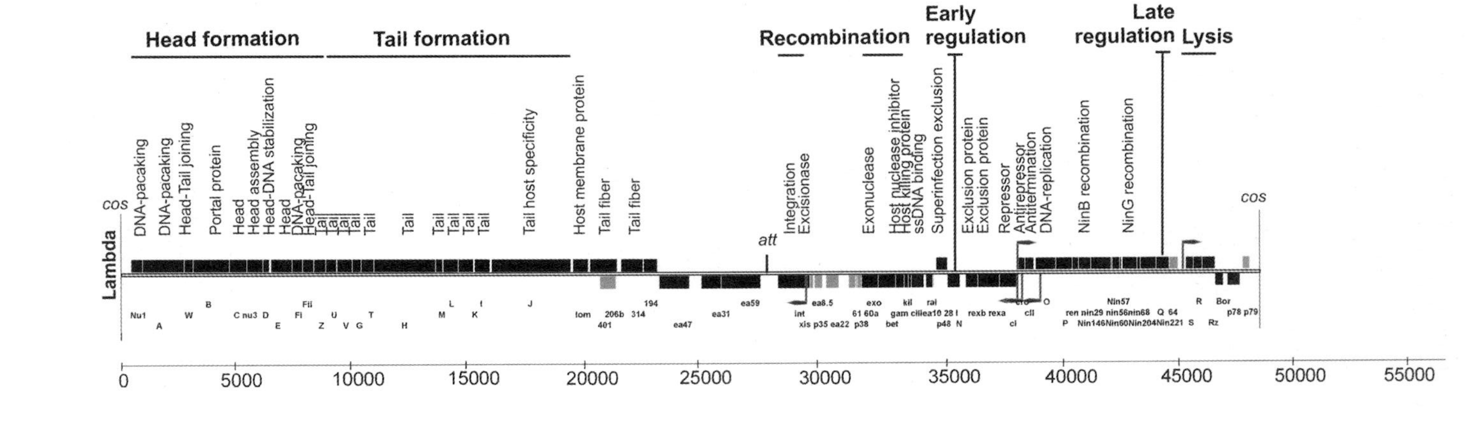
Head formation
Tail formation
Recombination
Early regulation
Late regulation
Lysis
cos
Lambda
DNA-pacaking
DNA-pacaking
Head-Tail joining
Portal protein
Head
Head assembly
Head-DNA stabilization
Head
DNA-pacaking
Head-Tail joining
Tail host specificity
Host membrane protein
Tail fiber
Tail fiber
att
Integration
Excisionase
Exonuclease
Host nuclease inhibitor
Host killing protein
ssDNA binding
Superinfection exclusion
Exclusion protein
Exclusion protein
Repressor
Antirepressor
Antitermination
DNA-replication
NinB recombination
NinG recombination
cos
0
5000
10000
15000
20000
25000
30000
35000
40000
45000
50000
55000

From its genome content, it is clear that phage D3 is another example of an evolved *Pseudomonas* phage, with a clear relationship to lambdoid coliphages. In addition, the altered position of the endolysin gene and the presence of a serotype conversion module provide evidence for a mosaic-based evolution – in part – by horizontal exchange.

10.2.4
Pseudomonas-Infecting *Podoviridae*

10.2.4.1 **gh-1: A T7-Like Member of the *Podoviridae***

Originally isolated on *P. putida* strain A.3.12 in 1966, bacteriophage gh-1 forms large, clear plaques (4–6 mm). Acriflavine agglutination and sodium dodecylsulfate–polyacrylamide gel electrophoresis analyses on the LPS of phage-resistant *P. putida* mutants revealed gh-1 requires the presence of an intact bacterial O-antigen for successful infection. Furthermore, gh-1 has an average burst size of 103 and a short latent period of 21 min, as shown in the original one-step growth curves [85]. gh-1 was originally used in environmental research and in comparison to the closely related and well-studied coliphage T7 [86,87].

The close relationship of gh-1 to the members of the T7-like phages was further shown upon comparison of the phages' genome sequences. The gh-1 genome contains regions highly similar to T7, interspersed with nonsimilar regions containing small unknown ORFs [88] (Figure 10.4). Nonetheless, it is somewhat more distantly related to T7 compared to phages T3 and ϕYeO3-12.

Like T7, transcription occurs from one strand only, using a phage-encoded single-subunit RNA polymerase, which was shown to be rifampicin resistant [89]. The gh-1 RNA polymerase recognizes conserved promoter sequences [90], located within the intergenic regions and comparable to T3 and ϕYeO3-12, both phages closely related to T7. Expression of the gh-1 RNA polymerase itself is presumed to be regulated by two host RNA polymerase recognition sites (σ^{70}-like promoters), located in the early region of the gh-1 genome.

Reviewing the encoded proteins, gh-1 contains 31 common gene products with the T7-like phages, while 12 putative gene products are unique. Although the temporal and functional distribution of gh-1 genes is very similar to that of the other members of the T7 group, allowing T7 gene nomenclature to be used, this phage is marked by the absence of ORFs prior to gene 1 (RNA polymerase). Nonetheless, the strong correlation in genome organization and its highly conserved replication strategy (transcription, conservation in phage promoter sequences) suggest this complex strategy evolved only once, followed by a differentiation in host specificity [88].

Figure 10.3 Genome organization overview of the *Siphoviridae* representative D3. The genome organization of D3 is shown in comparison with *E. coli* phage λ, with elements as described in Figure 10.2. The genome organization of phage D3 largely corresponds to that of phage λ. A typical cI/cro/cII life cycle regulation module, integrase genes, and a clearly defined head and tail formation cluster are present. In contrast to phage λ, the endolysin gene is not located near the right genome end. In addition, phage D3 contains a specific serotype conversion module, responsible for serotype conversion of *P. aeruginosa* after lysogenization. Four tRNA genes are predicted in phage D3. *attP*, attachment site of the phage; *cos*, cohesive end.

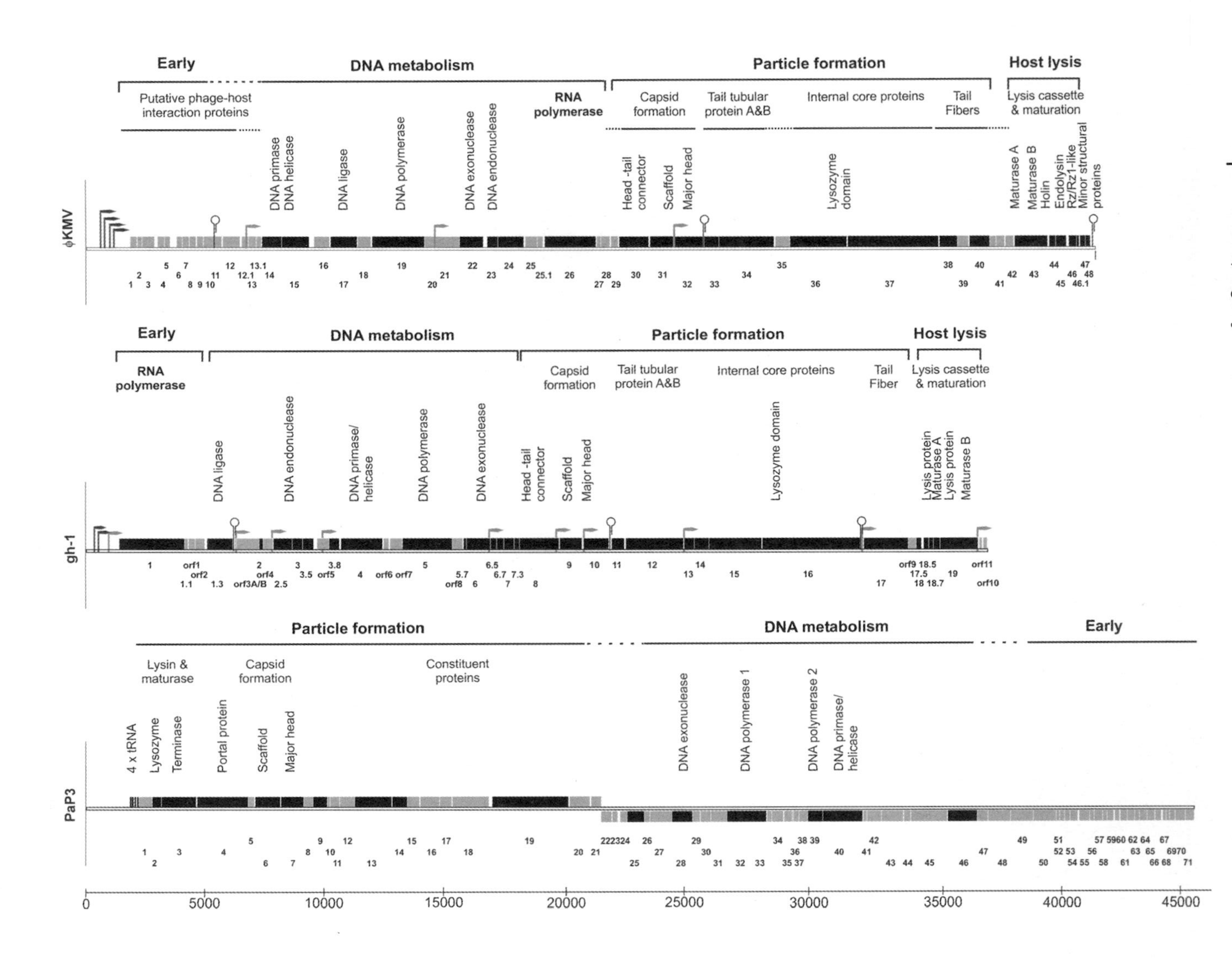
φKMV
Early
Putative phage-host interaction proteins
DNA metabolism
DNA primase
DNA helicase
DNA ligase
DNA polymerase
DNA exonuclease
DNA endonuclease
RNA polymerase
Particle formation
Capsid formation
Head -tail connector
Scaffold
Major head
Tail tubular protein A&B
Internal core proteins
Lysozyme domain
Tail Fibers
Host lysis
Lysis cassette & maturation
Maturase A
Maturase B
Holin
Endolysin
Rz/Rz1-like
Minor structural proteins
gh-1
Early
RNA polymerase
DNA metabolism
DNA ligase
DNA endonuclease
DNA primase/ helicase
DNA polymerase
DNA exonuclease
Particle formation
Capsid formation
Head -tail connector
Scaffold
Major head
Tail tubular protein A&B
Internal core proteins
Lysozyme domain
Tail Fiber
Host lysis
Lysis cassette & maturation
Lysis protein
Maturase A
Lysis protein
Maturase B
PaP3
Particle formation
Lysin & maturase
4 x tRNA
Lysozyme
Terminase
Portal protein
Capsid formation
Scaffold
Major head
Constituent proteins
DNA metabolism
DNA exonuclease
DNA polymerase 1
DNA polymerase 2
DNA primase/ helicase
Early
0
5000
10000
15000
20000
25000
30000
35000
40000
45000

10.2.4.2 **ϕKMV-Like Viruses**

The members of the newly established ϕKMV-like genus within the *Podoviridae* constitute a geologically widespread yet relatively tight-knit group of phages. This is illustrated by the isolation of over 15 ϕKMV-like phages across the different continents, with varying similarity towards ϕKMV, as confirmed by Southern hybridization experiments or (partial) genome sequencing [91].

Bacteriophage ϕKMV (Figure 10.1E) is most intensely studied, in combination with the recently sequenced bacteriophages LKD16 and LKA1 [92]. The structural proteins are identified [93], its DNA ligase and its lytic proteins are characterized [94–96], while its early genes and transcription scheme are currently under investigation [97].

The host range for the ϕKMV-like phages is narrow, but they have strong lytic properties as they form large clear plaques. The differences in host range, adhesion characteristics and infectivity parameters among the ϕKMV-like phages suggest a variation in bacterial adhesion factors.

These phages have linear dsDNA genomes with remarkably long direct terminal repeats of 414, 428 and 298 bp for ϕKMV, LKD16 and LKA, respectively. The LKD16 genome comprises 43200 bp and has an overall DNA identity of 83% with ϕKMV (42512 bp), while LKA1 has no significant DNA similarity whatsoever.

Functional annotation revealed a common architecture in the genomes of ϕKMV, LKD16 and LKA1, loosely corresponding to that of T7-like phages (Figure 10.4). However, contrary to other members of the T7-like phages, ϕKMV, LKD16 and LKA1 carry a single-subunit RNA polymerase gene adjacent to the structural genome region instead of in the early region, and they lack conserved phage promoters. Indeed, apart from a number of conserved, bacterial promoters at the left end of the genome, no clear, conserved phage promoters are present within these three genomes, in comparison to phage gh-1. These features suggest major differences in the transcription scheme for ϕKMV-like phages in which two functional genomic regions can be delineated. The first genomic region ends after the RNA polymerase

◄

Figure 10.4 Genome organization overview of the *Podoviridae* representatives. The genome organizations of ϕKMV, gh-1 and PaP3 are shown, with elements as described in Figure 10.2. Phages ϕKMV and gh-1 both belong to the T7 supergroup since the global genome organization is conserved, with all genes of the host infection, replication and particle formation/release transcribed from the plus strand. The gh-1 organization very closely matches that of bacteriophage T7. Major differences between gh-1 and ϕKMV include the large number of small unidentified genes in the early genome region of ϕKMV, absent in gh-1. A second crucial difference is the localization of the phage-encoded RNA polymerase, in the early region for gh-1, but adjacent to the particle formation genes in ϕKMV. This suggests a difference in the transcription scheme between both phages. The genome organization of PaP3 can only be described very tentatively compared to the detailed and specific outline of ϕKMV and gh-1, since so few PaP3 gene products show similarity to other phage proteins. The putative particle formation region and the DNA metabolism/early region are transcribed in opposite direction, divided by a bidirectional terminator. A limited number of constituent proteins and capsid formation proteins have been defined, linked to the lysis cassette. The DNA metabolism region boasts the presence of two DNA polymerase genes.

gene, and encompasses genes for host conversion ("early" genes) and DNA replication, while the second ("late") region comprises genes coding for structural and lysis proteins. In this organization, they have followed another evolutionary pathway compared with T7 *sensu stricto* phages like gh-1.

When comparing the ϕKMV-like phages, the early region of LKA1 has strongly diverged since similarity to other gene products (gp) in the databases is limited to gp3 similar to gp4 of both ϕKMV/LKD16 and to the unknown gp70 of phage PaP3. Between ϕKMV and LKD16, the overall DNA homology is slightly below average (58%) although the majority of genes are preserved between both phages.

The DNA replication genes are conserved among the three phages and functionally consistent to the corresponding proteins of the T7-like phages. In contrast to the T7-like genus, the predicted RNA polymerases of the ϕKMV-like phages are at the end of the DNA replication region, although the RNA polymerases are closely related. Within the RNA polymerase, the variation of the recognition and specificity loops between the T7-likes and ϕKMV-like phages reflects the proposed differences in promoter sequences.

From a structural perspective, homology among the ϕKMV-like phage structural proteins reflects their closely resembling morphology. LKA1, as the most distant member, shares an overall 30–50% protein similarity to ϕKMV and LKD16, and shows significant deletions in the genes encoding the scaffolding protein (180 bp) and an internal virion protein (999 bp). However, the major difference is observed in the tail fiber proteins, showing a markedly lower similarity between ϕKMV and LKD16, whereas LKA1 has a completely different tail fiber. This can be observed by electron microscopy imaging as well as by the occurrence of a single large putative tail fiber protein, resembling that of D3 gp27, located just upstream of the serotype conversion module [81]. This single gene contrasts the four small ORFs for ϕKMV and LKD16 in this region. These differences explain the discrepancies between the infectivity ranges of these phages. Indeed, preliminary analyses suggest a type IV pili-dependent infection mechanism for ϕKMV and LKD16, while the correlation between LKA1 and D3 suggests a possible LPS-dependent mechanism (Chibeu *et al.*, unpublished results).

Overall, the lysis cassette in the known ϕKMV-like phages seems conserved, despite limited sequence homology for LKA1, and consists of stop/start overlapping holin/lysin genes, followed by predicted out-of-frame internally overlapping Rz/Rz1-like genes. Interestingly, these phages have been suggested to use the SAR mechanism for lysis (see Section 10.1.1.2).

10.2.4.3 ***Pseudomonas*** **Phage PaP3**

Bacteriophage PaP3 is part of a collection of phages isolated from hospital sewage in Chongqing, China. Its semitransparent plaques and restriction analysis on its DNA have revealed it to be a temperate phage with a linear genome of 45.5 kbp and cohesive ends [98].

The PaP3 genome is organized in two regions, transcribed in opposite directions, divided by a bidirectional terminator (Figure 10.4). The left region (ORF1–21) comprises of a gene cluster (encoded on the Watson strand) predicted to encode the structural and lysis proteins, as well as the tRNA genes. ORF22–77 are located on the crick strand and encode replication proteins, and a large number of unknown predicted gene products. Intriguingly, the predicted replication gene products includes two different DNA polymerase, different in size, but homologous at the C-terminal.

Furthermore, genome analysis allowed establishing the exact host integration site (core site) for this phage, with its conserved, site-specific recombination region. From these data, a mechanism for the integration of the PaP3 genome could be extrapolated, similar to the model deduced by Campbell for bacteriophage λ [68,98]. Interestingly, the *attP* and *attB* integration sites are located within predicted tRNA genes encoded by the phage and host, respectively. This is not an uncommon feature among phages since the conserved secondary structure of tRNA genes might facilitate the integration reaction. Moreover, multiple copies of these genes are present within the bacterial genome and are conserved among different bacteria. This provides the means for multiple insertion sites and increased horizontal transfer to other bacteria [99–101].

10.2.4.4 *Pseudomonas* Phage F116

Originally isolated by B. W. Holloway in 1960, F116 was classified correctly as a *Podoviridae* member only in 1993 [102]. Biofilm-degrading properties have been established for this phage, allowing penetration to its host and infection through binding to the type IV pili of the cell [103]. This infection is achieved by subsequent retraction of the pili, allowing contact between the phage and the bacterial cell wall for injection of the phage DNA [104]. F116 is suggested to possess the unusual ability to replicate as a plasmid during its lysogenic replication cycle [105]. However, genome sequencing of its linear genome (terminally redundant, circularly permuted) revealed the presence of a possible integration (*int*) gene [106]. It is possible that F116, lacking specific *attB* sites in PAO1, has undergone evolutionary changes which permit replication in plasmid form. F116 rarely packages host DNA [106].

The mosaic nature of the genome of F116 (Figure 10.5) clearly contains evidence that it is not only the phages which impact their host's evolution, but the inverse is also true. Indeed 16 gene products of F116 show significant similarity to proteins in *P. aeruginosa*, *P. putida* and *P. syringae*, acquired by the phage through multiple and independent exchange events.

From a genome organization perspective, transcription of the DNA replication, lysis and morphogenic gene clusters seem to be regulated by the divergently transcribed *cI* and *cro* locus, suggesting that the F116 central regulatory system resembles that of coliphage λ and that of *Pseudomonas* phage D3 (see Section 10.2.3.2).

10.2.5 *Pseudomonas*-Infecting *Leviviridae*

Leviviridae have been isolated on various bacteria, including *P. aeruginosa*, and have been studied extensively from a replication, gene regulation and translation perspective (e.g. for model phages Qβ and MS2) [107]. Phages of the *Leviviridae*

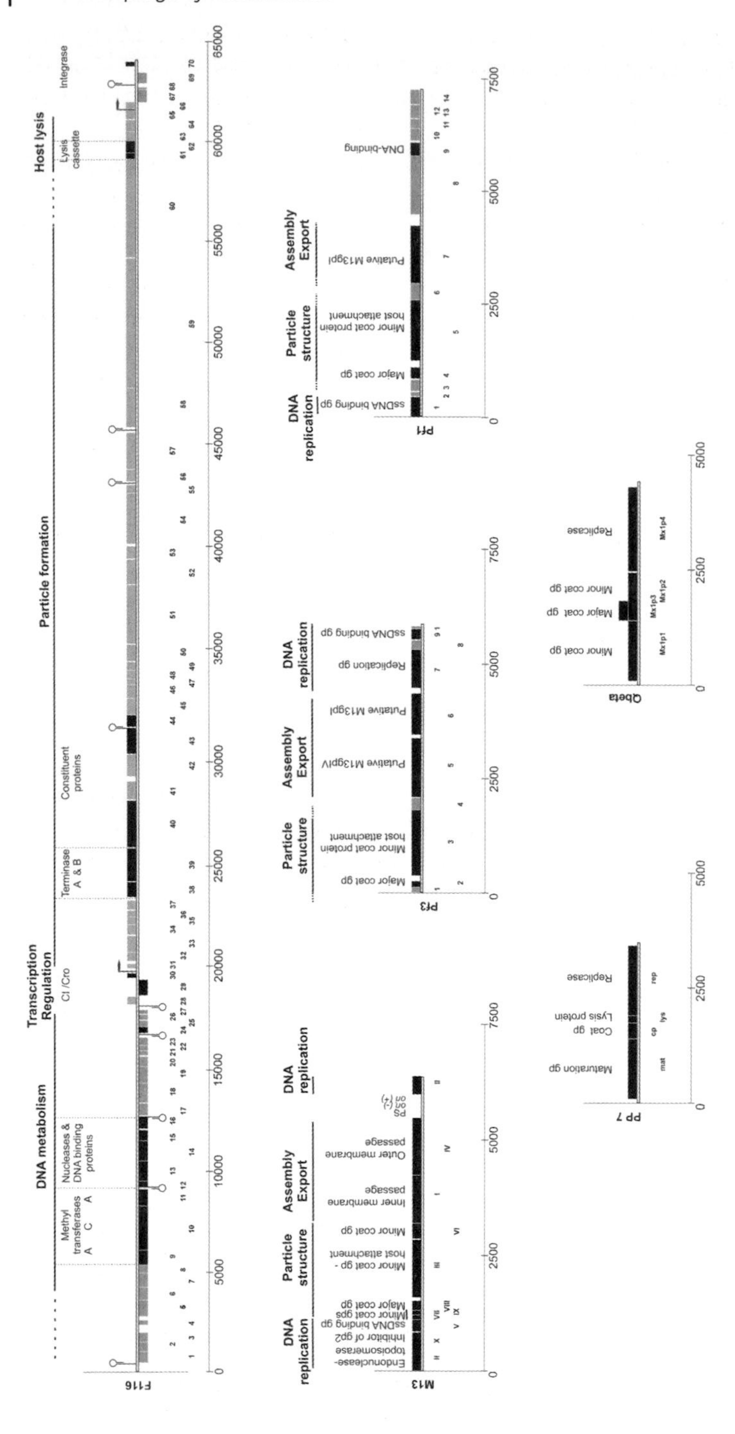

F116
DNA metabolism
Methyl transferases A C A
Nucleases & DNA binding proteins
Transcription Regulation
CI /Cro
Particle formation
Terminase A & B
Constituent proteins
Host lysis
Lysis cassette
Integrase
M13
DNA replication
Particle structure
Assembly Export
Endonuclease-topoisomerase
Inhibitor of gp2
ssDNA binding gp
Minor coat gps
Major coat gp
Minor coat gp - host attachment
Minor coat gp
Inner membrane passage
Outer membrane passage
Pf3
Major coat gp
Minor coat protein host attachment
Putative M13gpIV
Putative M13gpI
Replication gp
ssDNA binding gp
Pf1
ssDNA binding gp
Major coat gp
Minor coat protein host attachment
Putative M13gpI
DNA-binding
PP7
Maturation gp
Coat gp
Lysis protein
Replicase
Qbeta
Minor coat gp
Major coat gp
Minor coat gp
Replicase

family have a small RNA genome and are known to have very high mutation rates because the RNA polymerases have no proofreading capacity [108]. *Pseudomonas* phage PP7 was originally isolated in 1966 by Bradley [109], has an E1 (tailless) morphotype (Figure 10.1G) and forms clear plaques of about 0.5 mm. Its sequence was published in 1995 by Olsthoorn *et al.* [110] and consists of a positive, ssRNA genome of 3588 nucleotides, encoding four predicted ORFs. Based on its genome organization, it belongs to the Levivirus species and is similar to the MS2-like, group A coliphages, although DNA homology is limited.

Interestingly, PP7 contains many relatively well conserved RNA secondary structures associated with genome regulation, including the 5′-terminal hairpin, the gene start/stop-associated RNA folds, the replicase-binding sites and the 3′-untranslated region [110].

Based on protein similarity to MS2-like coliphages, the four encoded proteins are said to include a maturation protein, a coat protein, a replication protein and a predicted lysin (Figure 10.5).

Phage PP7 adsorbs to the side of polar pili of *P. aeruginosa*, inducing cleavage of the maturation protein, causing the release of RNA [111,112]. As such, the maturation protein performs multiple roles in the infection process and in RNA protection.

The coat protein (127 amino acids) of PP7 is arranged in trimeric modules to form the basic icosahedral unit of the phage particle [113]. For these subunits, the contacts formed around the 5- and 3-fold axes (i.e. the FG loops, a 15-residue loop between the F and G β-strands) are well conserved, in contrast to other small RNA phages. Contrarily, the most conserved residues among these phages lie within the region involved in RNA binding, suggesting a conserved function. As dimers, the coat proteins have an additional function in the translation repression of replicase, by binding an RNA stem–loop structure upstream from the viral replicase gene (as

Figure 10.5 Genome organization overview of several bacteriophage representatives. The genomes of Podovirus F116 (top), filamentous phages M13, Pf3 and Pf1 of the *Inoviridae* family (center), and *Leviviridae* PP7 and Qbeta (below) are shown. The genome of phage F116 is very large, compared to other *Podoviridae* members (Figure 10.4) and shows very little relationship to these strictly lytic phages. Indeed, among the *Podoviridae* F116 can be considered an orphan, based on its genome, showing only limited similarity to a small number proteins involved in metabolism, particle formation and lysis. Note the presence of a putative integration gene which contradicts the hypothesis which states that F116 is a nonintegrating phage. Pf1, Pf3 and M13 are filamentous phages belonging to the *Inoviridae* family. Both Pf1 and Pf3 have a similar genome organization as coliphage M13. Three major genome regions are delineated: DNA replication, particle formation and phage assembly/export. Bacteriophages PP7 and Qbeta represent two different classes within the *Leviviridae*: the Levivirus and the Allolevivirus, respectively. The differences in genome size have led to hypotheses of gene expansion and gene contraction, compared to a common ancestor. In PP7 (Levivirus) a clear lysis gene can be distinguished, whereas lysis in Qbeta (Allolevivirus) is mediated by the maturation protein. Also, the PP7-coding regions are initiated in different reading frames, unlike Qbeta, and the coat and lysis coding regions are overlapping.

shown for MS2) [114]. In addition, correct packaging and viral assembly are also influenced by this protein.

Homologous to coliphage Qβ, the PP7 replicase complex is an RNA-dependent RNA polymerase holoenzyme, consisting of four subunits, one of which is the phage-encoded replicase. This protein forms a complex with host-encoded elongation factors EF-Tu and EF-Ts (guanine-nucleotide exchange factor), the ribosomal protein S1, and possibly HF-I (a ribosome-associated host factor) [110,115]. Translation of the replicase gene is suggested to be coupled to the coat protein expression, similar to MS2 [116]. However, termination of replicase synthesis is mainly regulated by an interaction between coat protein dimers (see above) and a hairpin spanning the start of the replicase gene.

Contrary to the Alloleviviruses, the members of the Leviviruses like PP7 contain a lysis gene that overlaps with both the coat and replicase gene and which show translational coupling in the homologous MS2 phage [117]. This lysis protein of PP7 is predicted as a short (54 amino acids) L protein, containing a hydrophobic domain to cross the cytoplasmic membrane, thus destroying the membrane potential [118].

10.2.6
Pseudomonas-Infecting *Cystoviridae*

Cystoviridae are defined as phages with dsRNA in a procapsid, enveloped by a lipid/protein membrane (Figure 10.1F). They possess a segmented genome composing of a small (S), medium (M) and large (L) moiety [119,120]. This segmented dsRNA genome links the *Cystoviridae* to the *Reoviridae*, comprising mainly eukaryotic viruses. Indeed, this evolutionary link to eukaryotic viruses can be observed in several properties, discussed below.

The hallmark member of the genus Cystovirus within the *Cystoviridae* is ϕ6, which naturally infects *P. syringae* pv. *phaseolicola* and which has been thoroughly researched (genome sequence, replication cycle, viral structure). The isolation of eight additional Cystoviruses infecting *P. syringae* (ϕ7–ϕ14) revealed several closely and more distantly related phages, three of which have been sequenced (ϕ8, ϕ12 and ϕ13). For these phages, packaging and replication mechanisms are conserved, while entry and assembly mechanisms have diverged. These differences and the specificity in host range are reflected in structural variations in the virion.

10.2.6.1 **Bacteriophage ϕ6**

Phage ϕ6 infects *P. syringae*. Phage particles consist of proteins (70% of the weight) and membrane lipids (20%). Phages are adsorbed by specific receptors located on the type IV pili which retract and bring the virion into contact with the host outer membrane for membrane fusion [121].

The total genome size of ϕ6 is 13 × 379 bp, segmented in the L, M and S dsRNA segments with respective sizes of 6734 (encoding gene products P1, P2, P4, P7 and P14), 4063 (encoding P3, P6, P10, P13) and 2948 bp (P5, P8, P9, P12). Close relatives of this phage identified so far include ϕ7, ϕ9, ϕ10 and ϕ11, showing strong sequence similarity (80–85%) [122].

From a structural perspective, cryo-electron microscopy-based reconstruction has revealed the ϕ6 virion to be icosahedral, with 600 major (P8) and 72 minor (P4) proteins interacting with the membrane, with 2-nm spikes (P3) protruding from the membrane bilayer. Contrary to the *Caudovirales*, phage particles of the *Cystoviridae* have the capacity to fuse using a fusogenic protein in the envelope (as shown for ϕ6), resulting in a single enveloped particle containing two or more nucleocapsids [123].

The phage particle attaches to the host pilus using a spike protein (P3) which extends from the virion surface. Upon contact with its host, it uses a protein-triggered membrane fusion (P6) to penetrate the outer cell membrane with high efficacy [124]. Subsequently, a particle-associated endopeptidase (P5) degrades the peptidoglycan to allow further passage of the phage into the host. During this process, most of the nucleocapsid surface lattice protein is degraded in the cell, and only the core particle containing the genome and the RNA polymerase associated proteins are present within the cell. Interestingly, some of the entering core particles (25%) are recycled in progeny phage at the end infection cycle [125].

The genome replication and transcription mechanism in ϕ6 can clearly be linked to eukaryotic viruses with dsRNA and *Flaviviridae*. Indeed, replication/transcription in these viral species are catalyzed by an RNA-dependent RNA polymerase (P2 in ϕ6), which uses both ssRNA and dsRNA as a template. This process is dependent on the translocation of ssRNA into (packaging) and out of (transcription) the polymerase complex, and requires cognate nucleoside triphosphate (NTP), which acts as a primer to the initiation of the transcription reaction [126]. For this, a viral molecular motor (NTPase; P4) functionally similar to hexameric helices is present, actively translocating RNA during packaging and guiding RNA export. As such, P4 acts as a molecular switch between packaging and transcription, regulated by the NTPase activity within its polymerase core. A model for switching between RNA packaging, replication and semiconservative transcription is described in Kainov *et al.* [127].

Interestingly, this suggests that the RNA packaging system is highly similar to the nucleic translocation by hexameric helicases, as seen in cellular DNA helicase B and the T7 DNA helicase (gp4) [128]. The P4 hexamers are present as part of the procapsid, and can be seen on the 5-fold faces, which are pulled towards the inner core of the particle [129,130].

The particle-associated muralytic protein (P5) is not only responsible for peptidoglycan degradation during infection, but is also an essential enzyme for host lysis during progeny release. In addition, the small P10 protein, encoded on the M segment is necessary for successful lysis [131,132]. Indeed, P10 is predicted to contain a C-terminal transmembrane domain (amino acids 20–41), which may suggest a holin-like function for this protein, disrupting the inner cell membrane.

10.2.6.2 **Bacteriophages ϕ8, ϕ12 and ϕ13**

Bacteriophages ϕ8, ϕ12 and ϕ13 are the most distant relatives of ϕ6 within the *Cystoviridae* [122]. Although their genome segments are similar to that of ϕ6, no nucleotide similarity and limited protein similarity is observed. Differences in host range can be explained by the fact that ϕ8 and ϕ13 directly recognizes the rough LPS, extending its infection range to rough strains of the *Salmonella* family [133,134]. This is due to the presence of two host attachment proteins compared to a single protein in ϕ6. In addition, the host cell attachment proteins (P6, P3a–c) of ϕ12 and ϕ13 have marked similarity.

In addition, particle architecture differs significantly. The most significant difference is observed in the protein composition of the distinct protein shell (P8) around the procapsid for ϕ6, while in ϕ8, the P8 proteins are part of the membrane [133,135].

Despite these differences and the lack of sequence similarity the genomic organization is conserved. *In vivo* studies suggest that, at a molecular level, ϕ6 and ϕ8 have a similar, NTPase motor-based genomic packaging program [136]. In addition, the lysis mechanism can be expected to be conserved, since significant sequence similarity exists between the ϕ12 and ϕ6 lysis proteins.

10.2.7 *Pseudomonas*-Infecting *Inoviridae*

10.2.7.1 Introduction

The *Inoviridae* family encompasses the filamentous phages which contain a covalently closed ssDNA genome. The two *Inoviridae* phages infecting *P. aeruginosa* described are Pf1 (around 2000 nm × 7 nm) and Pf3 (around 700 × 7 nm) (Figure 10.1H). They roughly share the same morphology and gene arrangement with the well-studied Ff phages (f1, M13 and fd) infecting *E. coli* cells that harbor the F-factor. In Ff phages, the ssDNA genome is extended along the axis of the virion particle and encapsulated in a tube composed of around 2500 copies of the major capsid protein. A limited number of minor capsid proteins are localized at the extreme ends of the tubular capsid structure. These minor proteins are involved in phage assembly, infectivity and stability [7]. Filamentous phages infect bacteria bearing retractile pili and infection results in the continuous extrusion of progeny into the medium without causing cell lysis. This is a feature that distinguishes them from most other bacteriophages that accomplish their release through cell lysis.

The biology and life cycle of Ff phages has been documented in detail (reviewed in Ref. [7]). The *E. coli* phage Ff phage encodes 10 proteins, five of which are part of the virion structure (gp3, gp6, gp7, gp8 and gp9). Three are involved in phage DNA synthesis (gp2, gp10 and gp5), while two are involved in virion extrusion from the cell (gp1 and gp4) [137]. The propagation cycle of the Ff phage starts with adhesion of the g3p minor capsid protein to the tip of the bacterial F-pilus. After retraction of the pilus, gp3 interacts with the TolQRA complex in the cell wall of *E. coli*, leading to disassembly at the cytoplasmic membrane [138,139]. The DNA enters the cell and phage proteins are expressed. During DNA replication, phage proteins gp2 and gp10 and host proteins are required for the formation of the replicative form (RF) of the Ff genome. RF replication starts after expression of gp2 and stops once enough gp5 is produced. gp5 is a ssDNA-binding protein that strongly binds to ssDNA and consequently inhibits complementary strand synthesis and RF formation. With the exception of the formation of the gp5–ssDNA complex, all other assembly steps occur at or in a membrane [7]. This is in contrast with most bacteriophages, which assemble in the cytoplasm.

Upon formation of phage particles, gp5 is replaced by gp8 at the membrane [10]. gp1 spans the cytoplasmic membrane and presumably initiates genome packaging after interaction with the packaging signal. gp4 multimerizes in the outer membrane and forms a channel after interaction with gp1, allowing phage release [137,140].

10.2.7.2 Phages Pf1 and Pf3

The Ff phages belong to the class I filamentous phages, together with *E. coli* phage IKe. Phage IKe has 45% of DNA sequence similarity with the Ff phages [141]. In contrast, phage Xf of *Xanthomonas oryzae* and the best-studied filamentous phages infecting *Pseudomonas*, phages Pf1 and Pf3, are class II phages. This division is based on the structure of the protein coat in the phage particle, determined by X-ray diffraction [142]. Phage Pf3 specifically infects *P. aeruginosa* strain O harboring IncP1 plasmids [143], whereas phage Pf1 infects with *P. aeruginosa* strain K. Bacteriophage Pf1 interacts with the type IV PAK pilus, whereas bacteriophage Pf3 interacts with the conjugative RP4 pilus [144].

The genome of phage Pf1 consists of 7349 nucleotides and has 14 predicted ORFs [145]. This is large compared to the Pf3 genome which has 5833 nucleotides and nine predicted ORFs [143]. The G + C content of phage Pf1 (64.5%) is closer to that of its host, compared to the G + C content of Pf3 (45.4%). Pf1 and Pf3 do not share DNA nor significant amino acid sequence similarity with Ff filamentous (M13, fd and f1) and IKe of *E. coli*. There is, however, similarity in the overall genome organization and ORF size and succession of both Pf1 and Pf3 compared to other filamentous phage genomes (Figure 10.5). Therefore, their life cycle is thought to be similar to Ff phages infecting *E. coli*. Recently, amino acid similarity between Pf3 and filamentous phage B5 protein coat was reported [146]. Bacteriophage B5 is a filamentous phage infecting the Gram-positive bacterium *Propionibacterium feudenreichii*.

In contrast, the size and position of intergenic regions are less conserved and no large intergenic region is present in Ff phages (505 nucleotides), is present in Pf1 (reduced to 41 nucleotides) and Pf3 (reduced to 113 nucleotides) [143,145]. In Ff phages the large intergenic region contains the signals for the synthesis of (+) and (−) strands of DNA, the initiation of capsid formation (packaging signal and the termination of RNA synthesis [7]. Both Pf1 and Pf3 contain isolated stem–loop structures (hairpin structures), which may be involved in transcription termination, DNA replication and particle morphogenesis. Resemblance between predicted hairpin structures in Ff phages and Pf3 is observed [143]. Predicted promoter sequences present are related to known pseudomonad promoters and termination of transcription appears to be regulated by rho-independent terminators.

In phage Pf1, equivalents of Ff gene 8 and gene 5 are present [145], encoding the major capsid protein and the ssDNA-binding protein, respectively. Unlike its counterparts in Ff phages, phage IKe and phage Pf1, the major capsid protein of Pf3, is synthesized without a signal peptide [143,145], which is remarkable since upon release of M13 phages ssDNA-binding proteins are replaced at the cytoplasmic membrane by major capsid proteins. Chen *et al.* [147] showed that conversion of the membrane-associated Pf3 coat protein conformation into the transmembrane conformation is promoted by the membrane insertase protein YidC.

10.2.8 Prophage Sequences in *Pseudomonas* Bacterial Chromosomes

With the completion of various bacterial genome sequences the overall presence of prophage sequences in bacterial genomes became clear [2,3]. In the published genome sequence of *P. aeruginosa* PAO1, two prophage elements were reported [3,148]: (i) a prophage tail-derived bacteriocin, and (ii) a Pf1-like prophage of which the sequence in the bacterial host is remarkably flanked by a putative reverse transcriptase and an integrase. These genes are not present in the filamentous phage genome. The published genome sequence of *P. putida* contains four prophage sequences, among which two appear chimers of sequence modules of (at least two) phages, whereas the two others show homology with T7-like *Podoviridae* and phage D3, respectively [3,149].

Acknowledgments

The research contributions of the authors to this chapter were performed at the Laboratory of Gene Technology of the KULeuven (Belgium), headed by Guido Volckaert. Their bacteriophage research program runs in close collaboration with Vadim V. Mesyanzhinov and Viktor N. Krylov of the Shemyakin-Ovchinnikov Institute of Bioorganic Chemistry (Moscow) and the State Institute for Genetics and Selection of Industrial Microorganisms (Moscow), respectively. We gratefully acknowledge H.-W. Ackermann and V. V. Mesyanzhinov for providing phage electron micrographs and related information. We thank Pieter-Jan Ceyssens and Guido Volckaert for critical reading and constructive suggestions. K.H. and R.L. are postdoctoral fellows of the Fonds voor Wetenschappelijk Onderzoek-Vlaanderen, Belgium.

References

1 Guttman, B., Raya, R. and Kutter, E. (2005) Basic phage biology, In *Bacteriophages: Biology and Applications* (eds E. Kutter and A., Sulakvelidze), CRC Press, Boca Raton, FL, pp. 29–66.

2 Casjens, S., Hatfull, G. and Hendrix, R. (1992) *Semin Virol*, **3**, 383–397.

3 Canchaya, C., Proux, C., Fournous, G., Bruttin, A. and Brüssow, H. (2003) *Microbiol Mol Biol Rev*, **67**, 238–276.

4 Mann, N.H., Cook, A., Millard, A., Bailey, S. and Clokie, M. (2003) *Nature*, **424**, 741.

5 Letellier, L., Boulanger, P., Plançon, L., Jacquot, P. and Santamaria, M. (2004) *Front Biosci*, **9**, 1228–1339.

6 Mesyanzhinov, V.V. (2004) *Adv Virus Res*, **63**, 287–352.

7 Makowski, L. and Russel, M. (1997) Structure and assembly of filamentous

bacteriophages, in *Structural Biology of Viruses* (eds W. Chiu, R.M. Burnett and R. Garcea), Oxford University Press, Oxford, pp. 352–380.

8 Weigler, W. and Seitz, H. (2006) *FEMS Microbiol Rev,* **30**, 321–381.

9 Nakai, H., Doseeva, V. and Jones, J.M. (2001) *Proc Natl Acad Sci USA,* **98**, 8247–8254.

10 Pratt, D. and Erdahl, W.S. (1968) *J Mol Biol,* **37**, 181–200.

11 Linderoth, N.A., Ziermann, R., Haggård-Liungquist, E., Christie, G.E. and Calendar, R. (1991) *Nucleic Acids Res,* **19**, 7207–7214.

12 Sippy, J. and Feiss, M. (2004) *Mol Microbiol,* **52**, 501–513.

13 Chung, Y.B. and Hinkle, D.C. (1990) *J Mol Biol,* **216**, 911–926.

14 Son, M., Watson, R.H. and Serwer, P. (1993) *Virology,* **196**, 282–289.

15 Gründling, A., Manson, M.D. and Young, R. (2001) *Proc Natl Acad Sci USA,* **98**, 9348–9352.

16 Loessner, M.J., Maier, S.K., Daubek-Puza, H., Wendlinger, G. and Scherer, S. (1997) *J Bacteriol,* **179**, 2845–2851.

17 São-José, C., Parreira, R., Vieira, G. and Santos, M.A. (2000) *J Bacteriol,* **182**, 5823–5831.

18 Xu, X., Kashima, O., Saito, A., Azakami, H. and Kato, A. (2004) *Biosci Biotechnol Biochem,* **68**, 1273–1278.

19 Xu, M., Arulandu, A., Struck, D.K., Swanson, S., Sacchettini, J.C. and Young, R. (2005) *Science,* **307**, 113–117.

20 Young, I., Wang, I. and Roof, W.D. (2000) *Trends Microbiol,* **8**, 120–128.

21 Matthews, R.E.F. (1983) The history of viral taxonomy, in *A Critical Appraisal of Viral Taxonomy* (ed. R.E.F. Mattews), CRC Press, Boca Raton, FL, pp. 1–35.

22 Ackermann, H.-W. (2007) *Arch Virol,* **152**, 227–243.

23 Susskind, M. and Botstein, D. (1978) *Microbiol Rev,* **42**, 385–413.

24 Lawrence, J.G., Hatfull, G.F. and Hendrix, R.W. (2002) *J Bacteriol,* **184**, 4891–4905.

25 Rohwer, F. and Edwards, R. (2002) *J Bacteriol,* **184**, 4529–4535.

26 Ackermann, H.-W. (2005) Bacteriophage classification, in *Bacteriophages: Biology and Applications* (eds E. Kutter and A., Sulakvelidze), CRC Press, Boca Raton, FL, pp. 67–89.

27 Ackermann, H.-W. and DuBow, M.S. (1987) Phages of *Pseudomonas* and related bacteria, in *Viruses of Prokaryotes, Volume II, Natural Groups of Bacteriophages,* CRC Press, Boca Raton, FL, pp. 116–159.

28 Shaburova, O.V., Hertveldt, K., de la Cruz, D.M., Krylov, S.V., Pleteneva, E.A., Burkaltseva, M.V., Lavigne, R., Volcaert, G. and Krylov, V.N. (2006) *Genetika,* **42**, 1065–1074.

29 Sillankorva, S., Oliveira, R., Vieira, M.-J., Sutherland, I. and Azeredo, J. (2004) *FEMS Microbiol Lett,* **241**, 13–20.

30 Park, S.C., Shimamura, I., Fukunaga, M., Mori, K. and Nakai, T. (2000) *Appl Environ Microbiol,* **66**, 1416–1422.

31 Serwer, P., Hayes, S.J., Zaman, S., Lieman, K., Rolando, M. and Hardies, S.C. (2004) *Virology,* **329**, 412–424.

32 Lindberg, R.B., Latta, R.L., Brame, R.C. and Moncrief, J.A. (1964) *Bact Proc,* 81.

33 Lindberg, R.B. and Latta, R.L. (1974) *J Infect Dis,* **130**, S33–S43.

34 Ackermann, H.-W., Cartier, C., Slopek, S. and Vieu, J.-F. (1988) *Ann Inst Pasteur/Virol,* **139**, 389–404.

35 Sharp, R., Jansons, I.S., Gertman, E. and Kropinski, A.M. (1996) *Gene,* **177**, 47–53.

36 Hoang, T.T., Kutchma, A.J., Becher, A. and Schweizer, H.P. (2000) *Plasmid,* **43**, 59–72.

37 Darzins, A. and Casadaban, M.J. (1989) *J Bacteriol,* **17**, 3909–3916.

38 Sulakvelidze, A. and Kutter, E. (2005) Bacteriophage therapy in humans, in *Bacteriophages: Biology and Applications* (eds E. Kutter and A. Sulakvelidze), CRC Press, Boca Raton, FL, pp. 381–436.

39 Soothill, J.S. (1992) *J Med Microbiol,* **37**, 258–261.

40 Nakai, T. and Park, S.C. (2002) *Res Microbiol*, **153**, 13–18.
41 Park, S.C., Shimamura, I., Fukunaga, M., Mori, K. and Nakai, T. (2000) *Appl Environ Microbiol*, **66**, 1416–1422.
42 Soothill, J.S. (1994) *Burns*, **20**, 209–211.
43 Maniloff, J. and Ackermann, H.-W. (1998) *Arch Virol*, **143**, 2051–2063.
44 Maniloff, J. and Ackermann, H.-W. (2000), in: van Regenmortel, M.H.V., Fauquet, C.M., Bishop, D.H.L., Carstens, E.B., Estes, M.K., Lemon, S.M., Maniloff, J., McGeoch, D.J., Pringle, C.R., Wickner R.B. (Eds.), *Virus Taxonomy*, Seventh Report of the International Committee on Taxonomy of Viruses, Academic Press, San Diego.
45 Hertveldt, K., Lavigne, R., Pleteneva, E., Sernova, N., Kurochkina, L., Korchevskii, R., Robben, J., Mesyanzhinov, V., Krylov, V.N. and Volckaert, G. (2005) *J Mol Biol*, **354**, 536–545.
46 Mesyanzhinov, V.V., Robben, J., Grymonprez, B., Kostyuchenko, V.A., Burkal'tseva, M.V., Sykilinda, N.N., Krylov, V.N. and Volckaert, G. (2002) *J Mol Biol*, **317**, 1–19.
47 Nakayama, K., Kanaya, S., Ohnishi, M., Terawaki, Y. and Hayashi, T. (1999) *Mol Microbiol*, **31**, 399–419.
48 Hayashi, T., Baba, T., Matsumoto, H. and Terawaki, Y. (1990) *Mol Microbiol*, **4**, 1703–1709.
49 Baltch, A.L., Smith, R.P., Franke, M., Ritz, W., Michelsen, P., Bopp, L. and Lutz, F. (1994) *Toxicon*, **32**, 27–34.
50 Kageyama, M. (1964) *J Biochem*, **55**, 49–53.
51 Yokota, S., Hayashi, T. and Matsumoto, H. (1994) *J Bacteriol*, **176**, 5262–5269.
52 Hayashi, T., Matsumoto, H., Ohnishi, M. and Terawaki, Y. (1993) *Mol Microbiol*, **7**, 657–667.
53 Wang, Z., Xiong, G. and Lutz, F. (1995) *Mol Gen Genet*, **246**, 72–79.
54 Yu, A., Bertani, L.E. and Haggard-Ljungquist, E. (1989) *Gene*, **80**, 1–11.
55 Ziermann, R., Bartlett, B., Calendar, R. and Christie, G.E. (1994) *J Bacteriol*, **176**, 4974–4984.
56 Portelli, R., Dodd, I.B., Xue, Q. and Egan, B. (1998) *Virology*, **248**, 117–130.
57 Krylov, V.N., Dela Cruz, D.M., Hertveldt, K. and Ackermann, H.W. (2007) *Arch Virol*, **152**, 1955–1959.
58 Krylov, V.N. (2001) *Genetika*, **7**, 869–87.
59 Krylov, V.N., Smirnova, T.A., Minenkova, I.B., Plotnikova, T.G., Zhazikov, I.Z. and Khrenova, E.A. (1984) *Can J Microbiol*, **30**, 758–762.
60 Fokine, A., Kostyuchenko, V.A., Efimov, A.V., Kurochkina, L.P., Sykilinda, N.N., Robben, J., Volckaert, G., Hoenger, A., Chipman, P.R., Battisti, A.J., Rossmann, M.G. and Mesyanzhinov, V.V. (2005) *J Mol Biol*, **352**, 117–24.
61 Olson, N.H., Gingery, M., Eiserling, F.A. and Backer, T.S. (2001) *Virology*, **279**, 385–391.
62 Fokine, A., Chipman, P.R., Leiman, P.G., Mesyanzhinov, V.V., Rao, V.B. and Rossmann, M.G. (2004) *Proc Natl Acad Sci USA*, **101**, 6003–6008.
63 Miller, E.S., Kutter, E., Mosig, G., Arisaka, F., Kunisawa, T. and Rüger, W. (2003) *Microbiol Mol Biol Rev*, **67**, 86–156.
64 Akhverdyan, V.Z., Khrenova, E.A., Bogush, V.G., Gerasimova, T.V., Kirsanov, N.B. and Krylov, V.N. (1984) *Genetika*, **20**, 1612–1619.
65 Holloway, B.W., Egan, J.B. and Monk, M. (1960) *Aust J Exp Biol*, **39**, 321–330.
66 Budzik, J.M., Rosche, W.A., Rietsch, A. and O'Toole, G.A. (2004) *J Bacteriol*, **186**, 3270–3273.
67 Roncero, C., Darzins, A. and Casadaban, M.J. (1990) *J Bacteriol*, **172**, 1899–1904.
68 Campbell, A.M. (1962) *Adv Genet*, **11**, 101.
69 Morgan, G.J., Hatfull, G.F., Casjens, S. and Hendrix, R.W. (2002) *J Mol Biol*, **317**, 337–359.
70 Birge, E.A. (1994) *Bacterial and Bacteriophage Genetics*, Springer, Berlin.
71 Braid, M.D., Silhavy, J.L., Kitts, D.L., Cano, R.J. and Howe, M.M. (2004) *J Bacteriol*, **186**, 6560–6574.

72 Wang, P.W., Chu, L. and Guttman, D.S. (2004) *J Bacteriol*, **186**, 400–410.

73 Mit'kina, L.N. and Krylov, V.N. (1999) *Russ J Genet*, **35**, 1015–1022.

74 Marrs, C.F. and Howe, M.M. (1990) *Virology*, **174**, 192–203.

75 Salmon, K.A., Freedman, O., Ritchings, B.W. and DuBow, M.S. (2000) *Virology*, **272**, 85–97.

76 Autexier, C., Wragg-Legare, S. and DuBow, M.S. (1991) *Biochim Biophys Acta*, **1088**, 147–150.

77 Summer, E.J., Gonzalez, C.F., Carlisle, T., Mebane, L.M., Cass, A.M., Savva, C.G., LiPuma, J.J. and Young, R. (2004) *J Mol Biol*, **340**, 49–65.

78 Mathee, K. and Howe, M. (1990) *J Bacteriol*, **175**, 5314–5323.

79 Margolin, W., Rao, G. and Howe, M.M. (1989) *J Bacteriol*, **171**, 2003–2018.

80 Kuzio, J. and Kropinski, A.M. (1983) *J Bacteriol*, **155**, 203–212.

81 Newton, G.J., Daniels, C., Burrows, L.L., Kropinski, A.M., Clarke, A.J. and Lam, J.S. (2001) *Mol Microbiol*, **39**, 1237–1247.

82 Gilakjan, Z.A. and Kropinski, A.M. (1999) *J Bacteriol*, **181**, 7221–7227.

83 Cavenagh, M.M. and Miller, R.V. (1986) *J Bacteriol*, **165**, 448–452.

84 Kropinski, A.M. (2000) *J Bacteriol*, **182**, 6066–6074.

85 Lee, L.F. and Boezi, J.A. (1966) *J Bacteriol*, **92**, 1821–1827.

86 Chang, P.L. and Yen, T.F. (1985) *Appl Environ Microbiol*, **50**, 1545–1547.

87 Korsten, K.H., Tomkiewicz, C. and Hausmann, R. (1979) *J Gen Virol*, **43**, 57–73.

88 Kovalyova, I.V. and Kropinski, A.M. (2003) *Virology*, **311**, 305–315.

89 Jolly, J.F. (1979) *J Virol*, **30**, 771–776.

90 Lavigne, R., Sun, W.D. and Volckaert, G. (2004) *Bioinformatics*, **20**, 629–635.

91 Burkal'tseva, M.V., Pleteneva, E.A., Shaburova, O.V., Kadykov, V.A. and Krylov, V.N. (2006) *Genetika*, **42**, 33–38.

92 Ceyssens, P.J., Lavigne, R., Mattheus, W., Chibeu, A., Hertveldt, K., Mast, J., Robben, J. and Volckaert, G. (2006) *J Bacteriol*, **188**, 6924–6931.

93 Lavigne, R., Noben, J.P., Hertveldt, K., Ceyssens, P.J., Briers, Y., Dumont, D., Roucourt, B., Krylov, V.N., Mesyanzhinov, V.V., Robben, J. and Volckaert, G. (2006) *Microbiology*, **152**, 529–534.

94 Lavigne, R., Briers, Y., Hertveldt, K., Robben, J. and Volckaert, G. (2004) *Cell Mol Life Sci*, **61**, 2753–2759.

95 Lavigne, R., Roucourt, B., Hertveldt, K. and Volckaert, G. (2005) *Protein Pept Lett*, **12**, 645–648.

96 Briers, Y., Lavigne, R., Plessers, P., Hertveldt, K., Hanssens, I., Engelborghs, Y. and Volckaert, G. (2006) *Cell Mol Life Sci*, **63**, 1899–1905.

97 Roucourt, B., Chibeu, A., Lecoutere, E., Lavigne, R., Volckaert, G. and Hertveldt, K. (2007) *Arch Virol*, **152**, 1467–1475.

98 Tan, Y., Zhang, K., Rao, X., Jin, X., Huang, J., Zhu, J., Chen, Z., Hu, X., Shen, X., Wang, L. and Hu, F. (2007) *Cell Microbiol*, **9**, 479–91.

99 Freitas-Vieira, A., Anes, E. and Moniz-Pereira, J. (1998) *Microbiology*, **144**, 3397–3406.

100 McShan, W.M. and Ferretti, J.J. (1997) *J Bacteriol*, **179**, 6509–6511.

101 Campbell, A. (2003) *Res Microbiol*, **154**, 277–282.

102 Krylov, V.N., Tolmachova, T.O. and Akhverdian, V.Z. (1993) *Arch Virol*, **131**, 141–151.

103 Pemberton, J.M. (1973) *Virology*, **55**, 558–560.

104 Bradley, D.E. (1974) *Virology*, **58**, 149–163.

105 Miller, R.V., Pemberton, J.M. and Clark, A.J. (1977) *J Virol*, **22**, 844–847.

106 Byrne, M. and Kropinksi, A.M. (2005) *Gene*, **346**, 187–194.

107 Fiers, W., Contreras, R., Duerinck, F., Haegeman, G., Iserentant, D., Merregaert, J., Min, J.W., Molemans, F., Raeymaekers, A., Van den Berghe, A., Volckaert, G. and Ysebaert, M. (1976) *Nature*, **260**, 500–507.

108 Drake, J. (1993) *Proc Natl Acad Sci USA*, **90**, 4171–4175.

109 Bradley, D.E. (1966) *J Gen Microbiol*, **45**, 83–96.

110 Olsthoorn, R.C.L., Garde, G., Dayhuff, T., Atkins, J.F. and Duin, J. (1995) *Virology*, **206**, 611–625.

111 Bradley, D.E. (1972) *J Gen Microbiol*, **72**, 303–319.

112 Zinder, N.E. (ed.) (1975) *RNA Phages*, Cold Spring Harbor Laboratory Press, Cold Spring Harbor, NY.

113 Tars, K., Fridborg, K., Bundule, M. and Liljas, L. (2000) *Acta Crystallogr*, **D56**, 398–405.

114 Valegård, K., Murray, J.B., Stonehouse, N.J., van den Worm, S., Stockley, P.G. and Liljas, L. (1997) *J Mol Biol*, **270**, 724–738.

115 Kajitani, M., Kato, A., Wada, A., Inokuchi, Y. and Ishihama, A. (1994) *J Bacteriol*, **176**, 531–534.

116 van Himbergen, J., Van Geffen, B. and Van Duin, J. (1993) *Nucleic Acids Res*, **21**, 1713–1717.

117 Adhin, M.R. and van Duin, J. (1990) *J Mol Biol*, **213**, 811–818.

118 Young, R. (1993) *Microbiol Rev*, **56**, 431–481.

119 Vidaver, A.K., Koski, R.K. and Van Etten, J.L. (1973) *J Virol*, **11**, 799–805.

120 Semancik, J.S., Vidaver, A.K. and Van Etten, J.L. (1973) *J Mol Biol*, **78**, 617–25.

121 Romantschuk, M. and Bamford, D.H. (1985) *J Gen Virol*, **66**, 2461–2469.

122 Mindich, L., Qiao, X., Qiao, J., Onodera, S., Romantschuk, M. and Hoogstraten, D. (1999) *J Bacteriol*, **181**, 4505–4508.

123 Bamford, D.H., Romantschuk, M. and Somerharju, P.J. (1987) *EMBO J*, **6**, 1467–1473.

124 Olkkonen, V.M. and Bamford, D.H. (1989) *Virology*, **171**, 229–238.

125 Romantschuk, M., Olkkonen, V.M. and Bamford, D.H. (1988) *EMBO J*, **7**, 1821–1829.

126 Butcher, S.J., Grimes, J.M., Makeyev, E.V., Bamford, D.H. and Stuart, D.I. (2001) *Nature*, **410**, 235–240.

127 Kainov, D.E., Lisal, J., Bamford, D.H. and Tuma, R. (2004) *Nucleic Acids Res*, **32**, 3515–21.

128 Patel, S.S. and Picha, K.M. (2000) *Annu Rev Biochem*, **69**, 651–697.

129 de Haas, F., Paatero, A.O., Mindich, L., Bamford, D.H. and Fuller, S.D. (1999) *J Mol Biol*, **294**, 357–372.

130 Butcher, S.J., Dokland, T., Ojala, P.M., Bamford, D.H. and Fuller, S.D. (1997) *EMBO J*, **16**, 4477–4487.

131 Mindich, L. and Lehman, J. (1979) *J Virol*, **30**, 489–496.

132 Johnson, M.D. and Mindich, L. (1994) *J Virol*, **68**, 2331–2338.

133 Hoogstraten, D., Qiao, X., Sun, Y., Hu, A., Onodera, S. and Mindich, L. (2000) *Virology*, **272**, 218–24.

134 Qiao, X., Qiao, J., Onodera, S. and Mindich, L. (2000) *Virology*, **275**, 218–224.

135 Jäälinoja, H.T., Huiskonen, J.T. and Butcher, S.J. (2007) *Structure*, **15**, 157–167.

136 Qiao, J., Qiao, X. and Mindich, L. (2005) *BMC Microbiol*, 5.

137 Horabin, J.L. and Webster, R.E. (1988) *J Biol Chem*, **263**, 11575–11583.

138 Click, E.M. and Webster, R.E. (1997) *J Bacteriol*, **179**, 6464–6471.

139 Riechmann, L. and Holliger, P. (1997) *Cell*, **90**, 351–360.

140 Russel, M. (1994) *Science*, **265**, 612–614.

141 Peeters, B.P.H., Peters, R.M., Schoenmakers, J.G.G. and Konings, R.N.H. (1985) *J Mol Biol*, **181**, 27–39.

142 Marvin, D.A., Pigram, W.J., Wiseman, R.L. and Wachtel, E.J. (1974) *J Mol Biol*, **88**, 581–600.

143 Luiten, R.G., Putterman, D.G., Schoenmakers, J.G., Konings, R.N. and Day, L.A. (1985) *J Virol*, **56**, 268–76.

144 Holland, S.J., Sanz, C. and Perham, R.N. (2006) *Virology*, **345**, 540–548.

145 Hill, D.F., Short, N.J., Perham, R.N. and Petersen, G.B. (1991) *J Mol Biol*, **218**, 349–64.

146 Chopin, M.-C., Rouault, A., Ehrlich, S.D. and Gautier, M. (2002) *J Bacteriol*, **184**, 2030–2033.

147 Chen, M., Xie, K., Yuan, J., Yi, L., Facey, S.J., Pradel, N., Wu, L.F., Kuhn, A. and Dalbey, R.E. (2005) *Biochemistry*, **44**, 10741–10749.

148 Stover, C.K., Pham, X.Q., Erwin, A.L., Mizoguchi, S.D., Warrener, P., Hickey, M.J., Brinkman, F.S.L., Hufnagle, W.O., Kowalik, D.J., Lagrou, M., Garber, R.L., Goltry, L., Tolentino, E., Westbrock-Wadman, S., Yuan, Y., Brody, L.L., Coulter, S.N., Folger, K.R., Kas, A., Larbig, K., Lim, R., Smith, K., Spencer, D., Wong, G.K.-S., Wu, Z., Paulsen, I.T., Reizer, J., Saier, M.H., Hancock, R.E.W., Lory, S. and Olson, M.V. (2000) *Nature*, **406**, 959–964.

149 Nelson, K.E., Weinel, C., Paulsen, I.T., Dodson, R.J., Hilbert, H., Martins dos Santos, V.A., Fouts, D.E., Gill, S.R., Pop, M., Holmes, M., Brinkac, L., Beanan, M., DeBoy, R.T., Daugherty, S., Kolonay, J., Madupu, R., Nelson, W., White, O., Peterson, J., Khouri, H., Hance, I., Chris Lee, P., Holtzapple, E., Scanlan, D., Tran, K., Moazzez, A., Utterback, T., Rizzo, M., Lee, K., Kosack, D., Moestl, D., Wedler, H., Lauber, J., Stjepandic, D., Hoheisel, J., Straetz, M., Heim, S., Kiewitz, C., Eisen, J.A., Timmis, K.N., Düsterhöft, A., Tümmler, B. and Fraser, C.M. (2002) *Environ Microbiol*, **12**, 799–808.

11
Pseudomonas Plasmids

Christopher M. Thomas, Anthony S. Haines, Irina A. Kosheleva, and Alexander Boronin

11.1
Introduction

Our understanding of the genus *Pseudomonas* depends on knowledge of not only the chromosomal sequences of its major constituent species, but also the mobile genetic elements (MGEs) that carry additional genetic material within and between species. Plasmids are the most obvious MGEs since they replicate autonomously in the cytoplasm of their host and can be physically separated by a variety of techniques. Their autonomous replication allows establishment in a new host without the need for any sort of recombination so long as the plasmid's replication machinery is functional. Plasmids can vary in size from the smallest, encoding simply the ability to replicate (less than 1 kb), to the largest megaplasmids (greater than 600 kb), which can potentially encode enough genetic information for a whole bacterial cell. The genome consists of a scaffold of functions present in all members of a species or genus interspersed with extra DNA present in some strains but not others, which in many is due to acquisition of horizontally transferred DNA. The DNA associated with such elements across the many strains that comprise the species can equal or exceed another whole chromosome. The DNA that is accessible to the whole species due to association with MGEs, but only present in a few strains, is defined as the horizontal gene pool (HGP) [1]. This chapter deals with the plasmid-borne fraction of the HGP of the genus *Pseudomonas*.

Many plasmids are associated with important phenotypic properties, such as the ability to utilize a pollutant as a carbon source, to grow in the presence of one or more inhibitory substances, to invade a particular host or to cause disease [2]. The logic underlying which properties are plasmid associated has been discussed extensively [3,4]. However, since most genes can be found in both plasmid and nonplasmid locations, the phenotypic properties that a plasmid carries should not be seen as permanent or exclusive. Similarly, the phenotypic characteristics of plasmids do not generally relate to a constant parameter in the environment, but rather to variable factors, that the bacteria may or may not encounter. Alternatively it may be a property

Pseudomonas. Model Organism, Pathogen, Cell Factory. Edited by Bernd H.A. Rehm

ISBN: 978-3-527-31914-5

that existed previously in a different context and is spreading through a new microbial population or community where the phenotype may provide an advantage. Plasmid-associated genes also tend to exist in self-sufficient packages, so that they can confer their phenotype in a variety of hosts without depending very critically on specific interactions with the machinery of their host except for replication and gene expression. In some cases this means that the genetic segment conferring the trait is quite large (e.g. encoding a catabolic pathway).

Plasmids, by definition, contain a replication and maintenance system that makes them physically (but not necessarily biochemically) independent of the host chromosome for replication and stable inheritance. They can therefore carry their genetic cargo into the cytoplasm of any species in which they are able to replicate autonomously and thus avoid the constraints of homologous recombination. Each type of plasmid will have a certain host range where these generalizations apply. However, one can also have a situation in which the host range of the plasmid's transfer system and its replication/maintenance system are not identical, so that the plasmid may have the ability to spread into a host where establishment depends on some sort of integration into the chromosome, i.e. the plasmid naturally acts as a suicide vector. Indeed, screening available chromosomal sequences reveals the remnants of plasmid integration events demonstrating that integration and excision of plasmids can be an often repeated process [5].

The cataloguing of *Pseudomonas* plasmids has not proceeded uniformly. This is partly due to the complexity of functional incompatibility tests when plasmids and host already carry many selectable markers and the number of incompatibility groups grows. In addition, the replacement of functional incompatibility tests by hybridization to probe sequences in order to place plasmids in sequence families has reduced the need to determine incompatibility type, although for environmental studies, knowing which plasmids compete directly with each other is still of major importance. The ideal would be to know both sequence family and incompatibility grouping. It is pleasing to note that this is being recognized and in many cases incompatibility tests are being performed now to check whether plasmids related by sequence are compatible or incompatible. This chapter provides an update on a previous review [6]. That review has been used as the basis, but in addition to extensive revision and updating of the text, we have deleted sections where there has been little progress to report.

11.2
Plasmid Groups in *Pseudomonas*

Table 11.1 lists examples of plasmids from known groups in *Pseudomonas* species as well as plasmids that are documented as able to replicate in *Pseudomonas* whose replication systems are sequenced and which may represent the archetypes of new plasmid groups. The list includes plasmids across the whole size range – from pPST1 at 1.4 kb [7] to the megaplasmids of the pQBR series, the first of which to be sequenced has a size of 425 kb [8]. Although some plasmids carry no known phenotypic determinants (termed "cryptic" plasmids), others carry one or more traits such as antibiotic resistance, heavy metal resistance or degradative functions. There do not appear to be reported cases of naturally occurring plasmids carrying both antibiotic resistance and degradative determinants, but mercury resistance is certainly found on both sorts of plasmids and it would be surprising if there are no examples carrying both types of function, given the past widespread use of antibiotics in agricultural contexts.

The primary basis of classification is the plasmid incompatibility groups defined in the 1970s [9], and subsequently expanded slightly by work in the laboratories of George Jacoby and Alexander Boronin. At present there are 14 listed plasmid incompatibility groups in *Pseudomonas* [10,11], but Table 11.1 includes additional groups or individual plasmids that have not yet been given an IncP number. We propose not to add any more group numbers until the relation between unclassified plasmids and the defined groups is known. The basis for the main classification is still the functional incompatibility test for displacement of one plasmid by another where they carry different selectable or screenable markers. This is obviously a labor-intensive activity compared to hybridization or polymerase chain reaction (PCR) screening with specific primers, but such genetic tools are not yet available for all *Pseudomonas* plasmids. If the range of plasmids was the same across all bacterial genera then the tools available in one group of organisms should be applicable to others. However, in practice, tools developed for classification of *Escherichia coli* plasmids by isolation of minireplicons and DNA sequencing [12] are of only limited applicability to *Pseudomonas*. Apart from for plasmids with a known broad host range, the hybridization tools from the Couturier collection have largely failed to be useful for classifying plasmids from soil, marine or river samples [13–21]. This suggests that the plasmids of the *Pseudomonas* plasmid groups either belong to different phylogenetic families from those found in *E. coli* or are generally too distantly related to be detectable by cross-hybridization. Sequencing of plasmids from nonclinical environments, since hospitals tend to be the source of the plasmids for the Couturier collection, shows that most plasmids carry a replicon recognizable by its similarity to previously characterized systems, but often only with a low degree of sequence identity [22]. Therefore, the second of the two possibilities (that *Pseudomonas* plasmids generally are related, but only distantly, to replicons found in *E. coli*) seems more likely to be correct. There is thus a need for more nucleotide sequences of plasmids found in *Pseudomonas* to underpin screening and classification.

Currently complete sequences are available for archetypes of six *Pseudomonas* plasmid Inc groups: IncP-1; IncP-3; IncP-4, IncP-6, IncP-7 and IncP-9 (Table 11.1). For IncP-1 plasmids there are now more than 20 complete sequences. While the majority of these plasmids fall into the IncP-1β subgroup there appear to be five distinct divisions in the family, with the recently identified γ, δ and ε subgroups being represented by single plasmids pQKH54 [23] and pEST4011 [24], and the pair of plasmids pKJK5 [25] and pEMT3 [22], respectively.

For IncP-3 plasmids six complete sequences have become available recently. The replicon sequence from RA1 was identified some years ago when it was also demonstrated that the *E. coli* incompatibility groups IncA/C and IncC were united

Table 11.1 Representative plasmids of *Pseudomonas* incompatibility groups including all sequenced plasmids from these groups as well as other sequenced plasmids from *Pseudomonas*.

Incompatibility or homology group and group attributes	Subgroup	Plasmid	Size (kb)[e]	Phenotype; comments	Reference
IncP-1 (IncP)	IncP-1α	RK2 (RP1, RP4)	60	Cb Km Tc	[45][d]
Integrated regulation system					
Broad host range	IncP-1α	pTB11	69	Tc Km Ap Cc Ci Gm Km Sp Sm Tc Tb Tp	[149][d]
Conjugative					
Group divided into α, β, γ, δ, ε and possibly further subgroups	IncP-1α	pBS228	89	Ap Tc Sm Su Sp Tp	[150][d]
	IncP-1β	R751	53	Tp	[47][d]
	IncP-1β	pA1	47	cryptic	[157][d]
	IncP-1β	pA81	98	chlorobenzoic acid degradation; IncP-1β subgroup	[158][d]
	IncP-1β	pADP-1	96	atrazine catabolism	[160][d]
Acidovorax	IncP-1β	pAOVO02	64	unpublished	GenBank: CP000541
	IncP-1β	pB2	61	Ap Cm Sm Sp Su Tc	[153][d]
	IncP-1β	pB3	56	Ap Cm Sm Sp Su Tc	[153][d]
	IncP-1β	pB4	79	Ax Em Sm	[154][d]
	IncP-1β	pB8	57	Ax Sm Sp Su	[155][d]
	IncP-1β	pB10	64.5	Ap Sm Su Tc Hg	[143][d]
[Plasmid 1 from *Burkholderia cepacia*]	IncP-1β	pBC1	44	unpublished	GenBank: CP000443
	IncP-1β	pBP136	41	cryptic	[156][d]

	IncP-1β	pJP4	88	2,4-D degradation	[151][d]
	IncP-1β	pTP6	53	Hg	[58][d]
	IncP-1β	pUO1	67	haloacetate catabolism	[152][d]
	IncP-1γ	pQKH54	73	Hg	[23][d]
	IncP-1δ	pEST4011	80	2,4-D degradation	[24][d]
	IncP-1ε (proposed)	pKJK5	54		[159][d]
	IncP-1ε (proposed)	pEMT3		2,4-D degradation	[22]
IncP-2		pMG1	450	Bo Gm Hg Sm Su Te Uv; bacteriophage inhibition, pyocin inactivation	[161,162]
Conjugative, narrow host range					
Typically very large		pBS271	450	ε-caprolactam degradation; Te	[163]
All confer resistance to tellurite					
		pOZ176	450	bla (IMP-9) Te	[148]
IncP-3 (IncA/C)		pBS73	88	Cm Hg Km Sm Su Tc	[11]
Broad host range		RA1	130	Su Tc	[26,164]
Conjugative					
		R667	168	Ap Su	[165]
		pIP1202	183	Ap Cm Km Sm Sp Tc Mc Su	[27]
		pSN254	176	Ap Cm Sm Su Tc Cf Cx Gm	[27]
		pYR1	158	Sm Tc Tp Su	[27]
		pP91278	132		unpublished; GenBank: AB277724
IncP-4 (IncQ)	IncQ1α	RSF1010	8.7	Sm Su	[166][d]
Broad host range	IncQ1α	pIE1107	8.5	St Km	[167][d]
Mobilizable	IncQ1α	pIE1115	10.7	Sm Su	[55]
Strand displacement replication	IncQ1β	pDN1	5.1	cryptic	[142][d]

(Continued)

Table 11.1 (*Continued*)

Incompatibility or homology group and group attributes		Plasmid	Size (kb)[e]	Phenotype; comments	Reference
Small size	IncQ1β	pCCK381	10.9	Fm	[168][d]
High copy number	IncQ1γ	pIE1130	11	Cm Km Sm Su	[55][d]
Multiple essentially compatible subgroups[c]	IncQ2α IncQ2β IncQ2γ	pPTF-FC2	12	carries glutathioredoxin and MerR homologs	[169,170]
		pTC-F14	14	unknown	[171][d]
		pGNB2	8.5	Qn	[172][d]
IncP-5		Rms163	220	Bo Cm Su Tc	[9,173]
Large, conjugative, narrow host range					
IncP-6 (IncG/U)		Rms149	57	Cb Gm Sm Su; mobilizable	[174,175][d]
Broad host range					
		pRSB105	57	Em	[176][d]
		pFBAOT6	85	Tc (isolated from *Aeromonas punctata*)	[177][d]
IncP-7		Rms148	180	Sm	[9]
Narrow host range		pCAR1[a] pWW53 pND6–1	199	carbazole/dioxin metabolism	[73][d]
			108	toluene/xylene degradation	[146][d]
			102	naphthalene degradation	[178][d]
IncP-8		FP2	90	Hg Pm	[179]
IncP-9		pWW0 (TOL)	115	toluene catabolism; conjugative	[69][d]

Unstable in *P. aeruginosa* and *non-Pseudomonas* species at 43°C	pM3	75	Sm Tc; conjugative	[68]
	NAH7	82	naphthalene catabolism	[180][d]
	pDTG1	83	naphthalene catabolism; conjugative	[87][d]
IncP-10	R91	48	Cb; conjugative	[181,182]
Narrow host range				
IncP-11	RP1-1 (R18-1)		Cb; no detectable extrachromosomal DNA	[183]
	R151	33	Cb Gm Km Sm Su Tb	[184,185]
IncP-12	R716	165	Sm Hg	[186]
Narrow host range				
IncP-13	pMG25	99	Bo Cb Cm Gm Km Sm Su Tb	[10]
Narrow host range	pQM1	252	Hg Uv; conjugative	[52,187]
IncP-14	pBS222	17	Tc; conjugative	[11]
Broad host range				
pPT23A-like[b]	pPT23A	100	coronatine synthesis; conjugative	[188,189]
Narrow host range				
Compatible and largely homologous plasmids often found in one strain	pFKN	40	virulence (against plants)	[190][d]
pIP02-like	pIP02	40	cryptic, but carries multiple ORFs of unknown function	[31][d]
Conjugative, broad host range				
Unclassified *P. fluorescens* sugar beet group I	pQBR11	294	Hg; conjugative	[54]

(*Continued*)

Table 11.1 *(Continued)*

Incompatibility or homology group and group attributes	Plasmid	Size (kb)[e]	Phenotype; comments	Reference
	pQBR103	425	Hg; conjugative	[8][d]
Unclassified *P. fluorescens* sugar beet group II	pQBR24	307	Hg; conjugative	[54]
Unclassified *P. fluorescens* sugar beet group III	pQBR55	149	Hg; conjugative	[54]
Unclassified *P. fluorescens* sugar beet group IV	pQBR30	263	Hg; conjugative	[54]
Unclassified *P. fluorescens* sugar beet group V	pQBR61	64	Hg; conjugative	[54]
Unclassified (not P-1 or P-4)	pAM10.6	10.6	phenol catabolism	[35]
Narrow host range				
Unclassified	pKB740	8.1	2-amino-benzoate metabolism	[66] (GenBank: X66604)
Rolling circle replication(?)				
Unclassified Rolling circle replication	pPP8-1	2.53	cryptic	[36][d]

Unclassified	pPS10/pECB2	10.0/4.48	cryptic	[37,38]
Unclassified	pPST1	1.4	cryptic	[7]
Unclassified	pRA2	33	restriction-modification system, mobilizable	[131][d]
Unclassified	pRO1600	3	Cryptic	[191,192]
Unclassified	pVS1	30	Hg Su; mobilizable, narrow host range	[40,193][d]

Resistances: Ax = amoxycillin, Bc = borate resistance, Cb = carbenicillin, Cc = cefaclor, Ci = cefuroxime, Cf = cephalothin, Cm = chloramphenicol, Cx = ceftriaxone, Em = erythromycin, Ff = florfenicol, Gm = gentamycin, Hg = mercury, Km = kanamycin, Mc = minocycline, Pm = phenyl-mercuric acetate, Qc = quarternary ammonium compounds, Qn = quinolone, Sm = treptomycin, Sp = spectinomycin, St = streptothricin, Su = sulfonamides, Tb = tobramycin, Tc = tetracycline, Te = tellurite, Tp = trimethoprim, Uv = ultraviolet light.

[a] Typed by sequence homology.

[b] This grouping defined by homology; contains mutually compatible plasmids.

[c] Subgroups as given by Rawlings and Tietze [28].

[d] Paper reporting or finishing the complete plasmid sequence.

[e] Where no size estimate in kilobases could be found, size was calculated as 1.5 × (molecular weight in MDa).

by containing sequences that would hybridize to this replicon [26]. It was shown that the replicon would function in *P. aeruginosa*, confirming that this was the replicon identified also as IncP-3 in *Pseudomonas*. We have recently sequenced another IncA/C plasmid, R667, in collaboration with the Sanger Institute, Cambridge, (http://www.sanger.ac.uk/Projects/Plasmids). In parallel, two groups have sequenced three plasmids from *Yersinia pestis*, *Y. ruckeri* and *Salmonella enterica* [27], and used these to define an IncA/C (IncP-3) backbone. Two very closely related plasmids from *Photobacterium damselae* are also annotated in the databases, but not yet described in a publication. Interestingly, two putative Rep proteins encoded by the first megaplasmid, pQBR103, from *P. fluorescens* to be sequenced show similarity to the IncP-3 Rep [8].

For IncP-4-type plasmids there are more than 10 complete genome sequences, but these do not represent a single incompatibility group in contrast to the IncP-1 situation. There are at least three different incompatibility types within the same sequence family. It has been suggested [28] that the subdivisions in the family be referred to as IncP-4α and IncP-4β by analogy with the IncP-1 plasmids, although for IncP-1 plasmids at least some subgroups are incompatible with each other.

For IncP-6 we can now include three complete sequences of rather disparate plasmids. Although Rms149 was the only plasmid directly assigned to IncP-6, sequence analysis identified similarities in the putative Rep-protein binding sequences of Rms149 and plasmid pFBAOT6, identified as IncU in *E. coli* (Rms149 is IncG in *E. coli*). When cloning and testing of appropriate plasmids was carried out incompatibility was confirmed [29]. It was therefore suggested that the two groups are combined and designated IncU, and that these additional plasmids are conversely now included in IncP-6.

For IncP-7 there are four complete plasmid sequences and one partial sequence although the Rms148 sequence that we have determined has not been completely annotated or published. So far all three plasmids are tightly clustered. By contrast, for the IncP-9 group where there are three plasmids completely sequenced there are two main subdivisions each of which contain further subdivisions [30]. All exogenously isolated naphthalene degradative plasmids fall into the IncP-9δ subgroup, contained all the *nah* genes for complete naphthalene degradation and were very similar by their *Eco*RI restriction patterns to pDTG plasmid with exception of large restriction fragments containing catabolic genes (I.K., A.B. and colleagues, unpublished).

There are also complete sequences of a number of additional plasmids that either originate in *Pseudomonas* or have a sufficiently broad host range that they can exist either stably or semistably in *Pseudomonas* species, but have not been assigned to a particular group. The most notable of these are the plasmids pIP02, pSB102 and pXF51, which were recently identified as a new plasmid group, at least some of which appear to have a broad host range [31–33]. A separate broad-host-range plasmid that has been completely sequenced is RA2 [34]. These plasmids might belong to one of the known groups for which we do not have sequence information. The absence of DNA probes or DNA primers for PCR amplification for all the known groups makes it difficult to perform the classification easily. We have started a programme to create the tools to underpin the creation of such a catalog. This process will be helped by the existence of sequences for additional replicons, i.e. those of pAM10.6 [35], pPP8-1 [36], pPS10/pECB2 [37,38], pRO1600 [39] and pVS1 [40].

Some groups of quite closely related plasmids do not conform to the expectation that they should form an incompatibility group. Apart from the IncP-4-like plasmids, one example of this is the pT23A family which includes at least 47 plasmids [41,42]. Interestingly this family contains a range of compatible plasmids [43], although the level of sequence identity between apparently compatible plasmids is quite high and it is known that naturally occurring strains often carry more than one member of the family. These plasmids play an important role in the range of adaptive and virulence properties possessed by *P. syringae* [44].

11.3
Defining Characteristics of *Pseudomonas* Plasmids

Many plasmids play an important role in the horizontal gene pool of *Pseudomonas* species. However, the fact that a plasmid can be found in *Pseudomonas* does not automatically mean that it should be classed as a *Pseudomonas* plasmid. The genomes of different organisms have different characteristic G + C content, codon usage and other sequence motifs. Recently acquired horizontally transferred DNA can often be identified by its being atypical of the genome in which it is found, but with time such DNA is normally thought to acquire the signatures of its current host. Plasmids found in *Pseudomonas* can therefore be assessed with respect to whether they are typical in G + C content, codon usage and other sequence motifs.

One of the first analysis of a plasmid genome in this respect was carried out with the genomic sequence of IncP-1α plasmid RP4/RK2 [45,46]. The driving force behind the study was to test the idea that the IncP-1 plasmid backbone has been selected to be deficient in restriction endonuclease cleavage sites as the result of encountering restriction barriers during spread from one strain to another. What became clear was that the backbone was deficient in some palindromic hexamer sequences and very rich in others. To some extent this could be due to high G + C content in the plasmid genome (61.75%). However, randomly generated sequences with a similar G + C content gave a higher frequency of sites than appeared in RP4, suggesting that they have been selected against in RP4. Those hexamers that were underrepresented corresponded to recognition sequences for restriction endonucleases that have been cataloged in the *P. aeruginosa* cluster of *Pseudomonas* species. Thus, bias against the presence of such sites as a result of the selective pressure created by restriction barriers is a plausible explanation. In addition, the plasmid genome showed the same high frequency of GCCG and CGGC motifs that are characteristic of *P. aeruginosa*. It is therefore possible to deduce that RP4 evolved within *P. aeruginosa* or an organism very like it. In contrast, while the IncP-1β plasmids have a similar, if not higher, G + C content than the IncP-1α plasmids, they do not have the same preponderance of GCCG/CGGC motifs, suggesting that the evolutionary context in which they evolved was different [47].

Plasmids can be complex genomes and may consist of a core along with recently acquired DNA with different characteristics. We will take the approach that a typical *Pseudomonas* plasmid will have basic plasmid functions – replication, maintenance and transfer genes whose genomic signatures are characteristic of *Pseudomonas*. Its genetic cargo may have been acquired by forays into other genera, but unless the core has *Pseudomonas* characteristics then it should probably be regarded as a transient, even though this may simply mean that it has not been a resident long enough to acquire the genomic motifs. However, Table 11.1 does not exclude plasmids that do not fit this criterion – it lists plasmid families (or individual plasmids where there is not yet a family) that have been found in *Pseudomonas* or belong to families that are found in *Pseudomonas*. In general we have aimed to include any plasmids that can replicate in *Pseudomonas* or are belong to families that can replicate in *Pseudomonas* and have been sequenced. As discussed in Section 11.9, relatively small changes can be responsible for allowing a plasmid to replicate and be maintained efficiently in a new host, even if it was not stable to start with. It would therefore be possible for a plasmid to be reasonably well adapted to a new host, but not have the genomic characteristics of that host.

If we apply this analysis to the plasmid groups listed in Table 11.1, we can categorize the various plasmid groups. The plasmids that we would class as typical *Pseudomonas* plasmids on the basis of GtC content are: IncP-1 (61–64%), IncP-4 (61%), IncP-6 (59%), IncP-9 (59%); pIP02-like (61%), pPS10/pECB2 (60%), RA2 (60%) and pVS1 (60%). Interestingly the IncP-3, IncP-7 and pT23A replicons have G + C contents of 53–54%, much lower than would be expected. Plasmids pAM10.6 and pRO1600 have a slightly higher, but still not typical G + C content (56%). The IncW plasmids have in the past been classed as broad host range and included in the list of *Pseudomonas* plasmids, but despite their G + C content (the *rep* gene is 59%), they are not stably maintained in *Pseudomonas* in the absence of selection and therefore have not been included in our list. However, it is interesting to note the similarities with the *rep* genes of some of the *Pseudomonas* plasmids listed (see Section 11.6) and the similarity of organization of the transfer regions of IncP-9 plasmids to those of the IncWs. It may therefore be appropriate at some stage to give this group an IncP number, e.g. if an IncW plasmid that is stable in *Pseudomonas* is found. Clearly the relationships between plasmid replicons and their host genomes are of great interest, and the accumulating sequences should make this a productive area for bioinformatic studies, which is beyond the scope of this chapter.

11.4 Isolation of Plasmids

Although a very large number of plasmids have been isolated and at least partially characterized, it is not yet clear whether the plasmids that have been characterized are fully representative of those plasmids that carry important phenotypic determinants in natural environments. There is therefore a growing effort to sample new environmental niches and determine the nature of plasmids that are present.

Obviously the most direct approach (referred to as endogenous isolation) is to spread environmental samples on appropriate types of solid media and purify to single colonies bacteria that grow. Plasmid screening can then be performed either without purification of plasmid DNA, for example by the Eckhardt technique [48] in which bacteria are lysed in the gel well and plasmid DNA is induced to migrate away from the very-high-molecular-weight chromosomal DNA or after plasmid extraction by one of the standard methods that may be adapted to isolation of high-molecular-weight plasmids [49,50]. However, this relies on being able to grow a fully representative range of the bacteria that potentially harbor plasmids. It also is limited to finding bacteria that make up at least 0.1% or more of the population because screening thousands of cultures for plasmids may not be practical [18]. Attempts to use endogenous isolation to recover putative IncP-9 plasmids from pig manure samples that had tested strongly positive for these plasmids on the basis of PCR screening of total DNA extracted from manure proved negative [51].

Since it is known that many bacteria observed by microscopy in environmental samples are unculturable, approaches that do not rely on culturing the bacterial host have been devised – the so-called exogenous isolation. The first such approach, called biparental exogenous isolation, adds a new selectable recipient bacterial strain to the environmental samples, incubates the mixture for long enough to allow plasmid transfer to take place and then selects derivatives of the added recipient that have acquired a new marker from the environmental sample [52,53]. This procedure can in principle capture both conjugative plasmids and mobilizable plasmids that reside in a host that also carries a self-transmissible plasmid that can mobilize the plasmid carrying the selected trait. These studies are particularly relevant to this review because the hosts that have been used successfully for such isolations include *P. aeruginosa*, *P. putida* or *P. fluorescens*. Examples of acquired traits are resistance to mercury [54], antibiotic resistance [55] and ability to degrade petrol hydrocarbons such as naphthalene (I.K., A.B. and colleagues, unpublished) or herbicide [56]. This has been a useful approach, but is limited to plasmids that carry the selectable marker that has been chosen.

An alternative approach is to use what is termed triparental exogenous isolation [57]. In this approach two general plasmid properties are selected – the ability to transfer and the ability to mobilize a second plasmid. To the environmental sample are added a plasmid-free selectable recipient bacterial strain (strain A) as well as a second strain (strain B) carrying a mobilizable plasmid with a selectable marker. After enough time for plasmid transfer to take place the mixture is plated on medium to select for transfer of the mobilizable plasmid to strain A. This should only occur if a self-transmissible has transferred to strain B and allowed it to transfer to strain A. In many cases the new plasmid also transfers to strain B resulting in recipients carrying both plasmids. Recipients of the selected marker are then screened for acquisition of new environmental plasmids. Again *P. putida* has often been used as the recipient in such experiments. Isolation of pTP6 from contaminated sediment that had been screened by PCR and hybridization and shown to have IncP-1 plasmids was not achieved by direct plating or biparental mating, but was eventually achieved by triparental mating to mobilize IncQ plasmid pIE723 [58].

11.5 Use of PCR Primers and Associated Probes for *Pseudomonas* Plasmids

As mentioned above, one important aim is to develop sets of PCR primers that can be used to test whether a sample contains DNA belonging to a specific plasmid group. This could be applied to purified plasmid DNA, to total DNA from a pure culture or total DNA from an environmental sample that may contain not just a mixture of bacteria, but also DNA from other sources. This allows one to analyze a sample without the need to cultivate the bacteria carrying the plasmid. The power of this technique was originally demonstrated with IncP-1 (IncP), IncP-4 (IncQ), IncN and IncW plasmids [59]. Reconstruction experiments indicate that the detection limit of this method is somewhere in the region of one bacterium carrying a particular plasmid in 10^3–10^4 bacteria. Before such primers can be used it is important that they are shown to give a product reproducibly with complex samples containing DNA of a particular plasmid group and that a product is not obtained with DNA from plasmids of other groups. Although the presence of a PCR product of the right size can indicate the presence of the query sequence, in practice it is always necessary to use Southern blotting or DNA sequencing to check that the PCR product does correspond to a template of interest. It is surprising how often a PCR product of the right size is obtained which turns out to be spurious when it is cloned and sequenced. However, if a product is obtained that does not hybridize to the probe then it may be that the plasmid that gave the product is only distantly related to the archetype, i.e. it diverges too much to give cross-hybridization. Such a situation is interesting because it indicates the identification of novel relatives of known plasmids that may help to fill a gap in the continuity of a plasmid group.

Such primers are currently available for the groups IncP-1 [59,60], IncP-3/IncA/C [26], IncP-4 [59], IncP-9 [51,61], IncN and IncW [59], and pIP02-like [31]. In the near future we should be able to add probes for IncP-6 and IncP-7. The previously described primers have been used to screen the available environmental samples, and this will allow a comprehensive assessment of their distribution and abundance [51,55,59]. Interestingly, while the plasmids that we have designated as typical of *Pseudomonas* were found in a range of water and soil samples, the IncN and IncW plasmid signals were confined largely to manure samples, suggesting a closer link to enteric flora, reinforcing the categorization already suggested above [59].

11.6 Plasmid Replication

Autonomous replication and stable inheritance are the key properties of all plasmids, and these aspects will be a major focus of the rest of this chapter, with plasmid transfer being covered much less thoroughly. Plasmid copy number and size tend to be inversely related in all bacteria, and *Pseudomonas* species are no exception. Most attention has focused on the large, low-copy-number plasmids, but small, high-copy-number plasmids are known and some have been studied, particularly with a view to vector development. The aim of this section is not to give a detailed description of replication, but to illustrate the diversity and relationships between the systems in *Pseudomonas* plasmids. The reader is referred to previous reviews for more information on replication *per se* [62].

The smallest replicon that has been reported to replicate in *Pseudomonas* is only 640 bp in size and originated in *Acidothiobacillus ferrooxidans*, but can replicate in quite a number of Gram-negative genera including *Pseudomonas* [63]. Interestingly its G + C content is 59%, consistent with the criteria defined above as typical of the *Pseudomonas* genome. How it replicates is not known since it bears no resemblance to other replication systems and encodes no protein. It has only been studied as part of a shuttle vector and so it is conceivable that it is not capable of self-replication. Relevant to this point is the 300-bp replicon derived from the *P. fluorescens* megaplasmid pQBR11 that also encodes no protein [64] and seems very unlikely to be the *bona fide* replicon from such a large plasmid. An additional replicon was isolated in the same way on a 4.9-kb *Hin*dIII fragment from pQBR55 [65]. It has some of the characteristics of a replicon such as the presence of a *dnaB* family (helicase) gene, a GC skew inversion (which indicates the transition from one replication direction to another) and a gene related to *traA* of IncP-1α that make sense if it is the megaplasmid replicon. However, the complete DNA sequence of pQBR103 revealed a more probable replicon based on the gene(s) for a Rep protein with similarities to the IncP-3 Rep of RA1, and a series of possible interons for regulation and activation of the system [8].

A 2.5-kb rolling circle replication (RCR) plasmid was isolated from *P. putida* P8 [36]. The predicted Rep protein and the putative replication origin placed the replication system firmly in the pC194 family of RCR plasmids. Not only does the plasmid lack additional maintenance genes that would be needed if the plasmid became very large and forced the copy number down, but also the accumulation of single-stranded DNA for such an expanded plasmid would result in recombinational instability that would prevent growth in plasmid size. A second plasmid, pKB740 from *Pseudomonas* species [66], may also replicate via a RCR strategy [36], although this conclusion depends on similarity to the putative *rep* gene of a relatively poorly characterized bacteriophage [67]. It is interesting to note that the IncP-9 Rep protein has weak, but apparently significant similarity to a sequence group that includes RCR Rep proteins [68,69]. We can speculate that the IncP-9 Rep originally drove RCR, but evolved in such a way as to lose nicking activity and acquire the ability to recruit host replication functions.

The Rep of pT23A plasmids is related to that of ColE2, a small, medium copy number, DNA polymerase (Pol) I-dependent plasmid of *E. coli* [70]. The reason for the Pol I dependence is probably not that the Rep protein leads in DnaA, but rather processes a transcript going through the replication origin, creating an R-loop that allows Pol I to initiate replication, but would not allow Pol III to do so. Since pT23A plasmids are large and relatively low-copy-number plasmids, it is not clear to what extent the details of the replication system will match those of ColE2, despite the similarity of the Rep proteins.

All other groups of plasmids contain replicons which depend on at least one Rep protein to activate a replication origin containing multiple Rep-binding sites, termed iterons. We can group these replicons by the relationships between their Rep proteins. It is possible to put almost all of these proteins, including the ColE2 Rep, into a single alignment and build a phylogeny [71,72]. While some subdivisions as outlined below are very clear, the relationships between the less closely related are much more difficult to be sure of and so are not discussed here. The Rep proteins from IncP-1 plasmids (product of the *trfA* gene), the IncP-4 plasmids (product of the *repC* gene) and RA2 (product of the *repA* gene) each fall into separate groups found only on closely related plasmids. These plasmids all carry other stability functions belonging to clearly defined and broadly distributed families. The diversity of the Rep functions may thus reflect the drive of plasmids to avoid competition with each other by having a unique or at least not cross-reacting Rep–*oriV* system.

IncP-3 Rep is similar to Rep proteins of plasmids from *Buchnera aphidicola* and *Acidithiobacillus ferrooxidans*. IncP-7 Rep is related to that of many other plasmids [73] including *Pseudomonas* plasmids pPS10/pECB2 (which are very similar to each other) and pRO1600. This group also includes the Rep proteins from the FIA replicon of IncFI plasmids, RepE, for which there is a crystal structure [74], as well as the Pir protein of R6K and Rep from pSC101 that have also been studied in detail [75–78]. A general point that should be noted from this group is that a family of Rep proteins can be found in narrow-host-range plasmids from a range of different species that is much wider than the host ranges of the individual systems, i.e. the various members of the Rep family are adapted to different narrow groups of hosts. IncP-9 Rep shows similarity to the Rep proteins of pBBR1 from *Bordetella pertussis* [68,79] and pL10.6 from *P. fluorescens* [35]. This set seems to be an odd one out because sequence alignments place it in a group of Rep proteins from RCR plasmids, although admittedly the similarity is very low. The Rep proteins from the pIP02-like group are most closely related to the IncW Rep from plasmid pSa [31]. Finally, pVS1 shows similarity to Reps from plasmids of *Paracoccus alcaliphilus* [80] and *Erwinia stewartii* [81].

The general model for the action of these Rep proteins is that they bind to the replication origin and assist host proteins to activate it [62]. The normal sequence is DnaA binding along with the plasmid Rep, followed by localized unwinding. This can allow the DnaBCD helicase to enter and unwind the origin further, which enables primase to initiate leading strand synthesis. Replication origins normally contain an A + T-rich region that is proposed to be where the double-stranded DNA melts first. In most *oriV*s it is also possible to identify potential DnaA boxes and where it has been investigated directly DnaA has generally been confirmed at these sites. In the case of IncP-9, DnaA does not bind when present on its own under standard conditions, but does bind when Rep is present [82]. The plasmid-encoded Rep protein may also play additional roles during the assembly and action of the initiation complex – the Rep protein TrfA-2 (TrfA33) of IncP-1 plasmids contains a sequence motif that interacts with the β protein sliding clamp (DnaN) of DNA Pol III and may be involved in controlling overreplication of the plasmid [83].

The most complicated replicon is that of the multicopy IncP-4 system that is the only one to date known to require three plasmid-encoded proteins – an iteron-binding protein (RepC), a helicase (RepA) and a primase (RepB) [84]. The replication strategy employed is called strand displacement because there are no lagging strand origins to assemble the primosome that normally generates Okazaki fragments. Instead the replication origin has two divergent leading strand origins that can fire simultaneously or sequentially. The replisome which assembles continuously displaces the nonreplicated strand. A practical consequence of this is that the system seems to be inherently unsuited to replicating a large plasmid, because as the plasmid gets larger, the increased delay in converting the lagging strand to double-stranded DNA by the action of the other leading strand origin, results in accumulation of single-stranded DNA. This can be recombinogenic and so large derivatives of IncP-4 plasmids are structurally unstable, a fact that came to light most forcibly when cosmid cloning vectors based on the IncP-4 replicon, were found to lose inserted DNA (often up to 40 kb) during genomic library construction. There is the possibility that the IncP-7 replicon includes a helicase since there is a gene downstream of *rep*, and possibly cotranscribed, that shows sequence similarity to a helicase from *Salmonella enterica* [73], but as yet no experimental evidence either way is available.

As can be seen, therefore, apart from a few exceptions referred to at the start of this section, we can make predictions about the functions of most of the replication systems in *Pseudomonas* plasmids described to date. We expect that as other large plasmids are discovered and characterized, they will be found to have replicons related to those described above.

11.7 Plasmid Maintenance

The organization of *Pseudomonas* plasmids that are known to encode additional stability determinants are summarized in Figure 11.1. The clustering of replication and maintenance systems is a way of maximizing coinheritance during plasmid recombination and transfer [85]. The most extreme example of this is the IncP-1 plasmids where the majority of the plasmid is a core of coordinately regulated replication, maintenance and transfer functions [45,47,86]. Many other plasmids demonstrate this principle as well. For example, comparison of the pWW0 sequence [69] with that of the naphthalene degradation plasmids pDTG1 [87] and NAH7 [88], shows that the conserved IncP-9 core is the clustered replication, maintenance and transfer region, outside which the plasmids diverge almost completely. The three major categories of these factors are multimer resolution, active partitioning and postsegregational killing. For an overview of how these systems work the reader should consult more specialized reviews [89]. However, one must also not forget that the ability of a conjugative plasmid to reinfect a plasmid-free segregant may act as a system to prevent plasmid loss from a population and the power of this mechanism has been demonstrated with respect to IncP-1 plasmids in bacteria growing either under standard microbiological laboratory conditions [90] or in the rat gut [91].

Apart from RCR plasmids (that should generate monomeric products even from a dimeric template as a result of their mode of replication) plasmids are generally

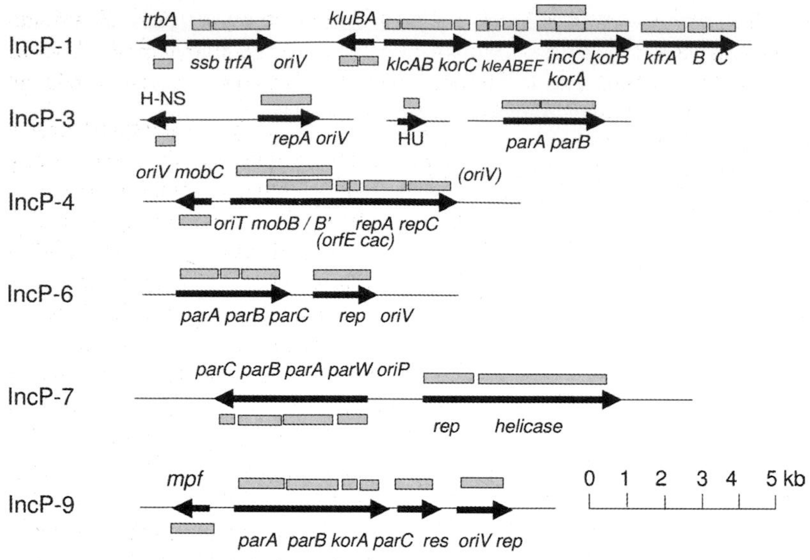

Figure 11.1 Summary of the organization of the replication and stable inheritance regions from major IncP plasmid groups discussed in this chapter. Only those genes implicated in stable inheritance by phenotype or coregulation with such genes are shown, except for IncP-1 *trbA* and IncP-9 *mpfR* that are involved in regulation of transfer genes, and the *mob* genes of IncP-4 that are integrally linked with the *rep* genes. For Inc P-1, *incC* and *korB* encode the plasmid's ParA and ParB proteins. The *incC* gene contains two translational starts and a regulatory protein KorA is encoded within the longer *incC* coding frame. For IncP-3 genes two nucleoid-associated proteins are shown because there is evidence that these proteins supplement host homologs and prevent the plasmid becoming an excessive burden. These IncP-3 functions seem to be the most dispersed on the genome. For IncP-4 the *mob*/*parB* gene contains two translational starts. The shorter product is required for replication; the longer is required for mobilization.

thought to need a multimer resolution system (*mrs*). This generates monomers from the dimers that arise by recombination [92]. That multimer resolution is vital for plasmid stable inheritance has been most extensively demonstrated for ColE1 through study of its *cer* system [93]. Resolution is generally achieved by site-specific recombinases, that prevent the accumulation of dimers and higher-order multimers. The most complicated *mrs* is encoded by the IncP-1α plasmids where three genes (*parCBA*) out of a set of five (*parABCDE*) transcribed from two divergent operons encode multimer resolution [94]. Interestingly, this block of genes is located physically away from the rest of the replication and stable inheritance functions, and appears to have been acquired after creation of the IncP-1 regulon, although the initial impression that it has never been present in the IncP-1β plasmids [47] may be wrong because recently isolated IncP-1β plasmids do have *parA* [95]. It is the ParA of this system that shows sequence identity to the well-characterized resolvase protein family and most other plasmids just seem to encode a homolog of ParA if they have an identifiable putative *mrs* function. That ParA alone can provide multimer-resolution activity is consistent with the observation that the RK2 *parA* plus its associated *cis*-acting site can give plasmid stabilization, at least in some strains [94], although greater stabilization was found when *parB* and *parC* were included. pIP02 also encodes a ParB homolog, close to but not directly next to the *parA*, the importance of either of these functions for pIP02 is unknown since *parA* was inactivated by the cassette that was inserted to label the cryptic plasmid [31]. Other plasmids with *res* homologs include IncP-9. Surprisingly, deletion of *res* from the IncP-9 minireplicon in pMT2 did not result in increased loss rate or apparent accumulation of multimers [68], even after ultraviolet irradiation that might be expected to elevate recombination rates. The plasmid pVS1 also encodes a resolvase which does not appear too necessary for stable inheritance [40]. The inconsistency in whether or not a resolvase gene is found leaves us wondering whether the need for such a function is actually such a universal property of plasmids.

The second type of general stability determinant is the active partitioning system. It is normally only the low-copy-number plasmids that carry active partition systems and the presence of such a system allows one to functionally define plasmids as low or high copy number, as well as by the yield of DNA obtained during screening. The IncP-4 plasmids appear to fall into the high-copy-number category and this was nicely demonstrated by inserting extra copies of the *oriV* iterons into the plasmid [96]. This causes copy number to be reduced and results in decreased segregational stability as expected if segregation is random.

The active partitioning systems that have been recognized on *Pseudomonas* plasmids all fall into the type I *par* family, all of which have a Walker-type ATPase as their ParA [89,97]. These systems can be subdivided, not just by sequence similarity, but also by organization of the *par* locus, the presence of a putative DNA-binding domain at the N-terminus of the ParA protein, the family to which the ParB protein belongs and the location of the *parS* sequence which acts as the centromere-like sequence for the partitioning process. The classic P1 and F systems are designated Ia, while those in which ParA does not have a DNA-binding domain, does not have a classic ParB protein and has its *parS* upstream rather than downstream of the *par* operon are referred to as Ib. The *par* systems of the IncP-1, IncP-7, IncP-9 and pIP02-like systems currently fall outside either of these two major groups both organizationally and phylogenetically. The IncP-1, IncP-9 and pIP02 systems are more closely related at the protein level to each other than they are to other systems, but the IncP-7 *par* system is quite distinct. The pRA2 system is a type Ib system as may be the system based on the *parA* gene of pT23A, but pVS1 is peculiar because knockout of its putative *parB* gene apparently has no destabilizing effect (cited in the database annotation). The ParA and ParB homologs in the IncP-1 system appear to be closer to their chromosomal counterparts, particularly from *P. aeruginosa* and *P. putida*, than to other plasmid-encoded Par proteins. Additional genes have been designated *par* on the basis of coregulation with *parA* and *parB* genes, but further evidence for a direct role in partitioning is lacking. This includes *parC* of pWW0 (*tolA* in pM3), and *parW* and *parC* in pL6.5 and pCAR1.

The repertoire of maintenance functions observed in the *Pseudomonas* plasmids is thus exactly what we would expect from our general knowledge of plasmids that have been studied in detail. However, there are still many associated genes whose role if any has not yet been defined. Of these, the role of the KfrA-like proteins that bear resemblance to the Smc proteins (structural maintenance of chromosomes) is most intriguing. Hopefully the accumulation of information on other *Pseudomonas* plasmids will add to our ability to identify additional gene families that are conserved across many groups and that this will justify further work to establish the role that they play.

11.8
Determinants of Incompatibility

Knowledge of what determines incompatibility is useful because it can help to focus attention on key areas during the analysis of new plasmid genomes. Incompatibility normally can be observed to arise between two plasmids in the same bacterium, where the plasmids are phenotypically distinguishable, but which are indistinguishable with respect to at least one of the key mechanisms that results in their stable inheritance: replication, postsegregational killing or active partitioning. If two plasmids share effectively identical replication and replication control systems, then random selection from a pool of plasmid molecules to be replicated will result in a bias towards one or other type which will be reinforced by subsequent selection until cell lines with only one type breed true. If they share a postsegregation killing system then it can only protect against loss of both plasmids from the cell, i.e. loss of just one of the plasmids is ignored because the other plasmid can still provide the antidote that prevents the toxin from killing the bacterial host. Thus, the shared *pks* negates the effectiveness against plasmid loss, for both plasmids. If the two plasmids share an active partition system based on a pairing step that is not locked into the replication cycle, then random selection from the pool of plasmids will again result in the loss of effectiveness of the partitioning system – daughter bacteria will always get one plasmid, but not necessarily both plasmids.

Does this background fit with what we find in practice in *Pseudomonas* plasmids? For the IncP-1 plasmids the replication origin and the iterons that it carries are the primary incompatibility determinant [98,99] with the partitioning and control functions of the *korA–incC–korB* operon playing a secondary role [99,100]. The contribution of the *parDE* region (which encodes a postsegregational killing system) to incompatibility has been assessed by placing the *parD* gene in *trans* and showing that this can destabilize RK2. However, the strength of the effect is strain-specific, reflecting the relative importance of the different stability determinants in different host backgrounds [90,101]. The *klc–kle* region has been shown to be important for stability in *P. aeruginosa* [102]. While studying this effect it was discovered that a series of direct repeats when present in both of a pair of plasmids caused incompatibility in *P. aeruginosa*, but not in *E. coli* [103].

For IncP-4 plasmids incompatibility appears to be due to the iterons that bind RepC in the replication origin. Some IncP-4-like plasmids carry an extra, nonfunctional copy of the *oriV* region, located downstream of *repC* sequences [28]. While the extra *oriV* is not activated by the RepC protein of the plasmid it is in, it does bind RepC from other IncP-4-like plasmids, thus interfering with the replication of the second plasmid. This may be a strategy to displace other IncP-4-like plasmids. This does cause a problem for the IncP-4 classification system, because is means that plasmids whose Rep proteins clearly fall into different specificity groups are placed in the same group because of the one way effect of the second *oriV* region. Where we understand the basis for such effects, we propose that the unidirectional incompatibility should not be used as the basis for primary classification, but should be regarded as a secondary effect.

For IncP-9 plasmids the replication origin region alone is not able to express incompatibility, but needs to be combined with the *rep* gene [68]. A second region corresponding to the putative centromere-like region associated with the *parAB* region of the minireplicon plasmid pMT2 from pM3 also expresses incompatibility very strongly [68]. This may correlate with the importance of this region in stable inheritance of the plasmid and the fact that the *oriV–rep* region is not able to function as an independent replicon. These observations show clearly that incompatibility is not due only to replication determinants, but can also be associated with stable maintenance functions and thus can be usefully used as an indication of regions that are important for stable inheritance.

11.9 Determinants of Host Range

The ability of a plasmid to carry genes to and from distantly related bacteria is perceived as an important characteristic of their potential to fuel the evolution of their hosts. Plasmids with a broad host range are often seen as having greater potential to drive the appearance of variant strains. However, the term "broad-host-range plasmid" has been used rather arbitrarily over the years, but with increasing awareness of the genetic relatedness of bacterial groups it has been proposed for plasmids of Gram-negative bacteria that this should mean at least that the plasmid is able to establish itself and then be maintained reasonably stably without selection in at least two of the subdivisions of proteobacteria. Thus, a plasmid that can shuttle between *P. putida* and *E. coli* may not necessarily fit this definition since these both are within the γ subdivision. On this basis IncP-9 plasmids do not count as having a broad host range, whereas a number of plasmid groups in *Pseudomonas* do have a broad host range and in general these are already included also in the plasmid classification system of *E. coli* (IncP-1 – IncP in *E. coli*, IncP-3 – IncA/C, IncP-4 – IncQ, IncP-6 – IncG/U). In general, host range has been determined by testing the ability of plasmids to transfer to and be maintained in test strains representing a selection of different bacterial groups. A recent development uses tagging of plasmids with the gene for fluorescent protein that is switched off in the donor bacteria, but is activated after transfer allowing individual transconjugant bacteria to be isolated by fluorescence activated cell sorting [104,105]. For IncP-1 plasmid pJP4 it was demonstrated that the plasmid could establish itself in bacteria of α, γ, γ–β and β proteobacteria, although the 2,4-dichlorophenoxyacetic acid (2,4-D) degradation genes were functional only in a limited sub-range of these bacteria [104]. Given the diversity of species within the *Pseudomonas* genus and the importance of transporting genes from other genera into *Pseudomonas*, the factors that determine whether a plasmid can establish itself in a *Pseudomonas* species and be maintained there stably is an important property. Unfortunately it appears from recent detailed comparison of IncP-1 plasmid stability in different species and strains that one should be wary of generalizing about plasmid behavior from a limited number of tests [106]. While IncP-1 plasmid pB10 is generally quite stable in *Pseudomonas* species single strains of each *P. korrensis* and *P. putida* were found in which the plasmid is lost rapidly in the absence of selection [106]. Conversely, the factors that limit the maintenance of *Pseudomonas* plasmids outside the genus are also of interest and the innovations in plasmid tagging and cell sorting referred to above has allowed the demonstration that IncP-1 plasmids can transfer efficiently to an unexpectedly wide range of bacteria including *Arthrobacter* which are Gram-positive bacteria belonging to the Actinobacteria [105].

That it is the replication and stable inheritance system that is particularly important in determining host range was shown by studies on hybrid plasmids created between IncP-1 plasmid RK2 and *E. coli* IncFI plasmid F [107]. By transferring the IncP-1 and IncFI replicons from *E. coli* to *P. aeruginosa* with either the IncP-1 or the IncFI transfer systems it was shown that although the F transfer system is not so efficient as the IncP-1 transfer system, the absolute limitation on host range is the replication system. The IncFI transfer system could mobilize the IncP-1 replicon to *P. aeruginosa*, but when the IncP-1 transfer was used to move the IncFI replicon efficiently from *E. coli* to *P. aeruginosa* it could not replicate at all. Recent biochemical studies showed that the defect in replication may be due to a failure of the plasmid RepE protein to form a stable and productive complex with DnaB protein from *Pseudomonas* species [108]. The possibility that poor transcription from the *repE* promoter may contribute to the replication defect was also considered, but replacement by a promoter that was known to function in *Pseudomonas* species failed to overcome the defect. Inclα plasmids could also be transferred to species in which it can not replicate [109] but the basis for the limit on its replication is not yet known. Similarly, the IncP-9 plasmid pWW0 could transfer to *Rhizobium* species as indicated by the detection of the *xylE* gene but was not stably maintained [110]. There is therefore considerable interest in determining the factors that limit the host range of a plasmid's replication and maintenance system as discussed below. However, the importance of the transfer process should not be totally ignored – since the range of species into which the IncP-1β plasmid pB10 spreads in an activated sludge microbial community depended on the initial donor in which it was introduced [111].

For the IncP-4 plasmid it seems clear that the major factor in its broad host range is the multiple Rep proteins that it encodes which give it a degree of independence from the machinery of its host. It not only provides the machinery to prime leading strand synthesis, but also dispenses with the need for lagging strand synthesis, by its strand displacement replication strategy [84]. It is more of a challenge to explain those plasmids that on the face of it are very similar to each other, possessing both a *rep* and *oriV*, but which have different host ranges. The IncP-1 plasmid has been studied in this respect as a broad-host-range plasmid, while pPS10 has been studied as a narrow host-range plasmid. For these plasmids it appears that it is the macromolecular interactions between the plasmid and host replication machinery that determine their host range.

The minimal replicon of IncP-1 plasmids consists of the *oriV* region and the *trfA* gene encoding proteins that in conjunction with host DNA replication proteins activate *oriV*. Clues about the host range came from the discovery that the *rep* gene *trfA* encoded two related Rep proteins, TrfA-1 and TrfA-2, from alternative translational starts in the same open reading frame [112,113]. Subsequently, it was shown that TrfA-2 was sufficient to drive plasmid replication in *E. coli* and *P. putida*, but not in *P. aeruginosa* [114,115]. Conversely the sequences needed to provide a functional *oriV* in different species were not identical. In *P. aeruginosa* a segment encoding only the minimal five TrfA-binding iterons, plus the adjacent A + T and G + C regions is sufficient for (albeit slightly less efficient) replication, whereas in *E. coli* and *P. putida* the region containing DnaA-binding sites next to the TrfA iterons is necessary [116]. Subsequent biochemical analysis has shown that both TrfA and DnaA are necessary to activate the replication origin, and that TrfA helps DnaA to recruit additional host factors necessary to activate the replication origin [117–120]. Further studies showed that it is the N-terminal domain of TrfA-1 (and absent from TrfA-2) that is responsible for recruiting DnaB in *P. aeruginosa* independently of DnaA [121]. Thus, the TrfA complex is flexible enough to cope with species differences, but this does depend on there being two forms of the TrfA protein – it may be that TrfA-1 does not work as well as TrfA-2 in some species.

The results with RK2 highlight the importance of interactions between the plasmid and host-encoded parts of the replication machinery. That the differences in the interactions necessary in different species can be quite small is suggested by studies on how a *Pseudomonas* plasmid of limited host range can extend its host range. Plasmid pPS10 from *P. syringae* can replicate in *E. coli* at 30°C, but not at higher temperature [122]. This may be because many *Pseudomonas* species exist mainly in environments where the average temperature is in the region of 10–25°C and so the systems are optimized for lower temperatures. For example, temperature adaptation has been documented for plasmid transfer in aquatic environments [53]. However, it may also reflect less robust interactions with host replication factors to which the plasmid is not adapted. Such plasmid temperature-sensitivity is not unique – it is also observed for the IncP-9 plasmid pM3 [123]. To understand the limitations on interactions with the host it was possible to isolate mutations that were able to suppress this temperature sensitivity. Such mutations could occur in either the plasmid or the host. Those in the plasmid mapped in the *rep* gene [122], while those in the host mapped to the *dnaA* gene [124]. These studies demonstrate that the host range of a plasmid is not fixed and can be altered by small changes in key proteins.

In addition to replication, the stability of a plasmid in different species can be an important determinant of its effectiveness as a carrier of genetic information in different host populations. The way that a host can modulate effectiveness of stable inheritance functions is illustrated again by the IncP-1 plasmids. For example, the effectiveness of the *parABC* gene varies from strain to strain in both *E. coli* and *P. aeruginosa* [94,124], and so one should expect that it may also vary from species to species as well. The stability conferred by the *kle* region of RK2 is also host specific – apparently being much more important in *P. aeruginosa* than in *E. coli* [102]. Another factor that may create host–specific variations in the effectiveness of replication and stable inheritance genes is the strength of gene expression signals. For example, not all bacteria recognize promoter sequences in the same way. Although the nature of the mutational changes at DNA sequence level were not reported, the IncP-1 plasmid pB10 was shown to adapt to a *P. putida* host in which it was unstable and this process was accelerated by frequent conjugative transfer [125]. Changes to the expression profile of plasmid genes were detected, but a full explanation of the adaptive effects is not yet available. This suggests that at least in some cases the IncP-1 regulatory circuits that use very strong promoters repressed by autoregulation to compensate for contexts where the "standard" signals work less effectively may not provide the robustness that the plasmid requires for all hosts [85].

The host range of a plasmid may not easily be linked to individual genetic components. Clearly where a plasmid is a chimera of two narrow-host-range replicons with different host ranges the host range of the complete plasmid will be broader than either component. However, other interactions may also extend the host range. Thus, in the IncP-9 plasmid pM3, the wild-type isolated *rep–oriV* region cannot replicate in *E. coli* and requires at least part of the *par* region to allow it to replicate in this species. The host range of the *rep–oriV* region alone can be extended by mutating the *rep* promoter to increase expression of *rep*, but the defect is normally suppressed by the action of ParB bound to its binding site parS that must be in *cis* to the *rep–oriV* region [126]. An example showing some similarity is *E. coli* IncX plasmid R6K whose replication is normally confined to *E. coli* and closely related bacteria, but which can be allowed to replicate in *Pseudomonas* species by overexpressing its *rep* gene [127]. An interesting third example is the pCTX-M3 plasmid that can transfer to and replicate in *Pseudomonas* species, but whose IncL/M replicon is apparently incapable of being introduced into *P. aeruginosa* when separated as a minireplicon. A second candidate *rep* gene was rule out by mutagenesis, so the basis of the broad host range of the complete plasmid is not clear [128]. Perhaps a locus adjacent to the IncL/M replicon is necessary to modulate the replicon behavior as found for IncP-9.

11.10
Plasmid Transfer

One of the key features of plasmids is their ability to spread from one bacterium to another either under their own volition or because they are assisted in some way. Transfer can occur by transformation, transduction or conjugation and the context determines which of these is dominant. With the large plasmids of Gram-negative bacteria it is generally expected that transfer will be by conjugation, but in fact transformation is more widespread a process than often realized [129,130]. This is illustrated by pRA2, which appears to transfer by a transformation process since it is blocked by DNase I [131]. It appears that DNA is released from donors and is then taken up by recipients in the immediate vicinity. Surprisingly, pRA2 DNA seemed to be much more efficiently acquired by the recipients than DNA of control plasmids, suggesting that pRA2 may promote its uptake through DNA structure or an expressed function.

The most studied mechanism for active transfer is conjugation which relies on two sets of functions – mating pair formation (Mpf), and DNA transfer and replication (Dtr) [132]. Many small plasmids carry only the Dtr functions which are then termed *mob*, for mobilizable, since they require a conjugative system to mobilize them. A coupling protein is needed to allow a Dtr system to harness a Mpf system and the specificity of such coupling proteins determines which plasmids are efficiently mobilized by which self-transmissible plasmids [133]. The best characterized interaction among the *Pseudomonas* plasmids is the efficient mobilization of the IncP-4 plasmids by IncP-1 plasmids. The most widespread gene set responsible for pilus biosynthesis and the Mpf apparatus belong to the *virB/trb* family that is a subgroup of the type IV secretion systems (T4SS) [134]. Such genes are found on the IncP-1, IncP-9 and pIP02-like plasmids. The equivalent apparatus of pCAR1, the putative IncP-7 plasmid, shows much more similarity to the transfer system of plasmids from Enterobacteriaceae, most notably Rts1 (IncT) [73]. The DNA processing systems associated with these Mpf systems are not always the same. Thus, the IncP-9 Dtr system is like the IncW system, which is quite distinct from the IncP-1 system. The Dtr systems include all those functions that are needed to transfer once a mating bridge has been formed – assembly of the relaxosome, association with the mating bridge, triggering of nicking and initiation of transfer rolling circle replication, and then reconstitution of the plasmid in the recipient cell by lagging strand synthesis and maturation. The transfer origin where nicking occurs, *oriT*, is the core of the Dtr system and there is recognizable sequence similarity between all *oriT* sequences [132]. All the self-transmissible or mobilizable *Pseudomonas* plasmids described to date have a recognizable *oriT* sequence.

One topical area of research is the visualization of biological processes proceeding in real-time. For the IncP-1β plasmid this has been achieved by tagging the plasmid with multiple copies of the *lac* operator. The plasmid is then allowed to transfer into a recipient that is expressing a fusion protein consisting of the *lac* repressor protein LacI and the green fluorescent protein from the jelly fish *Aequorea victoria* that is commonly used as a reporter [135]. The fluorescent protein is distributed evenly in the recipient, until the plasmid enters and then it is sequestered to where the plasmid is and forms a clear focus that eventually replicates and starts to segregate symmetrically. An alternative way of visualizing transfer more generally (without being able to see the plasmid foci inside the bacteria) involves tagging the plasmid with a *lacp–gfp* cassette so that green fluorescent protein production is switched off in donor bacteria by the presence of *lacI*^{q} in the chromosome [136,137]. Expression is then switched on in recipients after transfer due to the absence of *lacI* gene, allowing easy detection of the cells that have acquired the plasmid. This has been applied to pWW0 because it is the best-studied degradative plasmid and has been used as a model for manipulation of bioremediation strategies that involve environmental release of plasmid-bearing strains. When pWW0 transfer in microbial biofilm communities was studied [136] the results showed that transfer occurs when a donor interacts with a growing or established biofilm, but that transfer to endogenous bacteria only occurs at the interface between the donor and recipient cultures, and does not spread throughout the recipient population [138]. The importance of plasmid transfer systems in promoting biofilm formation by their hosts has been discovered recently [139] and this raises a number of important issues about key aspects of conjugation. The most important of these stems from the observation that F plasmid can stimulate biofilm formation even in the absence of recipient F^- bacteria. It appears that interaction between plasmid-positive bacteria is not inhibited under conditions of biofilm formation despite plasmid-encoded surface exclusion functions. Thus, plasmid transfer into plasmid-positive bacteria may occur at reasonable frequency, allowing homologous recombination between related plasmids to occur. Therefore, in natural environments conjugation may promote inter-plasmid recombination and thus favor rapid plasmid evolution. Evidence that this is indeed occurring may come from current projects to sequence multiple members of the same plasmid family. If recombination turns out to be a frequent event then it will be an important factor of the HGP.

11.11 Phenotypic Cargo of *Pseudomonas* Plasmids

Plasmids are major vehicles for gene movement between bacteria. Their genetic cargo is often acquired through the action of transposable elements or integrons [140] which may also facilitate loss of such DNA in the absence of selection. While the sequence of IncP-6 plasmid Rms149 from *P. aeruginosa* and the *P. fluorescens* plasmid pQBR103 appear to indicate that a plasmid can apparently maintain a very large amount of not selectively advantageous DNA, the isolation of cryptic IncP-1 [141] and IncP-4-like [142] plasmids containing little other than "backbone" functions (replication, maintenance and transfer/mobilization) is consistent with this clear distinction between "vehicle" and "cargo". Indeed, loss of unselected tetracycline resistance genes can be observed rapidly from IncP-1β plasmid pB10 where they are flanked by directly repeated copies of a truncated transposase gene [143]. A similar type of truncated insertion sequence (IS) element is responsible for recombination between different copies of pJP4 in the same cell resulting in derivatives with either duplications of the 2,4-D degradation genes or none at all [144], illustrating the plasticity of the plasmid genome. Resistance plasmid pB10 also contains a complete copy of IS1071, which is relatively unusual because this IS is most frequently associated with genes for catabolic functions [145].

Insertions tend to occur repeatedly in the same places probably because the initial insertion defines one of the places where the backbone can be disrupted without major impact on plasmid maintenance and also because the initial insertion may represent an expanded target itself into which further insertions will take place [85,88]. However, the suggestion that repeated sequence motifs present in the IncP-1β backbone where the majority of insertions occur may represent a hotspot for transposition was not borne out by the data on a collection of new insertions [88].

Making sense of the multiple insertion events can be done through bioinformatics approaches as illustrated for pWW0 [69]. This involved identifying transposases, the characteristic inverted repeat sequences associated with them and then the direct duplications (commonly 5 or 9 bp) associated with insertion events which allowed individual events to be defined. For some plasmids the number of events driven by quite closely related elements makes it very difficult to establish which transposase works on which transposon end sequences. However, expression of the different tranposase genes from IncP-7 plasmid pWW53 in *trans* to antibiotic resistance cassettes flanked by all possible combinations of end sequences defined which work together and allowed a model for creation of the plasmid including transposition of one element into the ends of another creating ends whose specificity had changed in subtle ways [146]. This provides one model for such studies.

An efficient way of identifying, cloning and assessing the nonbackbone part of plasmids belonging to a well-characterized plasmid group was applied to IncP-1 plasmids. After isolation of new IncP-1 plasmids, plasmid DNA was purified, subject to restriction, and the fragments separated by electrophoresis, Southern blotted and then probed with DNA corresponding to the known IncP-1 core. Restriction fragments that did not light up were then extracted from similar gels, and then cloned and sequenced [21]. In this particular study a number of functions not previously associated with IncP-1 were identified, including restriction/modification, superoxide dismutase and multidrug efflux system of the RND (resistance/nodulation/cell division) family.

An interesting phenotype is that conferred by the widespread genes originally identified as homologs of the *umuCD* genes of pKM101 that were associated with ultraviolet mutagenesis [147]. These genes have been shown to encode DNA Pol V which is error prone and expressed under stress conditions. It elevates the rate of mutation and increases the appearance of mutants that are able to survive the adverse selective pressure. While degradation functions are often the highest profile phenotypes associated with *Pseudomonas* plasmids, resistance in *P. aeruginosa* is a major problem. Hence the discovery of a new β-lactamase on an IncP-2 plasmid isolated repeatedly from clinical samples in China [148] is of significant importance. This is first time that such a resistance marker has been discovered on a megaplasmid and monitoring its spread will be important to establish the role of this location in dissemination of resistance.

11.12
Conclusions

The aim of cataloguing the plasmids of *Pseudomonas* is to provide a means of investigating the HGP of *Pseudomonas* and ordering the information that we have about it. The last half decade has seen an explosion of the number of plasmid sequences including those relevant to this book. While there are still quite a number of plasmid families requiring the first archetype to be sequenced, the fact that multiple sequences are becoming available for certain plasmid families reflects not only the ability to detect those plasmids for which we have molecular tools, but also the abundance of these dominant plasmid types in the environments tested. There is still a need for more data before we will have a clear picture of the diversity of the HGP in this genus. Concerted effort is now required to characterize the less common plasmids that may represent the vehicles for future spread of genes.

Acknowledgements

Recent work in this area in the author's laboratory was supported by a project grant from The Wellcome Trust (063083) as well as INTAS programmes (99-01487 and 2001-2383).

References

1 Thomas, C.M. (ed.), (2000) *The Horizontal Gene Pool: Bacterial Plasmids and Gene Spread*, Harwood Academic, Amsterdam.

2 Top, E. *et al.* (2000) Phenotypic traits conferred by plasmids, in *The Horizontal Gene Pool: Bacterial Plasmids and Gene Spread* (ed. C.M. Thomas), Harwood Academic, Amsterdam, pp. 249–285.

3 Thomas, C.M. (2004) Evolution and population genetics of bacterial plasmids, in *Plasmid Biology* (eds B. Funnell and G. Phillips), ASM Press, Washington, DC, pp. 509–528.

4 Eberhard, W.G. (1989) Why do bacterial plasmids carry some genes and not others? *Plasmid*, **21**, 167–174.

5 Chiu, C.-M. and Thomas, C.M. (2004) Evidence for past integration of IncP-1 plasmids into bacterial chromosomes. *FEMS Lett*, **241**, 163–169.

6 Thomas, C.M. and Haines, A.S. (2004) *Plasmids of Pseudomonas species, in Pseudomonas* (ed. J. Ramos), Kluwer/Plenum, London, pp. 197–231.

7 Fujita, M. *et al.* (1989) Identification and DNA sequencing of a new plasmid (pPST1) in *Pseudomonas stutzeri* MO-19. *Plasmid*, **22**, 271–274.

8 Tett, A. *et al.* (2007) Sequence-based analysis of pQBR103; a representative of a unique, transfer-proficient mega plasmid resident in the microbial community of sugar beet. *ISME J*, **1**, 331–340.

9 Sagai, H. *et al.* (1976) Classification of R plasmids by incompatibility in *Pseudomonas* aeruginosa. *Antimicrob Agents Chemother*, **10**, 573–578.

10 Jacoby, G.A. (1986) Resistance plasmids of *Pseudomonas*, in *The Bacteria* (ed. J.R. Sokatch), Academic Press, New York, pp. 265–293.

11 Boronin, A.M. (1992) Diversity of pseudomonas plasmids – to what extent? *FEMS Microbiol Lett*, **100**, 461–467.

12 Couturier, M. *et al.* (1988) Identification and classification of bacterial plasmids. *Microbiol Rev*, **52**, 375–395.

13 Kobayashi, N. and Bailey, M.J. (1994) Plasmids isolated from the sugar-beet phyllosphere show little or no homology to molecular probes currently available for plasmid typing. *Microbiology*, **140**, 289–296.

14 Top, E. *et al.* (1994) Exogenous isolation of mobilizing plasmids from polluted soils and sludges. *Appl Environ Microbiol*, **60**, 831–839.

15 Campbell, J.I.A., Jacobsen, C.S. and Sorensen, J. (1995) Species variation and plasmid Incidence among fluorescent *Pseudomonas* strains isolated from agricultural and industrial soils. *FEMS Microbiol Ecol*, **18**, 51–62.

16 Hill, K.E. *et al.* (1995) Retrotransfer of IncP1-like plasmids from aquatic bacteria. *Lett Appl Microbiol*, **20**, 317–322.

17 Dahlberg, C. *et al.* (1997) Conjugative plasmids isolated from bacteria in marine environments show various degrees of homology to each other and are not closely related to well-characterized plasmids. *Appl Environ Microbiol*, **63**, 4692–4697.

18 Sobecky, P.A. *et al.* (1997) Plasmids isolated from marine sediment microbial communities contain replication and incompatibility regions unrelated to those of known plasmid groups. *Appl Environ Microbiol*, **63**, 888–895.

19 Smit, E., Wolters, A. and van Elsas, J.D. (1998) Self-transmissible mercury resistance plasmids with gene-mobilizing capacity in soil bacterial populations: influence of wheat roots and mercury addition. *Appl Environ Microbiol*, **64**, 1210–1219.

20 van Elsas, J.D. *et al.* (1998) Isolation, characterization, and transfer of cryptic gene-mobilizing plasmids in the wheat rhizosphere. *Appl Environ Microbiol*, **64**, 880–889.

21 Droge, M., Puhler, A. and Selbitschka, W. (2000) Phenotypic and molecular characterization of conjugative antibiotic resistance plasmids isolated from bacterial communities of activated sludge. *Mol Gen Genet*, **263**, 471–482.

22 Gstalder, M.-E. *et al.* (2003) Replication functions of new broad host range plasmids isolated from polluted soils. *Res Microbiol*, **154**, 499–509.

23 Haines, A.S. *et al.* (2006) Plasmids from freshwater environments capable of IncQ retrotransfer are diverse and include pQKH54, a new IncP-1 subgroup archetype. *Microbiology*, **152**, 2689–2701.

24 Vedler, E., Vahter, M. and Heinaru, A. (2004) The completely sequenced plasmid pEST4011 contains a novel IncP1 backbone and a catabolic transposon harboring *tfd* genes for 2,4-dichlorophenoxyacetic acid degradation. *J Bacteriol*, **186**, 7161–7174.

25 Bahl, M.I. *et al.* (2007) The multiple antibiotic resistance IncP-1 plasmid pKJK5 isolated from a soil environment is phylogenetically divergent from members of the previously established α, β and δ sub-groups. *Plasmid*, **58**, 31–43.

26 Llanes, C. *et al.* (1994) Cloning and characterization of the IncA/C plasmid RA1 replicon. *J Bacteriol*, **176**, 3403–3407.

27 Welch, T.J. *et al.* (2007) Multiple antimicrobial resistance in plague: an emerging public health risk. *PLoS One*, **2**, e309.

28 Rawlings, D.E. and Tietze, E. (2001) Comparative biology of IncQ and IncQ-like plasmids. *Microbiol Mol Biol Rev*, **65**, 481–496.

29 Haines, A.S., Cheung, M. and Thomas, C.M. (2006) Evidence that IncG (IncP-6) and IncU plasmids form a single incompatibility group. *Plasmid*, **55**, 210–215.

30 Izmalkova, T.Y. *et al.* (2006) Molecular classification of IncP-9 naphthalene degradation plasmids. *Plasmid*, **56**, 1–10.

31 Tauch, A. *et al.* (2002) The complete nucleotide sequence and environmental distribution of the cryptic, conjugative, broad-host-range plasmid pIPO2 isolated from bacteria of the wheat rhizosphere. *Microbiology*, **148**, 1637–1653.

32 Schneiker, S. *et al.* (2001) The genetic organization and evolution of the broad host range mercury resistance plasmid pSB102 isolated from a microbial population residing in the rhizosphere of alfalfa. *Nucleic Acids Res*, **29**, 5169–5181.

33 Marques, M.V., da Silva, A.M. and Gomes, S.L. (2001) Genetic organization of plasmid pXF51 from the plant pathogen *Xylella fastidiosa*. *Plasmid*, **45**, 184–199.

34 Kwong, S.M. *et al.* (1998) Sequence analysis of plasmid pRA2 from *Pseudomonas alcaligenes* NCIB 9867 (P25X) reveals a novel replication region. *FEMS Microbiol Lett*, **158**, 159–165.

35 Peters, M. *et al.* (2001) Features of the replicon of plasmid pAM10.6 of *Pseudomonas fluorescens*. *Plasmid*, **46**, 25–36.

36 Holtwick, R. *et al.* (2001) A novel rolling-circle-replicating plasmid from *Pseudomonas putida* P8: molecular characterization and use as vector. *Microbiology*, **147**, 337–344.

37 Nieto, C. *et al.* (1992) Genetic and functional analysis of the basic replicon of pPS10, a plasmid specific for *Pseudomonas* isolated from *Pseudomonas syringae* pathovar *savastanoi*. *J Mol Biol*, **223**, 415–426.

38 Charnock, C. (1997) Characterization of the cryptic plasmids of the *Pseudomonas alcaligenes* type strain. *Plasmid*, **37**, 189–198.

39 Jansons, I. *et al.* (1994) Deletion and transposon mutagenesis and sequence-analysis of the pRO1600 *oriR* region found in the broad-host-range plasmids of the pQF Series. *Plasmid*, **31**, 265–274.

40 Heeb, S. *et al.* (2000) Small, stable shuttle vectors based on the minimal pVS1 replicon for use in Gram-negative, plant-associated bacteria. *Mol Plant Microbe Interact*, **13**, 232–237.

41 Sesma, A., Sundin, G.W. and Murillo, J. (2000) Phylogeny of the replication regions of pPT23A-like plasmids from *Pseudomonas syringae*. *Microbiology*, **146**, 2375–2384.

42 Ma, Z. *et al.* (2007) Phylogenetic analysis of the pPT23A plasmid family of *Pseudomonas syringae*. *Appl Environ Microbiol*, **73**, 1287–1295.

43 Sesma, A., Sundin, G.W. and Murillo, J. (1998) Closely related plasmid replicons coexisting in the phytopathogen *Pseudomonas* syringae show a mosaic organization of the replication region and altered incompatibility behavior. *Appl Environ Microbiol*, **64**, 3948–3953.

44 Vivian, A., Murillo, J., and Jackson, R.W. (2001) The roles of plasmids in phytopathogenic bacteria: mobile arsenals? *Microbiology*, **147**, 763–780.

45 Pansegrau, W. *et al.* (1994) Complete nucleotide sequence of Birmingham IncP-alpha plasmids – compilation and comparative analysis. *J Mol Biol*, **239**, 623–663.

46 Wilkins, B.M. *et al.* (1996) Distribution of restriction enzyme recognition sequences on broad host range plasmid RP4: molecular and evolutionary implications. *J Mol Biol*, **258**, 447–456.

47 Thorsted, P.B. *et al.* (1998) Complete sequence of the IncP beta plasmid R751: implications for evolution and organization of the IncP backbone. *J Mol Biol*, **282**, 969–990.

48 Eckhardt, T. (1978) A rapid method for the identification of plasmid desoxyribonucleic acid. *Plasmid*, **1**, 584–588.

49 Hansen, J.B. and Olsen, R.H. (1978) Isolation of large plasmids and characterization of the P2 incompatibility group plasmids pMG1 and p MG5. *J Bacteriol*, **135**, 227–238.

50 Kado, C.I. and Liu, S.T. (1981) Rapid procedure for detection and isolation of large and small plasmids. *J Bacteriol*, **145**, 1365–1373.

51 Krasowiak, R. *et al.* (2002) PCR primers for detection and characterisation of IncP-9 plasmids. *FEMS Microbiol Ecol*, **42**, 217–225.

52 Bale, M.J., Fry, J.C. and Day, M.J. (1987) Plasmid transfer between strains of *Pseudomonas aeruginosa* on membrane filters attached to river stones. *J Gen Microbiol*, **133**, 3099–3107.

53 Bale, M.J., Fry, J.C. and Day, M.J. (1988) Transfer and occurrence of large mercury resistance plasmids in river epilithon. *Appl Environ Microbiol*, **54**, 972–978.

54 Lilley, A.K. *et al.* (1996) Diversity of mercury resistance plasmids obtained by exogenous isolation from the bacteria of sugar beet in three successive years. *FEMS Microbiol Ecol*, **20**, 211–227.

55 Smalla, K. *et al.* (2000) Exogenous isolation of antibiotic resistance plasmids from piggery manure slurries reveals a high prevalence and diversity of IncQ-like plasmids. *Appl Environ Microbiol*, **66**, 4854–4862.

56 Top, E.M., Holben, W.E. and Forney, L.J. (1995) Characterization of diverse 2,4-dichlorophenoxyacetic acid-degradative plasmids isolated from soil by complementation. *Appl Environ Microbiol*, **61**, 1691–1698.

57 Hill, K.E., Weightman, A.J. and Fry, J.C. (1992) Isolation and screening of plasmids from the epilithon which mobilize recombinant plasmid pD10. *Appl Environ Microbiol*, **58**, 1292–1300.

58 Smalla, K. *et al.* (2006) Increased abundance of IncP-1beta plasmids and mercury resistance genes in mercury-polluted river sediments: first discovery of IncP-1beta plasmids with a complex *mer* transposon as the sole accessory element. *Appl Environ Microbiol*, **72**, 7253–7259.

59 Gotz, A. *et al.* (1996) Detection and characterization of broad-host-range plasmids in environmental bacteria by PCR. *Appl Environ Microbiol*, **62**, 2621–2628.

60 Thomas, C.M. and Thorsted, P. (1994) PCR probes for promiscuous plasmids. *Microbiology*, **140**, 1.

61 Greated, A. and Thomas, C.M. (1999) A pair of PCR primers for IncP-9 plasmids. *Microbiology*, **145**, 3003–3004.

62 Espinosa, M. *et al.* (2000) Plasmid replication and copy number control, in *The Horizontal Gene Pool: Bacterial Plasmids and Gene Spread* (ed. C.M. Thomas), Harwood Academic, Amsterdam, pp. 1–47.

63 Kalyaeva, E. *et al.* (2002) A broad host range plasmid vector that does not encode replication proteins. *FEMS Microbiol Lett*, **211**, 91–95.

64 Viegas, C.A. *et al.* (1997) Description of a novel plasmid replicative origin from a genetically distinct family of conjugative plasmids associated with phytosphere microflora. *FEMS Microbiol Lett*, **149**, 121–127.

65 Turner, S.L., Lilley, A.K. and Bailey, M.J. (2002) Two *dnaB* genes are associated with the origin of replication of pQBR55, an exogenously isolated plasmid from the rhizosphere of sugar beet. *FEMS Microbiol Ecol*, **42**, 209–215.

66 Altenschmidt, U., Eckerskorn, C. and Fuchs, G. (1990) Evidence that enzymes of a novel aerobic 2-amino-benzoate metabolism in denitrifying Pseudomonas are coded on a small plasmid. *Eur J Biochem*, **194**, 647–653.

67 Stassen, A.P.M. *et al.* (1992) Nucleotide sequence of the genome of the filamentous bacteriophage-I 2-2– module evolution of the filamentous phage genome. *J Mol Evol*, **34**, 141–152.

68 Greated, A. *et al.* (2000) The replication and stable-inheritance functions of IncP-9 plasmid pM3. *Microbiology*, **146**, 2249–2258.

69 Greated, A. *et al.* (2002) The complete nucleotide sequence of TOL plasmid p WW0. *Environ Microbiol*, **4**, 856–871.

70 Hiraga, S.-I., Sugiyama, T. and Itoh, T. (1994) Comparative analysis of the replication regions of eleven ColE2-related plasmids. *J Bacteriol*, **176**, 7233–7243.

71 del Solar, G. *et al.* (1998) Replication and control of circular bacterial plasmids. *Microbiol Mol Biol Rev*, **62**, 434–464.

72 Espinosa, M. *et al.* (1982) Interspecific plasmid transfer between *Streptococcus pneumoniae* and *Bacillus subtilis*. *Mol Gen Genet*, **188**, 195–201.

73 Maeda, K. *et al.* (2003) Complete nucleotide sequence of carbazole/dioxin-degrading plasmid pCAR1 in *Pseudomonas resinovorans* strain CA10 indicates its mosaicity and the presence of large catabolic transposon Tn4676. *J Mol Biol*, **326**, 21–33.

74 Komori, H. *et al.* (1999) Crystal structure of a prokaryotic replication initiator protein bound to DNA at 2.6 Å resolution. *EMBO J*, **18**, 4597–4607.

75 Kruger, R. and Filutowicz, M. (2000) Dimers of Pi protein bind the A + T-rich region of the R6K gamma origin near the leading-strand synthesis start sites: regulatory implications. *J Bacteriol*, **182**, 2461–2467.

76 Lu, Y.B., Datta, H.J. and Bastia, D. (1998) Mechanistic studies of initiator–initiator interaction and replication initiation. *EMBO J*, **17**, 5192–5200.

77 Fueki, T. and Yamaguchi, K. (2001) The structure and function of the replication initiator protein (Rep) of pSC101: an analysis based on a novel positive–selection system for the replication–deficient mutants. *J Biochem (Tokyo)*, **130**, 399–405.

78 Sharma, R., Kachroo, A. and Bastia, D. (2001) Mechanistic aspects of DnaA-RepA interaction as revealed by yeast forward and reverse two-hybrid analysis. *EMBO J*, **20**, 4577–4587.

79 Antoine, R. and Locht, C. (1992) Isolation and molecular characterization of a novel broad-host-range plasmid from *Bordetella bronchisceptica* with sequence similarity to plasmids from Gram-positive organisms. *Mol Microbiol*, **6**, 1785–1799.

80 Bartosik, D. *et al.* (2001) Characterization and sequence analysis of the replicator region of the novel plasmid pALC1 from *Paracoccus alcaliphilus*. *Plasmid*, **45**, 222–226.

81 Fu, J.F. *et al.* (1996) Characterization of the replicon of plasmid pSW500 of *Erwinia stewartii*. *Mol Gen Genet*, **250**, 699–704.

82 Krasowiak, R. *et al.* (2006) IncP-9 replication initiator protein binds to multiple DNA sequences in *oriV* and recruits host DnaA protein. *Plasmid*, **56**, 187–201.

83 Kongsuwan, K. *et al.* (2006) The plasmid RK2 replication initiator protein (TrfA) binds to the sliding clamp beta subunit of DNA polymerase III: implication for the toxicity of a peptide derived from the amino-terminal portion of 33-kilodalton TrfA. *J Bacteriol*, **188**, 5501–5509.

84 Scherzinger, E. *et al.* (1991) Plasmid RSF1010 DNA-replication *in vitro* promoted by purified RSF1010 RepA, RepB and RepC proteins. *Nucleic Acids Res*, **19**, 1203–1211.

85 Thomas, C.M. (2000) Paradigms of plasmid organization. *Mol Microbiol*, **37**, 485–491.

86 Adamczyk, M. and Jagura-Burdzy, G. (2003) Spread and survival of promiscuous IncP-1 plasmids. *Acta Biochim Pol*, **50**, 425–453.

87 Dennis, J.J. and Zylstra, G.J. (2004) Complete sequence and genetic organization of pDTG1, the 83 kilobase naphthalene degradation plasmid from *Pseudomonas putida* strain NCIB 9816-4. *J Mol Biol*, **341**, 753–768.

88 Sota, M. *et al.* (2007) Region-specific insertion of transposons in combination with selection for high plasmid transferability and stability accounts for the structural similarity of IncP-1 plasmids. *J Bacteriol*, **189**, 3091–3098.

89 Gerdes, K. *et al.* (2000) Plasmid maintenance systems, in *The Horizontal Gene Pool: Bacterial Plasmids and Gene Spread* (ed. C.M. Thomas), Harwood Academic Press, Amsterdam, pp. 49–85.

90 Sia, E.A. *et al.* (1995) Different relative importance of the *par* operons and the effect of conjugal transfer on the maintenance of intact promiscuous plasmid RK2. *J Bacteriol*, **177**, 2789–2797.

91 Bahl, M.I. *et al.* (2007) Conjugative transfer facilitates stable maintenance of IncP-1 Plasmid pKJK5 in *Escherichia coli* cells colonizing the gastrointestinal tract of the germfree rat. *Appl Environ Microbiol*, **73**, 341–343.

92 Summers, D.K., Beton, C.W.H. and Withers, H.L. (1993) Multicopy plasmid instability – the dimer catastrophe hypothesis. *Mol Microbiol* **8**, 1031–1038.

93 Guhathakurta, A. and Summers, D. (1995) Involvement of ArgR and PepA in the pairing of ColE1 dimer resolution sites. *Microbiology*, **141**, 1163–1171.

94 Easter, C.L., Schwab, H. and Helinski, D.R. (1998) Role of the parCBA operon of the broad-host-range plasmid RK2 in stable plasmid maintenance. *J Bacteriol*, **180**, 6023–6030.

95 Szczepanowski, R. *et al.* (2007) Novel macrolide resistance module carried by the IncP-1β resistance plasmid pRSB111, isolated from a wastewater treatment plant. *Antimicrob Agents Chemother*, **51**, 673–678.

96 Becker, E.C. and Meyer, R.J. (1997) Acquisition of resistance genes by the IncQ plasmid R1162 is limited by its high copy number and lack of a partitioning mechanism. *J Bacteriol*, **179**, 5947–5950.

97 Bignell, C. and Thomas, C.M. (2001) The bacterial ParA-ParB partitioning proteins. *J Biotechnol*, **91**, 1–34.

98 Thomas, C.M., Stalker, D.M. and Helinski, D.R. (1981) Replication and Incompatibility properties of segments of the origin region of replication of the broad host range plasmid Rk2. *Mol Gen Genet*, **181**, 1–7.

99 Thomas, C.M. *et al.* (1984) Analysis of copy number control elements in the region of the vegetative replication origin of the broad host-range plasmid RK2. *EMBO J*, **3**, 57–63.

100 Meyer, R.J. and Hinds, M. (1982) Multiple mechanisms for expression of incompatibility by broad host range plasmid RK2. *J Bacteriol*, **152**, 1078–1090.

101 Easter, C.L., Sobecky, P.A. and Helinski, D.R. (1997) Contribution of different segments of the *par* region to stable maintenance of the broad-host-range plasmid RK2. *J Bacteriol*, **179**, 6472–6479.

102 Wilson, J.W., Sia, E.A. and Figurski, D.H. (1997) The *kilE* locus of promiscuous IncP alpha plasmid RK2 is required for stable maintenance in *Pseudomonas aeruginosa*. *J Bacteriol*, **179**, 2339–2347.

103 Wilson, J.W. and Figurski, D.H. (2002) Host-specific incompatibility by 9-bp direct repeats indicates a role in the maintenance of broad-host-range plasmid RK2. *Plasmid*, **47**, 216–223.

104 Bathe, S. *et al.* (2004) High phylogenetic diversity of transconjugants carrying plasmid pJP4 in an activated sludge-derived microbial community. *FEMS Microbiol Lett*, **235**, 215–219.

105 Musovic, S. *et al.* (2006) Cultivation-independent examination of horizontal transfer and host range of an IncP-1 plasmid among Gram-positive and Gram-negative bacteria indigenous to the barley rhizosphere. *Appl Environ Microbiol*, **72**, 6687–6692.

106 De Gelder, L. *et al.* (2007) Stability of a promiscuous plasmid in different hosts: no guarantee for a long-term relationship. *Microbiology*, **153**, 452–463.

107 Guiney, D.G. (1982) Host range of the conjugation and replication functions of the *Escherichia coli* sex plasmid Flac. *J Mol Biol*, **162**, 699–703.

108 Zhong, Z.P., Helinski, D. and Toukdarian, A. (2005) Plasmid host-range: restrictions to F replication in *Pseudomonas*. *Plasmid*, **54**, 48–56.

109 Boulnois, G.J. *et al.* (1985) Transposon donor plasmids, based on ColIb-P9, for use in *Pseudomonas putida* and a variety of other Gram-negative bacteria. *Mol Gen Genet*, **200**, 65–67.

110 Jussila, M.M. *et al.* (2007) TOL plasmid transfer during bacterial conjugation *in vitro* and rhizoremediation of oil compounds *in vivo*. *Environ Pollut*, **146**, 510–524.

111 De Gelder, L. *et al.* (2005) Plasmid donor affects host range of promiscuous IncP-1 beta plasmid pB10 in an activated-sludge microbial community. *Appl Environ Microbiol*, **71**, 5309–5317.

112 Smith, C.A. and Thomas, C.M. (1984) Nucleotide sequence of the *trfA* gene of broad host-range plasmid RK2. *J Mol Biol*, **175**, 251–262.

113 Shingler, V. and Thomas, C.M. (1984) Analysis of the *trfA* region of broad host-range plasmid RK2 by transposon mutagenesis and identification of polypeptide products. *J Mol Biol*, **175**, 229–249.

114 Durland, R.H. and Helinski, D.R. (1987) The sequence encoding the 43-kilodalton TrfA protein is required for efficient replication or maintenance of minimal RK2 replicons in *Pseudomonas aeruginosa*. *Plasmid*, **18**, 164–169.

115 Shingler, V. and Thomas, C.M. (1989) Analysis of nonpolar insertion mutations in the *trfA* gene of IncP plasmid RK2 which affect its broad-host-range property. *Biochim Biophys Acta*, **1007**, 301–308.

116 Cross, M.A., Warne, S.R. and Thomas, C.M. (1986) Analysis of the vegetative replication origin of broad-host-range plasmid RK2 by transposon mutagenesis. *Plasmid*, **15**, 132–146.

117 Konieczny, I. and Helinski, D.R. (1997) Helicase delivery and activation by DnaA and TrfA proteins during the initiation of replication of the broad host-range plasmid RK2. *J Biol Chem*, **272**, 33312–33318.

118 Doran, K.S., Helinski, D.R. and Konieczny, I. (1999) Host-dependent requirement for specific DnaA boxes for plasmid RK2 replication. *Mol Microbiol*, **33**, 490–498.

119 Caspi, R. *et al.* (2000) Interactions of DnaA proteins from distantly related bacteria with the replication origin of the broad host range plasmid RK2. *J Biol Chem*, **275**, 18454–18461.

120 Caspi, R. *et al.* (2001) A broad host range replicon with different requirements for replication initiation in three bacterial species. *EMBO J*, **20**, 3262–3271.

121 Zhong, Z.P., Helinski, D. and Toukdarian, A. (2003) A specific region in the N terminus of a replication initiation protein of plasmid RK2 is required for recruitment of *Pseudomonas aeruginosa* DnaB helicase to the plasmid origin. *J Biol Chem*, **278**, 45305–45310.

122 Fernandez-Tresguerres, M.E. *et al.* (1995) Host growth temperature and a conservative amino-acid substitution in the replication protein of pPS10 influence plasmid host-range. *J Bacteriol*, **177**, 4377–4384.

123 Titok, M., Maksimava, N.P. and Fomichev, Y.K. (1991) Characteristics of the broad host range IncP-9 R plasmid p M3. *Mol Genet Microbiol Virol*, **8**, 18–23.

124 Maestro, B. *et al.* (2003) Modulation of pPS10 host range by plasmid-encoded RepA initiator protein. *J Bacteriol*, **185**, 1367–1375.

125 Heuer, H., Fox, R.E. and Top, E.M. (2007) Frequent conjugative transfer accelerates adaptation of a broad-host-range plasmid to an unfavorable *Pseudomonas putida* host. *FEMS Microbiol Ecol*, **59**, 738–748.

126 Sevastsyanovich, Y.R. *et al.* (2005) Ability of IncP-9 plasmid pM3 to replicate in *E coli* is dependent on both *rep* and *par* functions. *Mol Microbiol*, **57**, 819–833.

127 Wild, J. *et al.* (2004) Gamma origin plasmids of RISK lineage replicate in diverse genera of Gram-negative bacteria. *Ann Microbiol*, **54**, 471–480.

128 Mierzejewska, J., Kulinska, A. and Jagura-Burdzy, G. (2007) Functional analysis of replication and stability regions of broad-host-range conjugative plasmid CTX-M3 from the IncL/M incompatibility group. *Plasmid*, **57**, 95–107.

129 Lorenz, M.G. and Wackernagel, W. (1991) High-frequency of natural genetic-transformation of *Pseudomonas stutzeri* in soil extract supplemented with a carbon energy and phosphorus source. *Appl Environ Microbiol*, **57**, 1246–1251.

130 Sikorski, J. *et al.* (1998) Natural genetic transformation of *Pseudomonas stutzeri* in a non-sterile soil. *Microbiology*, **144**, 569–576.

131 Kwong, S.M. *et al.* (2000) Characterization of the endogenous plasmid from *Pseudomonas alcaligenes* NCIB 9867: DNA sequence and mechanism of transfer. *J Bacteriol*, **182**, 81–90.

132 Zechner, E. *et al.* (2000) Conjugative-DNA transfer processes, in *The Horizontal Gene Pool: Bacterial Plasmids and Gene Spread* (ed. C.M. Thomas), Harwood Academic Press, Amsterdam, pp. 87–174.

133 Cabezon, E., Sastre, J.I. and de la Cruz, F. (1997) Genetic evidence of a coupling role for the TraG protein family in bacterial conjugation. *Mol Gen Genet*, **254**, 400–406.

134 Christie, P.J. and Vogel, J.P. (2000) Bacterial type IV secretion: conjugation systems adapted to deliver effector molecules to host cells. *Trends Microbiol*, **8**, 354–360.

135 Lawley, T.D. *et al.* (2002) Functional and mutational analysis of conjugative transfer region 1 (Tra1) from the IncHI1 plasmid R27. *J Bacteriol*, **184**, 2173–2180.

136 Christensen, B.B., Sternberg, C. and Molin, S. (1996) Bacterial plasmid conjugation on semi-solid surfaces monitored with green fluorescent protein (GFP) from *Aequorea victoria* as a marker. *Gene*, **173**, 59–65.

137 Aspray, T.J., Hansen, S.K. and Burns, R.G. (2005) A soil-based microbial biofilm exposed to 2,4-D: bacterial community development and establishment of conjugative plasmid pJP4. *FEMS Microbiol Ecol*, **54**, 317–327.

138 Christensen, B.B. *et al.* (1998) Establishment of new genetic traits in a microbial biofilm community. *Appl Environ Microbiol*, **64**, 2247–2255.

139 Ghigo, J.M. (2001) Natural conjugative plasmids induce bacterial biofilm development. *Nature*, **412**, 442–445.

140 Bissonnette, L. and Roy, P.H. (1992) Characterization of In0 of *Pseudomonas aeruginosa* plasmid pVS1, an ancestor of integrons of multiresistance plasmids and transposons of Gram-negative bacteria. *J Bacteriol*, **174**, 1248–1257.

141 Kamachi, K. *et al.* (2006) Plasmid pBP136 from *Bordetella pertussis* represents an ancestral form of IncP-1 beta plasmids without accessory mobile elements. *Microbiology*, **152**, 3477–3484.

142 Whittle, G. *et al.* (2000) Identification and characterization of a native *Dichelobacter nodosus* plasmid, pDN1. *Plasmid*, **43**, 230–234.

143 Schluter, A. *et al.* (2003) The 64,508 bp IncP-1 beta antibiotic multiresistance plasmid pB10 isolated from a waste-water treatment plant provides evidence for recombination between members of different branches of the IncP-1 beta group. *Microbiology*, **149**, 3139–3153.

144 Larrain-Linton, J. *et al.* (2006) Molecular and population analyses of a recombination event in the catabolic plasmid pJP4. *J Bacteriol*, **188**, 6793–6801.

145 Dennis, J.J. (2005) The evolution of IncP catabolic plasmids. *Curr Opin Biotechnol*, **16**, 291–298.

146 Yano, H. *et al.* (2007) Complete sequence determination combined with analysis of transposition/site-specific recombination events to explain genetic organization of IncP-7 TOL plasmid pWW53 and related mobile genetic elements. *J Mol Biol*, **369**, 11–26.

147 Tark, M. *et al.* (2005) DNA polymerase V homolog encoded by TOL plasmid pWW0 confers evolutionary fitness on *Pseudomonas putida* under conditions of environmental stress. *J Bacteriol*, **187**, 5203–5213.

148 Xiong, J.H. *et al.* (2006) bla(IMP-9) and its association with large plasmids carried by *Pseudomonas aeruginosa* isolates from the People's Republic of China. *Antimicrob Agents Chemother*, **50**, 355–358.

149 Tennstedt, T. *et al.* (2005) Sequence of the 68,869 bp IncP-1 alpha plasmid pTB11 from a waste-water treatment plant reveals a highly conserved backbone, a *Tn*402-like integron and other transposable elements. *Plasmid*, **53**, 218–238.

150 Haines, A.S. *et al.* (2007) Sequence of plasmid pBS228 and reconstruction of the IncP 1α phylogeny. *Plasmid*, **58**, 76–83.

151 Trefault, N. *et al.* (2004) Genetic organization of the catabolic plasmid pJP4 from *Ralstonia eutropha* JMP134 (pJP4) reveals mechanisms of adaptation to chloroaromatic pollutants and evolution of specialized chloroaromatic degradation pathways. *Environ Microbiol*, **6**, 655–668.

152 Sota, M., Kawasaki, H. and Tsuda, M. (2003) Structure of the haloacetate-catabolic IncP-1beta plasmid pUO1 and genetic mobility of its residing haloacetate-catabolic transposon. *J Bacteriol*, **185**, 6741–6745.

153 Heuer, H. *et al.* (2004) The complete sequences of plasmids pB2 and pB3 provide evidence for a recent ancestor of the IncP-1 beta group without any accessory genes. *Microbiology*, **150**, 3591–3599.

154 Tauch, A. *et al.* (2003) The 79,370-bp conjugative plasmid pB4 consists of an IncP-1 beta backbone loaded with a chromate resistance transposon, the strA–strB streptomycin resistance gene pair, the oxacillinase gene bla(NPS-1), a tripartite antibiotic efflux system of the resistance–nodulation–division family. *Mol Genet Genomics*, **268**, 570–584.

155 Schluter, A. *et al.* (2005) Plasmid pB8 is closely related to the prototype IncP-1 beta plasmid R751 but transfers poorly to *Escherichia coli* and carries a new transposon encoding a small multidrug resistance efflux protein. *Plasmid*, **54**, 135–148.

156 Sota, M. *et al.* (2007) Region-specific insertion of transposons in combination with selection for high plasmid transferability and stability accounts for the structural similarity of IncP-1 plasmids. *J Bacteriol*, **189**, 3091–3098.

157 Harada, K.M. *et al.* (2006) Sequence and analysis of the 46.6-kb plasmid pA1 from *Sphingomonas* sp A1 that corresponds to the typical IncP-1 beta plasmid backbone without any accessory gene. *Plasmid*, **56**, 11–23.

158 Jencova, V. *et al.* (2004) Chlorocatechol catabolic enzymes from *Achromobacter xylosoxidans* A8. *Int Biodeterior Biodegrad*, **54**, 175–181.

159 Bahl, M.I., Hansen, L.H. and Sorensen, S.J. (2007) Impact of conjugal transfer on the stability of IncP-1 plasmid pKJK5 in bacterial populations. *FEMS Microbiol Lett*, **266**, 250–256.

160 Martinez, B. *et al.* (2001) Complete nucleotide sequence and organization of the atrazine catabolic plasmid pADP-1 from *Pseudomonas* sp. strain ADP. *J Bacteriol*, **183**, 5684–5697.

161 Jacoby, G.A. (1974) Properties of R-plasmids determining gentamicin resistance by acetylation in *Pseudomonas aeruginosa*. *Antimicrob Agents Chemother*, **6**, 239–252.

162 Jacoby, G.A. *et al.* (1983) Properties of IncP-2 plasmids of *Pseudomonas* spp. *Antimicrob Agents Chemother*, **24**, 168–175.

163 Boronin, A.M. *et al.* (1984) Plasmids specifying ε-caprolactam degradation in *Pseudomonas* strains. *FEMS Microbiol Lett*, **22**, 167–170.

164 Aoki, T. *et al.* (1971) Detection of resistance factors in fish pathogen *Aeromonas liquefaciens*. *J Gen Microbiol*, **65**, 343–349.

165 Hedges, R.W. (1974) R factors from Providence. *J Gen Microbiol*, **81**, 171-.

166 Scholz, P. *et al.* (1989) Complete nucleotide sequence and gene organization of the broad host range plasmid R SF1010. *Gene*, **75**, 271–288.

167 Tietze, E. (1998) Nucleotide sequence and genetic characterization of the novel IncQ-like plasmid p IE1107. *Plasmid*, **39**, 165–181.

168 Kehrenberg, C. and Schwarz, S. (2005) Plasmid-borne florfenicol resistance in *Pasteurella multocida*. *J Antimicrob Chemother*, **55**, 773–775.

169 Rawlings, D.E. *et al.* (1993) A molecular analysis of a broad-host-range plasmid isolated from *ThioBacillus ferrooxidans*. *FEMS Microbiol Rev*, **11**, 3–8.

170 Clennel, A.M., Johnston, B. and Rawlings, D.E. (1995) Structure and function of *Tn*5467, a *Tn*21-like transposon located on the *ThioBacillus ferrooxidans* broad-host-range plasmid pTF- FC2. *Appl Environ Microbiol*, **61**, 4223–4229.

171 Gardner, M.N., Deane, S.M. and Rawlings, D.E. (2001) Isolation of a new broad-host-range IncQ-like plasmid, pTC-F14, from the acidophilic bacterium *AcidithioBacillus caldus* and analysis of the plasmid replicon. *J Bacteriol*, **183**, 3303–3309.

172 Bonemann, G. *et al.* (2006) Mobilizable IncQ-related plasmid carrying a new quinolone resistance gene, *qnrS2*, isolated from the bacterial community of a wastewater treatment plant. *Antimicrob Agents Chemother*, **50**, 3075–3080.

173 Summers, A.O. and Jacoby, G.A. (1978) Plasmid-determined resistance to boron and chromium compounds in *Pseudomonas aeruginosa*. *Antimicrob Agents Chemother*, **13**, 637–640.

174 Hedges, R.W. and Jacoby, G.A. (1980) Compatibility and molecular properties of plasmid Rms149 in *Pseudomonas aeruginosa* and *Escherichia coli*. *Plasmid*, **3**, 1–6.

175 Haines, A.S. *et al.* (2005) The IncP-6 plasmid Rms149 consists of a small mobilizable backbone with multiple large insertions. *J Bacteriol*, **187**, 4728–4738.

176 Schluter, A. *et al.* (2007) Erythromycin resistance-conferring plasmid pRSB105, isolated from a sewage treatment plant, harbors a new macrolide resistance determinant, an integron-containing *Tn*402-like element, a large region of unknown function. *Appl Environ Microbiol*, **73**, 1952–1960.

177 Rhodes, G. *et al.* (2004) Complete nucleotide sequence of the conjugative tetracycline resistance plasmid pFBAOT6, a member of a group of IncU plasmids with global ubiquity. *Appl Environ Microbiol*, **70**, 7497–7510.

178 Li, W. *et al.* (2004) Complete nucleotide sequence and organization of the naphthalene catabolic plasmid pND6-1 from *Pseudomonas* sp. strainND6. *Gene*, **336**, 231–240.

179 Pemberton, J.M. and Clark, A.J. (1973) Detection and characterization of plasmids in *Pseudomonas aeruginosa* strains PAO. *J Bacteriol*, **114**, 424–433.

180 Sota, M. *et al.* (2006) Genomic and functional analysis of the IncP-9 naphthalene-catabolic plasmid NAH7 and its transposon *Tn*4655 suggests catabolic gene spread by a tyrosine recombinase. *J Bacteriol*, **188**, 4057–4067.

181 Jacoby, G.A. *et al.* (1978) An explanation for the apparent host specificity of Pseudomonas plasmid P91 expression. *J Bacteriol*, **136**, 1159–1164.

182 Moore, R.J. and Krishnapillai, V., (1982) *Tn*7 and *Tn*501 insertions into *Pseudomonas aeruginosa* plasmid R91-5: mapping of two transfer regions. *J Bacteriol* **149**, 276–183.

183 Ingram, L. *et al.* (1972) A transmissible resistance element from a strain of *Pseudomonas aeruginosa* containing no detectable extrachromosomal DNA. *J Gen Microbiol*, **72**, 269–279.

184 Bryan, L.E., Shahrabadi, M.S. and Van der Elzen, H.M. (1974) Gentamicin resistance in *Pseudomonas aeruginosa*: R-factor mediated resistance. *Antimicrob Agents Chemother*, **6**, 191–199.

185 Partridge, S.R., Collis, C.M. and Hall, R.M. (2002) Class 1 integron containing a new gene cassette *aadA10* associated with *Tn*1404 from R151. *Antimicrob Agents Chemother*, **46**, 2400–2408.

186 Bryan, L.E. *et al.* (1973) Characterization of R931 and other *Pseudomonas aeruginosa* R-factors. *Antimicrobial Agents Chemother*, **3**, 625–637.

187 Fry, J.C. and Day, M.J. (1990) Plasmid transfer in the epilithon, in *Bacterial Genetics in Natural Environments* (eds J.C. Fry and M.J. Day), Cambridge University Press, Cambridge, pp. 55–80.

188 Gibbon, M.J. *et al.* (1999) Replication regions from plant-pathogenic *Pseudomonas syringae* plasmids are similar to ColE2-related replicons. *Microbiology,* **145**, 325–334.

189 Bender, C.L., Malvick, D.K. and Mitchell, R.E. (1989) Plasmid-mediated production of the phytotoxin coronatine in *Pseudomonas syringae pv tomato. J Bacteriol,* **171**, 807–812.

190 Rohmer, L. *et al.* (2003) Nucleotide sequence, functional characterization and evolution of pFKN, a virulence plasmid in *Pseudomonas syringae* pathovar *maculicola. Mol Microbiol,* **47**, 1545–1562.

191 Olsen, R.H., Debusscher, G. and McCombie, W.R. (1982) Development of broad-host-range vectors and gene banks – self-cloning of the *Pseudomonas aeruginosa* PAO chromosome. *J Bacteriol,* **150**, 60–69.

192 West, S.E. *et al.* (1994) Construction of improved *Escherichia–Pseudomonas* shuttle vectors derived from pUC18/19 and sequence of the region required for their replication in *Pseudomonas aeruginosa. Gene,* **148**, 81–86.

193 Itoh, Y. *et al.* (1984) Genetic and molecular characterization of the *Pseudomonas* plasmid p VS1. *Plasmid,* **11**, 206–220.

12
Biosynthesis and Regulation of Phenazine Compounds in *Pseudomonas* spp.

Dmitri V. Mavrodi, Linda S. Thomashow, and Wulf Blankenfeldt

12.1
Introduction

Fluorescent *Pseudomonas* spp. are foremost among the Gram-negative bacteria in their ability to produce secondary metabolites – a diverse assemblage of compounds that often exhibit antibiotic activity, but have no obvious function in microbial growth. Many of these compounds are complex in structure and energetically costly to synthesize, and the reason for their production has long puzzled researchers. With the availability of the molecular techniques that have enabled contemporary microbiologists to bridge the gap between studies *in vitro* and those in the environment, however, the contribution of secondary metabolites to microbial activities in their native habitats is now becoming apparent. The phenazines, a class of heterocyclic, nitrogen-containing metabolites produced by fluorescent pseudomonads and members of a few other bacterial genera, have proven particularly interesting in this regard. Phenazines were recognized more than 150 years ago as a factor in human and animal infections, but knowledge of their synthesis and antibiotic activity in soil environments, and their importance to microbial competitiveness and survival under those conditions, is comparatively recent. Remarkably, only now are the complex regulatory mechanisms and the key reactions in phenazine biosynthesis, reviewed here, being revealed. Together with more classical laboratory studies of microbial physiology, these results provide insight and a more holistic understanding of the full range of microbial responses required for growth and survival under diverse environmental conditions.

12.2
Diversity of Phenazine Compounds Produced by *Pseudomonas* spp

Fluorescent pseudomonads are Gram-negative bacteria that belong to the γ subclass of Proteobacteria [1]. The ability to produce phenazines is well documented in three *Pseudomonas* species: soil-inhabiting *P. fluorescens*, *P. chlororaphis* and *P. aureofaciens* (now classified as *P. chlororaphis*), and the opportunistic human and

Pseudomonas. Model Organism, Pathogen, Cell Factory. Edited by Bernd H.A. Rehm

ISBN: 978-3-527-31914-5

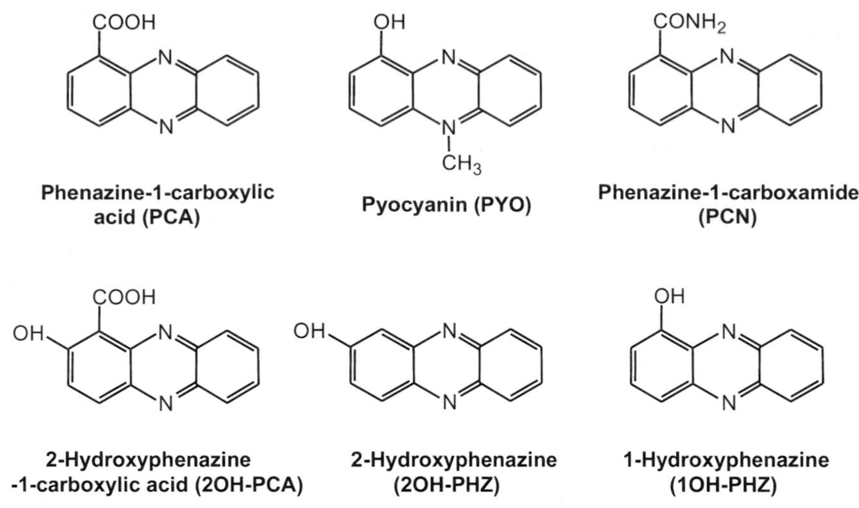

Figure 12.1 Phenazine compounds produced by fluorescent *Pseudomonas* spp.

animal pathogen *P. aeruginosa*. All pseudomonads except for *P. fluorescens*, which produces only phenazine-1-carboxylic acid (PCA) (Figure 12.1), have the capacity to synthesize multiple phenazine derivatives, and while the variety of compounds synthesized by given species appears to be specific, the relative amounts of possible products are affected by the growth conditions [2]. In addition to PCA, *P. chlororaphis* can produce phenazine-1-carboxamide [chlororaphin (PCN)] [3] or 2-hydroxyphenazine-1-carboxylic acid (2OH-PCA) and 2-hydroxyphenazine (2OH-PHZ) [4] (Figure 12.1).

P. aeruginosa was first isolated from pus colored with phenazines in 1882 [5]. It produces a characteristic blue–green phenazine pyocyanin [5-*N*-methyl-1-hydroxyphenazine (PYO)] (Figure 12.1), which is associated with the majority of *P. aeruginosa* isolates [2]. Phenazines are redox-active compounds capable of undergoing oxidation–reduction transformations (see Section 12.5) and a so-called "chameleon phenomenon" – in which cultures of *P. aeruginosa* change the color of the medium depending on the pH value and aeration – in fact reflects changes in the redox state of PYO [6]. In addition to PYO, *P. aeruginosa* also produces other phenazine compounds: PCA, PCN, 1-hydroxyphenazine (1OH-PHZ), and aeruginosins A (5-methyl-7-amino-1-carboxymethylphenazinium betaine) and B (5-methyl-7-amino-1-carboxy-3-sulphophenazinium betaine) [2,7].

Some recently discovered *Pseudomonas* strains that do not belong to the aforementioned phenazine-producing species also have been reported. For example, De Maeyer *et al.* [8] isolated two PCA- and PCN-producing *Pseudomonas* strains from the rhizosphere of cocoyam plants in Cameroon that subsequently were identified as possible new species. In a different study, Hu *et al.* [9] isolated a new phenazine-producing strain from the rhizosphere of sweet melon in China. The strain, *Pseudomonas* sp. M18, is closely related to *P. aeruginosa*, but does not grow at 41°C, and produces only PCA and the polyketide antibiotic pyoluteorin.

12.3
Biosynthesis of Phenazines and Evolution of Pathways

12.3.1
Core Biosynthetic Pathway and Assembly of the Phenazine Scaffold

The study of phenazine biosynthesis can be divided into different phases distinguished by the methodology employed and the questions addressed. An excellent overview of the earlier stages has been given by Turner and Messenger in 1986 [7], and only milestones will be repeated here.

A key finding of the first period, which started with the isolation of PYO from wound dressings by Fordos in the second half of the 19th century [10], was the observation that PYO production is linked to the presence of a microorganism [5] then named *Bacillus pyocyaneus* and now known as *P. aeruginosa*. It soon became obvious that the presence or absence of phenazine pigments depends on the composition of culture media [11] – a feature still used for the identification of *P. aeruginosa* in the clinic today [12]. A phenazine moiety in PYO was first suggested by Wrede and Strack [13], and the correct chemical structure was determined later by Hillemann [14]. Electrochemical properties of PYO subsequently were studied by Friedheim and Michaelis [15]. It was recognized that pigments purified from other strains also contain the phenazine nucleus, leading to the suggestion that the genus *Pseudomonas* be redefined to contain only phenazine-producing strains [16].

Research during the second stage mainly involved feeding experiments with isotope-labeled compounds to obtain a first glimpse at the underlying chemistry of phenazine biosynthesis. These studies identified shikimic acid as a precursor of PYO [17,18] and other phenazine derivatives (reviewed in Ref. [7]), indicating that they originate from the aromatic acid biosynthesis pathway. However, the exact branch point was not determined, and this problem was only resolved later, when two groups independently reported work with mutant strains of *P. aeruginosa*, leading to the proposition that chorismic acid is the common precursor of all phenazines [19,20]. Detailed insight into the mode of phenazine ring assembly was gained from experiments with specifically and not uniformly labeled precursors, first applied by Podojil and Gerber in 1970 [21]. This method ultimately revealed a symmetrical head-to-tail dimerization of two shikimate-derived molecules as a key step in phenazine biosynthesis [22]. Using various ^{15}N-labeled substrates, the side-chain of glutamine was identified as the source of the two phenazine ring nitrogen atoms [23]. A further problem presented itself in the large number of phenazine derivatives found in nature and in the observation that some strains produced several phenazines simultaneously, raising questions about the existence and identity of a common phenazine precursor from which this variety is generated. While the earlier literature suggests that this common precursor is phenazine-1,6-dicarboxylic acid (PDC) (reviewed in Ref. [7]), it has in the meantime become evident that, at least in *Pseudomonas*, a decarboxylation reaction precedes the formation of the aromatic ring system of phenazines [24]. Therefore, PCA seems to be the main precursor of strain-specific phenazine derivatives in pseudomonads, but other species that produce and incorporate both PCA and PDC have also been reported [25].

A new era in the investigation of phenazine biosynthesis began in the 1990s, when genes required for phenazine biosynthesis were first identified [26] and sequenced [27,28]. In *P. fluorescens* 2–79 this revealed the presence of a conserved gene cluster containing seven open reading frames termed the *phz* operon (Figure 12.2), which has since been found in all phenazine-producing *Pseudomonas* strains. It should be noted that *P. aeruginosa*, unlike other phenazine-producing species, carries two functionally homologous copies of the *phz* operon, but the reason for this redundancy is not fully understood [29]. The *phz* operon is responsible for the "core" phenazine biosynthesis pathway leading to PCA, whereas other genes that convert PCA to strain-specific derivatives are often located within or in the immediate vicinity of the operon [30]. Of the seven genes *phzA–G*, *phzC* encodes a type II 3-deoxy-D-arabino-heptulosonate 7-phosphate (DAHP) synthase, which catalyzes the first committed step of the shikimate pathway and thus acts upstream of phenazine biosynthesis. Interestingly, the activity of DAHP synthases is often controlled by allosteric mechanisms and it is thus conceivable that the presence of the *phzC* gene adds an additional layer of control to the mechanisms discussed in Section 12.4. Sequence alignment of PhzC with that of a type II DAHP synthase from *Mycobacterium tuberculosis* has revealed a large deletion in a region of the enzyme from *Pseudomonas* that has been implicated in feedback inhibition of these enzymes [31]. PhzC may therefore ensure a sufficient supply of metabolites into phenazine biosynthesis when other DAHP synthases are already inhibited. However, since

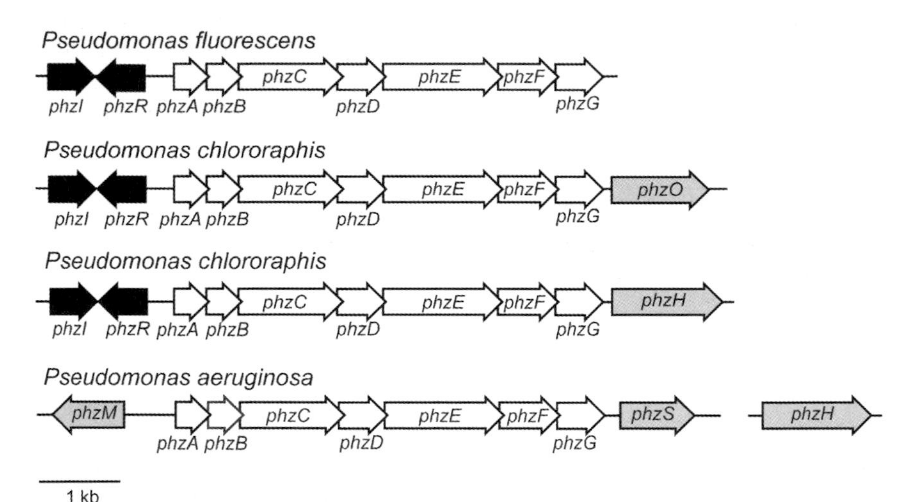

Figure 12.2 Comparison of phenazine biosynthetic loci from *P. fluorescens*, *P. chlororaphis* and *P. aeruginosa*. White, grey and black arrows indicate the position and direction of transcription of the core, phenazine-modifying and regulatory genes, respectively. The figure is drawn to scale.

the reaction catalyzed by PhzC is not unique to phenazine biosynthesis, it is not surprising that some phenazine producers lack *phzC* in their *phz* operons [30]. The remaining open reading frames have, with one notable exception [32], been found in all other phenazine-producing bacteria investigated to date and the following summary therefore seems valid for members of other genera in addition to *Pseudomonas*.

A fourth chapter in the elucidation of the biosynthetic reactions leading to phenazines was opened in 2001 and is still not closed today. Using whole-cell extracts of *Escherichia coli* harboring plasmids with different fragments of the *phz* operon of *P. fluorescens* 2–79, McDonald *et al.* [24] followed the conversion of chorismic acid and other substances to suggest a detailed reaction scheme for the final steps of PCA biosynthesis that was in line with predictions based on earlier DNA sequence analysis [28]. PhzE, a large enzyme related to anthranilate synthase, catalyzes the first committed step, and reacts chorismate and glutamine to 2-amino-2-deoxychorismate (ADIC), but instead of further converting ADIC to anthranilate, ADIC is released to be hydrolyzed to *trans*-2,3-dihydro-3-hydroxyanthranilate (DHHA). Hydrolysis is catalyzed by PhzD, an α/β-hydrolase that has been characterized in detail by crystal structure and biochemical analysis [33].

At variance with the pathway suggested by McDonald *et al.*, two independent reports indicate that DHHA is the substrate of PhzF, rather than of the FMN-dependent oxidase PhzG (Figure 12.3) [34,35]. PhzF belongs to a poorly characterized enzyme family and converts DHHA to ketone **1** by initiating a [1,5]-prototropic shift. The ketone cannot be isolated but reacts further, surprisingly and without further enzymatic catalysis, to form small amounts of PCA via several unstable intermediates. The current believe is that two ketone molecules undergo a condensation reaction to form a tricylic phenazine-like double imine intermediate similar to **2**. The face-to-face arrangement of the PhzF dimer, especially in the ligand-bound closed conformation [35], seems to support this reaction by increasing the probability of a ketone–ketone collision, as the newly generated ketones can only leave the two active centers via the PhzF dimer interface. The isomerization of DHHA and subsequent condensation of two ketone molecules are key steps of phenazine biosynthesis, explaining why PhzF was found to be indispensable for PCA formation in experiments with enzymes from *Pseudomonas* [24,34,35]. Interestingly, the *phz* operon from a strain of *Streptomyces cinnamonensis* was recently found to lack a *phzF* gene, implicating an alternative route to the phenazines synthesized by these Gram-positive bacteria [32].

It is likely that the efficiency of the ketone condensation step normally is enhanced by the participation of the first two gene products of the *phz* operon from *Pseudomonas*, i.e. PhzA and PhzB, which have been found to increase the yield of phenazine biosynthesis severalfold [24,34,35]. These two small proteins are highly homologous to one another and belong to the Δ^5-3-ketosteroid isomerase family, but residues of the active centers are not conserved, making prediction of their function difficult [36]. It has been speculated that PhzA/B are part of a multienzyme complex that protects unstable intermediates from breakdown [24,28]. In this multienzyme complex,

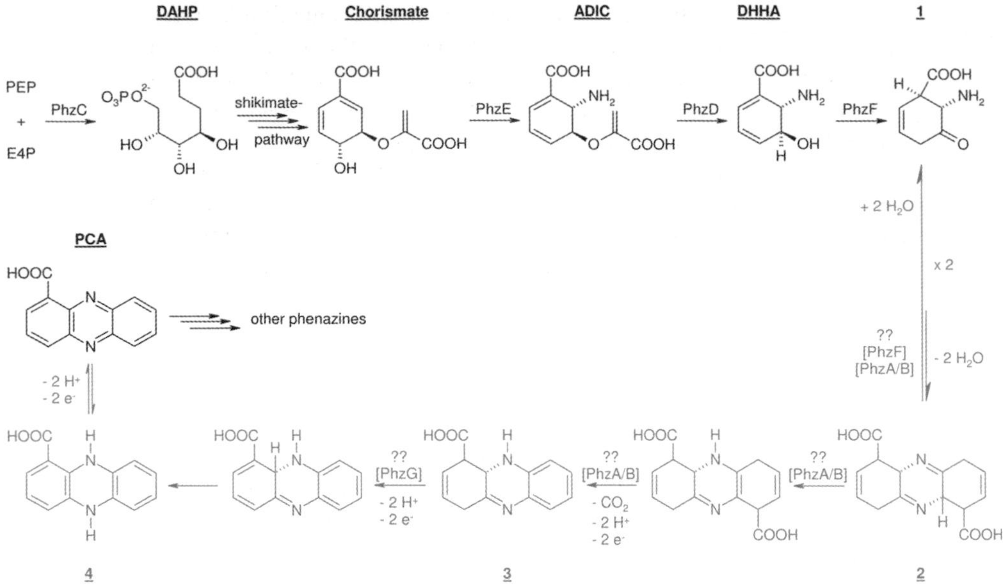

Figure 12.3 Present view of phenazine biosynthesis. Intermediates without definitive structural confirmation are shown in grey. Enzyme names in brackets indicate that involvement of the respective enzyme is assumed, but experimental proof is lacking. PEP, phosphoenolpyruvate; E4P, erythrose-4-phosphate; DAHP, 3-deoxy-D-arabino-heptulosonate-7-phosphate; ADIC, 2-amino-2-deoxychorismate; DHHA, *trans*-2;3-dihydro-3-hydroxyanthranilate; PCA, phenazine-1-carboxylic acid.

PhzA/B could catalyze ketone condensation itself or, alternatively, stabilize the condensation product, e.g. by rearrangement of the double bond system of **2** to protect against back-hydrolysis, which probably is the preferred reaction of **2** in an aqueous environment. Another possible role of PhzA/B is that of a cofactor-free oxygenase in catalysis of the following step in phenazine biosynthesis, which renders formation of the tricyclic phenazine backbone irreversible: mass spectrometric evidence indicates that the double imine **2** undergoes oxidative decarboxylation to form **3**. The driving force of this reaction probably stems from the stability gained by generating a partially aromatized intermediate and could explain why PCA is formed to some extent by all phenazine-producing bacteria investigated to date, since the reaction also can proceed without catalysis [35]. While the *phzA/B* genes are conserved in all phenazine-producing species, the apparent gene duplication is limited to *Pseudomonas* species [30], possibly explaining the efficiency with which pseudomonads produce phenazines and why PCA, rather than PDC, is the precursor for their strain-specific phenazine derivatives.

To arrive at PCA, the final product in the core pathway in *Pseudomonas*, intermediate **3** needs to undergo further oxidation. The available data suggest that these oxidations can proceed uncatalyzed, but the presence of PhzG, an FMN-dependent oxidase related to pyridoxine-5′-phosphate oxidase, increased the amount of PCA obtained in *in vitro* reactions [34,35]. This supports the involvement of PhzG in the final steps, e.g. by catalyzing the conversion of **3** to **4** as shown in Figure 12.3, and is in accordance with crystal structures of PhzG from *P. aeruginosa* and *P. fluorescens* showing that the solvent-exposed active site is large enough to bind tricyclic substrates like intermediate **3** [37]. Flavin cofactors accept two electrons, but four electrons must be removed from **3** to produce fully aromatized PCA. However, the prospective PhzG product **4**, 5,10-dihydro-PCA, is identical in structure to the reduced phenazine moiety thought to participate in physiological electron shuttling reactions of phenazine derivatives [38]. Therefore, core phenazine biosynthesis is likely complete at this oxidation state.

12.3.2
Phenazine-Modifying Genes

The core phenazine operons of all *Pseudomonas* spp. are functionally homologous and sufficient to enable synthesis of PCA. The diversity of phenazines produced by *Pseudomonas* spp. results from the presence and activity of dedicated modifying enzymes that derivatize PCA via hydroxylation, decarboxylation, methylation, trans-amidation or a combination of these reactions [30,39]. The corresponding phenazine-modifying genes are often linked to and coordinately regulated with the core operon [4,29,40].

In *P. chlororaphis*, PCA can be modified by two types of enzymes. Some strains, previously known as *P. aureofaciens*, carry a phenazine-modifying gene, *phzO*, situated immediately downstream of the core biosynthetic locus and encoding an unusual 55-kDa monooxygenase that uses $FADH_2$ as a cosubstrate rather than a cofactor [4,41]. PhzO probably functions together with an as yet unidentified flavin reductase that supplies it with reduced FAD to convert PCA to 2OH-PCA, which then spontaneously decomposes to 2OH-PHZ [4]. A second group of *P. chlororaphis* strains converts PCA to PCN with the aid of another phenazine-modifying gene, *phzH* [3,40,42], which encodes an enzyme with similarity to glutamine-dependent asparagine synthases. PhzH has an N-terminal class II glutamine amidotransferase domain that hydrolyzes Gln to yield ammonia, and a C-terminal domain that presumably mediates synthesis of an AMP-containing phenazine intermediate and its subsequent reaction with ammonia to form PCN, Glu, AMP and PPi [43]. The novel phenazine-producing strains of *Pseudomonas* recently discovered by De Maeyer *et al.* [8] also carry homologs of *phzH* and, like *P. chlororaphis* and *P. aeruginosa* (see below), have the capacity to produce PCN.

In addition to PhzH, *P. aeruginosa* has genes for PhzS and PhzM, which are involved in the synthesis of PYO from PCA and represent the best-studied phenazine-modifying enzymes, since both have been crystallized [29,44–46]. PYO is produced from PCA in two steps. PhzM, a dimeric *S*-adenosyl methionine-dependent N-methyltransferase, first converts PCA to 5-methyl-PCA betaine [44]. The second step is carried out by PhzS, a FAD-dependent dimeric oxidoreductase that uses NADH as an external electron donor and catalyzes hydroxylative decarboxylation of 5-methyl-PCA (or PCA) to PYO [44,45]. Interestingly, PhzM appears to be active only in the presence of PhzS, which suggests that the two might transiently interact, thus avoiding the premature release of a potentially unstable and deleterious intermediate, 5-methyl-PCA [44].

12.3.3 Evolution of Phenazine Pathways

Recent progress in microbial genome sequencing has resulted in the identification of novel phenazine biosynthesis gene clusters in non-*Pseudomonas* species including *Brevibacterium linens* BL2 (GenBank accession no. AAGP00000000), *Erwinia carotovora* ssp. *atroseptica* SCRI1043 [47] and *Burkholderia cepacia* 383 (GenBank accession no. NC_007511). A cursory examination of these novel core pathways reveals that *phzA, phzD, phzE, phzF* and *phzG* are highly conserved across all bacterial lineages, consistent with biochemical evidence that in *Pseudomonas*, the corresponding enzymes are absolutely required for phenazine synthesis. Unlike the operons in *Pseudomonas* and *B. cepacia*, however, the one in *E. carotovora* lacks *phzC*. The core operons of *B. cepacia, E. carotovora* and *B. linens* differ further from those of *Pseudomonas* in having only one copy of the *phzA* homolog. Another notable difference is the presence in *E. carotovora* of a small open reading frame upstream of the key phenazine biosynthetic gene *phzF*. This open reading frame in *B. cepacia* is fused with *phzF* and encodes a 120-residue N-terminal extension of unknown function that is absent from PhzF in *Pseudomonas*. Cluster analyses of sequences for 16S rDNA and PhzA/B and PhzF across several genera reveal no evidence of horizontal gene transfer of phenazine genes among fluorescent *Pseudomonas* spp., since the tree topology for PhzF closely follows that for 16S rDNA (Figure 12.4).

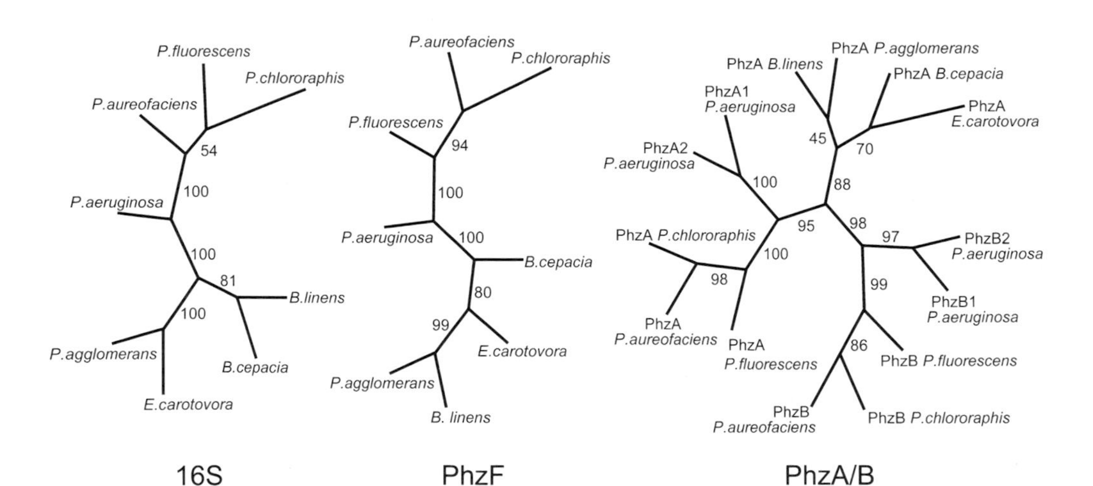

Figure 12.4 Bootstrapped neighbor-joining trees based on alignments of 16S rDNA, PhzF and PhzA/B sequences from *P. fluorescens, P. chlororaphis/aureofaciens, P. aeruginosa, B. cepacia, Pantoea agglomerans, E. carotovora* and *B. linens*.

Despite the fact that the PhzF and PhzA/B trees exhibit similar grouping patterns, the PhzA and PhzB proteins fall into separate clusters, suggesting that the corresponding genes arose from ancestral gene duplication and have evolved to carry out different enzymatic functions.

12.4 Regulation of Phenazine Production in *Pseudomonas* Spp.

12.4.1 Phenazine Biosynthesis and *N*-Acyl-Homoserine Lactone (Acyl-HSL)-Based Quorum Sensing (QS)

Phenazine production in all fluorescent *Pseudomonas* spp. occurs in the stationary phase of growth, and is regulated by two-component sensory-transduction elements and canonical QS, a phenomenon in which an accumulation of acyl-HSL signal molecules allows bacteria to coordinately modulate expression of certain genes in response to population density [30,48]. QS is based on the ability of dedicated LuxR-like activator proteins to recognize and bind cognate acyl-HSLs. This binding triggers a conformational change within the activator protein, enabling it to recognize specific promoter elements and thus activate the transcription of target genes [48]. In *P. fluorescens* and *P. chlororaphis*, phenazine synthesis is regulated by pathway-specific QS circuits consisting of an autoinducer synthase, PhzI, that produces acyl-HSL, and a cognate transcriptional activator, PhzR [42,49,50]. The corresponding regulatory genes are situated adjacent to the structural genes encoding enzymes for assembly of the phenazine scaffold and its modification. The structural genes are driven by a promoter containing an essential 18-bp almost-perfect inverted repeat, the *phz* box, which resembles the binding sites of other members of the LuxR family. *P. fluorescens* produces six acyl-HSLs, of which 3-OH-C6-HSL is the most abundant [49], while *P. chlororaphis* produces three acyl-HSLs with C6-HSL directly mediating phenazine biosynthesis [42,51]. *P. chlororaphis* also has a second QS circuit, CsaI/R, that is not involved in phenazine production but instead regulates cell surface properties and protease production [52].

The QS machinery in *P. aeruginosa* is more complex and consists of two QS circuits, Las and Rhl, comprising a hierarchical cascade, and the negative regulator QscR, which modulates the activity of LasR and RhlR via formation of inactive heterodimers [48,53,54]. The Las and Rhl circuits utilize as cognate autoinducers 3-oxo-C12-HSL and C4-HSL, respectively [48], and their activity is modulated by a third signaling system based on 2-heptyl-3-hydroxy-4-quinolone, also known as the *Pseudomonas* quinolone signal (PQS) [55]. Synthesis of PQS is controlled by the LysR-like transcription factor MvfR [56], and numerous independent studies have revealed that mutations in various components of the LasI/R, RhlI/R and MvfR–PQS machinery result in PYO-deficient phenotypes, presumably because of interactions between the PQS and *rhl* genes, and the fact that PQS directly regulates the expression of the *phz*1 operon [57–61]. A fourth LuxR-like protein, termed VsqR, is crucial for synthesis of both the major acyl-HSLs and PYO, and a *vsqR* mutation interferes with expression of the *phz2* operon [62]. Finally, a recent study by Dietrich *et al.* [63] suggests that phenazines themselves act as signaling compounds downstream of the PQS system, since PYO affects the expression of a small subset of genes. The study also revealed that the main regulatory networks form a temporal cascade, with the QS, PQS and *phz* genes being expressed in the early exponential, late exponential and early stationary phases of growth, respectively.

12.4.2
Global Regulation

There is mounting evidence that the production of phenazines, together with certain other secondary metabolites, is regulated globally in *Pseudomonas* spp. at both the transcriptional and posttranscriptional levels [64]. Transcriptional control is exerted through the induction of a stationary-phase σ factor (RpoS or σ^{38}), which causes global changes in gene expression patterns during the stationary phase of growth [64]. In *P. chlororaphis*, RpoS, together with a TetR-like regulatory protein PsrA, modulates the synthesis of both acyl-HSLs and phenazines [65].

Posttranscriptional control of phenazine production in pseudomonads is carried out by the highly conserved two-component signal transduction system GacA–GacS and small regulatory RNAs [43,64,66]. In response to yet unknown environmental cue(s), the GacA/S system activates the expression of small regulatory RNAs whose function is to titrate the RsmA-like translational repressor that otherwise obscures ribosomal-binding sites of transcripts from genes for the synthesis of secondary metabolites [64]. In *P. aeruginosa*, the GacA/S system, together with RsmA and the small regulatory RNA RsmZ, controls the *las* and *rhl* systems, and the synthesis of PYO, hydrogen cyanide (HCN) and the lectin PA-IL [64,67]. Inactivation of the *gacA* gene results in delayed and significantly decreased production of PYO [68,69]. GacA also negatively regulates the QS circuitry through the σ factor RpoN (σ^{54}) [70], which in turn affects the production of elastase, rhamnolipids, HCN and PYO [71]. Finally, the synthesis of phenazines and other virulence factors is controlled in *P. aeruginosa* by a cAMP receptor protein homolog termed Vfr [72]. Inactivation of this regulator has pleiotropic effects, and interferes with QS, twitching motility and production of PYO.

12.5
Biological Activity and Role of Phenazine Compounds from *Pseudomonas* Spp.

12.5.1
Spectrum of Biological Activity of Phenazines

It has long been known that phenazines produced by pseudomonads have broad cross-phylum inhibitory activity [2,7,39]. Hence, these compounds are often considered to be antibiotics. A comprehensive analysis of the antimicrobial properties carried out by Smirnov and Kiprianova [2] revealed that all *Pseudomonas*-derived phenazines exhibit some degree of bacteriostatic activity, with PYO and the hydroxylated phenazines from *P. aeruginosa* and *P. chlororaphis* being particularly active against Gram-positive bacteria. Similarly, numerous studies have documented the antifungal properties of phenazines from rhizosphere-inhabiting strains of *P. chlororaphis* and *P. fluorescens*, and their contribution to the ability of these bacteria to antagonize economically important soil-borne phytopathogenic fungi including *Gaeumannomyces graminis*, *Fusarium oxysporum*, *Rhizoctonia solani*, *Pythium splendens* and *Sclerotinia slerotiorum* (reviewed in Refs. [30,73]). Furthermore, phenazines produced in the environment can affect the functional diversity of microbial communities. For example, the production *in situ* of PYO plays an important role in the ability of *P. aeruginosa* to compete with *Staphylococcus aureus* [74], *Candida albicans* and *Aspergillus fumigatus* [75] during colonization of the lungs of patients with cystic fibrosis. The presence of PYO-producing strains in oil-contaminated sites can reduce microbial community diversity and the ability to degrade polycyclic aromatic hydrocarbons [76]. Phenazines derived from *Pseudomonas* spp. also are active against plants and were reported to inhibit the growth of the noxious weed downy brome (*Bromus tectorum*) [77], and to inhibit cell growth and cause root tissue necrosis in winter wheat [78]. PCA and 2-OH-PCA have even been patented as herbicides for control of the weeds timothy (*Phleum pratense*) and garden cress (*Nasturnium officinale*), and as potential algicides for the blue–green algae *Microcystis aeruginosa* and *Anabaena flos-aquae*, which cause fresh water blooms [79]. Finally, multiple studies focused on the virulence mechanisms employed by the opportunistic pathogen *P. aeruginosa* to infect various hosts have revealed the broad toxicity of PYO. In all tested model systems, PYO is a nonspecific pathogenicity factor required for the generation of pathogenic symptoms in plants and effective killing of *Caenorhabditis elegans*, *Drosophila melanogaster* and burned mice [68,80–82].

12.5.2
Mode of Action and Mechanisms of Self-Defense

Naturally occurring phenazines include well over 50 known compounds, many of which are biologically active [7,30]. However, most "biological activity" reports describe only simple antagonism *in vitro* and it is difficult to systematically derive structure–activity relationships from these published results because of the diverse targets and assay conditions employed [39]. Models including DNA intercalation and topoisomerase inhibition have been suggested to explain the mode of action of phenazines, but recent studies with *P. aeruginosa* indicate that much of their biological activity is due to their ability to cause oxidative stress [39,83]. Studies *in vitro* utilizing cell cultures and purified PYO have revealed that phenazines can be reduced by NADH forming an unstable free radical which reacts with molecular oxygen to form toxic superoxide ($O_2^{-}\cdot$) and hydrogen peroxide (H_2O_2) [84]. The latter two can oxidize glutathione and interact with protease-cleaved Fe-transferrin and the siderophore pyochelin, with formation of the highly cytotoxic hydroxyl radical

(·OH) [85,86]. PYO also can directly oxidize glutathione [87–89] and inhibit antioxidant enzymes [90], thus causing activation of proinflammatory processes in human lung epithelial cells and neutrophil-mediated tissue damage [91–94]. It also has been reported that PCA can stimulate the influx and activation of neutrophils, and the release of reactive oxygen species (ROS) in human lung epithelium, resulting in lung tissue damage [95]. The proposed role of phenazines as inducers of oxidative stress was confirmed by Angell *et al.* [96], who carried out microarray profiling of *Saccharomyces cerevisiae* that had been treated with PYO. The majority of genes in which expression was affected were involved in defense against oxidative stress and DNA damage, and in cell wall and protein synthesis, drug resistance, transport, and the cell cycle.

The broad cross-phylum activity of phenazines suggests that they target conserved cellular pathways, and Ran *et al.* [69] employed a functional genomic approach to identify the eukaryotic pathways disrupted by PYO. They screened a deletion library of *S. cerevisiae* and identified 50 potential targets, most of which have human orthologs. These targets encompassed genes involved in electron transport and respiration, oxidative stress, the cell cycle, protein sorting, vesicle transport, and the assembly of vacuolar ATPase machinery [69]. Vacuolar ATPases (V-ATPases) are members of a diverse family of ATP-driven proton pumps that are localized in organelle and plasma membranes, and function to acidify lumen of vacuoles, and lysosomes, endosomes and secretory vesicles [97]. PYO severely inhibited V-ATPase activity in cultured human lung epithelial cells, thus reducing the optimal expression and functioning of cystic fibrosis conductance regulator (CFTR), an important cAMP-dependent chloride ion cannel [98]. It has been proposed that the inactivation of V-ATPase activity also may account for the disruption of ion transport, calcium homeostatsis and ciliary function observed in epithelial cells treated with PYO [69,99].

Although the molecular mechanisms underlying the activity of phenazines towards bacteria and plants remain unexplored, the redox properties of PYO have been implicated in the ability of *P. aeruginosa* to induce systemic resistance (ISR) in plants to pathogenic fungi. Two studies focused on *P. aeruginosa* 7NSK2 suggest that the balanced production of PYO and the synergistic interaction of the latter with pyochelin are crucial to the ability of this strain to trigger ISR in beans, tobacco, tomato and rice [100,101]. Experiments with purified PYO in rice revealed that it modulated transient production of ROS in the plant tissues, placing them into an ISR-induced state and priming them for faster reaction to pathogen attack [101].

All these findings raise obvious questions as to how phenazine producers protect themselves from phenazine toxicity. Oxidative stress imposed by PYO or other phenazine compounds results in the intracellular accumulation of toxic hydrogen peroxide, hydroxyl radicals and superoxide, with a concomitant increase in the cellular activity of catalase and superoxide dismutase to detoxify these compounds. In *E. coli*, two redox-sensing global regulators, SoxR and OxyR, control stress responses to superoxide and hydrogen peroxide, respectively [102]. *P. aeruginosa* encodes homologs of both regulators [103] and while the global role of OxyR appears to be preserved [104], the function of SoxR is reduced exclusively to the control of a small group of genes induced by PYO [63] or the redox-active compound paraquat [105]. Most strongly induced by these compounds are genes encoding a putative monooxygenase and components of a proton-driven pump, MexGHI–OpmD, that belongs to the resistance–nodulation–division (RND) superfamily and may be involved in active efflux of PYO [63]. Among other mechanisms contributing to the high resistance to PYO in *P. aeruginosa* are low intracellular levels of NADPH, lack of NADPH:PYO oxidoreductase, and high levels of superoxide dismutase and catalase activity [106]. In *P. aeruginosa*, the manganese- and iron-cofactored cytoplasmic superoxide dismutases protect the cell under pyocyanogenic conditions, and corresponding mutant strains are either entirely PYO deficient or produce markedly decreased amounts of phenazines [107].

12.5.3
Redox Homeostasis and Other Physiological Roles

It was long thought that phenazines, like other secondary metabolites, were of little or no importance in nature [7]. To date, the topic remains largely unexplored, but a handful of studies that have addressed the role of phenazines in the environment all demonstrated a positive correlation between the ability to produce phenazines and the competitiveness and long-term survival of the producers. For example, in a study by Mazzola *et al.* [108], phenazine-producing strains of *P. aureofaciens* and *P. fluorescens* were more competitive and survived longer in the plant rhizosphere than did the isogenic phenazine-deficient mutants. Phenazines have been detected at concentrations up to 10^{-4} M in the sputa of patients with cystic fibrosis [109] and PYO is critical for lung infection by *P. aeruginosa* in mice. Lau *et al.* [83] demonstrated that PYO-deficient mutants were less virulent and less competitive than wild-type *P. aeruginosa* PAO1 in the mouse acute and chronic pneumonia models of infection. When infected with 1×10^7 bacteria, CD-1 mice were unable to clear the wild-type strain and viable counts had increased by around 1.4 log after 16 h; in contrast, populations of strains mutated in *phzM* and *phzS* were attenuated by 4.47 and 4.15 log, respectively. In competitive mixed infection assays, *phzM* and *phzS* mutants were only 18 and 25% as competitive as PAO1.

The crucial role of phenazines can be explained by the emerging physiological role of these metabolites in *Pseudomonas* spp. It is well established that both saprophytic and pathogenic pseudomonads often exist in natural habitats in the form of complex surface-attached communities called biofilms [110,111], and a recent study by Maddula *et al.* [112] revealed that phenazine production is crucial for establishment of biofilms by *P. chlororaphis* 30–84. It has been suggested that under the oxygen-limited conditions present in mature biofilms, phenazines may help bacteria to generate energy for growth by acting as electron acceptors for re-oxidation of NADH [38]. Finally, the redox potentials of some phenazines are sufficiently low under environmental conditions that they can function as electron shuttles to reductively dissolve poorly crystalline iron and manganese oxides in soil, making these and other nutrients associated with mineral phases biologically accessible [38]. Hernandez *et al.* [113] showed that phenazines act as electron shuttles to promote microbial mineral reduction, and speculated that they could make iron and other nutrients in soil more accessible to microorganisms. Indeed, PYO produced by *P. aeruginosa* in fuel cells markedly stimulated growth and power output not only by the producer strain, but also by other bacteria in the mixed communities inhabiting microbial fuel cells [114,115].

12.6 Conclusions

Within *Pseudomonas* spp., the ability to produce phenazines is found among strains of three well-characterized species, i.e. *P. fluorescens*, *P. chlororaphis* and *P. aeruginosa*. These strains carry a homologous seven-gene core operon that encodes enzymes for the synthesis of a common precursor, PCA, which can be converted by dedicated modifying enzymes into species-specific phenazine products. Phenazine production is tightly regulated in *Pseudomonas* spp. at transcriptional and posttranscriptional levels by dedicated QS and global regulatory circuits, in response to changes in population density and environmental cues. Phenazines produced by *Pseudomonas* spp. contribute to redox homeostasis, bacterial virulence, plant disease suppression, microbial fuel cell efficiency and soil mineral reduction, and can act as intra- and intercellular molecular signals.

Acknowledgments

W.B. thanks Roger S. Goody for his continuous support of this project. Research of the authors summarized in this review was sponsored in part by the Max Planck Society (to W.B.) and the Deutsche Forschungsgemeinschaft (grant BL 587 to W.B.).

References

1 Stackebrandt, E., Murray, G.E. and Truper, H.G. (1988) *Proteobacteria classis* nov., a name for the phylogenetic taxon that includes the "purple bacteria and their relatives" *Int J Syst Bacteriol*, **38**, 321–325.

2 Smirnov, V. and Kiprianova, E. (1990) *Bacteria of Pseudomonas Genus*, Naukova Dumka, Kiev.

3 Chin-A-Woeng, T.F.C., Bloemberg, G.V., van der Bij, A.J., van der Drift, K.M.G.F., Schripsema, J., Kroon, B., Scheffer, R.J., Keel, C., Bakker, P.A.H.M., Tichy, H.V., de Bruijn, F.J., Thomas-Oates, J.E. and Lugtenberg, B.J.J. (1998) Biocontrol by phenazine-1-carboxamide-producing *Pseudomonas chlororaphis* PCL1391 of tomato root rot caused by *Fusarium oxysporum* f. sp. *radicis-lycopersici*. *Mol Plant Microbe Interact*, **11**, 1069–1077.

4 Delaney, S.M., Mavrodi, D.V., Bonsall, R.F. and Thomashow, L.S. (2001) *phzO*, a gene for biosynthesis of 2-hydroxylated phenazine compounds in *Pseudomonas aureofaciens* 30–84. *J Bacteriol*, **183**, 318–327.

5 Gessard, C. (1882) Sur les colorations bleue et verte des lignes a pansements. *CR Acad Sci D*, **94**, 536–568.

6 Britton, G. (1983) *Biochemistry of Natural Pigments*, Cambridge University Press, Cambridge.

7 Turner, J.M. and Messenger, A.J. (1986) Occurrence, biochemistry and physiology of phenazine pigment production. *Adv Microb Physiol*, **27**, 211–275.

8 De Maeyer, K., Perneel, M. and Hofte, M. (2006) Phenazine biosynthesis and its regulation in *Pseudomonas* CMR12A and CMR5C. In *Proceedings of the 7th International Workshop on Plant Growth Promoting Rhizobacteria* Noordwijkerhout.

9 Hu, H., Xu, Y., Chen, F., Zhang, X. and Ki Hur, B. (2005) Isolation and characterization of a new fluorescent *Pseudomonas* strain that produces both phenazine-1-carboxylic acid and pyoluteorin. *J Microbiol Biotechnol*, **15**, 86–90.

10 Fordos, J. (1859) *Receuil des Travaux de la Societe d'Emulation pour les Sciences Pharmaceutiques*. Vol. 3.

11 Wasserzug, E. (1887) Sur la formation de la matiere colorante chez le bacillus pyocyaneus. *Ann Inst Pasteur*, **1**, 581–591.

12 King, E.O., Ward, M.K. and Raney, D.E. (1954) Two simple media for the demonstration of pyocyanin and fluorescein. *J Lab Clin Med*, **44**, 301–307.

13 Wrede, F. and Strack, E. (1929) Pyocyanin, the blue pigment of *Bacillus pyocyaneus* IV. The constitution and synthesis of pyocyanin. *Hoppe-Seylers Z Physiol Chem*, **181**, 58–76.

14 Hillemann, H. (1938) Phenazine.III. Position of the methyl groups in pyocyanine and attempts to synthesize isopyocyanine. *Ber Deutsch Chem Ges B*, **71**, 46–52.

15 Friedheim, E. and Michaelis, L. (1931) Potentiometric study of pyocyanine. *J Biol Chem*, **91**, 355–368.

16 Tobie, W.C. (1945) A Proposed biochemical basis for the genus *Pseudomonas*. *J Bacteriol*, **49**, 459–462.

17 Millican, R.C. (1962) Biosynthesis of pyocyanine. Incorporation of [^{14}C] shikimic acid. *Biochim Biophys Acta*, **57**, 407–409.

18 Ingledew, W.M. and Campbell, J.J. (1969) Evaluation of shikimic acid as a precursor of pyocyanine. *Can J Microbiol*, **15**, 535–541.

19 Longley, R.P., Halliwell, J.E., Campbell, J.J. and Ingledew, W.M. (1972) The branchpoint of pyocyanine biosynthesis. *Can J Microbiol*, **18**, 1357–1363.

20 Calhoun, D.H., Carson, M. and Jensen, R.A. (1972) The branch point metabolite for pyocyanine biosynthesis in *Pseudomonas aeruginosa*. *J Gen Microbiol*, **72**, 581–583.

21 Podojil, M. and Gerber, N.N. (1970) Biosynthesis of 1,6-phenazinediol 5,10-dioxide (iodinin). Incorporation of shikimic acid. *Biochemistry*, **9**, 4616–4618.

22 Etherington, T., Herbert, R.B., Holliman, F.G. and Sheridan, J.B. (1979) The biosynthesis of phenazines: incorporation of [^{2}H]shikimic acid. *J Chem Soc Perkin Transact*, **1**, 2416–2419.

23 Römer, A. and Herbert, R.B. (1982) Further observations on the source of nitrogen in phenazine biosynthesis. *Z Naturforsch C*, **37**, 1070–1074.

24 McDonald, M., Mavrodi, D.V., Thomashow, L.S. and Floss, H.G. (2001) Phenazine biosynthesis in *Pseudomonas fluorescens*: branchpoint from the primary shikimate biosynthetic pathway and role of phenazine-1,6-dicarboxylic acid. *J Am Chem Soc*, **123**, 9459–9460.

25 Giddens, S.R., Feng, Y.J. and Mahanty, H.K. (2002) Characterization of a novel phenazine antibiotic gene cluster in *Erwinia herbicola* Eh1087. *Mol Microbiol*, **45**, 769–783.

26 Pierson, L.S., 3rd and Thomashow, L.S. (1992) Cloning and heterologous expression of the phenazine biosynthetic locus from *Pseudomonas aureofaciens*

30–84. *Mol Plant Microbe Interact*, **5**, 330–339.

27 Pierson, L.S., 3rd Gaffney, T., Lam, S. and Gong, F. (1995) Molecular analysis of genes encoding phenazine biosynthesis in the biological control bacterium *Pseudomonas aureofaciens* 30–84. *FEMS Microbiol Lett*, **134**, 299–307.

28 Mavrodi, D.V., Ksenzenko, V.N., Bonsall, R.F., Cook, R.J., Boronin, A.M. and Thomashow, L.S. (1998) A seven-gene locus for synthesis is of phenazine-1-carboxylic acid by *Pseudomonas fluorescens* 2–79. *J Bacteriol*, **180**, 2541–2548.

29 Mavrodi, D.V., Bonsall, R.F., Delaney, S.M., Soule, M.J., Phillips, G. and Thomashow, L.S. (2001) Functional analysis of genes for biosynthesis of pyocyanin and phenazine-1-carboxamide from *Pseudomonas aeruginosa* PAO1. *J Bacteriol*, **183**, 6454–6465.

30 Mavrodi, D.V., Blankenfeldt, W. and Thomashow, L.S. (2006) Phenazine compounds in fluorescent *Pseudomonas* spp.: biosynthesis and regulation. *Annu Rev Phytopathol*, **44**, 417–445.

31 Webby, C.J., Baker, H.M., Lott, J.S., Baker, E.N. and Parker, E.J. (2005) The structure of 3-deoxy-D-arabino-heptulosonate 7-phosphate synthase from *Mycobacterium tuberculosis* reveals a common catalytic scaffold and ancestry for type I and type II enzymes. *J Mol Biol*, **354**, 927–939.

32 Haagen, Y., Gluck, K., Fay, K., Kammerer, B., Gust, B. and Heide, L. (2006) A gene cluster for prenylated naphthoquinone and prenylated phenazine biosynthesis in *Streptomyces cinnamonensis* DSM 1042. *Chembiochem*, **7**, 2016–2027.

33 Parsons, J.F., Calabrese, K., Eisenstein, E. and Ladner, J.E. (2003) Structure and mechanism of *Pseudomonas aeruginosa* PhzD, an isochorismatase from the phenazine biosynthetic pathway. *Biochemistry*, **42**, 5684–5693.

34 Parsons, J.F., Song, F., Parsons, L., Calabrese, K., Eisenstein, E. and Ladner, J.E. (2004) Structure and function of the phenazine biosynthesis protein PhzF from *Pseudomonas fluorescens* 2–79. *Biochemistry*, **43**, 12427–12435.

35 Blankenfeldt, W., Kuzin, A.P., Skarina, T., Korniyenko, Y., Tong, L., Bayer, P., Janning, P., Thomashow, L.S. and Mavrodi, D.V. (2004) Structure and function of the phenazine biosynthetic protein PhzF from *Pseudomonas fluorescens*. *Proc Natl Acad Sci USA*, **101**, 16431–16436.

36 Ahuja, E.G., Mavrodi, D.V., Thomashow, L.S. and Blankenfeldt, W. (2004) Overexpression, purification and crystallization of PhzA, the first enzyme of the phenazine biosynthesis pathway of *Pseudomonas fluorescens* 2–79. *Acta Crystallogr*, **D60**, 1129–1131.

37 Parsons, J.F., Calabrese, K., Eisenstein, E. and Ladner, J.E. (2004) Structure of the phenazine biosynthesis enzyme PhzG. *Acta Crystallogr*, **D60**, 2110–2113.

38 Price-Whelan, A., Dietrich, L.E. and Newman, D.K. (2006) Rethinking "secondary" metabolism: physiological roles for phenazine antibiotics. *Nat Chem Biol*, **2**, 71–78.

39 Laursen, J.B. and Nielsen, J. (2004) Phenazine natural products: biosynthesis, synthetic analogues, and biological activity. *Chem Rev*, **104**, 1663–1685.

40 Chin-A-Woeng, T.F.C., Thomas-Oates, J.E., Lugtenberg, B.J.J. and Bloemberg, G.V. (2001) Introduction of the *phzH* gene of *Pseudomonas chlororaphis* PCL1391 extends the range of biocontrol ability of phenazine-1-carboxylic acid-producing *Pseudomonas* spp. strains. *Mol Plant Microbe Interact*, **14**, 1006–1015.

41 Gisi, M.R. and Xun, L. (2003) Characterization of chlorophenol 4-monooxygenase (TftD) and NADH:flavin adenine dinucleotide oxidoreductase (TftC) of *Burkholderia cepacia* AC1100. *J Bacteriol*, **185**, 2786–2792.

42 Chin-A-Woeng, T.F.C., van den Broek, D., de Voer, G., van der Drift, K.M.G.M., Tuinman, S., Thomas-Oates, J.E., Lugtenberg, B.J.J. and Bloemberg, G.V.

(2001) Phenazine-1-carboxamide production in the biocontrol strain *Pseudomonas chlororaphis* PCL1391 is regulated by multiple factors secreted into the growth medium. *Mol Plant Microbe Interact*, **14**, 969–979.

43 Chin-A-Woeng, T.F.C., Bloemberg, G.V. and Lugtenberg, B.J.J. (2003) Phenazines and their role in biocontrol by *Pseudomonas* bacteria. *New Phytologist*, **157**, 503–523.

44 Parsons, J.F., Greenhagen, B.T., Shi, K., Calabrese, K., Robinson, H. and Ladner, J.E. (2007) Structural and functional analysis of the pyocyanin biosynthetic protein PhzM from *Pseudomonas aeruginosa*. *Biochemistry*, **46**, 1821–1828.

45 Gohain, N., Thomashow, L.S., Mavrodi, D.V. and Blankenfeldt, W. (2006) The purification, crystallization and preliminary structural characterization of FAD-dependent monooxygenase PhzS, a phenazine-modifying enzyme from *Pseudomonas aeruginosa*. *Acta Crystallogr*, **F62**, 989–992.

46 Gohain, N., Thomashow, L.S., Mavrodi, D.V. and Blankenfeldt, W. (2006) The purification, crystallization and preliminary structural characterization of PhzM, a phenazine-modifying methyltransferase from *Pseudomonas aeruginosa*. *Acta Crystallogr*, **F62**, 887–890.

47 Bell, K.S., Sebaihia, M., Pritchard, L., Holden, M.T., Hyman, L.J., Holeva, M.C., Thomson, N.R., Bentley, S.D., Churcher, L.J., Mungall, K., Atkin, R., Bason, N., Brooks, K., Chillingworth, T., Clark, K., Doggett, J., Fraser, A., Hance, Z., Hauser, H., Jagels, K., Moule, S., Norbertczak, H., Ormond, D., Price, C., Quail, M.A., Sanders, M., Walker, D., Whitehead, S., Salmond, G.P., Birch, P.R., Parkhill, J. and Toth, I.K. (2004) Genome sequence of the enterobacterial phytopathogen *Erwinia carotovora* subsp. *atroseptica* and characterization of virulence factors. *Proc Natl Acad Sci USA*, **101**, 11105–11110.

48 Fuqua, C., Parsek, M.R. and Greenberg, E.P. (2001) Regulation of gene expression by cell-to-cell communication: acyl-homoserine lactone quorum sensing. *Annu Rev Genet*, **35**, 439–468.

49 Khan, S.R., Mavrodi, D.V., Jog, G.J., Suga, H., Thomashow, L.S. and Farrand, S.K. (2005) Activation of the phz operon of *Pseudomonas fluorescens* 2–79 requires the LuxR homolog PhzR, *N*-(3-OH-hexanoyl)-L-homoserine lactone produced by the LuxI homolog PhzI, and a *cis*-acting *phz* box. *J Bacteriol*, **187**, 6517–6527.

50 Wood, D.W. and Pierson, L.S. 3rd (1996) The *phzI* gene of *Pseudomonas aureofaciens* 30–84 is responsible for the production of a diffusible signal required for phenazine antibiotic production. *Gene*, **168**, 49–53.

51 Wood, D.W., Gong, F., Daykin, M.M., Williams, P. and Pierson, L.S. 3rd (1997) *N*-acyl-homoserine lactone-mediated regulation of phenazine gene expression by *Pseudomonas aureofaciens* 30–84 in the wheat rhizosphere. *J Bacteriol*, **179**, 7663–7670.

52 Zhang, Z. and Pierson, L.S. 3rd (2001) A second quorum-sensing system regulates cell surface properties but not phenazine antibiotic production in *Pseudomonas aureofaciens*. *Appl Environ Microbiol*, **67**, 4305–4315.

53 Chugani, S.A., Whiteley, M., Lee, K.M., D'Argenio, D., Manoil, C. and Greenberg, E.P. (2001) QscR, a modulator of quorum-sensing signal synthesis and virulence in *Pseudomonas aeruginosa*. *Proc Natl Acad Sci USA*, **98**, 2752–2757.

54 Lee, J.H., Lequette, Y. and Greenberg, E.P. (2006) Activity of purified QscR, a *Pseudomonas aeruginosa* orphan quorum-sensing transcription factor. *Mol Microbiol*, **59**, 602–609.

55 Pesci, E.C., Milbank, J.B., Pearson, J.P., McKnight, S., Kende, A.S., Greenberg, E.P. and Iglewski, B.H. (1999) Quinolone signaling in the cell-to-cell communication system of *Pseudomonas aeruginosa*. *Proc Natl Acad Sci USA*, **96**, 11229–11234.

56 Deziel, E., Lepine, F., Milot, S., He, J.X., Mindrinos, M.N., Tompkins, R.G. and Rahme, L.G. (2004) Analysis of *Pseudomonas aeruginosa* 4-hydroxy-2-alkylquinolines (HAQs) reveals a role for 4-hydroxy-2-heptylquinoline in cell-to-cell communication. *Proc Natl Acad Sci USA*, **101**, 1339–1344.

57 Cao, H., Krishnan, G., Goumnerov, B., Tsongalis, J., Tompkins, R. and Rahme, L.G. (2001) A quorum sensing-associated virulence gene of *Pseudomonas aeruginosa* encodes a LysR-like transcription regulator with a unique self-regulatory mechanism. *Proc Natl Acad Sci USA*, **98**, 14613–14618.

58 Gallagher, L.A., McKnight, S.L., Kuznetsova, M.S., Pesci, E.C. and Manoil, C. (2002) Functions required for extracellular quinolone signaling by *Pseudomonas aeruginosa*. *J Bacteriol*, **184**, 6472–6480.

59 Diggle, S.P., Winzer, K., Chhabra, S.R., Chhabra, S.R., Worrall, K.E., Camara, M. and Williams, P. (2003) The *Pseudomonas aeruginosa* quinolone signal molecule overcomes the cell density-dependency of the quorum sensing hierarchy, regulates *rhl*-dependent genes at the onset of stationary phase and can be produced in the absence of LasR. *Mol Microbiol*, **50**, 29–43.

60 McGrath, S., Wade, D.S. and Pesci, E.C. (2004) Dueling quorum sensing systems in *Pseudomonas aeruginosa* control the production of the *Pseudomonas* quinolone signal (PQS). *FEMS Microbiol Lett*, **230**, 27–34.

61 McKnight, S.L., Iglewski, B.H. and Pesci, E.C. (2000) The *Pseudomonas* quinolone signal regulates *rhl* quorum sensing in *Pseudomonas aeruginosa*. *J Bacteriol*, **182**, 2702–2708.

62 Juhas, M., Wiehlmann, L., Huber, B., Jordan, D., Lauber, J., Salunkhe, P., Limpert, A.S., von Gotz, F., Steinmetz, I., Eberl, L. and Tummler, B. (2004) Global regulation of quorum sensing and virulence by VqsR in *Pseudomonas aeruginosa*. *Microbiology*, **150**, 831–841.

63 Dietrich, L.E.P., Price-Whelan, A., Petersen, A., Whiteley, M., Newman, D.K. (2006) The phenazine pyocyanin is a terminal signalling factor in the quorum sensing network of *Pseudomonas aeruginosa*. *Mol Microbiol*, **61**, 1308–1321.

64 Haas, D. and Keel, C. (2003) Regulation of antibiotic production in root-colonizing *Pseudomonas* spp. and relevance for biological control of plant disease. *Annu Rev Phytopathol*, **41**, 117–153.

65 Girard, G., van Rij, E.T., Lugtenberg, B.J. and Bloemberg, G.V. (2006) Regulatory roles of *psrA* and *rpoS* in phenazine-1-carboxamide synthesis by *Pseudomonas chlororaphis* PCL1391. *Microbiology*, **152**, 43–58.

66 Chin-A-Woeng, T.F.C., van den Broek, D., Lugtenberg, B.J.J. and Bloemberg, G.V. (2005) The *Pseudomonas chlororaphis* PCL1391 sigma regulator *psrA* represses the production of the antifungal metabolite phenazine-1-carboxamide. *Mol Plant Microbe Interact*, **18**, 244–253.

67 Heurlier, K., Williams, F., Heeb, S., Dormond, C., Pessi, G., Singer, D., Camara, M., Williams, P. and Haas, D. (2004) Positive control of swarming, rhamnolipid synthesis, and lipase production by the posttranscriptional RsmA/RsmZ system in *Pseudomonas aeruginosa* PAO1. *J Bacteriol*, **186**, 2936–2945.

68 Rahme, L.G., Tan, M.W., Le, L., Wong, S.M., Tompkins, R.G., Calderwood, S.B. and Ausubel, F.M. (1997) Use of model plant hosts to identify *Pseudomonas aeruginosa* virulence factors. *Proc Natl Acad Sci USA*, **94**, 13245–13250.

69 Ran, H.M., Hassett, D.J. and Lau, G.W. (2003) Human targets of *Pseudomonas aeruginosa* pyocyanin. *Proc Natl Acad Sci USA*, **100**, 14315–14320.

70 Heurlier, K., Denervaud, V., Pessi, G., Reimmann, C. and Haas, D. (2003) Negative control of quorum sensing by

RpoN (σ^{54}) in *Pseudomonas aeruginosa* PAO1. *J Bacteriol*, **185**, 2227–2235.

71 Thompson, L.S., Webb, J.S., Rice, S.A. and Kjelleberg, S. (2003) The alternative sigma factor RpoN regulates the quorum sensing gene *rhlI* in *Pseudomonas aeruginosa*. *FEMS Microbiol Lett*, **220**, 187–195.

72 Beatson, S.A., Whitchurch, C.B., Sargent, J.L., Levesque, R.C. and Mattick, J.S. (2002) Differential regulation of twitching motility and elastase production by Vfr in *Pseudomonas aeruginosa*. *J Bacteriol*, **184**, 3605–3613.

73 Raaijmakers, J.M., Vlami, M. and de Souza, J.T. (2002) Antibiotic production by bacterial biocontrol agents. *Antonie Van Leeuwenhoek*, **81**, 537–547.

74 Voggu, L., Schlag, S., Biswas, R., Rosenstein, R., Rausch, C. and Gotz, F. (2006) Microevolution of cytochrome *bd* oxidase in Staphylococci and its implication in resistance to respiratory toxins released by *Pseudomonas*. *J Bacteriol*, **188**, 8079–8086.

75 Kerr, J.R., Taylor, G.W., Rutman, A., Hoiby, N., Cole, P.J. and Wilson, R. (1999) *Pseudomonas aeruginosa* pyocyanin and 1-hydroxyphenazine inhibit fungal growth. *J Clin Pathol*, **52**, 385–387.

76 Norman, R.S., Moeller, P., McDonald, T.J. and Morris, P.J. (2004) Effect of pyocyanin on a crude-oil-degrading microbial community. *Appl Environ Microbiol*, **70**, 4004–4011.

77 Gealy, D.R., Gurusiddaiah, S., Ogg, J. and Kennedy, A.C. (1995) Metabolites from *Pseudomonas fluorescens* strain D7 inhibit downy brome (*Bromus tectorum*) seedling growth. *Weed Technol*, **10**, 282–287.

78 Berestetskiy, O.A., Patyka, V.F., Mochalov, Y.M. and Grab, T.A. (1975) Studies of phytotoxic activity of *Pseudomonas putida* strain 2181. *Physiol Biochem Basis Plant Interact Phytocen*, **6**, 96–99.

79 Nelson, C.D. and Toohey, J.I. (1968) Methods of controlling the growth of noxious plants. US Patent 3,367,765.

80 Mahajan-Miklos, S., Tan, M.W., Rahme, L.G. and Ausubel, F.M. (1999) Molecular mechanisms of bacterial virulence elucidated using a *Pseudomonas aeruginosa–Caenorhabditis elegans* pathogenesis model. *Cell*, **96**, 47–56.

81 Lau, G.W., Goumnerov, B.C., Walendziewicz, C.L., Hewitson, J., Xiao, W., Mahajan-Miklos, S., Tompkins, R.G., Perkins, L.A. and Rahme, L.G. (2003) The *Drosophila melanogaster* Tol pathway participates in resistance to infection by the Gram-negative human pathogen *Pseudomonas aeruginosa*. *Infect Immun*, **71**, 4059–4066.

82 Rahme, L.G., Ausubel, F.M., Cao, H., Drenkard, E., Goumnerov, B.C., Lau, G.W., Mahajan-Miklos, S., Plotnikova, J., Tan, M.W., Tsongalis, J., Walendziewicz, C.L. and Tompkins, R.G. (2000) Plants and animals share functionally common bacterial virulence factors. *Proc Natl Acad Sci USA*, **97**, 8815–8821.

83 Lau, G.W., Hassett, D.J., Ran, H.M., Kong, F.S. (2004) The role of pyocyanin in *Pseudomonas aeruginosa* infection. *Trends Mol Med*, **10**, 599–606.

84 Hassan, H.M. and Fridovich, I. (1980) Mechanism of the antibiotic action of pyocyanine. *J Bacteriol*, **141**, 156–163.

85 Miller, R.A., Rasmussen, G.T., Cox, C.D. and Britigan, B.E. (1996) Protease cleavage of iron-transferrin augments pyocyanin-mediated endothelial cell injury via promotion of hydroxyl radical formation. *Infect Immun*, **64**, 182–188.

86 Britigan, B.E., Rasmussen, G.T. and Cox, C.D. (1997) Augmentation of oxidant injury to human pulmonary epithelial cells by the *Pseudomonas aeruginosa* siderophore pyochelin. *Infect Immun*, **65**, 1071–1076.

87 Kanthakumar, K., Taylor, G., Tsang, K.W.T., Cundell, D.R., Rutman, A., Smith, S., Jeffery, P.K., Cole, P.J. and Wilson, R. (1993) Mechanisms of action of *Pseudomonas aeruginosa* pyocyanin on human ciliary beat *in vitro*. *Infect Immun*, **61**, 2848–2853.

88 Muller, M. (2002) Pyocyanin induces oxidative stress in human endothelial cells and modulates the glutathione redox cycle. *Free Radic Biol Med*, **33**, 1527–1533.

89 O'Malley, Y.Q., Reszka, K.J., Spitz, D.R., Denning, G.M. and Britigan, B.E. (2004) *Pseudomonas aeruginosa* pyocyanin directly oxidizes glutathione and decreases its levels in airway epithelial cells. *Am J Physiol*, **287**, L94–L103.

90 O'Malley, Y.Q., Reszka, K.J., Rasmussen, G.T., Abdalla, M.Y., Denning, G.M. and Britigan, B.E. (2003) The *Pseudomonas* secretory product pyocyanin inhibits catalase activity in human lung epithelial cells. *Am J Physiol*, **285**, L1077–L1086.

91 Ginn-Pease, M.E. and Whisler, R.L. (1998) Redox signals and NF-κB activation in T cells. *Free Radic Biol Med*, **25**, 346–361.

92 Rahman, I. (2000) Regulation of nuclear factor κB, activator protein-1, and glutathione levels by tumor necrosis factor alpha and dexamethasone in alveolar epithelial cells. *Biochem Pharmacol*, **60**, 1041–1049.

93 Denning, G.M., Wollenweber, L.A., Railsback, M.A., Cox, C.D., Stoll, L.L. and Britigan, B.E. (1998) *Pseudomonas* pyocyanin increases interleukin-8 expression by human airway epithelial cells. *Infect Immun*, **66**, 5777–5784.

94 Lauredo, I.T., Sabater, J.R., Ahmed, A., Botvinnikova, Y. and Abraham, W.M. (1998) Mechanism of pyocyanin- and 1-hydroxyphenazine-induced lung neutrophilia in sheep airways. *J Appl Physiol*, **85**, 2298–2304.

95 Denning, G.M., Iyer, S.S., Reszka, K.J., O'Malley, Y., Rasmussen, G.T. and Britigan, B.E. (2003) Phenazine-1-carboxylic acid, a secondary metabolite of *Pseudomonas aeruginosa*, alters expression of immunomodulatory proteins by human airway epithelial cells. *Am J Physiol*, **285**, L584–L592.

96 Angell, S., Bench, B.J., Williams, H. and Watanabe, C.M. (2006) Pyocyanin isolated from a marine microbial population: synergistic production between two distinct bacterial species and mode of action. *Chem Biol*, **13**, 1349–1359.

97 Kane, P.M. (2006) The where, when, and how of organelle acidification by the yeast vacuolar H^+-ATPase. *Microbiol Mol Biol Rev*, **70**, 177–191.

98 Kong, F., Young, L., Chen, Y., Ran, H., Meyers, M., Joseph, P., Cho, Y.H., Hassett, D.J. and Lau, G.W. (2006) *Pseudomonas aeruginosa* pyocyanin inactivates lung epithelial vacuolar ATPase-dependent cystic fibrosis transmembrane conductance regulator expression and localization. *Cell Microbiol*, **8**, 1121–1133.

99 Denning, G.M., Railsback, M.A., Rasmussen, G.T., Cox, C.D. and Britigan, B.E. (1998) *Pseudomonas* pyocyanine alters calcium signaling in human airway epithelial cells. *Am J Physiol*, **274**, L893–L900.

100 Audenaert, K., Pattery, T., Cornelis, P. and Hofte, M. (2002) Induction of systemic resistance to *Botrytis cinerea* in tomato by *Pseudomonas aeruginosa* 7NSK2: role of salicylic acid, pyochelin, and pyocyanin. *Mol Plant Microbe Interact*, **15**, 1147–1156.

101 De Vleesschauwer, D., Cornelis, P. and Hofte, M. (2006) Redox-active pyocyanin secreted by *Pseudomonas aeruginosa* 7NSK2 triggers systemic resistance to *Magnaporthe grisea* but enhances *Rhizoctonia solani* susceptibility in rice. *Mol Plant Microbe Interact*, **19**, 1406–1419.

102 Pomposiello, P.J. and Demple, B. (2001) Redox-operated genetic switches: the SoxR and OxyR transcription factors. *Trends Biotechnol*, **19**, 109–114.

103 Stover, C.K., Pham, X.Q., Erwin, A.L., Mizoguchi, S.D., Warrener, P., Hickey, M.J., Brinkman, F.S., Hufnagle, W.O., Kowalik, D.J., Lagrou, M., Garber, R.L., Goltry, L., Tolentino, E., Westbrock-Wadman, S., Yuan, Y., Brody, L.L., Coulter, S.N., Folger, K.R., Kas, A., Larbig, K., Lim, R., Smith, K., Spencer, D., Wong, G.K., Wu, Z., Paulsen, I.T., Reizer, J., Saier, M.H., Hancock, R.E., Lory, S. and Olson,

M.V. (2000) Complete genome sequence of *Pseudomonas aeruginosa* PA01, an opportunistic pathogen. *Nature*, **406**, 959–964.

104 Ochsner, U.A., Vasil, M.L., Alsabbagh, E., Parvatiyar, K. and Hassett, D.J. (2000) Role of the *Pseudomonas aeruginosa oxyR–recG* operon in oxidative stress defense and DNA repair: OxyR-dependent regulation of *katB–ankB, ahpB,* and *ahpC–ahpF*. *J Bacteriol*, **182**, 4533–4544.

105 Palma, M., Zurita, J., Ferreras, J.A., Worgall, S., Larone, D.H., Shi, L., Campagne, F. and Quadri, L.E. (2005) *Pseudomonas aeruginosa* SoxR does not conform to the archetypal paradigm for SoxR-dependent regulation of the bacterial oxidative stress adaptive response. *Infect Immun*, **73**, 2958–2966.

106 Hassett, D.J., Charniga, L., Bean, K., Ohman, D.E. and Cohen, M.S. (1992) Response of *Pseudomonas aeruginosa* to pyocyanin: mechanisms of resistance, antioxidant defenses, and demonstration of a manganese-cofactored superoxide dismutase. *Infect Immun*, **60**, 328–336.

107 Hassett, D.J., Schweizer, H.P. and Ohman, D.E. (1995) *Pseudomonas aeruginosa sodA* and *sodB* mutants defective in manganese- and iron-cofactored superoxide dismutase activity demonstrate the importance of the iron-cofactored form in aerobic metabolism. *J Bacteriol*, **177**, 6330–6337.

108 Mazzola, M., Cook, R.J., Thomashow, L.S., Weller, D.M. and Pierson, L.S. 3rd (1992) Contribution of phenazine antibiotic biosynthesis to the ecological competence of fluorescent pseudomonads in soil habitats. *Appl Environ Microbiol*, **58**, 2616–2624.

109 Wilson, R., Sykes, D.A., Watson, D., Rutman, A., Taylor, G.W. and Cole, P.J. (1988) Measurement of *Pseudomonas aeruginosa* phenazine pigments in sputum and assessment of their contribution to sputum sol toxicity for respiratory epithelium. *Infect Immun*, **56**, 2515–2517.

110 Morris, C.E. and Monier, J.M. (2003) The ecological significance of biofilm formation by plant-associated bacteria. *Annu Rev Phytopathol*, **41**, 429–453.

111 Hall-Stoodley, L. and Stoodley, P. (2005) Biofilm formation and dispersal and the transmission of human pathogens. *Trends Microbiol*, **13**, 7–10.

112 Maddula, V.S., Zhang, Z., Pierson, E.A. and Pierson, L.S. 3rd (2006) Quorum sensing and phenazines are involved in biofilm formation by *Pseudomonas chlororaphis (aureofaciens)* strain 30–84. *Microb Ecol*, **52**, 289–301.

113 Hernandez, M.E., Kappler, A. and Newman, D.K. (2004) Phenazines and other redox-active antibiotics promote microbial mineral reduction. *Appl Environ Microbiol*, **70**, 921–928.

114 Rabaey, K., Boon, N., Denet, V., Verhaege, M., Hofte, M. and Verstraete, W. (2004) Bacteria produce and use redox mediators for electron transfer in microbial fuel cells. *Abstr Papers Am Chem Soc*, **228**, U622–U622.

115 Rabaey, K., Boon, N., Hofte, M. and Verstraete, W. (2005) Microbial phenazine production enhances electron transfer in biofuel cells. *Environ Sci Technol*, **39**, 3401–3408.

13
Pseudomonas–Plant Interactions

S.H. Miller, G.L. Mark, A. Franks, and F. O'Gara

13.1
Introduction

Pseudomonas is a diverse genus containing a large number of species with a variety of catabolic and metabolic abilities. This diversity allows for the colonization of an array of environmental niches and interactions with a wide range of eukaryotic hosts as saprotrophs, endophytes, commensals, plant pathogens and opportunistic human pathogens [1–5]. *Pseudomonas* species are of great interest to researchers in the fields of plant, soil and human-associated microbiology, and as such the genomic sequences of a range of *Pseudomonas* strains have been determined or are currently in progress (Table 13.1). *Pseudomonas*–plant interactions are found ubiquitously in nature and encompass a growing number of plant species, many of which are important to commercial horticulture. These interactions fall into two general groups: those that are beneficial and those that are detrimental to the host plants health (Figure 13.1). *Pseudomonas* sp. which successfully interact with plant hosts ultimately gain an advantage over the competing microorganisms in the immediate environment. This is usually achieved by occupying an exclusive niche provided by the plants surfaces and obtaining growth substrates exuded from the plant roots or released by disease-related breakdown of plant cells in the case of pathogenic interactions. The development of such interactions between *Pseudomonas* and host plants usually involves avoidance, subversion and sometimes even stimulation of the host plant's defenses. In return the plant host can respond to the presence of the *Pseudomonas* sp. by modulating defense mechanisms, and engaging in signaling between itself and the infecting bacterium in an attempt to further promote beneficial interactions or suppress pathogenic ones. The molecular game of cat and mouse between the partners, with each attempting to benefit from the other, has evolved over millions of years and provides us now with a vast array of *Pseudomonas*–plant interaction mechanisms to unravel. Indeed, genetic and physiological similarities

Pseudomonas. Model Organism, Pathogen, Cell Factory. Edited by Bernd H.A. Rehm

ISBN: 978-3-527-31914-5

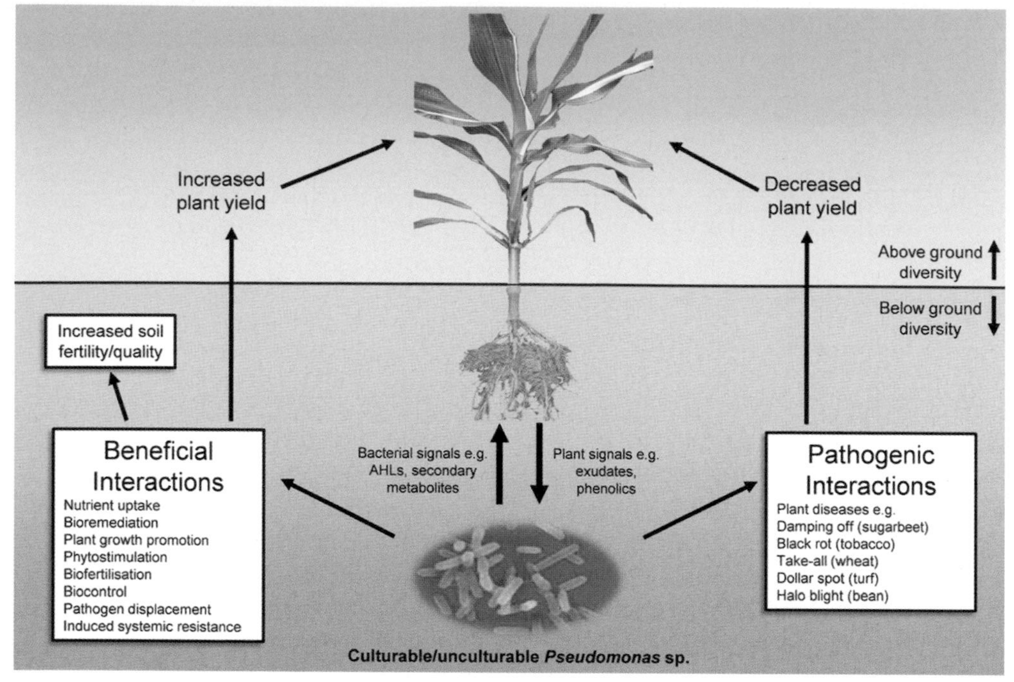

Figure 13.1 Examples of the signal exchange and various beneficial and pathogenic interactions between *Pseudomonas* and plant hosts.

between the *Pseudomonas* sp. which belong to the plant growth-promoting and the pathogenic disease-forming groups suggest that in actuality a continuum exists between these two extremes of interaction. Pathogenic strains can be close relatives of nonpathogenic strains and beneficial strains can share genetic traits with pathogens that are known to be involved in pathogenicity [3]. The study of *Pseudomonas*–plant interactions can be broken down into three interrelated aspects: (i) identification of the mechanisms involved in plant growth promotion and health by *Pseudomonas* sp., (ii) identification of the mechanisms which cause plant disease by *Pseudomonas* sp., and (iii) elucidation of the plant and bacterial signaling involved in modulating these interactions and determining the resulting type of interaction produced.

13.2
Beneficial *Pseudomonas*–Plant Interactions

It has long been recognized that there are many naturally occurring soil bacteria with intrinsic characteristics that make them ideal candidates for sustainable agricultural and industrial applications. Of these, *Pseudomonas* have attracted significant interest as “keystone species” that play a vital role in maintaining soil quality, plant yield and ecosystem function due to the ability of specific strains to colonies the rhizosphere at

Table 13.1 Examples of plant-beneficial, associative and pathogenic *Pseudomonas* strains.

	Species	Strain	Origin	Genome sequence status
Beneficial	*P. fluorescens*	F113	sugarbeet, Ireland [175]	–
	P. fluorescens	CHA	tobacco, Switzerland [16]	–
	P. fluorescens	PF01	tobacco, Switzerland [176]	completed[a] (unpublished)
	P. fluorescens	PF5	cotton, Texas [20]	completed [177][b]
	P. fluorescens	Q2-87	wheat, Washington [178]	–
	P. fluorescens	WCS365	tomato [179]	–
	P. chlororaphis	PCL1391	tomato [180]	in progress[c]
	P. aureofaciens	30–84	wheat, Kansas [181]	–
Associative	*P. fluorescens*	SBW25	sugarbeet [182]	annotation in progress[d]
	P. putida	KT2440	[183,184]	completed [13][e]
	P. aeruginosa	PA01	B. Holloway	completed [185][f]
Pathogenic	*P. syringae*	pv. *tomato* DC3000	bacterial speck tomato and *Arabidopsis* [186]	completed [187][b]
	P. syringae	pv. *syringae* B728a	bacterial brown spot in bean [188]	completed [189][b]
	P. syringae	pv. *phaseolicola* 1448A	halo blight in bean [190]	completed [191][g]
	P. aeruginosa	PA-14	wilting in *Arabidopsis*	completed [192][h]

Genome Sequencing Center:
[a] DOE Joint Genome Institute Microbial Genomics.
[b] The Institute for Genomic Research (TIGR).
[c] Leiden University/Greenomics.
[d] Sanger institute.
[e] TIGR/German Consortium.
[f] PathoGenesis Corporation/Cystic Fibrosis Foundation/University of Washington Genome Center.
[g] TIGR/Cornell University.
[h] National Center for Biotechnology Information.

high densities, produce secondary metabolites with powerful antifungal activities, produce phytostimulatory compounds, solubilize ecological significant compounds such as phosphates and degrade toxic contaminants (reviewed in Refs. [6–12]). Examples of *Pseudomonas* strains that have been shown to play roles in beneficial interactions in the rhizosphere such as phytostimulation, biofertilization, bioremediation and biocontrol are detailed in Table 13.2. The release of the genome sequence of beneficial *Pseudomonas* strains such as *P. putida* has demonstrated the catabolic and metabolic diversity of the soil-borne, nonpathogenic pseudomonads, and provided further insight to their potential application in agriculture [13].

Table 13.2 Examples of beneficial *Pseudomonas*–plant interactions.

Interaction and *Pseudomonas* strain	**Trait(s)**	**Reference**
Biocontrol	Biocontrol via inhibition of...	
P. fluorescens F113	*Pythium ultimum*, causal agent of damping-off in sugarbeet	[15,193,194]
	Erwinia carotovora ssp. *atroseptica*, causal agent of potato tuber rot	[23]
	Polymyxa betae, causal agent with BNYV *Rhizomania* in sugarbeet	[22]
	Globodera rostochiensis, potato cyst nematode; increased egg hatch and decreased juvenile motility	[23]
P. fluorescens CHA0	*Thielaviopsis basicola*, causal agent of black root rot of tobacco; *F. oxysporum* f.sp. *radicis lycopersici*, causal agent of tomato root and foot rot	[19,195]
P. fluorescens Pf-5	*P. ultimum*, causal agent of damping-off in cotton, cucumber and pea	[196,197]
	Rhizoctonia solani, causal agent of seedling disease in cotton	[20]
	Pyrenophora tritici-repentis, causal agent of tan spot in wheat	[198]
	Sclerotonia homoeocarpa and *Dreschslera poae*, causal agents of dollar spot and leaf spot in turfgrass respectively	[199]
	F. oxysporum f.sp. *radicis lycopersici*, causal agent of tomato root and foot rot	[195]
	E. carotovora subsp. *atroseptica*, causal agent of potato tuber rot	[200]
P. fluorescens Q2-87	*Gaeumannomyces graminis* var. *tritici*, causal agent of take-all in wheat	[201]
P. fluorescens WCS365	*F. oxysporum* f.sp. *radicis lycopersici*, causal agent of tomato crown and foot rot	[202]
P. chlororaphis PC1391	*F. oxysporum* f.sp. *radicis lycopersici*, causal agent of tomato root and foot rot	[203]
P. aureofaciens TX-1	*S. homoeocarpa*, causal agent of dollar spot	[204]
ISR	ISR in...	
P. fluorescens WCS417r	*Arabidposis thaliana*, *Dianthus caryophyllus*, *Raphanus sativus*, *Lycopersicon esculentum* L.	[33–35,38,39,205]
P. fluorescens CHA	*A. thaliana*	[24]

P. aeruginosa 7NSK	*Oryza sativa*	[206]
Biofertilization	Solubilization/recycling of nutrients	
P. fluorescens NCIMB 11764	cyanide degrading activity into CO_2, formic acid, formamide and ammonium	[207]
P. fluorescens GRS1, PRS9, ATCC13525	tricalcium phosphate solubilization	[61]
Pseudomonas sp.	phorate	[48,208]
Phytostimulation	Production of plant growth promoting hormones...	
Pseudomonas sp.	IAA	[66]
P. fluorescens HP72	IAA	[69]
P. fluorescens CHA0	IAA	[67]
Bioremediation	Degradation of...	
P. fluorescens F113 derivatives	polychlorinated biphenyls	[209,210]
P. fluorescens KP12	polychlorinated biphenyls	[211]
P. fluorescens CG5	oxadiazon	[212]
P. fluorescens	chlorinated pesticides (2,4-D and DDT)	[213]
P. fluorescens ATCC 17400 derivative	biodegradation of 2,4-dinitrotoluene	[214]
Pseudomonas sp.	naphthalene degradation	[215]
Pseudomonas sp. ONBA-17	*o*-nitrobenzaldehyde	[216]
Pseudomonas sp. NRRL B-12227	cyanuric acid	[217]
Pseudomonas sp. ADP	atrazine and phenol	[218]
P. putida KT2440	toluene	[219]
P. putida ZWL73	4-chloronitrobenzene	[220]
P. putida PCL1444	naphthalene	[91]
P. alcaligenes KP2	polychlorinated biphenyls	[211]

13.2.1 Biocontrol

Bacterial strains belonging to several *Pseudomonas* lineages including *P. putida*, *P. fluorescens*, *P. chlororaphis* and *P. aureofaciens* possess intrinsic biocontrol activities against a range of economically devastating plant pathogens (see examples in Table 13.2). This occurs through the secretion of a range of antibiotic metabolites including pyoverdin, phenazines, pyoluteorin, pyrrolnitrin, hydrogen cyanide, cyclic lipopeptides and phloroglucinols [7]. Biocontrol ability is often the result of a combination of the factors listed above and the successful commercial application in agriculture of strains which produce these metabolites can often depend on a range of mitigating factors. These factors include the ability of the microbial inoculant to survive in the rhizosphere and compete with the indigenous microbial populations as well as the ability to produce effective concentrations of the antibiotic metabolites at the time and site of disease [14].

The model strain *P. fluorescens* F113 has attracted attention in recent years as a potential commercial biocontrol inoculant due to its ability to successfully colonies the rhizosphere and produce high levels of the secondary metabolite 2,4-diacetylphloroglucinol (Phl). Genetic and molecular analysis has established a link between the production of this polyketide and the biocontrol efficacy of *P. fluorescens* F113 [15,16]. Phl is a significant secondary metabolite with a broad spectrum of antifungal, antirhizomania, antimicrobial and antihelminthic activities [15,17–24]. It is synthesized by condensation of acetyl-CoA molecules with malonyl-CoA to produce the precursor monoacetylphloroglucinol which is then transacetylated to produce Phl. It has been shown to cause membrane disruption in *Pythium* sp. [25]; however, biocontrol activity of Phl may also be achieved through inducing systemic (disease) resistance (ISR) in plants (see Section 13.2.2). Regulation of Phl production in *P. fluorescens* F113 is governed by complex regulatory cascades that function at both the transcriptional and posttranscriptional levels [26]. Extracellular environmental signals such as the presence of minerals and carbon sources have been shown to influence production of secondary metabolites such as Phl, and molecular signals have also been shown to enhance the production in *P. fluorescens* strains [27,28]. Interestingly, Phl can be perceived as a positive signal in a mixed bacterial population and give rise to increased expression of the Phl biosynthetic genes in the rhizosphere [29]. Questions about the influence of external microbial, plant and environmental signals still remain, and future investigations into the regulation of secondary metabolites such as Phl will hopefully provide strategies to aid in increasing their production and persistence in the rhizosphere.

Despite the success of *Pseudomonas* strains in reducing disease in plant hosts in soil-based microcosm and greenhouse systems, limitations still exist in the production of microbial inoculants that have the following traits: (i) a broad spectrum of biocontrol activity against a range of plant pathogens, (ii) competitive colonization and consistency in biocontrol efficacy under field conditions, and (iii) viability economically in competition with chemical-based disease control methods. There is now a preference to move towards mixed inoculant consortia to enhance biocontrol efficacy (see Section 13.4.5). For example, Dekkers *et al.* showed that mixed inoculation with *P. fluorescens* WCS365 and *P. chlororaphis* PCL1391 exhibited increased biocontrol ability in a *Fusarium oxysporum* f. sp. *radicis-lycopersici-tomato* system compared to single-strain inoculants, presumably by concomitant activation of ISR and production of phenazine-1-carboxamide [30]. This strategy, together with enhanced secondary metabolite production, could potentially increase both the biocontrol efficacy and economic viability for the commercial production of *Pseudomonas* biocontrol strains.

13.2.2 ISR

Several *Pseudomonas* species have been shown to elicit ISR responses against a range of pathogens in a number of plant hosts [24,31–36]. Induction of ISR in the plant is phenotypically similar to the pathogen-induced systemic acquired resistance [37]; however, it is likely that different pathways of induction occur for these two forms of resistance response. The tendency has been to investigate the ISR response elicited by *Pseudomonas* against leaf pathogens [38,39]; however, using split root systems this response can also be shown against root pathogens [40]. ISR induction can often be plant ecotype/cultivar specific. For example, *P. fluorescens* can induce ISR in some *Arabidopsis thaliana* ecotypes, but not in others [37,38]. Cultivar-specific ISR elicitation by *Pseudomonas* has also been shown in carnation and cucumber [33,41]. Despite advances in determining the molecular mechanisms underpinning bacterial induction of ISR, the signals responsible have not yet been fully identified. Potential inducers include lipopolysaccharide (LPS, specifically the O-antigen of LPS), Phl and iron-chelating siderophores [24,32].

13.2.3 Biofertilization

In most soils, nitrogen and phosphate are invariably the two most limiting nutrients for plant growth. Many soil bacteria and fungi have the ability to act as plant biofertilizers by increasing the pool of bioavailable nitrogen and phosphate via nitrogen fixation, inorganic phosphate solubilization and organic phosphate mineralization [8,42]. Several *Pseudomonas* sp. have been identified, usually from the screening of environmental bacterial isolates, with the ability to solubilize various forms of insoluble inorganic phosphate (P_i) found in the soil/rhizosphere such as di- and tri-calcium phosphates, hydrates, hydroxy- and fluroapatites, and crystalline forms of Al-P and Fe-P. [43–51]. It is generally accepted that the mechanism by which most *Pseudomonas* sp. solubilize inorganic phosphate involves the synthesis of organic acids, in particularly gluconic and 2-keto-gluconic acid, via the extracellular direct oxidative metabolism of glucose [52,53]. This causes acidification of the area surrounding the cell and subsequent release of P_i by ligand exchange reactions, and chelation of Ca^{2+}, Al^{3+} and Fe^{3+} cations [54]. There have been very few studies investigating the mechanisms and genetic regulation of *Pseudomonas*-mediated phosphate solubilization, as most of the research regarding the glucose dehydrogenase and gluconate dehydrogenase genes responsible for gluconic and 2-ketogluconic acid production has focused on alternative bacterial species [55]. One exception was the identification by Babu-Khan *et al.* of the *gabY* gene in *Pseudomonas cepacia* E-37, which is speculated to encode a membrane-bound transport protein which may play a role in expression and/or regulation of the direct oxidation pathway of glucose metabolism in this strain [56]. Organic phosphate (P_o) is also a major reservoir of immobilized phosphate found in soils and the ability to mineralize P_o forms such as phytates, phosphomono-, di- and triesters, and organophosphonates via the actions of various phosphatase, phytase and phosphonatase enzymes has been studied in several *Pseudomonas* sp. [57–60]. Expression of these enzymes in *Pseudomonas* sp. has been investigated in some detail, where it has been determined that the principle factor controlling phosphatase gene regulation is activation in response to low environmental bioavailable P_i concentrations acting through the *pho* regulon, similar to that characterized in *Escherichia coli* [60]. In addition, growth temperature was found to regulate acidic phosphatase expression in *P. fluorescens* MF3 [58]. The impact of microbial phosphate solubilization/mineralization on plant growth has been studied in several *Pseudomonas*–plant interactions, with the general conclusion being that inoculation leads to an increase in plant root and shoot growth, as well as increased plant phosphate uptake in some cases [61–65]. However, the ability of many *Pseudomonas* sp. to also produce other metabolites known to promote plant growth such as antimicrobials, siderophores and phytohormones may act to mask the beneficial effects of biofertilization. Indeed, in most cases plant growth stimulation has only been demonstrated when applying much higher levels of bacteria and substrates (e.g. phytate) for phosphate solubilization to the system than are generally found in the soil. Further study is required in this area to gain a more accurate understanding of the potential agricultural benefits of *Pseudomonas*-mediated biofertilization.

13.2.4
Phytostimulation

Auxins and cytokinins are hormones that play significant roles in plant physiological processes such as cell elongation/division, tissue differentiation and apical dominance [66]. They are primarily produced by plants; however, microorganisms such as plant pathogenic and beneficial *Pseudomonas* have also been shown to synthesis them [66–69] (see Table 13.2 for examples). Production of the auxin indole-3-acetic acid (IAA) by bacteria has been directly linked to the development of the plant host root systems and IAA production by *Pseudomonas* sp. can result in increased plant root growth and branching [70]. Increased root branching has been correlated with avoidance mechanisms of salinity stress in alfalfa, increased plant yield and promotion of soil nutrient acquisition [71–73]. In *P. fluorescens* HP72, synthesis of IAA causes development of shorter, branched plant root systems in creeping bentgrass and may be important in root colonization, but it is not involved in increased biocontrol efficacy against brown patch disease [69]. *Pseudomonas* can produce IAA either via the indole-3-acetamide pathway or indole-3-pyruvic pathway and beneficial *Pseudomonas* tend to synthesize IAA via the latter pathway [70]. *Pseudomonas* sp. are also able to conjugate free auxins and hydrolyze conjugated forms, adding another level to their abilities at plant phytohormone manipulation [66]. Identification of factors which affect the production of IAA during plant growth promotion by *Pseudomonas* is an area of growing interest. IAA production has been shown to be dependant on tryptophan, the precursor molecule for IAA in *P. putida*, and in *P. chlororaphis* IAA production appears to be under the control of the GacA/S global regulatory system [68,74].

13.2.5 Rhizoremediation

Industrial development has drastically increased the spread of anthropogenic organic pollutants in the environment [75]. Rhizoremediation can be defined as the utilization of biological organisms such as plants (phytoremediation) or microbes (bioremediation) to facilitate the elimination of hazardous contaminants from the environment. This technology can either be applied *in situ* or *ex situ*, and has long been recognized as having many economical and environmental benefits over conventional chemical remediation processes. Degradation of contaminants by indigenous microorganisms can often occur, but usually at a slow rate, particularly at sites polluted by multiple organic compounds. Due to their excellent root-colonizing ability and ecological competence, pseudomonads are ideal candidates for bioremediation. Several species of *Pseudomonas* have been identified as carrying the abilities to degrade a number of environmental pollutants (Table 13.2). Genetic engineering technology has the potential to increase the bioremediation ability of microorganisms in the degradation of three important organic pollutants, i.e. polycyclic aromatic hydrocarbons (PAHs), polychlorinated biphenyls (PCBs) and pesticides. Despite advances in biomolecular engineering, constraints in the use of bioremediation still exist, particularly with regard to regulation by environmental bodies both in the USA and Europe. Hence, few bioremediation inoculants have reached field application status. Genetically modified *P. fluorescens* HK44, which degrades naphthalene, is the only bacterial inoculant approved for bioremediation field testing in the USA [76]. In order to gain public acceptance and regulatory approval certain questions as to the safety of genetically modified organisms have to be addressed. In particular, the question of toxicity of bio-accumulated toxic compounds or toxic intermediates of the biodegradation process which may be more soluble and therefore more likely to leach into the groundwater. An approach such as using microorganisms that only function in the presence of specific compounds or in association with higher plants seems promising. Molecular strategies are being developed to (in)directly monitor the presence or absence of pollutants and to monitor the rate of biodegradation by bioremediation inoculants.

13.3
Pseudomonas–Plant Signaling

The rhizosphere is a site of intense interaction between microbes and plants, and although knowledge is well established on the biology and biochemistry of this dynamic niche, still little is known of the full extent of molecular interactions that take place between plants and microbes, and among the microbes themselves. The molecular signaling that occurs plays a fundamental role both in the establishment of beneficial interactions and pathogenesis of detrimental interactions between the plant–microbe partners. An understanding of these signaling processes and of the functions regulated by them may have profound implications for the design of new strategies to combat disease or to promote interactions of benefit to the plant. Understanding how microbes respond to plant signals and the role that plant signaling has in determining interaction specificity or modulating bacterial population selection is therefore central to reaping the benefits of plant–microbe interactions.

It is well established that different plant species can actively select for distinct microbial populations in the rhizosphere [77–82]. Using a culture-independent approach Sessitch *et al.* also reported the existence of plant host-specific endophytic bacterial communities in field-based trials [83]. This plant host selection can also occur within varieties or cultivars of the same plant species. In field-based studies, Mark *et al.* established that different varieties of sugarbeet can select for genetically and functionally diverse populations of resident culturable fluorescent pseudomonads in the rhizosphere [84]. Plant-mediated selection relies in part on the activation of bacterial gene expression in response to molecular signals exuded by the host roots.

13.3.1
Plant Root Exudates

Plant roots exude large amounts of nutrients including sugars, amino acids, various dicarboxylic acids and aromatic compounds which act as chemoattractants for bacteria [85–89]. In particular, carbohydrates and amino acids in root exudates have been implicated in stimulating bacterial chemotaxis to the root surface [90]. Metabolites exuded by the plant root may also act as signals to influence the ability of bacterial strains to colonies the root, and to compete and persist in the rhizosphere [3,91–93]. For example, compounds in tomato root exudates have been shown to elicit a chemotactic response in *P. fluorescens* and mutants in flagellar-driven chemotaxis have reduced ability in competitive colonization [94]. Plant root exudates therefore play a significant role in plant–microbial interactions as one of the earliest and most important signaling events between the partners [3,95]. A number of factors can influence chemical composition of exudates such as plant age, physiological status, soil type and nutrient availability [96]. Root exudates can also act as metal chelators and thus increase the availability of micronutrients such as Fe, Mn, Cu and Zn to bacteria in the rhizosphere [97]. In turn, bacteria can influence root exudate composition by altering root cell leakage, cell metabolism and plant nutrition status [77]. Experiments have suggested that Phl produced by fluorescent pseudomonad species can increase root cell permeability and result in improved root exudation by the plant [98]. Naesby *et al.* [99] also showed that inoculation with the Phl-producing *P. fluorescens* F113 increased the available carbon in the rhizosphere. This was also observed in work carried out by Meharg and Killham, where they showed extracellular metabolites from pseudomonads increased carbon exudation by the plant host [100]. Mechanisms behind how microbes alter plant root exudation not only have implications for both partners in the interaction, but also for nutrient cycling and availability in the soil. Plants cycle amino acids across root cell membranes and the net efflux is defined as exudation. The degree by which plants compete effectively with rhizosphere-colonizing microbes depends on the ability of the root to uptake the released compounds. Phillips *et al.* reported that treating roots with Phl increased the total net efflux of amino acids from plant roots [101]. This in theory can stimulate the availability of these compounds for the entire soil food web and thus have strong implications for soil fertility (Figure 13.2).

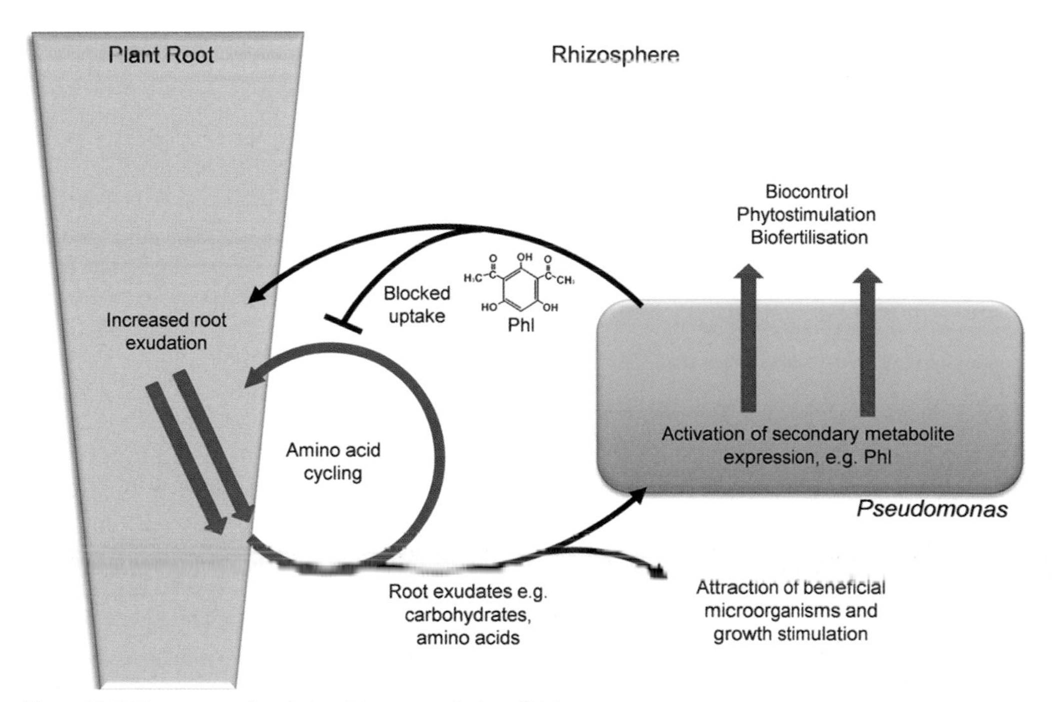

Figure 13.2 The proposed cycle involving mutually beneficial stimulatory mechanisms between a plant growth-promoting *Pseudomonas* strain and a plant host. Production of secondary metabolites such as Phl stimulates root exudate efflux and/or blocks amino acid uptake. This in turn enhances the growth and activity of beneficial microorganisms in the rhizosphere and increases soil fertility.

13.3.2
Plant-Derived Signals and Bacterial Gene Expression

Despite the general acceptance that plant-derived extracellular signals can influence the behavior of bacteria in the rhizosphere, very little is known about the effects of these signals on the patterns of bacterial gene expression and of the role of those genes with altered expression in the microbe–plant interaction. In order to address this, Mark *et al.* tested the concept that plant root exudates from two varieties of sugarbeet, which select for genetically distinct pseudomonad populations in the rhizosphere, differentially modulate expression of bacterial genes in the *P. aeruginosa* PA01 transcriptome [86]. The gene expression response to the two exudates showed only a partial overlap and the majority of those genes with altered expression were regulated in response to only one of the two exudates. Genes with altered expression included those with functions previously implicated in microbe–plant interactions such as aspects of metabolism, chemotaxis and type III secretion, and a new subset with putative or unknown function. Use of a panel of mutants with targeted disruptions allowed for the identification of novel genes with roles in the competitive ability of *P. aeruginosa* in the rhizosphere [86]. Homologs of the genes identified occur in the genomes of both beneficial and pathogenic root-associated bacteria, suggesting that this strategy may help to elucidate the basis of molecular interactions that are important for both plant growth promotion and for plant pathogenesis. Another approach for identification of rhizosphere specific bacterial genes is *in vivo* expression technology (IVET), now established as a method of analyzing bacterial gene expression in the rhizosphere [102–105]. One clear advantage of IVET is that experiments can be carried out *in situ* and there is no need for extrapolation from *in vitro* model systems. However, the method excludes genes that show significant expression *in vitro* in minimal medium and identifies only those that are upregulated in the rhizosphere.

13.3.3
N-Acyl-Homoserine Lactone (AHL) Signaling

Modulation of gene expression in response to cell densities is termed quorum sensing (QS). Bacterial density or sensing of restricted space for growth is based on small diffusible signal molecules. In Gram-negative bacteria, AHLs are a common QS molecule. They are comprised of a homoserine lactone with a variable fatty acid moiety which confers specificity [106,107]. On a genetic level, QS systems involve a LuxI-type AHL synthetase and a cognate LuxR-type transcriptional regulator. Binding of an appropriate AHL causes activation of the LuxR protein and this complex then typically binds as a dimer to a 20-bp *lux* box promoter element to activate transcription of target genes. The LuxI and LuxR QS gene pairs are often found adjacent to each other on the chromosome and are regulated in a positive feedback mechanism [106–108]. QS has been shown to effect a wide range of secondary metabolites and control a wide variety of functions, including those potentially important for plant interactions such as biofilm formation and attachments to surfaces [109].

In *P. aureofaciens*, phenazine antibiotics are produced under positive QS control by the binding of the PhzI product, *N*-hexanoyl-homoserine lactone (HHL), to the LuxR-type regulator, PhzR [110–114]. As with other AHL systems, the PhzI/PhzR system is autoinducible and expressed under higher cell densities [115]. Interestingly, other extracellular products of these strains, including HCN and exoproteases, are not subject to control by the PhzR/PhzI system [116]. Phenazine production in *P. aeruginosa* PAO1 is under the control of three LuxR-type regulators, the two activators LasR and RhlR, and the negative control of QscR [106,117,118]. LasR acts at a higher level in a regulatory cascade to RhlR and they each respond to different AHLs, *N*-(3-oxo-dodecanoyl)-homoserine lactone and *N*-butyryl-homoserine lactone respectively. It seems that the LasI product, *N*-(3-oxo-dodecanoyl)-homoserine lactone, also activates QscR, and that this regulator has a number of overlaps in gene expression control with LasR and RhlR [119]. *P. aureofaciens* 30–84 also contains CsrR/CsrI – a secondary QS system. This system does not influence phenazine regulation, but does appear to regulate cell surface properties and rhizosphere competence [116]. Homologs in other rhizosphere-associated pseudomonads have been identified, but are not as extensively researched. Mutants in the PupI/PupR QS system in the plant growth-promoting strain *P. putida* IsoF produce structured biofilms containing microcolonies and water channels rather than the homogenous unstructured biofilms of the wild-type [120]. The biofilm formed by the mutant lost structure when the native AHL was added back to the biofilm. The root-colonizing *P. fluorescens* NCIMB 10586 also contains the QS system MupI/MupR, which is quite homologous to the PupI/PupR system of *P. putida* [121]. This system controls the biosynthesis of the polyketide antibiotic mupirocin. Using different AHL biosensors a number of other pseudomonads, including *P. savastanol*, *P. tomato*, *P. helianthi* and *P. corrugate*, were demonstrated to produce AHL signal molecules. In a screen of 80 rhizosphere-isolated *P. putida* strains using AHL biosensors, 44% were found to give a positive result [122]. Interestingly, AHL production appeared to be more common among plant-associated than soil-borne *Pseudomonas* sp. [123,124].

The proportion of rhizosphere bacteria in which AHLs have been detected with available assays only represents a small fraction of the culturable rhizosphere organisms. Cross-talk between AHL-producing rhizosphere organisms is thought to be complex, and may give rise to both positive and negative effects. Bacterial communities may stimulate QS-controlled phenotypes by production of compatible AHLs, or by interfering with the quorum signals through degradation or production of AHL antagonists. For example, the phenazine-producing strain *P. chlororaphis* PCL1391 loses its ability to protect tomato plants against *Fusarium oxysporum* var. *lycopersici* in the presence of other rhizobacteria that are able to degrade AHLs through the production of lactonases that cleave the homoserine lactone ring [125]. It is unclear then whether manipulation of quorum signals within the rhizosphere will lead to increased plant growth promotion capabilities, as AHL signals are also used as signals by many plant pathogenic species as well.

13.3.4
Pathogen-Associated Molecular Patterns (PAMPs)

PAMPs, also known as general elicitors, are highly conserved and ubiquitous molecules that elicit defense responses in plants, and are widely distributed amongst microbial species in both pathogenic and nonpathogenic organisms [126,127]. A number of bacterial-derived compounds have been shown to independently activate plant defense responses, including flagellin, cold-shock proteins, LPS and elongation factor Tu (EF-Tu). Plants have developed systems to recognize PAMPs by using pattern recognition receptors (PRRs). Activation of these signals in turn activates, through signal transduction, a range of basal defense mechanisms, such as ethylene production, oxidative bursts, callose deposition, induction of defense-related genes and the hypersensitive response (HR) [128–132]. PAMP-triggered immunity in the plant is the first active response to microbial perception, in some cases causing changes in plant gene expression within a few minutes [133]. Different plant species react differently to different PAMPs, e.g. tobacco responds to bacterial cold-shock proteins, whereas *Arabidopsis* does not [128].

Bacterial LPS is the most widely studied PAMP in relation to plant pathogens. It has been shown to initiate plant synthesis of antimicrobial products and to suppress the development of programmed cell death associated with the HR [134,135]. Variations in the composition of LPS contribute to evasion and suppression of plant responses by plant pathogenic pseudomonads and the activation of plant defense responses by many plant growth-promoting pseudomonads. A central molecule required to elicit this defense response is the lipid-A core oligosaccharide structure that is common to many bacteria. Induction of plant defense mechanisms by plant growth-promoting pseudomonad LPS may enhance local responses to pathogenic bacteria, but more importantly it may also have a role in priming the systemic expression of plant defense responses [134]. Induction of antimicrobial production in plants exposed to a pathogen was found to be much greater in plants pretreated with LPS from certain plant growth-promoting pseudomonads. For the *Pseudomonas* sp. in particular, flagellin and LPS are two of the most well-studied PAMPs eliciting defense responses in plants. In *Arabidopsis* a membrane-associated kinase with an extracellular leucine-rich domain, FLS-2, mediates recognition of flagellin [136]. Purified flagellin from *P. syringae, P. aeruginosa* or *P. fluorescens*, and a flagellin peptide of 22 conserved amino acids, elicit an oxidative burst response, callose deposition and synthesis of antimicrobial proteins in plant cells [133]. Recognition of flagellin by plants is host and strain specific. In *Arabidopsis* the Ws-0 ecotype does not respond to *Pseudomonas* flagellin. Commonly, pathogens will not elicit a response in a susceptible species, e.g. *P. syringae* pvs. *tabaci* purified flagellin elicits a defense response in tobacco, whereas the *P. syringae* pvs. *tomato* and *glycinea* do not [137]. Flagella are known to be important in the colonization of plant roots and leaves, but not for endophytic multiplication [3], which makes it possible that flagellin regulation may be an important mechanism for some *Pseudomonas* species to evade the plant defense response.

13.4 Current and Future Prospects for the Study of *Pseudomonas*–Plant Interactions

Studies of *Pseudomonas*–plant interactions have now entered the "-omics" era with the availability of the complete genome sequences of many pseudomonads (see Table 13.1). This is coupled with the sequencing of several of the commonly studied plant hosts of both pathogenic and beneficial *Pseudomonas* species. The molecular bases of the interactions between plants and microbes are now being explored to new and exciting levels thought impossible in the pre-omics era.

13.4.1 Transcriptomics

The development of oligonucleotide microarrays of both bacteria and their plant hosts encompassing all mRNA transcripts in the cell has allowed for the analysis of global gene expression of either partner during plant–microbe interactions. Microarrays of the model plant host *A. thaliana* have been extensively used for identifying plant genes involved in interactions with both pathogenic [138–142] and beneficial *Pseudomonas* sp. [143,144]. Microarrays based on soybean (Glycine max) [145,146] and tomato [147,148] have also been used to study pathogenic *Pseudomonas* interactions. Comparisons of the transcriptome profiles of wild-type *P. syringae* with those of mutants in genes known to be important in plant virulence have uncovered previously unidentified genes involved in plant–host interaction and provided a better understanding of the complex regulation of virulence genes [149,150]. Sarkar *et al.* used a microarray-based comparative genomics approach to identify virulence genes in *P. syringae* isolates associated with the particular host plant they were isolated from [151]. As described in Section 13.3.2, the influence of plant root exudates on global gene expression in *P. aeruginosa* PA01 was also examined by transcriptome profiling [86].

13.4.2 Proteomics

One of the major drawbacks of analysis of plant and microbe transcriptomes is that they do not provide information about what changes are occurring in the cell at the posttranscriptional level. The use of proteomics to measure the levels of actively expressed proteins in the cell is one way to overcome this limitation. *Arabidopsis* has once again been the most common plant host used to examine changes in the proteome during interactions with plant pathogens such as *P. syringae* [152,153]. A more directed approach involves examination of the phosphoproteome, i.e. the proteins that have become phosphorylated in response to a plant pathogen. Jones *et al.* have used this technique to discover proteins activated in *A. thaliana* in response to attack by *P. syringe* pv. *tomato* [154].

13.4.3
Metabolomics

Analysis of the metabolome can involve both unbiased (metabolic fingerprinting) and biased approaches (focused target compound analysis, metabolite profiling of predefined targets) towards the goal of identifying the metabolic state of the plant or microbe of interest [155]. Metabolomic studies hold an advantage over transcriptomics and proteomics in that it is assumed changes in the metabolome more closely reflect the activities within the cell. A functional screen of the secretome of *P. syringae* by Guttman *et al.* identified several genes whose products are specifically secreted during infection of *A. thaliana* [156].

As additional genomic sequences of both plants and plant-associated *Pseudomonas* are completed, the application of the "-omics" technologies listed above will generate large amounts of information regarding changes in gene expression, protein production and metabolite production during interactions between the two partners. Each approach has its disadvantages, as they each tend to represent a biased view of the effect the other partner in the interaction has on cellular processes. The future of "-omics" studies will therefore rely on the ability to integrate the analyses from transcriptomic, proteomic and metabolomic studies to obtain a more detailed, global understanding of these complex interactions. The use of integrated system biology strategies such as the SYSTOMONAS website (www.systomonas.de) [157] will greatly help to promote consolidation and analysis of the data produced in this field.

13.4.4
In Situ Analysis of *Pseudomonas*–Plant Interactions

While the ability to measure genome-wide regulation of genes and global protein/ metabolite synthesis using the above "-omics" technologies can provide an overview of cellular activities during *Pseudomonas*–plant interactions, the destructive nature of these sampling methods means that they are limited in their ability to provide information regarding spatial or temporal variations in cellular activity during the interaction processes. The ability to analyze gene regulation and expression in environmentally relevant, *in situ* conditions in real-time is a key issue underpinning the future understanding of microbe–plant interactions at the most fundamental level.

Reporter gene technology, i.e. the fusion of the promoter from a gene of interest to a reporter gene for the study of *in situ* expression, has advanced considerably over the past 20 years. Early methods utilizing reporters such as *lacZ*, *gusA* and *luxAB* were limited in their application due to the requirement of external substrates, cellular energy (ATP) and additional culturing of cells in order to measure expression. To an extent they have since been superseded by the discovery and consequent widespread application of green fluorescent protein (GFP) and the subsequently developed autofluorescent protein (AFP) derivatives, including unstable variants useful for monitoring short-term gene expression and variants with a variety of different spectral characteristics [158]. When exposed to light of a certain wavelength, AFPs will emit light at a longer wavelength without the requirement for external substrates or cellular energy. Emissions can then be measured by a number of methods including epifluorescent microscopy and confocal laser scanning microscopy in real-time without the need to destructively process the sample. This technique is particularly useful for studying spatial distribution of bacteria during plant–microbe interactions.

AFPs have been used most simply to "tag" a *Pseudomonas* strain of interest in order to monitor its interaction with a host plant [159–164]. The dual labeling of strains with autofluorescent proteins which emit light at different spectra [165], as well as combinations of AFPs with other expression systems, can be useful in identifying the effects of multiple bacterial interactions. This technique has been successfully used for monitoring the distribution of phase variants of *P. brassicacearum* during colonization of wheat [166], for monitoring relationships between the localization and metabolic activity of a *Pseudomonas* strain during plant interaction [167], and for monitoring the competitive ability of mutant strains versus wild-type. When an autofluorescent protein is combined with a promoter that is induced in the presence of certain plant/microbial signals or environmental conditions (e.g. AHLs, plant exudates, C, N or P, temperature, osmolarity, iron availability and pH) this creates a biosensor strain, which can provide useful information about biological conditions within the soil/rhizosphere and the types of interactions occurring [168–170]. The increase or decrease of expression of specific genes during plant–*Pseudomonas* interactions can also be monitored [171] and once again dual labeling can be used to monitor activity of several genes in the same strain. In this way reporter gene studies can be used as a method of confirming the results of genome-wide transcriptomic analysis in a functional way.

13.4.5 Plant-Beneficial Microbial Consortia

Plant-beneficial microbes are often found to be compatible with one another, i.e. they do not inhibit the survival or plant growth promotion activity of each other. This fortuitous occurrence has paved the way for the combination of bacterial and fungal strains with differing plant growth promotion traits to form inoculation consortia. It has been demonstrated that the use of such consortia can often have a synergistic effect on host plant growth [172–174]. Incorporation of beneficial *P. fluorescens* strains carrying specific desirable traits such as biocontrol/phytostimulation metabolite production and phosphate solubilization into such plant growth promotion consortia is currently being investigated for application in the maize rhizosphere as part of the EU Sixth Framework project "Micromaize" (cordis.europa.eu/fp6/dc/index.cfm). It is hoped that inclusion of *P. fluorescens* in consortia with nitrogen-fixing *Azospirillum* and abuscular mycorrhiza fungi of the *Glomus* genus will create a system where mutual beneficial feedback of nutrients and growth signals between the plant and microbes and from microbe to microbe will allow for production of conventional yields of maize with reduced inputs of nitrogen and phosphorus fertilizers.

13.5 Conclusions

Even thought remarkable progress has been made in elucidation of the signaling and interaction processes between pathogenic and beneficial *Pseudomonas* sp. and plants, it is clear that there is still much to learn to unravel the full extent of the complex interactions that occur. It is clear that the variety of types of interaction that can occur is vast, and many more interesting associations will likely be discovered as studies of rhizosphere and soil ecology continue to progress. Many of these interactions are of great importance to the human race, be they pathogenic or beneficial, and thus there is a strong desire to gain a greater understanding as to how and why they occur. The specificity that exists between the eukaryotic and prokaryotic partners in these interactions is a particular area of future interest. Each of the interactions investigated thus far have evolved over a long period of time, with the plant host constantly battling to maintain equilibrium between allowing beneficial *Pseudomonas* sp. to interact in close association while mounting a defense against pathogens, thus the mechanisms involved have often been highly fine tuned from an evolutionary standpoint. We have much to learn from the study of the ecologically diverse and highly adapted *Pseudomonas*, and increasing our understanding of their ability to fit into a large variety of niche environments, including the association with plant hosts, will ultimately provide us with knowledge of great benefit to the human race, particularly in the field of increasing crop production.

Acknowledgments

The work in the BIOMERIT Research Centre is supported by Science Foundation of Ireland Principal Investigator Award 02/IN.1/B1261 (to F. O'G.), the Irish Research Council for Science, Engineering and Technology Post-doctoral fellowship (to A. F.), Science Foundation of Ireland Basic Research Grant 04/BR/B0597, the European Union (Pseudomics QLK5-CT-2002-0091 and Micromaize 036314), the Irish Department of Agriculture and Food Agricultural Research Stimulus Fund Programme (grants RSF 06 321 and RSF 06 377), and the Higher Education Authority (Ireland) PRTLI cycle 3 (Env Biotech).

References

1 Thomashow, L.S. (1996) *Curr Opin Biotechnol*, **7**, 343–347.
2 Cao, H., Baldini, R.L. and Rahme, L.G. (2001) *Annu Rev Phytopathol*, **39**, 259–284.
3 Lugtenberg, B.J.J. and Dekkers, L. and Bloemberg, G.V. (2001) *Ann Rev Phytopathol*, **39**, 461–490.
4 Preston, G.M., Haubold, B. and Rainey, P.B. (1998) *Curr Opin Microbiol*, **1**, 589–597.
5 Preston, G.M. (2004) *Philos Trans R Soc Lond B Biol Sci*, **359**, 907–918.
6 Mark, G., Morrissey, J.P., Higgins, P. and O'Gara, F. (2006) *FEMS Microbiol Ecol*, **56**, 167–177.

7 Haas, D. and Défago, G. (2005) *Nat Rev Microbiol*, **3**, 307–319.

8 Bloemberg, G.V. and Lugtenberg, B.J. (2001) *Curr Opin Plant Biol*, **4**, 343–350.

9 de Lorenzo, V. (2001) *EMBO Rep*, **2**, 357–359.

10 Walsh, U.F., Morrissey, J.P. and O'Gara, F. (2001) *Curr Opin Biotechnol*, **12**, 289–295.

11 Walsh, U.F., O'Gara, F., Economidis, I. and Hogan, S. (2000) *Harnessing the Potential of Genetically Modified Microorganisms and Plants*, European Commission, Brussels.

12 Haas, D., Blumer, C. and Keel, C. (2000) *Curr Opin Biotechnol*, **11**, 290–297.

13 Nelson, K.E. *et al.* (2002) *Environ Microbiol*, **4**, 799–808.

14 Chin-A-Woeng, T.F.C. *et al.* (2001) *Mol Plant Microbe Interact*, **14**, 969–979.

15 Fenton, A.M., Stephens, P.M., Crowley, J., O'Callaghan, M. and O'Gara, F. (1992) *Appl Environ Microbiol*, **58**, 3873–3878.

16 Keel, C. *et al.* (1992) *Mol Plant Microbe Interact*, **5**, 4–13.

17 Delaney, I. *et al.* (2001) *Plant Soil*, **232**, 195–205.

18 Duffy, B., Keel, C. and Défago, G. (2004) *Appl Environ Microbiol*, **70**, 1836–1842.

19 Laville, J. *et al.* (1992) *Proc Natl Acad Sci USA*, **89**, 1562–1566.

20 Howell, C.R. and Stipanovic, R.D. (1979) *Phytopathology*, **69**, 480–482.

21 Marchand, P.A., Weller, D.M. and Bonsall, R.F. (2000) *J Agric Food Chem*, **48**, 1882–1887.

22 Resca, R. *et al.* (2001) *Plant Soil*, **232**, 215–226.

23 Cronin, D. *et al.* (1997) *Appl Environ Microbiol*, **63**, 1357–1361.

24 Iavicoli, A., Boutet, E., Buchala, A. and Métraux, J.P. (2003) *Mol Plant Microbe Interact*, **16**, 851–858.

25 de Souza, J.T. *et al.* (2003) *Phytopathology*, **93**, 966–975.

26 Abbas, A. *et al.* (2002) *J Bacteriol*, **184**, 3008–3016.

27 Abbas, A. *et al.* (2004) *Microbiology*, **150**, 2443–2450.

28 Duffy, B.K. and Défago, G. (1999) *Appl Environ Microbiol*, **65**, 2429–2438.

29 Maurhofer, M., Baehler, E., Notz, R., Martinez, V. and Keel, C. (2004) *Appl Environ Microbiol*, **70**, 1990–1998.

30 Dekkers, L.C. *et al.* (2000) *Mol Plant Microbe Interact*, **13**, 1177–1183.

31 Audenaert, K., Pattery, T., Cornelis, P. and Höfte, M. (2002) *Mol Plant Microbe Interact*, **15**, 1147–1156.

32 van Loon, L.C., Bakker, P.A. and Pieterse, C.M. (1998) *Annu Rev Phytopathol*, **36**, 453–483.

33 van Peer, R., Niemann, G.J. and Schippers, B. (1991) *Phytopathology*, **81**, 728–734.

34 Leeman, M. *et al.* (1995) *Phytopathology*, **85**, 1021–1027.

35 Duijff, B.J., Alabpourette, C. and Lemanceau, P. (1996) *IOBC-WPRS Bull*, **19**, 120–124.

36 Ongena, M. *et al.* (2004) *Mol Plant Microbe Interact*, **17**, 1009–1018.

37 Ton, J., Pieterse, C.M. and Van Loon, L.C. (1999) *Mol Plant Microbe Interact*, **12**, 911–918.

38 van Wees, S.C. *et al.* (1997) *Mol Plant Microbe Interact*, **10**, 716–724.

39 Pieterse, C.M., van Wees, S.C., Hoffland, E., van Pelt, J.A. and van Loon, L.C. (1996) *Plant Cell*, **8**, 1225–1237.

40 Chen, C.Q., Bélanger, R.R., Benhamou, N., Paulitz, T.C. (1998) *Eur J Plant Pathol*, **104**, 877–886.

41 Liu, L., Kloepper, J.W. and Tuzun, S. (1995) *Phytopathology*, **85**, 695–698.

42 Rodríguez, H. and Fraga, R. (1999) *Biotechnol Adv*, **17**, 319–339.

43 Illmer, P. and Schinner, F. (1992) *Soil Biol Biochem*, **24**, 389–395.

44 Das, K., Katiyar, V. and Goel, R. (2003) *Microbiol Res*, **158**, 359–362.

45 Nautiyal, C.S. (1999) *FEMS Microbiol Lett*, **170**, 265–270.

46 Rodríguez, H., Gonzalez, T. and Selman, G. (2000) *J Biotechnol*, **84**, 155–161.

47 Bano, N. and Musarrat, J. (2003) *Lett Appl Microbiol*, **36**, 349–353.

48 Bano, N. and Musarrat, J. (2004) *FEMS Microbiol Lett*, **231**, 13–17.

49 Hameeda, B., Reddy, Y.H., Rupela, O.P., Kumar, G.N. and Reddy, G. (2006) *Curr Microbiol*, **53**, 298–302.

50 Pandey, A., Trivedi, P., Kumar, B. and Palni, L.M. (2006) *Curr Microbiol*, **53**, 102–107.

51 Ahmad, F., Ahmad, I. and Khan, M.S. (2006) *Microbiol Res*, in press.

52 Lessie, T.G. and Phibbs, P.V. Jr. (1984) *Annu Rev Microbiol*, **38**, 359–388.

53 Fuhrer, T., Fischer, E. and Sauer, U. (2005) *J Bacteriol*, **187**, 1581–1590.

54 Richardson, A.E. (2001) *Aus J Plant Physiol*, **28**, 897–906.

55 Rodríguez, H., Fraga, R., Gonzalez, T. and Bashan, Y. (2006) *Plant Soil*, **287**, 15–21.

56 Babu-Khan, S. *et al.* (1995) *Appl Environ Microbiol*, **61**, 972–978.

57 McGrath, J.W., Wisdom, G.B., McMullan, G., Larkin, M.J. and Quinn, J.P. (1995) *Eur J Biochem*, **234**, 225–230.

58 Burini, J.F., Gügi, B., Merieau, A. and Guespin-Michel, J.F. (1994) *FEMS Microbiol Lett*, **122**, 13–18.

59 Richardson, A.E. and Hadobas, P.A. (1997) *Can J Microbiol*, **43**, 509–516.

60 Monds, R.D., Newell, P.D., Schwartzman, J.A. and O'Toole, G.A. (2006) *Appl Environ Microbiol*, **72**, 1910–1924.

61 Katiyar, V. and Goel, R. (2003) *Microbiol Res*, **158**, 163–168.

62 Lifshitz, R. *et al.* (1987) *Can J Microbiol*, **33**, 390–395.

63 Hameeda, B., Harini, G., Rupela, O.P., Wani, S.P., Reddy and G. (2006) *Microbiol Res*, in press.

64 Richardson, A.E., Hadobas, P.A. and Hayes, J.E. (2000) *Plant Cell Environ*, **23**, 397–405.

65 Richardson, A.E., Hadobas, P.A., Hayes, J.E., O'Hara, C.P. and Simpson, R.J. (2001) *Plant Soil*, **229**, 47–56.

66 Costacurta, A. and Vanderleyden, J. (1995) *Crit Rev Microbiol*, **21**, 1–18.

67 Oberhänsli, T., Défago, G. and Haas, D. (1991) *J Gen Microbiol*, **137**, 2273–2279.

68 Kang, B.R. *et al.* (2006) *Curr Microbiol*, **52**, 473–476.

69 Suzuki, S., He, Y. and Oyaizu, H. (2003) *Curr Microbiol*, **47**, 138–143.

70 Patten, C.L. and Glick, B.R. (2002) *Appl Environ Microbiol*, **68**, 3795–3801.

71 Vaughan, L.V., MacAdam, J.W., Smith, S.E. and Dudley, L.M. (2002) *Crop Sci*, **42**, 2064–2071.

72 Lynch, J. and van Beem, J.J. (1993) *Crop Sci*, **33**, 1253–1257.

73 Saindon, G., Michaud, R. and St Pierre, C.A. (1991) *Can J Plant Sci*, **71**, 727–735.

74 Patten, C.L. and Glick, B.R. (2002) *Can J Microbiol*, **48**, 635–642.

75 Ang, E.L., Zhao, H.M. and Obbard, J.P. (2005) *Enz Microb Technol*, **37**, 487–496.

76 Ripp, S. *et al.* (2000) *Environ Sci Technol*, **34**, 846–853.

77 Yang, C.H. and Crowley, D.E. (2000) *Appl Environ Microbiol*, **66**, 345–351.

78 Smalla, K. *et al.* (2001) *Appl Environ Microbiol*, **67**, 4742–4751.

79 Kuske, C.R. *et al.* (2002) *Appl Environ Microbiol*, **68**, 1854–1863.

80 Kowalchuk, G.A., Buma, D.S., de Boer, W., Klinkhamer, P.G. and van Veen, J.A. (2002) *Antonie Van Leeuwenhoek*, **81**, 509–520.

81 Landa, B.B. *et al.* (2002) *Appl Environ Microbiol*, **68**, 3226–3237.

82 Wieland, G., Neumann, R. and Backhaus, H. (2001) *Appl Environ Microbiol*, **67**, 5849–5854.

83 Sessitsch, A., Reiter, B. and Berg, G. (2004) *Can J Microbiol*, **50**, 239–249.

84 Mark, G.L., Morrissey, J., Baysse, C., Sweeney, P. and O'Gara, F. (2004) Molecular ecology strategies to investigate the behavior of GM and non-modified biocontrol inoculants in the rhizosphere. Presented at the *11th International Congress on Molecular Plant–Microbe Interactions, St Petersburg*.

85 Ashby, A.M., Watson, M.D., Loake, G.J. and Shaw, C.H. (1988) *J Bacteriol*, **170**, 4181–4187.

86 Mark, G.L. *et al.* (2005) *Proc Natl Acad Sci USA*, **102**, 17454–17459.

87 Soby, S., Kirkpatrick, B. and Kosuge, T. (1991) *Appl Environ Microbiol*, **57**, 2918–2920.

88 Sood, S.G. (2003) *FEMS Microbiol Ecol*, **45**, 219–227.

89 Yao, J. and Allen, C. (2006) *J Bacteriol*, **188**, 3697–3708.

90 Somers, E., Vanderleyden, J. and Srinivasan, M. (2004) *Crit Rev Microbiol*, **30**, 205–240.

91 Kuiper, I., Bloemberg, G.V. and Lugtenberg, B.J. (2001) *Mol Plant Microbe Interact*, **14**, 1197–1205.

92 Simons, M., Permentier, H.P., deWeger, L.A., Wijffelman, C.A. and Lugtenberg, B.J.J. (1997) *Mol Plant Microbe Interact*, **10**, 102–106.

93 Lugtenberg, B.J., Kravchenko, L.V. and Simons, M. (1999) *Environ Microbiol*, **1**, 439–446.

94 de Weert, S. *et al.* (2002) *Mol Plant Microbe Interact*, **15**, 1173–1180.

95 Bais, H.P., Prithiviraj, B., Jha, A.K., Ausubel, F.M. and Vivanco, J.M. (2005) *Nature*, **434**, 217–221.

96 Brady, N.C. and Weil, R.R. (1999) *The Nature and Property of Soils*, Prentice Hall, Upper Saddle Hall NJ.

97 Dakora, F.D. and Phillips, D.A. (2002) *Plant Soil*, **245**, 35–47.

98 Barber, D.E. and Martin, J.K. (1976) *New Phytopathol*, **76**, 69–80.

99 Naseby, D.C., Pascual, J.A. and Lynch, J.M. (1999) *J Appl Microbiol*, **87**, 173–181.

100 Meharg, A.A. and Killham, K. (1995) *Plant Soil*, **170**, 345–349.

101 Phillips, D.A., Fox, T.C., King, M.D., Bhuvaneswari, T.V. and Teuber, L.R. (2004) *Plant Physiol*, **136**, 2887–2894.

102 Silby, M.W. and Levy, S.B. (2004) *J Bacteriol*, **186**, 7411–7419.

103 Ramos-González, M.I., Campos, M.J. and Ramos, J.L. (2005) *J Bacteriol*, **187**, 4033–4041.

104 Rainey, P.B. (1999) *Environ Microbiol*, **1**, 243–257.

105 Rediers, H., Rainey, P.B., Vanderleyden, J. and De Mot, R. (2005) *Microbiol Mol Biol Rev*, **69**, 217–261.

106 Fuqua, C., Parsek, M.R. and Greenberg, E.P. (2001) *Annu Rev Genet*, **35**, 439–468.

107 Whitehead, N.A., Barnard, A.M.L., Slater, H., Simpson and N.J.L. and Salmond, G.P.C. (2001) *FEMS Microbiol Rev*, **25**, 365–404.

108 Swift, S. *et al.* (2001) *Adv Microb Physiol*, **45**, 199–270.

109 Webb, J.S., Givskov, M. and Kjelleberg, S. (2003) *Curr Opin Microbiol*, **6**, 578–585.

110 Jude, F. *et al.* (2003) *J Bacteriol*, **185**, 3558–3566.

111 Pierson, L.S., 3rd, Keppenne, V.D., Wood, D.W. (1994) *J Bacteriol*, **176**, 3966–3974.

112 Pierson, L.S., 3rd, Wood, D.W., Pierson, E.A. (1998) *Ann Rev Phytopathol*, **36**, 207–225.

113 Pierson, L.S., 3rd, Wood, D.W., Pierson, E.A. and Chancey, S.T. (1998) *Eur J Plant Pathol*, **104**, 1–9.

114 Tateda, K. *et al.* (2001) *Antimicrob Agents Chemother*, **45**, 1930–1933.

115 Chancey, S.T., Wood and D.W., Pierson, L.S. 3rd, (1999) *Appl Environ Microbiol*, **65**, 2294–2299.

116 Zhang, Z.G. and Pierson, L.S. 3rd, (2001) *Appl Environ Microbiol*, **67**, 4305–4315.

117 Chugani, S.A. *et al.* (2001) *Proc Natl Acad Sci USA*, **98**, 2752–2757.

118 Whiteley, M., Lee, K.M. and Greenberg, E.P. (1999) *Proc Natl Acad Sci USA*, **96**, 13904–13909.

119 Lequette, Y., Lee, J..-H., Ledgham, F., Lazdunski, A. and Greenberg, E.P. (2006) *J. Bacteriol.*, **188**, 3365–3370.

120 Steidle, A. *et al.* (2002) *Appl Environ Microbiol*, **68**, 6371–6382.

121 El-Sayed, A.K., Hothersall, J. and Thomas, C.M. (2001) *Microbiology*, **147**, 2127–2139.

122 Berg, G. *et al.* (2002) *Appl Environ Microbiol*, **68**, 3328–3338.

123 Dumenyo, C.K., Mukherjee, A., Chun, W. and Chatterjee, A.K. (1998) *Eur J Plant Pathol*, **104**, 569–582.

124 Elasri, M. *et al.* (2001) *Appl Environ Microbiol*, **67**, 1198–1209.

125 Molina, L. *et al.* (2003) *FEMS Microbiology Ecology*, **45**, 71–81.

126 Nürnberger, T. and Lipka, V. (2005) *Mol Plant Pathol*, **6**, 335–345.

127 Parker, J.E. (2003) *Trends Plant Sci*, **8**, 245–247.

128 Felix, G. and Boller, T. (2003) *J Biol Chem*, **278**, 6201–6208.

129 Felix, G., Duran, J.D., Volko, S. and Boller, T. (1999) *Plant J*, **18**, 265–276.

130 Ingle, R.A., Carstens, M. and Denby, K.J. (2006) *Bioessays*, **28**, 880–889.

131 Kunze, G. *et al.* (2004) *Plant Cell*, **16**, 3496–3507.

132 Nürnberger, T. *et al.* (1994) *Cell*, **78**, 449–460.

133 Gómez-Gómez, L., Felix, G. and Boller, T. (1999) *Plant J*, **18**, 277–284.

134 Dow, M., Newman, M.A. and von Roepenak, E. (2000) *Ann Rev Phytopathol*, **38**, 241.

135 Newman, M.A., von Roepenack-Lahaye, E., Parr, A., Daniels, M.J. and Dow, J.M. (2002) *Plant J*, **29**, 487–495.

136 Gómez-Gómez, L. and Boller, T. (2002) *Trends Plant Sci*, **7**, 251–256.

137 Taguchi, F. *et al.* (2003) *Plant Cell Physiol*, **44**, 342–349.

138 Mohr, P.G. and Cahill, D.M. (2007) *Funct Integr Genomics*, **7**, 181–191.

139 Scheideler, M. *et al.* (2002) *J Biol Chem*, **277**, 10555–10561.

140 de Torres-Zabala, M. *et al.* (2007) *Embo J*, **26**, 1434–1443.

141 Sato, M. *et al.* (2007) *Plant J*, **49**, 565–577.

142 Thilmony, R., Underwood, W. and He, S.Y. (2006) *Plant J*, **46**, 34–53.

143 Verhagen, B.W. *et al.* (2004) *Mol Plant Microbe Interact*, **17**, 895–908.

144 Wang, Y., Ohara, Y., Nakayashiki, H., Tosa, Y. and Mayama, S. (2005) *Mol Plant Microbe Interact*, **18**, 385–396.

145 Zabala, G. *et al.* (2006) *BMC Plant Biol*, **6**, 26.

146 Zou, J. *et al.* (2005) *Mol Plant Microbe Interact*, **18**, 1161–1174.

147 Cohn, J.R. and Martin, G.B. (2005) *Plant J*, **44**, 139–154.

148 Mysore, K.S. *et al.* (2002) *Plant J*, **32**, 299–315.

149 Ferreira, A.O. *et al.* (2006) *Mol Plant Microbe Interact*, **19**, 1167–1179.

150 Lu, S.E., Wang, N., Wang, J., Chen, Z.J. and Gross, D.C. (2005) *Mol Plant Microbe Interact*, **18**, 324–333.

151 Sarkar, S.F., Gordon, J.S., Martin, G.B. and Guttman, D.S. (2006) *Genetics*, **174**, 1041–1056.

152 Jones, A.M., Thomas, V., Bennett, M.H., Mansfield, J. and Grant, M. (2006) *Plant Physiol*, **142**, 1603–1620.

153 Jones, A.M. *et al.* (2004) *Phytochemistry*, **65**, 1805–1816.

154 Jones, A.M., Bennett, M.H., Mansfield, J.W. and Grant, M. (2006) *Proteomics*, **6**, 4155–4165.

155 Weckwerth, W. (2003) *Annu Rev Plant Biol*, **54**, 669–689.

156 Guttman, D.S. *et al.* (2002) *Science*, **295**, 1722–1726.

157 Choi, C. *et al.* (2007) *Nucleic Acids Res*, **35**, D533–537.

158 Larrainzar, E., O'Gara, F. and Morrissey, J.P. (2005) *Annu Rev Microbiol*, **59**, 257–277.

159 Gau, A.E., Dietrich, C. and Kloppstech, K. (2002) *Environ Microbiol*, **4**, 744–752.

160 Bloemberg, G.V., O'Toole, G.A., Lugtenberg, B.J. and Kolter, R. (1997) *Appl Environ Microbiol*, **63**, 4543–4551.

161 Normander, B., Hendriksen, N.B. and Nybroe, O. (1999) *Appl Environ Microbiol*, **65**, 4646–4651.

162 Tombolini, R., van der Gaag, D.J., Gerhardson, B. and Jansson, J.K. (1999) *Appl Environ Microbiol*, **65**, 3674–3680.

163 Tombolini, R., Unge, A., Davey, M.E., deBruijn, F.J. and Jansson, J.K. (1997) *FEMS Microbiol Ecol*, **22**, 17–28.

164 Villacieros, M. *et al.* (2003) *Plant Soil*, **251**, 47–54.

165 Bloemberg, G.V., Wijfjes, A.H., Lamers, G.E., Stuurman, N. and Lugtenberg, B.J. (2000) *Mol Plant Microbe Interact*, **13**, 1170–1176.

166 Achouak, W., Conrod, S., Cohen, V. and Heulin, T. (2004) *Mol Plant Microbe Interact*, **17**, 872–879.

167 Unge, A. and Jansson, J. (2001) *Microb Ecol*, **41**, 290–300.

168 Joyner, D.C. and Lindow, S.E. (2000) *Microbiology*, **146**, 2435–2445.

169 Gantner, S. *et al.* (2006) *FEMS Microbiol Ecol*, **56**, 188–194.

170 Steidle, A. *et al.* (2001) *Appl Environ Microbiol*, **67**, 5761–5770.

171 Ramos, C., Molbak, L. and Molin, S. (2000) *Appl Environ Microbiol*, **66**, 801–809.

172 Marschner, P. and Timonen, S. (2005) *Appl Soil Ecol*, **28**, 23–36.

173 Russo, A. *et al.* (2005) *Biol Fertil Soils*, **41**, 301–309.

174 Mar Vázquez, M., César, S., Azcón, R. and Barea, J.M. (2000) *Appl Soil Ecol*, **15**, 261–272.

175 Shanahan, P., Osullivan, D.J., Simpson, P., Glennon, J.D. and O'Gara, F. (1992) *Appl Environ Microbiol*, **58**, 353–358.

176 Keel, C. *et al.* (1996) *Appl Environ Microbiol*, **62**, 552–563.

177 Paulsen, I.T. *et al.* (2005) *Nat Biotechnol*, **23**, 873–878.

178 Weller, D.M. (1984) *Appl Environ Microbiol*, **48**, 897–899.

179 Dekkers, L.C. Leiden University, 1997.

180 Chin-A-Woeng, T.F.C. *et al.* (1998) *Mol Plant Microbe Interact*, **11**, 1069–1077.

181 Pierson, L.S., 3rd, and Thomashow, L.S. (1992) *Mol Plant Microbe Interact*, **5**, 330–339.

182 Rainey, P.B. and Bailey, M.J. (1996) *Mol Microbiol*, **19**, 521–533.

183 Bagdasarian, M. *et al.* (1981) *Gene*, **16**, 237–247.

184 Regenhardt, D. *et al.* (2002) *Environ Microbiol*, **4**, 912–915.

185 Stover, C.K. *et al.* (2000) *Nature*, **406**, 959–964.

186 Whalen, M.C., Innes, R.W., Bent, A.F. and Staskawicz, B.J. (1991) *Plant Cell*, **3**, 49–59.

187 Buell, C.R. *et al.* (2003) *Proc Natl Acad Sci USA*, **100**, 10181–10186.

188 Beattie, G.A. and Lindow, S.E. (1994) *Appl Environ Microbiol*, **60**, 3799–3808.

189 Feil, H. *et al.* (2005) *Proc Natl Acad Sci USA*, **102**, 11064–11069.

190 Taylor, J.D., Teverson, D.M., Allen, D.J. and Pastor-Corrales, M.A. (1996) *Plant Pathology*, **45**, 469–478.

191 Joardar, V. *et al.* (2005) *J Bacteriol*, **187**, 6488–6498.

192 Lee, D.G. *et al.* (2006) *Genome Biol*, **7**, R90.

193 Carroll, H., Moënne-Loccoz, Y., Dowling, D.N. and O'Gara, F. (1995) *Appl Environ Microbiol*, **61**, 3002–3007.

194 Delany, I. *et al.* (2000) *Microbiology*, **146**, 537–543.

195 Sharifi-Tehrani, A., Zala, M., Natsch, A., Moënne-Loccoz, Y. and Défago, G. (1998) *Eur J Plant Pathol*, **104**, 631–643.

196 Howell, C.R. and Stipanovic, R.D. (1980) *Phytopathology*, **70**, 712–715.

197 Kraus, J. and Loper, J.E. (1992) *Phytopathology*, **82**, 264–271.

198 Pfender, W.F., Kraus, J. and Loper, J.E. (1993) *Phytopathology*, **83**, 1223–1228.

199 Rodriguez, F. and Pfender, W.F. (1997) *Phytopathology*, **87**, 614–621.

200 Xu, G.W. and Gross, D.C. (1986) *Phytopathology*, **76**, 414–422.

201 Bangera, M.G. and Thomashow, L.S. (1996) *Mol Plant Microbe Interact*, **9**, 83–90.

202 de Weert, S., Kuiper, I., Lagendijk, E.L., Lamers, G.E. and Lugtenberg, B.J. (2004) *Mol Plant Microbe Interact*, **17**, 1185–1191.

203 Chin-A-Woeng, T.F.C., Bloemberg, G.V., Mulders, I.H., Dekkers, L.C. and Lugtenberg, B.J. (2000) *Mol Plant Microbe Interact*, **13**, 1340–1345.

204 Powell, J.F., Vargas, J.M., Nair, M.G., Detweiler, A.R. and Chandra, A. (2000) *Plant Disease*, **84**, 19–24.

205 Pieterse, C.M. and van Loon, L.C. (1999) *Trends Plant Sci*, **4**, 52–58.

206 De Vleesschauwer, D., Cornelis, P. and Hofte, M. (2006) *Mol Plant Microbe Interact*, **19**, 1406–1419.

207 Kunz, D.A., Wang, C.S. and Chen, J.L. (1994) *Microbiology*, **140**, 1705–1712.

208 Kuklinsky-Sobral, J. *et al.* (2004) *Environ Microbiol*, **6**, 1244–1251.

209 Villacieros, M. *et al.* (2005) *Appl Environ Microbiol*, **71**, 2687–2694.

210 Brazil, G.M. *et al.* (1995) *Appl Environ Microbiol*, **61**, 1946–1952.

211 Totevová, S., Prouza, M., Burkhard, J., Demnerová, K. and Brenner, V. (2002) *Folia Microbiol (Praha)*, **47**, 247–254.

212 Garbi, C., Casasús, L., Martinez-Álvarez, R., Ignacio Robla, J. and Martín, M. (2006) *Water Res*, **40**, 1217–1223.

213 Santacruz, G., Bandala, E.R. and Torres, L.G. (2005) *J Environ Sci Health B*, **40**, 571–583.

214 Monti, M.R., Smania, A.M., Fabro, G., Alvarez, M.E. and Argaraña, C.E. (2005) *Appl Environ Microbiol*, **71**, 8864–8872.

215 Gomes, N.C., Kosheleva, I.A., Abraham, W.R. and Smalla, K. (2005) *FEMS Microbiol Ecol*, **54**, 21–33.

216 Fang-Bo, Y., Biao, S. and Shun-Peng, L. (2006) *Curr Microbiol*, **53**, 457–461.

217 Shiomi, N. *et al.* (2006) *J Biosci Bioeng*, **102**, 206–209.

218 Neumann, G. *et al.* (2004) *Appl Environ Microbiol*, **70**, 1907–1912.

219 Aranda-Olmedo, I., Marín, P., Ramos, J.L. and Marqués, S. (2006) *Appl Environ Microbiol*, **72**, 7418–7421.

220 Xiao, Y. *et al.* (2006) *Appl Microbiol Biotechnol*, **73**, 166–171.

14
Biotechnological Relevance of Pseudomonads

Bernd H.A. Rehm

14.1
Introduction

Pseudomonads represent a versatile group of Gram-negative bacteria capable of synthesizing a variety of secondary metabolites and biopolymers that are of biotechnological relevance (Figure 14.1). In addition to their enormous biosynthesis capacity, pseudomonads are able to degrade various recalcitrant compounds such as chlorinated aromatic hydrocarbons that can be found as pollutants in the environment. Thus, pseudomonads and their degradation pathways in engineered microorganisms have been conceived for the clean-up of polluted sites. The ability to metabolize a variety of diverse nutrients combined with the ability to form biofilms enables these bacteria to survive in a variety of habitats. Pseudomonads have been considered for the following biotechnological applications: bioremediation, production of polymers, biotransformation, low-molecular-weight compound production, recombinant protein production and biocontrol agents. The relevance of pseudomonads in the particular biotechnological field will be compiled and discussed below.

14.2
Biodegradation

Some members of the genus *Pseudomonas* are able to metabolize chemical pollutants in the environment and as a result can be used for bioremediation. Bioremediation has been defined as any process that involves microorganisms, fungi, plants or enzymes derived from living systems to remove contaminants from the environment, i.e. to return the environment to the original conditions that were present before it was altered by contaminants. This usually implies that recalcitrant contaminants, e.g. chlorinated hydrocarbons, are degraded by the activity of enzymes

Pseudomonas. Model Organism, Pathogen, Cell Factory. Edited by Bernd H.A. Rehm

ISBN: 978-3-527-31914-5

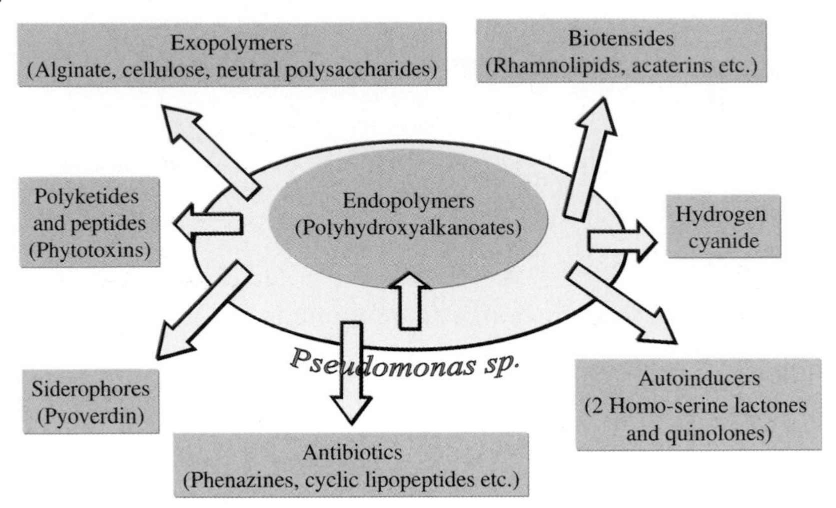

Figure 14.1 Biosynthesis capacity of pseudomonads. Representative compounds produced by various *Pseudomonas* species are shown.

and are usually fully oxidized to harmless molecules via the microbial metabolism. If the end product is carbon dioxide then this process has been defined as mineralization. The use of microorganisms to be added to polluted sites to speed up bioremediation has been designated as bioaugmentation. Bioaugmentation has been successfully applied for bioremediation of oil spills. One example is the *Exxon Valdez* oil spill in 1989, which was cleaned up by implementing bioremediation/ bioaugmentation. Since microorganisms require nitrogen and phosphorus to convert hydrocarbons to biomass, formulations containing these oil degraders must also contain adequate nutrients. Bioaugmentation is often used in terrestrial systems to kick-start the degradation process.

Among the genus *Pseudomonas*, the following species have been demonstrated to be particularly suitable for metabolizing chemical pollutants in the environment and thus can be used for bioremediation:

- *P. alcaligenes* [polycyclic aromatic hydrocarbons (PAHs)] [1]
- *P. mendocina* (toluene) [2]
- *P. pseudoalcaligenes* (cyanide) [3]
- *P. resinovorans* (carbazole) [4]
- *P. veronii* (simple aromatic compounds) [5,6]
- *P. putida* (organic solvents, e.g. toluene) [7]
- *P. stutzeri* (chlorinated hydrocarbons) [8–10]

14.2.1
Microbial Degradation of PAHs

Recalcitrant PAHs (Figure 14.2) are one of the most challenging toxic environmental pollutants that are widely and abundantly produced due to human activities (e.g.

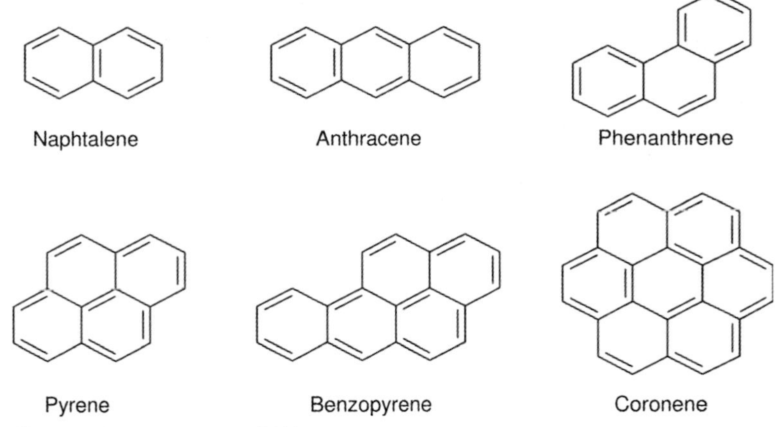

Figure 14.2 Representative PAHs.

incomplete combustion of organic matter, automobile exhausts, coal-fired electricity-generating power plants, agricultural burning, etc.). This group of aromatic compounds comprises highly toxic and carcinogenic substances, and continuous microbial degradation is an important environmental aspect. It is of particular relevance to avoid entrance of these substances into the food chain. Thus, bioremediation of these substances has become a priority of the US Environmental Protection Agency.

Some pseudomonads, such as *P. putida, P. fluorescens, P. paucimobilis, P. vesicularis, P. cepacia, P. testosteroni* and *P. aeruginosa,* are capable of degrading various PAHs. The degradation pathway starts with a chemical reaction catalyzed by a multicomponent dioxygenase enzyme system, which incorporates oxygen atoms at two carbon atoms of a benzene ring of a PAH, resulting in the formation of a *cis*-dihydrodiol [11]. The rearomatization is catalyzed by dehydrogenases, leading to dihydroxylated intermediates, which subsequently undergo ring cleavage (*ortho* or *meta* cleavage) catalyzed by dioxygenases to produce citric acid cycle intermediates (Figure 14.3) [11]. The genetics and biochemistry of PAH degradation pathways found in pseudomonads have been recently summarized by Habe and Omori [12].

14.2.2
Metabolic Engineering of Strains for Biodegradation

In the late 1970s research commenced to identify and clone bacterial genes encoding catabolic enzymes for recalcitrant compounds such as PAHs. The seminal work by Gunsalus and Chakrabarty on the genetics of degradation of a variety of recalcitrant compounds by *Pseudomonas* strains resulted in 1981 in a granted patent (US Patent 425944) for engineered strains that could degrade camphor, octane, salicylate and naphthalene. The emergence of these novel carbon sources based on human activities resulted in the evolution of the respective degradation pathways via natural genetic variation [13]. The respective degradation gene clusters are often localized on

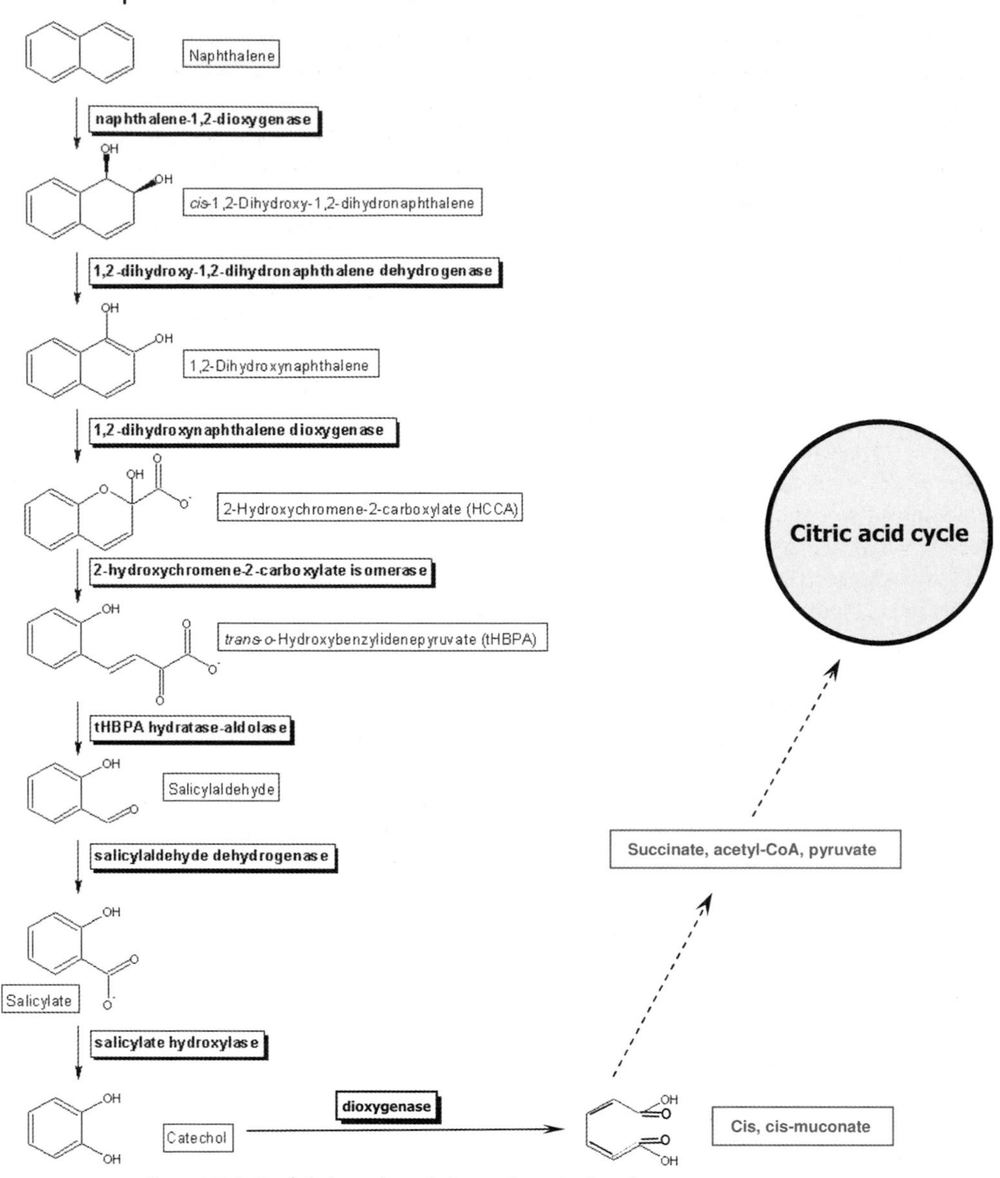

Figure 14.3 Naphthalene degradation pathway in *Pseudomonas*.

plasmids and are mostly inducible, i.e. expression of the genes happens only in the presence of the substrate [14]. To speed up these natural evolution processes leading to emergence of novel degradation gene clusters, genetic engineering of degradation pathways has been employed in order to obtain "super bugs" for the removal of recalcitrant compounds from polluted sites. The many published *Pseudomonas* genome sequences (nine genomes were published in May 2007) as well as other species (mycobacteria, streptomycetes, etc.) relevant for biodegradation provide an enormous resource of biodegradation genes that can be combined in hybrid gene clusters as well as individual genes, encoding key degradation enzymes, that can be subjected to mutagenesis in order to accelerate biodegradation of the targeted pollutant. The ground-breaking work on the degradation of petroleum components and chloroaromatic compounds [15,16] laid the foundation for the genetic engineering approach towards the biodegradation of recalcitrant pollutants. In fact, the use of engineered pseudomonads for bioremediation was strongly enhanced after the discovery of degradation gene clusters [17]. Genes encoding degradation enzymes from various species were combined to establish hybrid degradation pathways in *Pseudomonas* strains, which were capable of degrading very recalcitrant compounds, e.g. mixtures of chloro-benzoates and alkyl-benzoates [18–20]. The use of a strong substrate-responsive promoter embedded in a suitable regulatory network, which enables expression of the respective degradation genes, turned out to be crucial for *in situ* bioremediation. Currently, the behavior of genetically engineered degrading microorganisms is hard to predict in a certain environment and *in situ* bioremediation remains a challenge. However, exciting approaches using metagenomics to provide access to novel biocatalytic genes from nonculturable microorganisms represent a further and very promising opportunity.

Apart from the bioremediation aspect of the conversion of recalcitrant pollutants, many of the catalyzed conversion reactions leading to potentially valuable substances have been exploited by the biocatalysis/biotransformations sector.

The past successes and failures of applying genetically engineered microorganisms for bioremediation have been recently discussed by Cases and de Lorenzo [21]. However, bioremediation of polluted sites in the environment via the use of genetically engineered pseudomonads remains a very promising outlook.

14.3 Biotransformation/Bioconversion

Bioconversion (or biotransformation), i.e. the use of biological systems as living catalysts for the production of valuable metabolites, is a very promising approach for industrial applications. Currently, the number of biotransformations used at the industrial scale is doubling every decade, which indicates the enormous biotechnological potential of the use of cells/enzymes as biocatalysts.

Most biotransformation products are fine chemicals, which are produced at a scale from below 100 to up to 10 000 tons per annum. High-value biotransformation products are chiral compounds serving as intermediates or precursors for the synthesis of pharmaceuticals. Many enzymes isolated from pseudomonads as well as some whole *Pseudomonas* cells are currently used as biocatalysts in industrial biotransformations. If enzymes are used as biocatalysts it is often beneficial to retain the enzyme for continuous catalysis of conversion. In almost 50% of all enzyme-based industrial biotransformations the catalyzing enzyme is immobilized to enable recycling. Recently, a new *in vivo* enzyme immobilization strategy was developed that enables the one-step production of the relevant enzyme already immobilized to an intracellular polymer particle [22]. Examples of *Pseudomonas* enzymes currently applied in various industrial biotransformation reactions are:

- Aminopeptidase (DSM, The Netherlands)
- Aryl alcohol dehydrogenase (Pfizer, USA)
- Aspartate β-decarboxylase (Tanabe Seiyaku, Japan)
- Benzaldehyde dehydrogenase (Pfizer, USA)
- Benzoate dioxygenase (ICI, UK)
- Carbamoylase (Dr. Vig Medicaments, India)
- Dehalogenase (Astra Zeneca, UK)
- Glutaryl amidase (Asahi Kasei Chemical Industry, Japan)
- Haloalkane dehydrogenase (Daiso, Japan)
- Hydantoinase (Dr. Vig Medicaments, India)
- Lipase (Bristol Myers Squibb, USA; Pfizer, USA)
- Malease (DSM, The Netherlands)
- Monooxygenases (Pfizer, USA; Lonza, Switzerland)
- Nitrile hydratase (DuPont, USA)
- Oxidase (DSM, The Netherlands)

Biocatalysis employing whole cells has emerged as an important tool in the industrial synthesis of bulk chemicals, pharmaceuticals and agrochemical intermediates. Nevertheless, the applicability of whole cells is impaired by issues related to strain stability, narrow educt range as well as low productivity. Thus, whole-cell biocatalysts are only used in a few industrial applications [23]. Due to their enormous metabolic capacity and versatility, pseudomonads have been found to convert a variety of organic molecules into valuable compounds, e.g. ferulic acid to vanillin or indene to indandiol [24,25]. The use of whole cells as biocatalysts is of particular advantage when expensive cofactors need to be recycled for catalysis of the respective chemical reaction as well as when multiple reactions are involved in conversion. Indandiol is the precursor of *cis*-aminoindanol, a key chiral precursor to the HIV protease inhibitor Crixivan, which can be derived from indandiol by (2*R*) stereochemistry. Since living systems inherently enable enantioselective conversion reactions, their application for the synthesis of drug precursor enantiomers is superior to chemical synthesis reactions. Here, the conversion of eugenol (plant extract) to vanillin (food flavor) will be discussed as a model conversion using whole *Pseudomonas* cells as biocatalysts.

Most of the currently available vanillin is produced via a synthetic process with a market price of US\$15 kg^{-1} [26]. In contrast, the value of vanillin extracted from vanilla pods, which represents only 0.5% of the vanillin market, ranges between US\$1200 and 4000 kg^{-1} [26]. This high value of naturally nonchemically produced vanillin provides an opportunity for biotechnology to enable natural vanillin production in an economically competitive process. Based on their short generation times and accessibility through molecular genetics, microorganisms are ideal candidates to convert a more abundant natural as well as cheap vanillin precursor into vanillin. Eugenol, a principal aromatic constituent of clove oil with a market price of around

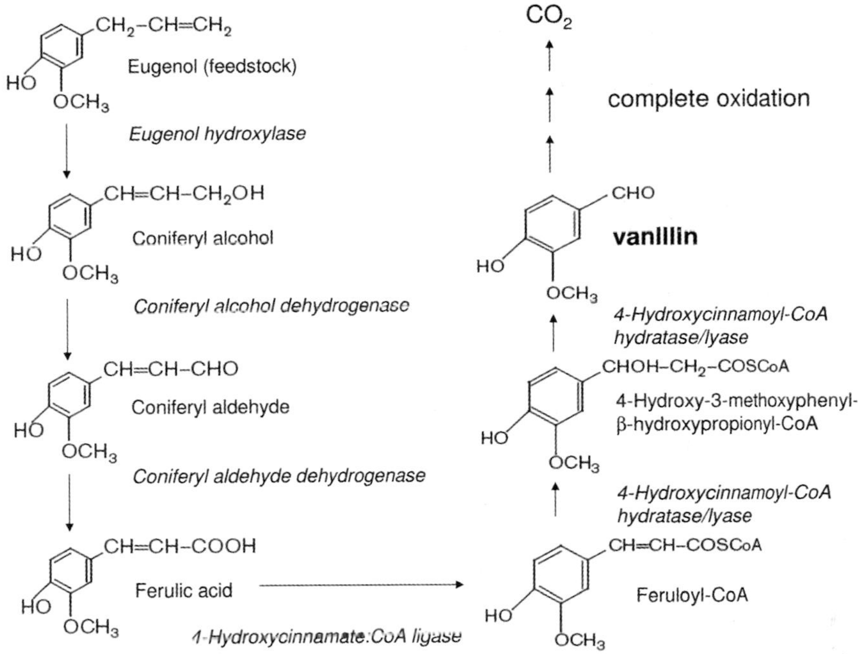

Figure 14.4 Conversion of eugenol into vanillin by *Pseudomonas* sp. HR199.

US$5 kg^{-1}, was considered as an economically realistic feedstock. The conversion of the feedstock eugenol to ferulic acid has been characterized in strain *Pseudomonas* sp. HR199 [27–31] (Figure 14.4). The reactions are successively catalyzed by eugenol hydroxylase (*ehyA* and *ehyB* genes), coniferyl alcohol dehydrogenase (*calA*), and coniferyl aldehyde dehydrogenase (*calB*). Eugenol hydroxylase is a heterodimeric flavocytochrome *c* enzyme; *ehyA* encodes the cytochrome *c* subunit. The respective enzyme, designated eugenol dehydrogenase, from *P. fluorescens* E118 has been purified and biochemically characterized [32]. Although the eugenol conversion pathway has been unraveled in pseudomonads, which also enables development of genetically engineered microorganisms, e.g. knock-out mutants incapable of further vanillin degradation that accumulate vanillin, this biotransformation process has not yet passed the regulatory hurdles [27].

14.4 Low-molecular-weight Compounds: Microbial Production of Biosurfactants

Pseudomonads are characterized by an enormous biosynthesis capacity and versatility. This group of bacteria is capable of synthesizing numerous exoproducts, e.g. various exoenzymes, polyketides, pyocyanine, various exopolysaccharides, homoserine lactones and rhamnolipids, as well as the intracellular storage polymer

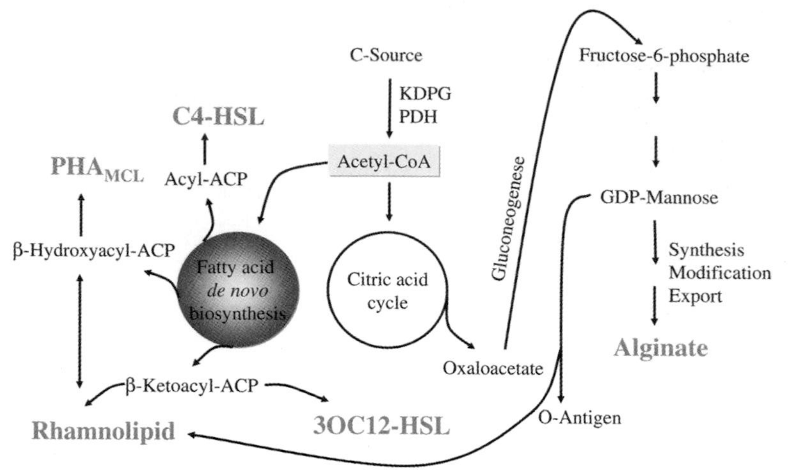

Figure 14.5 Metabolic network in *P. aeruginosa* involved in biosynthesis of PHA, alginate, rhamnolipid and the homoserine lactones (HSL). KDPG, 2-keto-3-deoxy-6-phosphogluconate; PDH, pyruvate dehydrogenase; PHA$_{MCL}$, medium-chain-length PHAs.

polyhydroxyalkanoate (PHA). The diversity of products synthesized by pseudomonads is outlined in Figure 14.1.

Pseudomonads can be used as cell factories harnessing nature's capacity to synthesize valuable compounds from cheap unrelated carbon sources. A carbon source, e.g. glucose, will be oxidized to acetyl-CoA, which serves as a precursor for the synthesis of various sugar and fatty acid derivatives. Then multiple, often species-specific, biosynthesis reactions are employed for the production of the various high- and low-molecular-weight compounds. The metabolic biosynthesis pathways are often highly regulated and metabolically interconnected [33,34] (Figure 14.5). Here, the focus will be on low-molecular-weight compounds such as rhamnolipids (biotensides).

14.4.1 Rhamnolipids as Biosurfactants

Most of the surface-active compounds on the market are produced by chemical synthesis. Biosurfactants, e.g. rhamnolipids, exhibit low toxicity and are fully biodegradable, which strongly increases their ecological acceptance. Biosurfactant activities can be measured as a decrease of surface tension enhancing, e.g. the solubility of hydrophobic molecules in water. Most of the known biosurfactants are glycolipids, which are composed of carbohydrates in combination with long-chain aliphatic acids or hydroxyaliphatic acids. Rhamnolipids (Figure 14.6) are some of the best known glycolipids.

Potential applications of biosurfactants in the cosmetic, food, healthcare, pulp and paper processing, coal, ceramic, and metal industries have been conceived. Even

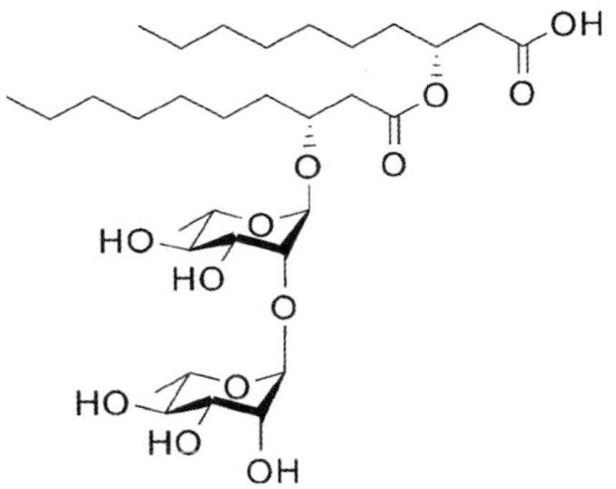

Figure 14.6 Chemical structure of the biosurfactant, di-rhamnolipid.

more promising are applications considering the cleaning of oil-contaminated tankers, oil spill removal, transportation of crude oil, recovery of crude oil from soil, facilitated oil recovery as well as bioremediation of sites contaminated with hydrocarbons, heavy metals and other pollutants [35]. The following biosurfactants have been reported to be produced by pseudomonads:

- Rhamnolipids (*P. aeruginosa, P. putida, P. chlororaphis*) [36,37]
- Viscosin (*P. fluorescens*) [38]
- Acaterin (*Pseudomonas* sp. A92) [39]
- Carbohydrate–protein–lipid (*P. fluorescens* 378) [40]
- Protein PA (*P. aeruginosa*) [41]

Rhamnolipids are composed of one or two molecules of rhamnose linked to one or two molecules of β-hydroxydecanoic acid. Rhamnolipids are the best-studied glycolipids (Figure 14.6).

14.4.1.1 Rhamnolipid Biosynthesis

The first description of rhamnose-containing glycolipids produced by *P. aeruginosa* was published by Jarvis and Johnson [42]. The first genes (*rhlAB*), encoding the rhamnosyltransferase, involved in rhamnolipid biosynthesis were identified and functionally assigned by Ochsner *et al.* [43]. Shortly after the discovery of these first rhamnolipid biosynthesis genes experimental evidence for cell density-dependent regulation [quorum sensing (QS)] of the *rhlAB* genes was provided [44]. Thus, at high bacterial densities diffusible compounds called autoinducers bind to specific transcriptional regulators. The regulatory network involved in QS-dependent regulation of rhamnolipid production plays an important role for the development of rhamnolipid production strains [45]. This transcriptional regulation was dependent on the *rhlR* and *rhlI* genes, encoding a regulator protein and the autoinducer synthetase, respectively. AlgC, a phosphomannomutase involved in alginate biosynthesis, was found to be also required for the synthesis of deoxythymidine-diphospho-L-rhamnose (dTDP-L-rhamnose) (Figure 14.5). dTDP-L-rhamnose is a substrate of the rhamnosyltransferase and thus a direct precursor of rhamnolipid [46]. The hydrophobic fatty acid moiety of the rhamnolipid is most likely diverted from fatty acid *de novo* biosynthesis by the activity of the NADPH-dependent β-ketoacyl-acyl carrier protein (β-ketoacyl-ACP) reductase (RhlG), which catalyzes conversion of β-ketoacyl-ACP to (*R*)-3-hydroxyacyl-ACP [47] (Figure 14.5). It was assumed that (*R*)-3-hydroxyacyl-ACP could be converted to (*R*)-3-hydroxyacyl-CoA (3HA) by the activity of PhaG, a (*R*)-3-hydroxyacyl-ACP:CoA transacylase, but this could not be confirmed experimentally [34,48,49]. A second rhamnosyltransferase (RhlC) catalyzes the transfer of a second rhamnose unit to the mono-rhamnolipid leading to the di-rhamnolipid [50]. It still remains intriguing how the fatty acid moiety 3-hydroxydecanoyl-3-hydroxydecanoate is synthesized.

14.5 Biopolymer Production

P. aeruginosa has emerged as a model organism to study the biosynthesis of the polymers alginate and PHA. *P. aeruginosa* is capable of producing at least three exopolysaccharides (alginate, cellulose and neutral exopolysaccharide) and one intracellular storage polymer (PHA) [51,52].

14.5.1 Alginate Biosynthesis

Alginates are nonrepeating copolymers of β-D-mannuronic acid (M) and α-L-guluronic acid (G), linked by 1–4 linkages. Alginates are produced by two bacterial genera, i.e. *Pseudomonas* and *Azotobacter*, as well as by brown seaweeds. In *Azotobacter* alginate is an important constituent of the outer layer of the dormant cyst, whereas in *Pseudomonas* it contributes to the formation of differentiated biofilms. The material properties of naturally produced alginates range from viscous solutions to pseudoplastic materials. The comonomers can be present in blocks of M residues (M blocks), G residues (G blocks) or alternating residues (MG blocks) [53] (Figure 14.7). The M residues in bacterial alginates can be acetylated to at positions O-2 and/or O-3. The acetylation degree and the comonomer arrangement strongly affect the material properties of the alginate. Control of this variability as well as knowledge about the alginate structure–function relationship would enable the production of tailor-made alginates with material properties relevant for certain applications. Since the 1920s, alginates have been commercial products and have been industrially produced from brown seaweeds.

The biosynthesis of alginate starts with the conversion of fructose-6-phosphate into mannose-6-phosphate (Figure 14.5). This reaction is catalyzed by the phosphomannose isomerase-guanosine diphosphomannose pyrophosphorylase (PMI-GMP encoded by *algA*), which is a bifunctional protein catalyzing the initial and third steps of alginate synthesis. Mannose-6-phosphate is then converted to mannose-1-phosphate catalyzed by phosphomannomutase (AlgC). The GMP activity of the bifunctional enzyme (AlgA) catalyzes the conversion of mannose-1-phosphate into GDP-mannose. GDP-mannose is then oxidized to GDP-mannuronic acid catalyzed by the guanosine

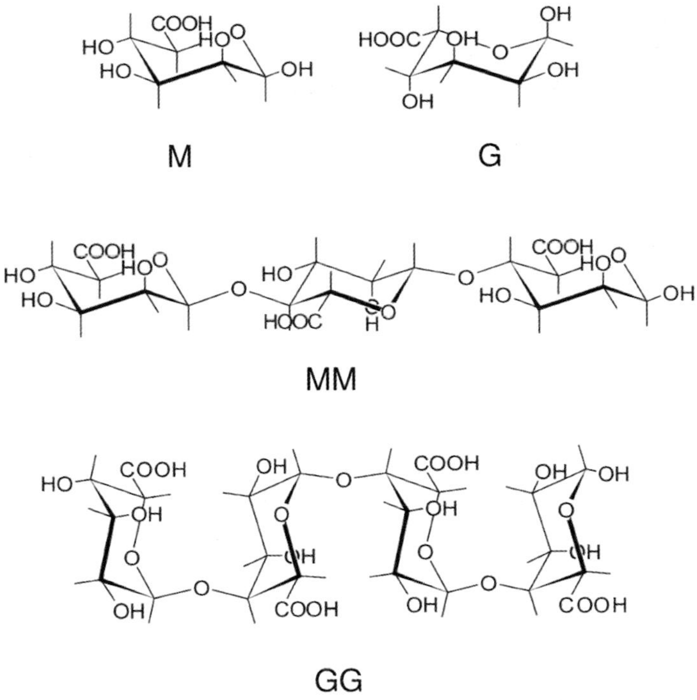

Figure 14.7 Chemical structure of alginate. M, mannuronic acid; G, guluronic acid; MM, secondary structure of alginate composed of M residues; GG, secondary structure of alginate composed of G residues.

diphosphomannose dehydrogenase (GMD encoded by *algD*). Excluding *algC* (encoding phosphomannomutase), all other alginate biosynthesis genes in bacteria are organized in a cluster (for review, see Refs. [54,55]). GDP-mannuronic acid is the activated precursor of alginate biosynthesis. The gene product of *alg8*, encoding a putative glycosyltransferase, presumably representing the catalytic subunit the alginate polymerase, was proposed to catalyze the polymerization reaction. The gene products of *alg44*, *algK* and *algX* are further candidates for being subunits of the alginate polymerase [55–58]. The polymerase, presumably localized as a multiprotein complex in the envelope, could enable alginate formation and secretion [53]. Secretion of the alginate through the outer membrane might be facilitated by an alginate-specific channel protein [59–62]. Modifications of bacterial alginate are introduced after polymerization and presumably in the periplasm. The enzymes involved in this modification, which have been characterized in some detail, are the transacetylase (AlgI, AlgJ and AlgF), the epimerase (AlgG) and the lyase (AlgL) [63–65].

Regulation of alginate biosynthesis becomes increasingly complex, and involves various specific and globally acting gene products [66]. Alginate biosynthesis is regulated based on environmental stimuli and/or by a genotypic switch comprising a mutation inactivating the anti-σ factor MucA [67]. The entire alginate biosynthesis gene cluster of *P. aeruginosa* is transcribed under control of the *algD* promoter, which represent the major target for transcriptional regulation [68].

14.5.1.1 Applications

There are currently no commercial applications of bacterial alginates. However, the possibility to produce tailor-made alginates by manipulation of alginate-producing bacteria opens up new commercial opportunities. Tailor-made alginates can be achieved by controlling the expression of genes encoding alginate-modifying enzymes. The application of highly purified alginates in the medical field is already the subject of strong research and development activities. Highly purified alginates could be used for immunoisolation of allogenic and xenogenic cells, enabling functional implantation into the human body. This was demonstrated for allogenic parathyroid tissue which was implanted into patients with severe hypoparathyroidism. Natural and synthetic polymers are used as biomaterials for wound dressings, dental implants, tissue engineering and microencapsulation of drugs aiming at controlled drug delivery. The current crucial problems in purity, biocompatibility and batch-to-batch reproducibility, especially in the case of alginates derived from varying algae populations, must be addressed in the future to enable application in medicine. Polyphenolic compounds and proteins, which are difficult to remove, contaminate alginates derived from harvested seaweeds and result in bioincompatibility. Thus, intensive purification of algal alginates is required for implantation and other medical applications. These circumstances certainly favor the biotechnological production of alginates using bacteria in a defined production processes (for review, see Ref. [51]).

14.5.2 Elastomeric Bioplastics: Medium-chain-length PHAs

PHAs composed of medium-chain length (*R*)-3-hydroxyfatty acids (6–14 carbon atoms) are mainly produced by fluorescent pseudomonads (Figure 14.8). These polymers serve as a carbon and energy reserve, but might also function as an electron sink as well as having an impact on biofilm formation [33]. Thermoplastic properties, biodegradability and biocompatibility make these materials suitable for several applications in the packaging, medicine, pharmacy, agriculture and food industries (Figure 14.8).

14.5.2.1 Biosynthesis of Medium-chain-length PHAs

Medium-chain-length PHAs are synthesized by converting intermediates of fatty acid metabolism to 3HA, which is substrate of the PHA synthase. The PHA synthase is the key biosynthesis enzyme and catalyzes the enantioselective polymerization of 3HA to PHA [69]. If the carbon source is oxidized to acetyl-CoA excluding the fatty acid β-oxidation pathway, then fatty acid *de novo* biosynthesis intermediates are diverted towards PHA biosynthesis catalyzed by the transacylase PhaG [48,49]. This specific transacylase catalyzes the transfer of the (*R*)-3-hydroxyacyl moiety of the respective ACP thioester to CoA [48,49]. If the carbon source is oxidized through the fatty acid β-oxidation pathway, then the (*R*)-specific enoyl-CoA hydratase (PhaJ)

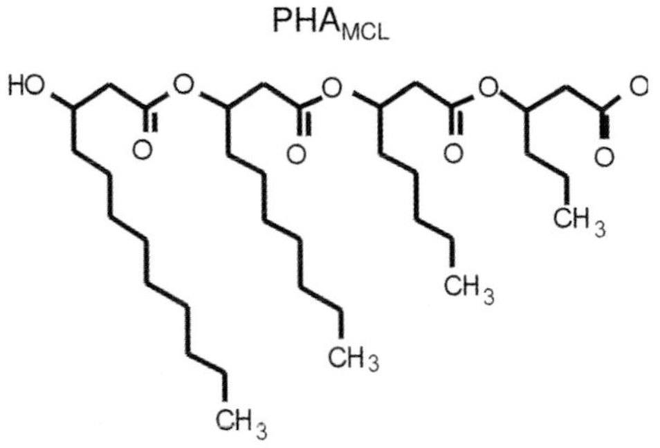

Properties	PHA_{MCL}	PP
T_m (°C)	61	176
T_g (°C)	-36	-10
Crystallinity (%)	30	60
Elongation at break (%)	300	400

Figure 14.8 Chemical structure of medium-chain-length PHAs (PHA_{MCL}) and material properties compared with the oil-based polypropylene (PP). T_m, melting temperature; T_g, glass transition temperature.

catalyzes the oxidation of enoyl-CoA to 3HA [70]. 3HA is a substrate for the polyester synthase (PhaC) and the direct precursor of PHA biosynthesis (Figure 14.5). The genes encoding for the metabolically linking enzymes (PhaG and PhaJ) are not colocalized with the polyester synthase gene, but are coregulated [49,71,72]. Metabolic engineering of medium-chain-length PHA biosynthesis pathways was successfully applied in order to obtain PHAs with a particular composition exerting material properties relevant for specific applications [73].

14.5.2.2 Potential Applications of Medium-chain-length PHAs

The elastomeric bioplastics have only very limited potential applications as commodity polymers mainly because of the rather low melting temperature (low crystallinity) as well as currently high production costs when compared to oil-based polymers (Figure 14.8). Production costs become irrelevant if medical applications are envisaged where performance is the key driver. The inherent high rate of biodegradability of medium-chain-length PHAs combined with their biocompatibility make these polymers ideal candidates for medical applications such as tissue engineering, implantation, microencapsulation and wound dressing [74].

14.6 Recombinant Protein Production

One of the most commercially used protein production hosts is *Escherichia coli* with the famous example of recombinant human insulin production. However, with a rapidly increasing number of therapeutic proteins, alternative microbial hosts might be beneficial to overcome *E. coli*-specific protein production problems such as the formation of inactive inclusion bodies, which require tedious refolding procedures. Recently, *P. fluorescens* has been discovered and exploited as a protein production host, resulting in the development of the so-called Pfenex (Dowpharma, USA) expression technology. Pfenex is built around specially modified strains of *P fluorescens* that increase cellular expression while maintaining critical solubility and activity characteristics of recombinantly expressed therapeutic proteins. In comparison to *E. coli* and other expression platforms, it uses different pathways in the metabolism of certain critical sugars. It was assumed that these differences result in reduced production of metabolic by-products that have been shown to negatively impact cell growth. In addition, the company has developed specific gene promoter and regulating systems that it claims further improve productivity.

14.7 Biocontrol Agents: Plant Growth Promotion

Pseudomonas species which inhabit the rhizosphere of plants have been already applied to cereal seeds or directly to soils in order to prevent growth of crop pathogens [75]. The use of microorganisms enabling the control of growth of pathogenic microorganisms/fungi has been designated as biocontrol. *P. fluorescens* strains (e.g. CHA0 or Pf-5) are currently the best understood biocontrol agents. However, it is still not completely understood how the plant growth-promoting properties of *P. fluorescens* are established. The current hypotheses imply that these rhizobacteria might induce systemic resistance to the pathogen in the host plant and/or the respective bacteria strongly impair the growth and establishment of pathogenic organisms living in the same habitat. Pseudomonads are characterized by the production of the fluorescent siderophore, pyoverdin, which chelates iron with high affinity (Figure 14.1). The respective siderophore hypothesis postulates that these plant growth-promoting bacteria show plant growth promotion by depriving pathogens of iron [76,77]. This hypothesis was supported by studies using addition of iron to soil, which led to reversal of plant growth promotion by reestablishment of pathogenic fungi [78]. Further support for this hypothesis was obtained when pyoverdin-negative mutants of fluorescent pseudomonads showed less protection in some systems. Another protective mechanism could rely on the capability of many pseudomonads to produce compounds antagonistic to other soil microorganisms/fungi, such as phenazine-type antibiotics or hydrogen cyanide. There is experimental evidence to support these hypotheses, in certain systems. A recent comprehensive review of the topic was written by Haas and Defago [75]. A few strains of *P. fluorescens*, *P. putida* and *P. chlororaphis* (Cedomon; BioAgri, Uppsala, Sweden) [79,80] are commercially available as biocontrol rhizobacteria. These biocontrol bacteria can be applied as dry products (granules or powders), cell suspensions (with or without microencapsulation) or seed coatings [80].

14.8 Outlook

Pseudomonads represent a group of extremely versatile bacteria, which are increasingly being investigated with respect to various biotechnological applications. Recent success towards the understanding of the molecular mechanism related to the biosynthesis of compounds as well as degradation of pollutants opens up new avenues for the use of this group of bacteria for synthesis of tailor-made high-value compounds, e.g. biopolymers for various medical applications. The increasing commercial use (therapeutic protein production host, biocontrol agent, enzyme producer, etc.) of pseudomonads is a strong indicator of the biotechnological potential of this group of bacteria. It is likely that the number of commercially viable *Pseudomonas*-based processes will further increase in the future.

References

1 Weissenfels, W.D., Beyer, M. and Klein, J. (1990) Degradation of phenanthrene, fluorine and fluoranthene by pure bacterial cultures. *Appl Microbiol Biotechnol*, **32**, 479–484.

2 Stoffels, M., Amann, R., Ludwig, W., Hekmat, D. and Schleifer, K.H. (1998) AT Bacterial community dynamics during start-up of a trickle-bed bioreactor degrading aromatic compounds. *Appl Environ Microbiol*, **64**, 930–939.

3 Huertas, M.J., Luque-Almagro, V.M., Martinez-Luque, M., Blasco, R., Moreno-Vivian, C., Castillo, F. and Roldan, M.D. (2006) Cyanide metabolism of *Pseudomonas pseudoalcaligenes* CEC: T5344: role of siderophores. *Biochem Soc Trans*, **34**, 152–155.

4 Widada, J., Nojiri, H., Yoshida, T., Habe, H. and Omori, T. (2002) Enhanced degradation of carbazole and 2,3-dichlorodibenzo-*p*-dioxin in soils by *Pseudomonas resinovorans* strain CA10. *Chemosphere*, **49**, 485–491.

5 Hong, H.B., Nam, I.H., Murugesan, K., Kim, Y.M. and Chang, Y.S. (2004) Biodegradation of dibenzo-*p*-dioxin, dibenzofuran, and chlorodibenzo-*p*-dioxins by *Pseudomonas veronii* PH-03. *Biodegradation*, **15**, 303–313.

6 Nam, I.H., Chang, Y.S., Hong, H.B. and Lee, Y.E. (2003) A novel catabolic activity of *Pseudomonas veronii* in biotransformation of pentachlorophenol. *Appl Microbiol Biotechnol*, **62**, 284–290.

7 Shim, H. and Yang, S.T. (1999) Biodegradation of benzene, toluene, ethylbenzene, and *o*-xylene by a coculture of *Pseudomonas putida* and *Pseudomonas* fluorescens immobilized in a fibrous-bed bioreactor. *J Biotechnol*, **67**, 99–112.

8 Lewis, T.A., Paszczynski, A., Gordon-Wylie, S.W., Jeedigunta, S., Lee, C.H. and Crawford, R.L. (2001) Carbon tetrachloride dechlorination by the bacterial transition metal chelator pyridine-2,6-bis (thiocarboxylic acid). *Environ Sci Technol*, **35**, 552–559.

9 Lewis, T.A., Cortese, M.S., Sebat, J.L., Green, T.L., Lee, C.H. and Crawford, R.L. (2000) A *Pseudomonas stutzeri* gene cluster encoding the biosynthesis of the CCl_4-dechlorination agent pyridine-2,6-bis (thiocarboxylic acid). *Environ Microbiol*, **2**, 407–416.

10 Sepulveda-Torres, L.C., Rajendran, N., Dybas, M.J. and Criddle, C.S. (1999) Generation and initial characterization of *Pseudomonas stutzeri* KC mutants with impaired ability to degrade carbon

tetrachloride. *Arch Microbiol*, **171**, 424–429.

11 Kanaly, R.A. and Harayama, S. (2000) Biodegradation of high-molecular-weight polycyclic aromatic hydrocarbons by bacteria. *J Bacteriol*, **182**, 2059–2067.

12 Habe, H. and Omori, T. (2003) Genetics of polycyclic aromatic hydrocarbon metabolism in diverse aerobic bacteria. *Biosci Biotechnol Biochem*, **67**, 225–243.

13 van der Meer, J.R. (1997) Evolution of novel metabolic pathways for the degradation of chloroaromatic compounds. *Antonie Van Leeuwenhoek*, **71**, 159–178.

14 Tropel, D. and van der Meer, J.R. (2004) Bacterial transcriptional regulators for degradation pathways of aromatic compounds. *Microbiol Mol Biol Rev*, **68**, 474–500.

15 Harvey, S., Elashvili, I., Valdes, J.J., Kamely, D. and Chakrabarty, A.M. (1990) Enhanced removal of *Exxon Valdez* spilled oil from Alaskan gravel by a microbial surfactant. *Biotechnology (NY)*, **8**, 228–230.

16 Haugland, R.A., Schlemm, D.J., Lyons, R.P., III, Sferra, P.R. and Chakrabarty, A.M. (1990) Degradation of the chlorinated phenoxyacetate herbicides 2,4-dichlorophenoxyacetic acid and 2,4,5-trichlorophenoxyacetic acid by pure and mixed bacterial cultures. *Appl Environ Microbiol*, **56**, 1357–1362.

17 Ramos, J.L., Diaz, E., Dowling, D., de Lorenzo, V., Molin, S., O'Gara, F., Ramos, C. and Timmis, K.N. (1994) The behavior of bacteria designed for biodegradation. *Biotechnology (NY)*, **12**, 1349–1356.

18 Rojo, F., Pieper, D.H., Engesser, K.H., Knackmuss, H.J. and Timmis, K.N. (1987) Assemblage of ortho cleavage route for simultaneous degradation of chloro- and methylaromatics. *Science*, **238**, 1395–1398.

19 Kaschabek, S.R., Kasberg, T., Muller, D., Mars, A.E., Janssen, D.B. and Reineke, W. (1998) Degradation of chloroaromatics: purification and characterization of a novel type of chlorocatechol 2,3-dioxygenase of *Pseudomonas putida* GJ31. *J Bacteriol*, **180**, 296–302.

20 Reineke, W. (1998) Development of hybrid strains for the mineralization of chloroaromatics by patchwork assembly. *Annu Rev Microbiol*, **52**, 287–331.

21 Cases, I. and de Lorenzo, V. (2005) Genetically modified organisms for the environment: stories of success and failure and what we have learned from them. *Int Microbiol*, **8**, 213–222.

22 Peters, V. and Rehm, B.H.A. (2006) *In vivo* enzyme immobilization by use of engineered polyhydroxyalkanoate synthase. *Appl Environ Microbiol*, **72**, 1777–1783.

23 Schoemaker, H.E., Mink, D. and Wubbolts, M.G. (2003) Dispelling the myths – biocatalysis in industrial synthesis. *Science*, **299**, 1694–1697.

24 O'Brien, X.M., Parker, J.A., Lessard, P.A. and Sinskey, A.J. (2002) Engineering an indene bioconversion process for the production of *cis*-aminoindanol: a model system for the production of chiral synthons. *Appl Microbiol Biotechnol*, **59**, 389–399.

25 Walton, N.J., Narbad, A., Faulds, C. and Williamson, G. (2000) Novel approaches to the biosynthesis of vanillin. *Curr Opin Biotechnol*, **11**, 490–496.

26 Lomascolo, A., Stentelaire, C., Asther, M. and Lesage-Meessen, L. (1999) Basidiomycetes as new biotechnological tools to generate natural aromatic flavours for the food industry. *Trends Biotechnol*, **17**, 282–289.

27 Schrader, J., Etschmann, M.M., Sell, D., Hilmer, J.M. and Rabenhorst, J. (2004) Applied biocatalysis for the synthesis of natural flavor compounds – current industrial processes and future prospects. *Biotechnol Lett*, **26**, 463–472.

28 Priefert, H., Rabenhorst, J. and Steinbuchel, A. (2001) Biotechnological production of vanillin. *Appl Microbiol Biotechnol*, **56**, 296–314.

29 Overhage, J., Kresse, A.U., Priefert, H., Sommer, H., Krammer, G., Rabenhorst, J. and Steinbuchel, A. (1999) Molecular characterization of the genes *pcaG* and *pcaH*, encoding protocatechuate 3,4-dioxygenase, which are essential for vanillin catabolism in *Pseudomonas* sp. strain HR199. *Appl Environ Microbiol*, **65**, 951–960.

30 Overhage, J., Priefert, H., Rabenhorst, J. and Steinbuchel, A. (1999) Biotransformation of eugenol to vanillin by a mutant of *Pseudomonas* sp. strain HR199 constructed by disruption of the vanillin dehydrogenase (*vdh*) gene. *Appl Microbiol Biotechnol*, **52**, 820–828.

31 Priefert, H., Rabenhorst, J. and Steinbuchel, A. (1997) Molecular characterization of genes of *Pseudomonas* sp. strain HR199 involved in bioconversion of vanillin to protocatechuate. *J Bacteriol*, **179**, 2595–2607.

32 Furukawa, H., Wieser, M., Morita, H., Sugio, T. and Nagasawa, T. (1998) Purification and characterization of eugenol dehydrogenase from *Pseudomonas fluorescens* E118. *Arch Microbiol*, **171**, 37–43.

33 Pham, T.H., Webb, J.S. and Rehm, B.H.A. (2004) The role of polyhydroxyalkanoate biosynthesis by *Pseudomonas aeruginosa* in rhamnolipid and alginate production as well as stress tolerance and biofilm formation. *Microbiology*, **150**, 3405–3413.

34 Rehm, B.H.A., Mitsky, T.A. and Steinbuchel, A. (2001) Role of fatty acid *de novo* biosynthesis in polyhydroxyalkanoic acid (PHA) and rhamnolipid synthesis by pseudomonads: establishment of the transacylase (PhaG)-mediated pathway for PHA biosynthesis in *Escherichia coli*. *Appl Environ Microbiol*, **67**, 3102–3109.

35 Desai, J.D. and Banat, I.M. (1997) Microbial production of surfactants and their commercial potential. *Microbiol Mol Biol Rev*, **61**, 47–64.

36 Tuleva, B.K., Ivanov, G.R. and Christova, N.E. (2002) Biosurfactant production by a new *Pseudomonas putida* strain. *Z Naturforsch C*, **57**, 356–360.

37 Guerra-Santos, L., Kappeli, O. and Fiechter, A. (1984) *Pseudomonas aeruginosa* biosurfactant production in continuous culture with glucose as carbon source. *Appl Environ Microbiol*, **48**, 301–305.

38 Nielsen, T.H., Christophersen, C., Anthoni, U. and Sorensen, J. (1999) Viscosinamide, a new cyclic depsipeptide with surfactant and antifungal properties produced by *Pseudomonas fluorescens* DR54. *J Appl Microbiol*, **87**, 80–90.

39 Naganuma, S., Sakai, K., Hasumi, K. and Endo, A. (1992) Acaterin, a novel inhibitor of acyl-CoA: cholesterol acyltransferase produced by *Pseudomonas* sp. A92. *J Antibiot (Tokyo)*, **45**, 1216–1221.

40 Persson, A., Molin, G. and Weibull, C. (1990) Physiological and morphological changes induced by nutrient limitation of *Pseudomonas fluorescens* 378 in continuous culture. *Appl Environ Microbiol*, **56**, 686–692.

41 Hisatsuk, K., Yamada, K. and Nakahara, T. (1972) Protein-like activator for alkane oxidation by *Pseudomonas aeruginosa* S7 B1. *Agric Biol Chem*, **36**, 1361–1369.

42 Jarvis, F.G. and Johnson, M.J. (1949) A glyco-lipid produced by *Pseudomonas aeruginosa*. *J Am Chem Soc*, **71**, 4124–4126.

43 Ochsner, U.A., Fiechter, A. and Reiser, J. (1994) Isolation, characterization, and expression in *Escherichia coli* of the *Pseudomonas aeruginosa* rhlAB genes encoding a rhamnosyltransferase involved in rhamnolipid biosurfactant synthesis. *J Biol Chem*, **269**, 19787–19795.

44 Ochsner, U.A. and Reiser, J. (1995) Autoinducer-mediated regulation of rhamnolipid biosurfactant synthesis in *Pseudomonas aeruginosa*. *Proc Natl Acad Sci USA*, **92**, 6424–6428.

45 Maier, R.M. and Soberon-Chavez, G. (2000) *Pseudomonas aeruginosa* rhamnolipids: biosynthesis and potential applications. *Appl Microbiol Biotechnol*, **54**, 625–633.

46 Olvera, C., Goldberg, J.B., Sanchez, R. and Soberon-Chavez, G. (1999) The *Pseudomonas aeruginosa* algC gene product participates in rhamnolipid biosynthesis. *FEMS Microbiol Lett*, **179**, 85–90.

47 Campos-Garcia, J., Caro, A.D., Najera, R., Miller-Maier, R.M., Al-Tahhan, R.A. and Soberon-Chavez, G. (1998) The *Pseudomonas aeruginosa rhlG* gene encodes an NADPH-dependent beta-ketoacyl reductase which is specifically involved in rhamnolipid synthesis. *J Bacteriol*, **180**, 4442–4451.

48 Hoffmann, N., Amara, A.A., Beermann, B.B., Qi, Q., Hinz, H.J. and Rehm, B.H.A. (2002) Biochemical characterization of the *Pseudomonas putida* 3-hydroxyacyl ACP: CoA transacylase, which diverts intermediates of fatty acid *de novo* biosynthesis. *J Biol Chem*, **277**, 42926–42936.

49 Rehm, B.H.A., Kruger, N. and Steinbuchel, A. (1998) A new metabolic link between fatty acid *de novo* synthesis and polyhydroxyalkanoic acid synthesis. The PHAG gene from *Pseudomonas putida* KT2440 encodes a 3-hydroxyacyl-acyl carrier protein-coenzyme a transferase. *J Biol Chem*, **273**, 24044–24051.

50 Rahim, R., Ochsner, U.A., Olvera, C., Graninger, M., Messner, P., Lam, J.S. and Soberon-Chavez, G. (2001) Cloning and functional characterization of the *Pseudomonas aeruginosa rhlC* gene that encodes rhamnosyltransferase *2*, an enzyme responsible for di-rhamnolipid biosynthesis. *Mol Microbiol*, **40**, 708–718.

51 Rehm, B.H.A. (2005) Biosynthesis and applications of alginates, in *Encyclopedia of Biomaterials and Biomedical Engineering*, (eds G.L. Bowlin and G. Wnek), Dekker, New York, pp. 1–9.

52 Rehm, B.H.A. (2006) Biopolyester particles produced by microbes or using polyester synthases: self-assembly and potential applications, in *Microbial Bionanotechnology: Biological Self-Assembly Systems and Biopolymer-Based Nanostructures*, (ed. B.H.A. Rehm), Horizon, Norwich, pp. 1–34.

53 Rehm, B.H.A. and Valla, S. (1997) Bacterial alginates: biosynthesis and applications. *Appl Microbiol Biotechnol*, **48**, 281–288.

54 Rehm, B.H.A. (2002) Alginates from bacteria, in *Bioploymers, Vol. 5: Polysaccharides I – Polysaccharides from Prokaryotes*, (eds A. Steinbüchel and S. De Baets). Wiley-VCH, Weinheim, pp. 179–212.

55 Remminghorst, U. and Rehm, B.H.A. (2006) Bacterial alginates: from biosynthesis to applications. *Biotechnol Lett*, **28**, 1701–1712.

56 Remminghorst, U. and Rehm, B.H.A. (2006) Alg*44*, a unique protein required for alginate biosynthesis in *Pseudomonas aeruginosa*. *FEBS Lett*, **580**, 3883–3888.

57 Gutsche, J., Remminghorst, U. and Rehm, B.H.A. (2006) Biochemical analysis of alginate biosynthesis protein AlgX from *Pseudomonas aeruginosa*: purification of an AlgX-MucD (AlgY) protein complex. *Biochimie*, **88**, 245–251.

58 Jain, S. and Ohman, D.E. (1998) Deletion of *algK* in mucoid *Pseudomonas aeruginosa* blocks alginate polymer formation and results in uronic acid secretion. *J Bacteriol*, **180**, 634–641.

59 Rehm, B.H.A. (1996) The *Azotobacter vinelandii* gene *algJ* encodes an outer-membrane protein presumably involved in export of alginate. *Microbiology*, **142**, 873–880.

60 Rehm, B.H.A., Grabert, E., Hein, J. and Winkler, U.K. (1994) Antibody response of rabbits and cystic fibrosis patients to an alginate-specific outer membrane protein of a mucoid strain of *Pseudomonas aeruginosa*. *Microb Pathog*, **16**, 43–51.

61 Rehm, B.H.A., Boheim, G., Tommassen, J. and Winkler, U.K. (1994) Overexpression of *algE* in *Escherichia coli*: subcellular localization, purification, and ion channel properties. *J Bacteriol*, **176**, 5639–5647.

62 Chu, L., May, T.B., Chakrabarty, A.M. and Misra, T.K. (1991) Nucleotide sequence and expression of the *algE* gene involved in alginate biosynthesis by *Pseudomonas aeruginosa*. *Gene*, **107**, 1–10.

63 Franklin, M.J. and Ohman, D.E. (1996) Identification of *algI* and *algJ* in the *Pseudomonas aeruginosa* alginate biosynthetic gene cluster which are required for alginate O acetylation. *J Bacteriol*, **178**, 2186–2195.

64 Franklin, M.J., Chitnis, C.E., Gacesa, P., Sonesson, A., White, D.C. and Ohman, D.E. (1994) *Pseudomonas aeruginosa* AlgG is a polymer level alginate C5-mannuronan epimerase. *J Bacteriol*, **176**, 1821–1830.

65 Schiller, N.L., Monday, S.R., Boyd, C.M., Keen, N.T. and Ohman, D.E. (1993) Characterization of the *Pseudomonas aeruginosa* alginate lyase gene (*algL*): cloning, sequencing, and expression in *Escherichia coli*. *J Bacteriol*, **175**, 4780–4789.

66 Schmitt-Andrieu, L. and Hulen, C. (1996) [Alginates of *Pseudomonas aeruginosa*: a complex regulation of the pathway of biosynthesis]. *C R Acad Sci III*, **319**, 153–160.

67 Schurr, M.J., Yu, H., Martinez-Salazar, J.M., Boucher, J.C. and Deretic, V. (1996) Control of AlgU, a member of the sigma E-like family of stress sigma factors, by the negative regulators MucA and MucB and *Pseudomonas aeruginosa* conversion to mucoidy in cystic fibrosis. *J Bacteriol*, **178**, 4997–5004.

68 Chitnis, C.E. and Ohman, D.E. (1993) Genetic analysis of the alginate biosynthetic gene cluster of *Pseudomonas aeruginosa* shows evidence of an operonic structure. *Mol Microbiol*, **8**, 583–593.

69 Rehm, B.H.A. (2003) Polyester synthases: natural catalysts for plastics. *Biochem J*, **376**, 15–33.

70 Fukui, T., Shiomi, N. and Doi, Y. (1998) Expression and characterization of (*R*)-specific enoyl coenzyme A hydratase involved in polyhydroxyalkanoate biosynthesis by *Aeromonas caviae*. *J Bacteriol*, **180**, 667–673.

71 Hoffmann, N. and Rehm, B.H.A. (2005) Nitrogen-dependent regulation of medium-chain length polyhydroxyalkanoate biosynthesis genes in pseudomonads. *Biotechnol Lett*, **27**, 279–282.

72 Hoffmann, N. and Rehm, B.H.A. (2004) Regulation of polyhydroxyalkanoate biosynthesis in *Pseudomonas putida* and *Pseudomonas aeruginosa*. *FEMS Microbiol Lett*, **237**, 1–7.

73 Madison, L.L. and Huisman, G.W. (1999) Metabolic engineering of poly(3-hydroxyalkanoates): from DNA to plastic. *Microbiol Mol Biol Rev*, **63**, 21–53.

74 Zinn, M., Witholt, B. and Egli, T. (2001) Occurrence, synthesis and medical application of bacterial polyhydroxyalkanoate. *Adv Drug Deliv Rev*, **53**, 5–21.

75 Haas, D. and Defago, G. (2005) Biological control of soil-borne pathogens by fluorescent pseudomonads. *Nat Rev Microbiol*, **3**, 307–319.

76 Kloepper, J.W., Schroth, M.N., Leong, J. and Teintze, M. (1980) Enhanced plant-growth by siderophores produced by plant growth-promoting rhizobacteria. *Abstr Papers Am Chem Soc*, **180**, 147-BIOL.

77 Schroth, M.N. and Hancock, J.G. (1981) Selected topics in biological control. *Annu Rev Microbiol*, **35**, 453–476.

78 Kloepper, J.W., Schroth, M.N., Leong, J., Teintze, M. and Theis, J.F. (1980. *Pseudomonas* siderophores – a mechanism explaining disease-suppressive soils. *Abstr Papers Am Chem Soc*, **180**, 146-BIOL.

79 Paulitz, T.C. and Belanger, R.R. (2001) Biological control in greenhouse systems. *Annu Rev Phytopathol*, **39**, 103–133.

80 Schisler, D.A., Slininger, R.J., Behle, R.W. and Jackson, M.A. (2004) Formulation of *Bacillus* spp. for biological control of plant diseases. *Phytopathology*, **94**, 1267–1271.

Index

Pseudomonas. Model Organism, Pathogen, Cell Factory. Edited by Bernd H.A. Rehm

ISBN: 978-3-527-31914-5

g

h

i

l

m

n

q